Drinfeld Modules, Modular Schemes and Applications

Proceedings of the Workshop on

Drinfeld Modules, Modular Schemes and Applications

Alden-Biesen 9–14 September 1996

Editors

E.-U. Gekeler
University of Saarbrücken, Germany

M. van der Put
University of Groningen, The Netherlands

M. Reversat
University of Toulouse, France

J. Van Geel
University of Gent, Belgium

World Scientific
Singapore • New Jersey • London • Hong Kong

Published by

World Scientific Publishing Co. Pte. Ltd.

5 Toh Tuck Link, Singapore 596224

USA office: 27 Warren Street, Suite 401-402, Hackensack, NJ 07601

UK office: 57 Shelton Street, Covent Garden, London WC2H 9HE

British Library Cataloguing-in-Publication Data
A catalogue record for this book is available from the British Library.

DRINFELD MODULES, MODULAR SCHEMES AND APPLICATIONS

ISBN-13 978-981-02-3067-8
ISBN-10 981-02-3067-2

PREFACE

Elliptic modules (nowadays called Drinfeld modules) were invented by V.G. Drinfeld in the early seventies and introduced to the mathematical world in his seminal paper "Elliptic modules", Math. Sbornik 94, 1974. Apparently, his construction evolved from trying to extend the well-known analogy between number fields and function fields to that part of number theory that deals with modular forms, modular curves etcetera. The theory of Drinfeld modules soon turned out to be a powerful tool that allowed succesful resolution of various "classical" problems in the arithmetic of function fields, e.g. global and local Langlands conjectures, the Stark conjectures and explicit class field theory. But it also shed new light on several questions of number theory in the strict sense, and of course raised many new and intriguing problems.

Unfortunately, despite its relevance and importance, there do not exist many references in the literature that might serve as a quick and comprehensive introduction. The workshop that took place in September 1996 in Alden-Biesen, Belgium, aimed to fill that gap. It was addressed to young mathematicians having some background in number theory and algebraic geometry. The present refereed proceedings volume contains the twelve instructional talks and most of the research papers presented at Alden-Biesen.

We will now describe the contents of, and interrelations between the various lectures.

The first instructional lecture gives a general introduction to Drinfeld modules and level structures. Some basic properties of the moduli schemes for Drinfeld modules with a level structure are stated and proved in lecture 2. Here and elsewhere, we have attempted to give elementary proofs and constructions whenever possible, often restricting ourselves to the standard case where the base ring A has the form $\mathbf{F}_q[T]$. For example, Drinfeld's involved proof of the smoothness of the moduli scheme is replaced by an elementary proof for the standard case.

The connection between lattices and Drinfeld modules is given in the third lecture. The analytic theory involved in this, is worked out in more detail in the later lectures 6 and 7.

A survey of class field theory and rank one Drinfeld modules is provided in lecture 4. The standard case provides a non-trivial example for this theory. Complete details are given in lecture 9.

A survey of Drinfeld modules over finite fields is given in lecture 5. The endomorphism algebra of such Drinfeld modules are studied. Further zeta functions and the theory of supersingular Drinfeld modules is presented.

The subsequent lectures focus on Drinfeld modules of rank two, modular curves, modular forms etcetera. A basic analytic tool is "the upper half plane" Ω and its associated Bruhat-Tits tree $\mathcal{T}$. Its rigid analytic structure is explained in lectures 6 and 7. The affine modular curve, associated to rank two Drinfeld modules with a given level, can be interpreted in the analytic category as the quotient $\Gamma\backslash\Omega$, for a suitable subgroup Γ of $Gl(2, A)$. This analytic curve has an obvious compactification, explained in lecture 8. The new points are, as usual, called "cusps". The universal rank two Drinfeld module with level structure lives above $\Gamma\backslash\Omega$. An attempt to extend this universal object above the cusps leads to a new concept, namely a generalized rank two Drinfeld module with prescribed level. This concept can be seen as an analogue of the Néron model of Tate's elliptic curve, used for the compactification of classical modular curve.

In lecture 9, the algebraic version of Tate-Drinfeld modules and generalized Drinfeld modules of rank two is worked out. For the standard case $A = \mathbf{F}_q[T]$, the compactification of the Drinfeld modular curve is constructed by comparison with the analytic theory. Moreover, it is shown that the (compactified) modular curve, corresponding to Drinfeld modules with a level (f)-structure, is connected and smooth over $A[1/f]$.

Modular forms are introduced in lecture 8. A survey of the subject is given in lecture 10. The typical phenomena in characteristic p, e.g. the non-existence of cusps forms of weight one, are emphasized. The theme of modular forms returns in lecture 12.

Lecture 11 aims to provide the material for understanding and proving Drinfeld's reciprocity law (DRL), which relates two-dimensional Galois representations and automorphic representations. Etale cohomology for rigid spaces is explained and the structure of the first l-adic cohomology group of the projective limit of all modular curves with a fixed base ring A is analyzed. The relevant theory of admissible representations is summarized and many results from that theory are taken as input for the proof of DRL.

The aim of lecture 12 is to describe and partially prove the link between DRL and the Shimura-Taniyama-Weil correspondence (STW) for function fields. Let E be an elliptic curve over a function field, that has split multiplicative reduction at ∞. Then DRL associates to E a certain automorphic representation. Unlike the classical case, this does not correspond to a uniquely determined modular form. One has to introduce theta functions to prove the STW statement, i.e., the elliptic curve is a quotient of the Jacobian variety of a suitable Drinfeld modular curve.

The research papers deal with applications to the arithmetic of function fields: construction of class field towers, explicit reciprocity laws, unit index calculations, various questions around modular curves and forms, with a view towards a Gross-Zagier formula for function fields.

We would like to thank the people who made this conference into a success. First of all, the local organizers and, in particular G. Cornelissen and the staff of Alden-Biesen. Thanks are due to our two sponsors FWO and CAGE. Last but not least, we thank the participants and the lecturers for their contributions.

E.-U. Gekeler
M. van der Put
M. Reversat
J. Van Geel

1. J. Van Geel
2. H. Oukhaba
3. G. Cornelissen
4. J. Top
5. M. L. Brown
6. H. G. Rück
7. R. Vidunas
8. B. Snyder
9. G. Van Steen
10. I. Rust
11. U. Tipp
12. L. Van Hamme
13. M. van der Put
14. S. Hoeppner
15. J. Denef
16. N. Rebinguet
17. M. Reversat
18. E. U. Gekeler
19. J. Scholten
20. V. Mauduit
21. B. Angles
22. B. Wack
23. F. Recher
24. C. Michaux
25. D. Zagier
26. O. Scheja
27. H. Matzat
28. M. Saidi
29. Y. Hellegouarch
30. D. Ulmer
31. A. Alvarez
32. B. Toen
33. B. Lopez

CONTENTS

Research Papers

Drinfeld Modules, Modular Schemes and Applications

Lectures

Lecture 1

Introduction to Drinfeld Modules

B. Heinrich Matzat
Interdisziplinaeres Zentrum fuer Wissenschaftliches
Rechnen der Universitaet Heidelberg
Im Neuenheimer Feld 368, D-69120 Heidelberg Germany
matzat@iwr.uni-heidelberg.de

Introduction

In his pioneering paper [4] Drinfeld introduced the notion of an elliptic module of rank r, which for $r = 2$ generalizes the concept of an elliptic curve. Such elliptic modules are usually named Drinfeld modules now.

The main difference between elliptic curves and Drinfeld modules of rank 2 is that for Drinfeld modules the base ring $\mathbb{Z}$ is substituted by a Dedekind domain A inside a global field F of positive characteristic, so that the group acting on the new objects becomes $GL_2(F)$ instead of $GL_2(\mathbb{Q})$. Nevertheless the reader will see that most of the basic results and notions for Drinfeld modules look very similar to the corresponding results for elliptic curves. This covers results concerning isogenies, endomorphism rings and algebras, torsion, Tate modules, level structures, etc.

This lecture gives a very brief introduction to the basic results about Drinfeld modules over fields, which are necessary for the understanding of the definition of Drinfeld modules over schemes at the end of this lecture, and some closely related results, which are needed for the construction of moduli schemes for Drinfeld modules in the subsequent lecture by Saïdi [10].

Almost all results presented here go back to Drinfeld in his original paper [4]. Some simplifications are taken from the existing survey papers of Deligne, Husemöller [3], and Hayes [7], as well as from the books of Gekeler [5], Goss [6], Laumon [8], and Blum, Stuhler [1], which also are highly recommended for further reading.

1 Endomorphisms of the Additive Group

1.1. Let B be a commutative ring with unity of finite characteristic p. We assume therefore the existence of a canonical homomorphismus from $\mathbb{F}_p$ into B.

The *additive algebraic group over* B is denoted by $\mathbb{G}_{a,B}$, or $\mathbb{G}_a$ for short. By definition this is the affine scheme $\mathrm{Spec}(B[T])$ over B together with the group law

$$\varphi : \mathbb{G}_a \times \mathbb{G}_a \longrightarrow \mathbb{G}_a$$

induced by the B-homomorphism

$$f : B[T] \longrightarrow B[T] \otimes_B B[T], \qquad T \longmapsto T \otimes 1 + 1 \otimes T.$$

Then the zero and the inverse of $\mathbb{G}_a$ are induced by the endomorphism

$$f_e : B[T] \longrightarrow B[T],\ T \longmapsto eT$$

with $e = 0$ and $e = -1$, respectively.
1.2. A morphism $\beta : \mathbb{G}_{a,B} \longrightarrow \mathbb{G}_{a,B}$ of the underlying B-schemes is induced by an endomorphism $b : B[T] \longrightarrow B[T]$, given for example by the image $b(T)$ of T. The morphism β is a group endomorphism (with respect to φ) of the group scheme $\mathbb{G}_{a,B}$ if and only if

$$b(T \otimes 1 + 1 \otimes T) = b(T \otimes 1) + b(1 \otimes T).$$

This condition is equivalent to $b(T)$ is an *(absolutely) additive polynomial.* Such additive polynomials are easily characterized.

Proposition. *Let B be a commutative ring of finite characteristic p. Then $b(T) \in B[T]$ is an additive polynomial if and only if there exist $b_i \in B$ such that $b(T) = \sum_{i \geq 0} b_i T^{p^i}$.*

The proof is by comparing the coefficients of $b(T_1 + T_2) \in B[T_1, T_2]$ and $b(T_1) + b(T_2)$.
1.3. Denoting the *Frobenius endomorphism of* $B[T]$ by $\tau : T \longmapsto T^p$ and taking the composition of polynomials as a product, we get the following very convenient description of the ring of group endomorphisms of $\mathbb{G}_a$ over B.

Corollary. *The ring of group endomorphisms* $\mathrm{End}_B(\mathbb{G}_a)$ *is the skew polynomial ring $B\{\tau\}$ over B given by the commutation rule $\tau \cdot b = b^p \cdot \tau$ for $b \in B$.*

1.4. In the following proposition we want to normalize the endomorphisms of $\mathbb{G}_a$ by conjugating with suitable automorphisms.

Proposition. *Let* $\beta = \sum_{i=0}^{n} b_i \tau^i \in B\{\tau\} \cong \mathrm{End}_B(\mathbb{G}_a)$.
(a) The element $\beta \in B\{\tau\}$ is an automorphism of $\mathbb{G}_a$ if and only if b_0 is invertible, i.e., $b \in B^\times$ and the b_i are nilpotent for $i > 0$.
(b) Assume $b_d \in B^\times$ and b_i nilpotent for $i > d$. Then there exists a unique

$\alpha = \sum_{i\geq 0} a_i\tau^i \in \mathrm{Aut}_B(\mathbb{G}_a)$ *with* $a_0 = 1$ *such that*

$$\alpha \cdot \beta \cdot \alpha^{-1} = \sum_{i=0}^{d} c_i\tau^i \qquad \text{where } c_d \in B^\times.$$

Proof. Let $\mathfrak{b}$ be the ideal in B generated by $b_1, ..., b_n$. Then there exists an $s \in \mathbb{N}$ with $\mathfrak{b}^s = 0$ if and only if $b_1, ..., b_n$ are nilpotent. This already proves (a).

To prove the existence of α in (b), it is sufficient to show that if $b_i = 0$ for all $i > n$ with $n > d$ an integer, the coefficient of τ^i in

$$\left(1 - \frac{b_n}{b_d^{p^{n-d}}}\tau^{n-d}\right) \cdot \beta \cdot \left(1 - \frac{b_n}{b_d^{p^{n-d}}}\tau^{n-d}\right)^{-1} \in K\{\tau\}$$

is zero for all $i \geq n$. But this is an obvious computation. Finally, the uniqueness follows by decreasing induction on n modulo the powers of the nilpotent ideal $\mathfrak{c} \triangleleft B$ generated by a_i for $i > 0$ and b_i for $i > d$ (see [8], Lemma 1.1.2, for the details). □

1.5. Now let $B = K$ be a field of characteristic p and let $K\{\tau\} = \mathrm{End}_K(\mathbb{G}_a)$. Then we can define a *degree map* and a *height map*

$$\deg : K\{\tau\} \longrightarrow \mathbb{N} \cup \{-\infty\}, \quad \sum_{i=h}^{d} b_i\tau^i \longmapsto d \text{ if } b_d \neq 0 \text{ with } \deg(0) := -\infty,$$

$$\mathrm{ht} : K\{\tau\} \longrightarrow \mathbb{N} \cup \{\infty\}, \quad \sum_{i=h}^{d} b_i\tau^i \longmapsto h \text{ if } b_h \neq 0 \text{ with } \mathrm{ht}(0) := \infty,$$

as well as the *value at the origin* (sometimes called *derivative*)

$$\partial_0 : K\{\tau\} \longrightarrow K, \quad \sum_{i=h}^{d} b_i\tau^i \longmapsto b_0.$$

Obviously, the degree and height maps have the following properties:

$$\deg(\alpha \cdot \beta) = \deg(\alpha) + \deg(\beta), \quad \deg(\alpha + \beta) \leq \max\{\deg(\alpha), \deg(\beta)\},$$
$$\mathrm{ht}(\alpha \cdot \beta) = \mathrm{ht}(\alpha) + \mathrm{ht}(\beta), \quad \mathrm{ht}(\alpha + \beta) \geq \min\{\mathrm{ht}(\alpha), \mathrm{ht}(\beta)\}.$$

Further, from the characterisation of inseparable polynomials over K we get the following result. Here $\beta(\tau) = \sum_{i=0}^{n} b_i\tau^i$ is called *separable* if and only if the corresponding polynomial $b(T) = \sum_{i=0}^{n} b_iT^{p^i} \in K[T]$ is separable.

Proposition. *For every $\alpha \in K\{\tau\}$ there exists a unique $h \in \mathbb{N}$ and a unique separable polynomial $\beta \in K\{\tau\}$ such that $\alpha = \beta \cdot \tau^h$.*

1.6. Before proceeding further it is very useful to state the following elementary results concerning $K\{\tau\}$.

Proposition. *Let K be a field of characteristic $p > 0$.*
(a) The skew polynomial ring $K\{\tau\}$ is right Euclidean, i.e. for each pair $\alpha, \beta \in K\{\tau\}$ there exists a unique pair $\gamma, \delta \in K\{\tau\}$ such that

$$\alpha = \gamma \cdot \beta + \delta \quad \textit{with } \deg(\delta) < \deg(\beta).$$

(b) The ring $K\{\tau\}$ is a principal ideal domain with respect to multiplication from the left.

Proof. Let $\alpha = \sum_{i=0}^{m} a_i \tau^i$, $\beta = \sum_{j=0}^{n} b_j \tau^j$ and $\gamma = \sum_{k=0}^{m-n} c_k \tau^k$ then for a suitably chosen $c_k \in K$ we obtain

$$\sum_{i=0}^{m} a_i \tau^i = \sum_{i=0}^{m} \Big(\sum_{j+k=i} c_k b_j^{p^k} \Big) \tau^i + \sum_{i=0}^{n-1} d_i \tau^i .$$

This already proves (a), from which (b) follows immediately. □

1.7. Proposition 1.6 can be used to obtain the following results on the structure of finite subgroup schemes of $\mathbb{G}_a$.

Proposition. *Let K be a field of positive characteristic, K^{alg} an algebraic closure of K and let E be a finite subgroup scheme of $\mathbb{G}_{a,K}$. Then the following holds:*
(a) There exists a unique monic $\epsilon \in K\{\tau\}$ such that $E = \mathrm{Ker}(\epsilon)$ (as a subgroupscheme) and $\#E(K^{alg}) = p^{\deg(\epsilon)}$.
(b) For each $\beta \in K\{\tau\}$ annihilating E, there exists a unique $\gamma \in K\{\tau\}$ with $\beta = \epsilon \cdot \gamma$.

Proof. Define $e(T) := \prod_{x \in E(K^{alg})} (T - x)$. Then with a further variable T_0,

$$e_0(T) := e(T + T_0) - e(T_0) \in K(T_0)[T]$$

is a polynomial annihilating $E(K^{alg})$ since $E(K^{alg})$ is a subgroup of $(K^{alg}, +)$. Because moreover the degrees of $e_0(T)$ and $e(T)$ coincide and both polynomials are monic, $e_0(T)$ equals $e(T)$. Therefore $e(T) \in K^{alg}[T]$ is additive, i.e. $e(T) =$

$\sum_{i\geq 0} e_i T^{p^i}$, and the corresponding $\epsilon = \sum_{i\geq 0} e_i\tau^i \in K\{\tau\}$ has the properties claimed in (a).

To prove (b) we use Proposition 1.6(a). By Euclidean division we get $\beta = \gamma\cdot\epsilon+\delta$, with $\deg(\delta) < \deg(\epsilon)$. Since with β and ϵ also δ annihilates $E(K^{alg})$, either $p^{\deg(\delta)} \geq \#E(K^{alg})$ in contradiction to $\deg(\delta) < \deg(\epsilon)$, or $\delta = 0$. □

1.8. *Remark.* In sections 1.2 to 1.7 we used the property that $b(T) \in K[T]$ is additive, which coincides with $\mathbb{F}_p$-linear. Assuming instead that $b(T)$ is $\mathbb{F}_q$-linear we would obtain a skew polynomial ring $K\{\tau\}$ with Frobenius endomorphism $\tau : T \longmapsto T^q$ and the commutation rule $\tau\cdot b = b^q\cdot\tau$. Obviously, under this stronger hypothesis Propositions 1.4 to 1.7 remain true with p replaced by q.

2 Definition of Drinfeld Modules over a Field

2.1. We use the following standard notations:

$\mathbb{F}_q$ is the finite field of q elements with $\mathrm{char}(\mathbb{F}_q) = p$,
$\mathcal{C}$ is an absolutly irreducible projective and smooth curve over $\mathbb{F}_q$,
∞ is a fixed closed point of $\mathcal{C}$ of degree $\deg(\infty) := [\mathbb{F}_q(\infty) : \mathbb{F}_q]$,
A is the ring $\Gamma(\mathcal{C}-\infty, \mathcal{O}_{\mathcal{C}})$ of functions on $\mathcal{C}$ which are regular outside ∞,
F is the function field $\mathbb{F}_q(\mathcal{C})$ of $\mathcal{C}$ over $\mathbb{F}_q$,
$|\cdot|_\infty$ is the normalized valuation of F defined by ∞ with valuation ring $\mathfrak{o}_\infty$ and valuation ideal $\mathfrak{p}_\infty$,
F_∞ is the completion of F with respect to $|\cdot|_\infty$.

Remark. By definition A is a Dedekind domain and F is a global field of characteristic $p > 0$.
Example. In the case $\mathcal{C} = \mathbb{P}^1$ (projective line) and $\mathcal{C} - \infty = \mathbb{A}^1$ (affine line), we obtain $A = \mathbb{F}_q[T]$, $F = \mathbb{F}_q(T)$, $|\cdot|_\infty = q^{\deg(\cdot)}$, and $F_\infty = \mathbb{F}_q((T^{-1}))$.

2.2. Definition. Let K be a field of characteristic p endowed with an $\mathbb{F}_q$-homomorphism $\gamma : A \longrightarrow K$ which gives K the structure of an A-algebra. Then a ring homomorphism

$$\phi : A \longrightarrow K\{\tau\} = \mathrm{End}_K(\mathbb{G}_a), \qquad a \longmapsto \phi_a$$

with the q-power Frobenius endomorphism τ (induced by $T \mapsto T^q$) is called a *Drinfeld A-module over K* if both:

$$\phi(A) \not\subseteq K \quad \text{and} \quad \partial_0 \circ \phi = \gamma.$$

In this case the prime ideal

$$\mathfrak{p}_0 := \mathrm{Ker}(\partial_0 \circ \phi) = \mathrm{Ker}(\gamma) \lhd A$$

is called the *characteristic* of ϕ and is denoted by $\mathrm{char}(\phi)$. The Drinfeld module ϕ has *general characteristic* if $\mathfrak{p}_0 = (0)$ and *finite characteristic* otherwise.

2.3. In order to define the rank and the height of a Drinfeld module, we first state the following preliminary result:

Proposition. *Let K be a field of positive characteristic p and let ϕ be a Drinfeld A-module over K. Then*
(a) the homomorphism $\phi : A \longrightarrow K\{\tau\}$ is injective,
(b) there exists an $r \in \mathbb{R}_{>0}$ such that for all $a \in A$

$$\deg(\phi_a) = \log_q(|a|_\infty^r) = -r \,\mathrm{ord}_{\mathfrak{p}_\infty}(a) \deg(\mathfrak{p}_\infty),$$

(c) in the case of finite characteristic there exists an $h \in \mathbb{R}_{>0}$ such that

$$\mathrm{ht}(\phi_a) = h \,\mathrm{ord}_{\mathfrak{p}_0}(a) \deg(\mathfrak{p}_0) \qquad \text{for all } a \in A.$$

Proof. By definition there exists an $a \in A$ with $\deg(\phi_a) > 0$. From this it follows by 1.5 that $\#\phi(A) = \infty$. Now assume that $\mathfrak{p} := \mathrm{Ker}(\phi) \neq (0)$. Then the residue field $A/\mathfrak{p}$ is finite, contradicting $\#\phi(A) = \infty$.

To prove (b), we only have to state that $q^{\deg(\cdot)}$ defines a valuation equivalent to $|\cdot|_\infty$. Finally for (c), $q^{-\mathrm{ht}(\cdot)}$ defines a valuation which is equivalent to the normalized valuation $|\cdot|_0$ belonging to $\mathfrak{p}_0$. □

In the sequel we shall sometimes use ord_∞ and ord_0 as abbreviations for $\mathrm{ord}_{\mathfrak{p}_\infty}$ and $\mathrm{ord}_{\mathfrak{p}_0}$, respectively.

The positive real number r in Proposition 2.3 is called the *rank* of the Drinfeld A-module ϕ and is denoted by $\mathrm{rank}(\phi)$. Similarly, in the case of finite characteristic the positive real number h is called the *height* of ϕ and is denoted by $\mathrm{height}(\phi)$. We will see in Section 2.6 that the rank and the height of a Drinfeld module are in fact positive integers.

2.4. Definition. Let K be a field of positive characteristic, ϕ a Drinfeld A-module and B a K-algebra. Then B becomes an *A-module via ϕ*, which means an A-module with respect to the action

$$a.b := \phi_a \cdot b \qquad \text{for } a \in A,\ b \in B$$

where the product sign on the right hand side is used to denote the substitution $\tau \longmapsto b^q$. Now let $\mathfrak{a}$ be an ideal in A. Then also

$$\mathrm{Ker}(\phi_{\mathfrak{a}})(B) := \{b \in B | \phi_a \cdot b = 0 \text{ for all } a \in \mathfrak{a}\}$$

defines an A-module via ϕ, in fact an $A/\mathfrak{a}$-module, called the *$\mathfrak{a}$-torsion submodule of B*, and

$$\mathrm{Ker}(\phi_{\mathfrak{a}}) : K\text{-}\mathcal{A}lg \longrightarrow A\text{-}\mathcal{M}od, \qquad B \longmapsto \mathrm{Ker}(\phi_{\mathfrak{a}})(B)$$

is a functor from the category of K-algebras into the category of A-modules called the *$\mathfrak{a}$-torsion functor of ϕ.*

Remark. Using the fact that by Proposition 1.6(b) the ring $K\{\tau\}$ is left principal, we obtain a unique monic skew polynomial $\phi_{\mathfrak{a}} \in K\{\tau\}$ with the property:

$$\mathrm{Ker}(\phi_{\mathfrak{a}})(B) = \{b \in B | \phi_{\mathfrak{a}}(b) = 0\}.$$

2.5. Theorem (Drinfeld). *Let K be a field of positive characteristic with algebraic closure K^{alg}, and let ϕ be a Drinfeld A-module over K of rank r and of height h (in the case of finite characteristic). Further denote by $\mathfrak{a}$ an ideal of A.*
(a) For $\mathfrak{a} \not\subseteq \mathfrak{p}_0 = \mathrm{char}(\phi)$, the $\mathfrak{a}$-torsion submodule of K^{alg} is a free $A/\mathfrak{a}$-module of rank r.
(b) For $\mathfrak{a} = \mathfrak{p}_0^m$, the $\mathfrak{a}$-torsion submodule $\mathrm{Ker}(\phi_{\mathfrak{a}})(K^{alg})$ is a free $A/\mathfrak{a}$-module of rank $r - h$.

Proof. First let $\mathfrak{a}$ be a prime ideal $\mathfrak{a} = \mathfrak{p} \triangleleft A$. Since A is a Dedekind domain there exists an $m \in \mathbb{N}$ such that $\mathfrak{b} := \mathfrak{p}^m$ is principal, $(b) = \mathfrak{b}$ say. By Proposition 1.6 for $l \in \mathbb{N}$

$$\#\mathrm{Ker}(\phi_{\mathfrak{b}^l})(K^{alg}) = \#\mathrm{Ker}(\phi_{\mathfrak{b}^l})(K^{alg}) = q^{l(\deg(\phi_b) - \delta \mathrm{ht}(\phi_b))}, \qquad (1)$$

where $\delta := \delta_{\mathfrak{p},\mathfrak{p}_0}$ is the Kronecker symbol. Hence the A-module $\mathrm{Ker}(\phi_{\mathfrak{b}^l})(K^{alg})$ is finite and therefore a finitely generated torsion A-module. From the structure theorem for torsion modules over Dedekind domains (see for example [2], Ch.VII, Prop.23) we obtain

$$\mathrm{Ker}(\phi_{\mathfrak{b}^l})(K^{alg}) \cong \bigoplus_{i=1}^{l \cdot m} (A/\mathfrak{p}_i)^{s_i}$$

for some $s_i \in \mathbb{N}$. Because of

$$\phi_b \cdot \mathrm{Ker}(\phi_{\mathfrak{b}^l})(K^{alg}) = \mathrm{Ker}(\phi_{\mathfrak{b}^{l-1}})(K^{alg})$$

equation (1) at least implies

$$\mathrm{Ker}(\phi_{\mathfrak{p}^n})(K^{alg}) \cong (A/\mathfrak{p}^n)^s \quad \text{for } m/n. \qquad (2)$$

Here by (1) and the definition of rank and height of ϕ, the exponent s coincides with $r - \delta \cdot h$, where $r := \mathrm{rank}(\phi)$ and $h := \mathrm{height}(\phi)$. By using the filtration $\mathrm{Ker}(\phi_{\mathfrak{p}^n})(K^{\mathrm{alg}}) \leq \mathrm{Ker}(\phi_{\mathfrak{p}^{n+1}})(K^{\mathrm{alg}})$ for $n \in \mathbb{N}$ and action by an element $a \in \mathfrak{p} \setminus \mathfrak{p}^2$, we derive the validity of (2) for all $n \in \mathbb{N}$, which proves (b). Part (a) finally follows by prime decomposition of $\mathfrak{a}$. □

2.6. The following Corollary is immediate from Theorem 2.5:

Corollary. *The rank and height (in the case of finite characteristic) of a Drinfeld A-module are positive integers.*

2.7. The standard example

Take $A = \mathbf{F}_q[T]$ and let K be a field provided with an A-module structure via a homomorphism $\gamma : A \to K$. A Drinfeld module (over K) of rank r is a homomorphism $\phi : A \to K\{\tau\}$ of a certain type (defined above). This homomorphism is determined by $\phi_T = \gamma(T) + c_1\tau + \cdots + c_r\tau^r$ with $c_1, \ldots, c_r \in K$ and $c_r \in K^*$. On the other hand, any choice of the $c_1, \ldots, c_r$ (with $c_r \in K^*$) defines a Drinfeld module over K of rank r. Two such Drinfeld modules ϕ and ϕ' are isomorphic if and only if there is an element $a \in K^*$ such that $\phi' = a^{-1}\phi a$, or what is the same $\phi'_T = a^{-1}\phi_T a$. Thus $\phi'_T = \gamma(T) + a^{q-1}c_1\tau + a^{q^2-1}c_2\tau + \ldots a^{q^r-1}c_r\tau^r$. For an algebraically closed field K the set of isomorphy classes of Drinfeld modules of rank r is therefore isomorphic to K^{r-1}. In this isomorphism, the class of ϕ, with $\phi_T = \gamma(T) + c_1\tau + \cdots + c_r\tau^r$, is mapped to the point $(\frac{c_1^{(q^r-1)/(q-1)}}{c_r}, \frac{c_2^{(q^r-1)/(q^2-1)}}{c_r}, \frac{c_{r-1}^{(q^r-1)/(q^{r-1}-1)}}{c_r}) \in K^{r-1}$.
For $r = 1$ one finds that there is only one isomorphism class of Drinfeld modules. The standard choice, called *the Carlitz module*, for a rank one Drinfeld is given by $\phi_T = \gamma(T) + \tau$. For $r = 2$, the isomorphism class of the Drinfeld module $\phi_T = \gamma(T) + c_1\tau + c_2\tau^2$ is determined by its *j-invariant* $j(\phi) = \frac{c_1^{q+1}}{c_2}$.

For a polynomial $f \in A$ of degree n the element $\phi_f \in K\{\tau\}$ has the form $\gamma(f) + *\tau + \cdots + *\tau^{2n}$, where the coefficient of τ^{2n} lies in K^*. In other words, ϕ_f is a polynomial in the (non-commutative) variable τ of degree $2n$. The expression $\phi_f(z)$ is a polynomial in the ordinary variable z of degree q^{2n}. The equation $\phi_f(z) = 0$ is a polynomial equation of degree q^{2n}. The derivative (with respect to z) of $\phi_f(z)$ is $\gamma(f)$. Thus the equation is separable if and only if $\gamma(f)$ is nonzero.

Let K^{sep} denote the separable algebraic closure of K. Suppose that that $\gamma(f) \neq 0$ then $Ker(\phi_f)(K^{sep}) = \{z \in K^{sep} | \phi_f(z) = 0\}$ is a set with cardinality q^{2n}. The map ϕ induces on K, K^{sep} and $Ker(\phi_f)(K^{sep})$ another structure as A-module. For this stucture $Ker(\phi_f)(K^{sep})$ is isomorphic (as A-module) with $(A/fA)^r$. The Galois group $Gal(K^{sep}/K)$ acts in an obvious way on $Ker(\phi_f)(K^{sep})$.

Suppose now that $\gamma(T) = 0$. Write, as before, $\phi_T = \gamma(T) + c_1\tau + \cdots + c_r\tau^r$. Then the height of ϕ is h precisely when $c_1, \ldots c_{h-1}$ are zero and $c_h \in K^*$. In other words, h is the largest integer such that $\phi_T(z)$ is a polynomial in z^{q^h}. Clearly, the height h can be any integer with $1 \leq h \leq r$. One defines the height to be zero if $\gamma(T) \neq 0$.

3 Isogenies of Drinfeld Modules

3.1. Definition. Let ϕ, ψ be Drinfeld A-modules over a field K endowed with an $\mathbb{F}_q$-homomorphism $\gamma : A \longrightarrow K$ as in Section 2.2. Then a skew polynomial $\lambda \in K\{\tau\}$ is called a *homomorphism from* ϕ *to* ψ and is denoted by $\lambda : \phi \to \psi$ if and only if

$$\lambda \cdot \phi_a = \psi_a \cdot \lambda \qquad \text{for all } a \in A.$$

Obviously the set

$$\mathrm{Hom}_A(\phi, \psi) := \{\lambda \in K\{\tau\} | \lambda : \phi \longrightarrow \psi \text{ homomorphism}\}$$

of all homomorphisms from ϕ to ψ becomes an A-module via ψ.

A non-trivial element $\lambda \in \mathrm{Hom}_A(\phi, \psi)$ is called an *isogeny,* and an *isomorphism* if in addition λ is invertible, i.e., $\deg(\lambda) = 0$. Then the respective Drinfeld A-modules are called *isogeneous* or *isomorphic,* respectively.

3.2. Proposition. *Let ϕ and ψ be Drinfeld A-modules over a field K and $\lambda \in \mathrm{Hom}_A(\phi, \psi)$ an isogeny. Then there exist an $a \in A$ and an isogeny $\mu \in \mathrm{Hom}_A(\psi, \phi)$ with $\mu \cdot \lambda = \phi_a$.*

Proof. Obviously, $\mathrm{Ker}(\lambda)(K) := \{b \in K | \lambda \cdot b = 0\}$ with $\lambda \cdot b = \lambda(b^q) \in K$ is a finite torsion A-module via ϕ. Therefore it is annihilated by ϕ_a for some $a \in A$. Now by Proposition 1.7(b) there exists a $\mu \in K\{\tau\}$ with $\phi_a = \mu \cdot \lambda$. Using $\lambda \in \mathrm{Hom}_A(\phi, \psi)$ we obtain

$$\mu \cdot \psi(b) \cdot \lambda = \mu \cdot \lambda \cdot \phi(b) = \phi(a) \cdot \phi(b) = \phi(b) \cdot \phi(a) = \phi(b) \cdot \mu \cdot \lambda$$

in $K\{\tau\}$ and therefore $\mu \cdot \psi(b) = \phi(b) \cdot \mu$, which proves $\mu \in \mathrm{Hom}_A(\psi, \phi)$. □

This proposition immediatly implies the following

Corollary. *The isogeny relation between Drinfeld A-modules is an equivalence relation which in addition preserves the rank and the characteristic.*

3.3. Now let $\mathfrak{p} \triangleleft A$ be a prime ideal. Then by Theorem 2.5 the $\mathfrak{p}^n$-torsion submodule $\mathrm{Ker}(\phi_{\mathfrak{p}^n})(K^{alg})$ is a free $A/\mathfrak{p}^n$-module of rank $r - \delta_{\mathfrak{p},\mathfrak{p}_0} \cdot h$ for each $n \in \mathbb{N}$. The union of all these submodules is denoted by $\mathrm{Ker}(\phi_{\mathfrak{p}^\infty})(K^{alg})$. Further let $A_\mathfrak{p}$ be the completion of A with respect to $\mathfrak{p}$. Then the $A_\mathfrak{p}$-module

$$T_\mathfrak{p}(\phi) := \mathrm{Hom}_{A_\mathfrak{p}}(K_\mathfrak{p}/A_\mathfrak{p}, \mathrm{Ker}(\phi_{\mathfrak{p}^\infty})(K^{alg}))$$

of continuous $A_\mathfrak{p}$-homomorphisms is called the $\mathfrak{p}$-*adic Tate-module of* ϕ. Note that

$$T_\mathfrak{p}(\phi) \cong \varprojlim(\mathrm{Ker}(\phi_{\mathfrak{b}^l})(K^{alg})),$$

where $\mathfrak{p}^m = \mathfrak{b} = (b)$ as in the proof of Theorem 2.5, so that in the case of principal ideals the above definition of a Tate-module coincides with the usual one known from the theory of elliptic curves.

Finally writing $F_\mathfrak{p} := \mathrm{Quot}(A_\mathfrak{p})$, we obtain an $F_\mathfrak{p}$-algebra

$$V_\mathfrak{p}(\phi) := T_\mathfrak{p}(\phi) \otimes_{A_\mathfrak{p}} F_\mathfrak{p},$$

named the $\mathfrak{p}$-*adic Tate-algebra of* ϕ.

3.4. Proposition. *Let ϕ and ψ be Drinfeld A-modules over a field K. Then for all prime ideals $\mathfrak{p} \triangleleft A$ different from $\mathfrak{p}_\infty$ and $\mathfrak{p}_0$, the natural homomorphism*

$$\mathrm{Hom}_A(\phi, \psi) \otimes_A A_\mathfrak{p} \longrightarrow \mathrm{Hom}_{A_\mathfrak{p}}(T_\mathfrak{p}(\phi), T_\mathfrak{p}(\psi))$$

is injective.

Proof. Since A has finite class number we can always find an $a \in A$ with $|a|_\mathfrak{p} < 1$ and $|a|_\mathfrak{q} = 1$ for $\mathfrak{q} \notin \{\mathfrak{p}, \mathfrak{p}_\infty\}$. Then first for all $\mathfrak{p} \notin \{\mathfrak{p}_0, \mathfrak{p}_\infty\}$ the natural homomorphism

$$\mathrm{Hom}_A(\phi, \psi) \otimes_A (A/(a)) \longrightarrow \mathrm{Hom}_{A_\mathfrak{p}}(T_\mathfrak{p}(\phi), T_\mathfrak{p}(\psi)) \otimes_{A_\mathfrak{p}} (A_\mathfrak{p}/(a)) \quad (1)$$

is injective. To see (1), we take a $\lambda \in \mathrm{Hom}(\phi, \psi)$ annihilating $\mathrm{Ker}(\phi_a)(K^{alg})$. Then by Proposition 1.7(b) there exists a $\mu \in K\{\tau\}$ with $\lambda = \mu \cdot \phi_a$, which by

$$\mu \cdot \phi_b \cdot \phi_a = \mu \cdot \phi_a \cdot \phi_b = \lambda \cdot \phi_b = \psi_b \cdot \lambda = \psi_b \cdot \mu \cdot \phi_a$$

even belongs to $\mathrm{Hom}_A(\phi, \psi)$. This shows $\lambda = \mu \cdot \phi_a = 0$ on the left hand side of (1). Now the assertion is derived by taking the inverse limit. □

3.5. Theorem (Drinfeld). *Let ϕ, ψ be Drinfeld A-modules of rank r over a field K. Then* $\mathrm{Hom}_A(\phi, \psi)$ *is a projective A-module of rank equal to or less than r^2.*

Proof. We denote the A-module $\mathrm{Hom}_A(\phi, \psi)$ via ψ by M, and the vector space $M \otimes_A F_\infty$ over F_∞ by V_∞. Then with

$$||a \cdot \lambda|| := |a|_\infty q^{\deg(\lambda)} \qquad \text{for } a \in F_\infty, \lambda \in M,$$

V_∞ becomes a normed space over F_∞. Because of $M \subseteq K\{\tau\}$, the A-module M is torsion free and hence flat, and M is discrete in V_∞. Now let W be a finite dimensional subspace of V. Then the A-module $M_W := M \cap W$ is finitely generated and torsion free, which implies that M_W is projective (see [2], Ch.VI, Prop.22). Since by Proposition 3.4 the natural map

$$M \otimes_A A_\mathfrak{p} \longrightarrow \mathrm{Hom}_{A_\mathfrak{p}}(T_\mathfrak{p}(\phi), T_\mathfrak{p}(\psi))$$

is injective for $\mathfrak{p} \notin \{\mathfrak{p}_0, \mathfrak{p}_\infty\}$, the $A_\mathfrak{p}$-module on the right hand side has projective dimension less than or equal to r^2 by Theorem 2.5. This implies that for increasing subspaces $W_1 \leq W_2 \leq \ldots$ of V the corresponding projective A-modules M_{W_i} finally become constant, which proves that M is projective of $\mathrm{rank}(M) \leq r^2$. □

3.6. From Theorem 3.5 of Drinfeld we immediately derive the following results about the endomorphism rings $\mathrm{End}_A(\phi) := \mathrm{Hom}_A(\phi, \phi)$ of Drinfeld A-modules.

Corollary. *Let ϕ be a Drinfeld A-module of rank r over a field K.*
(a) The endomorphism algebra $\mathrm{End}_F(\phi) := \mathrm{End}_A(\phi) \otimes_A F$ *of ϕ over F is a skew field of degree less than or equal to r^2 over F.*
(b) In the case of general characteristic, $\mathrm{End}_F(\phi)$ *is a field of degree less than or equal to r over F.*

Sketch of proof. By Proposition 3.2 each $0 \neq \lambda \in \mathrm{End}_A(\phi)$ is invertible in $\mathrm{End}_F(\phi)$. Therefore $\mathrm{End}_F(\phi)$ is a skew field of degree $[\mathrm{End}_F(\phi) : F] \leq r^2$ because of $\mathrm{rank}(\mathrm{End}_A(\phi)) \leq r^2$. This already proves (a).

From the analytic construction of Drinfeld modules (presented for example in [9]) it will follow that in general characteristic $\mathrm{End}_A(\phi)$ is commutative. This then will imply that $\mathrm{End}_F(\phi)$ is a field with $[\mathrm{End}_F(\phi) : F] \leq r$. □

4 Definition of Drinfeld Modules over a Scheme

4.1. Definition. Let S be a scheme over A with the canonical ring homomorphism $\gamma : A \longrightarrow \Gamma(S, \mathcal{O}_S)$ and let $r \in \mathbb{N}$. A pair (G, ϕ) consisting of a group scheme G over S and a ring homomorphism

$$\phi : A \longrightarrow \mathrm{End}_S(G), \qquad a \longmapsto \phi_a$$

from A into the ring of S-group endomorphisms of G is called a *Drinfeld A-module of rank r over S* if the following conditions are satisfied. The group scheme is locally isomorphic to the additive group $\mathbb{G}_{a,S}$ over S (i.e., G is a line bundle). For all affine open subsets U of S, with $U \cong \mathrm{Spec}(B)$ say, there exists an isomorphism of group schemes $\psi_U : G_U \longrightarrow \mathbb{G}_{a,U}$ such that for all $a \in A \setminus \{0\}$ the endomorphism ϕ_a has the property

$$\psi_U \cdot \phi_a \cdot \psi_U^{-1} = \sum_{n \geq 0} b_n(a)\tau^n \in B\{\tau\},$$

where τ denotes the q-power Frobenius endomorphism, and

$$b_0(a) = \gamma(a), \;\; b_n(a) \in B^\times \text{ for } n = d, \;\; b_n(a) \in B \text{ is nilpotent for } n > d$$

with $d := -r \,\mathrm{ord}_\infty(a) \deg(\mathfrak{p}_\infty)$ (compare Proposition 1.4).

Further there is a unique way to define a *derivative*

$$\partial_0 : \mathrm{End}_S(G) \longrightarrow \Gamma(S, \mathcal{O}_S),$$

which locally coincides with the value at the origin defined in Section 1.5. This derivative has the additional property $\partial_0 \cdot \phi = \gamma$. Finally the induced morphism

$$\vartheta := \mathrm{Spec}(\partial_0 \cdot \phi) : S \longrightarrow \mathrm{Spec}(A)$$

is called the *characteristic* of (G, ϕ).

4.2. *Remark.* It follows from Proposition 1.4 that every Drinfeld A-module (G, ϕ) over a scheme is isomorphic to one of the same rank with the stronger property that $b_n(a) = 0$ for $n > d$. Using the fact that G is a line bundle L over S with associated invertible sheaf $\mathcal{L}$, the defining properties of ϕ then can be simplified to

$$\phi_a = \sum_{n=0}^{d} \beta_n(a) \cdot \tau^n.$$

In this expression $\beta_n(a) \cdot \tau^n$ denotes the endomorphism of $\mathcal{L}$ given by $\tau^n : \mathcal{L} \to \mathcal{L}^{\otimes q^n}$, which is locally defined by $b \mapsto b^{\otimes q^n}$, and $\beta_n(a) \in \Gamma(S, L^{\otimes(1-q^n)})$, denotes the morphism of multiplication by the global section $\beta_n(a)$, with the additional properties

$$\beta_0(a) = \gamma(a) \quad \text{ and } \quad \beta_d(a) \text{ is locally invertible.}$$

A Drinfeld A-module (L,ϕ) of this type is called a *standard Drinfeld A-module over S* (see the subsequent lecture of Saïdi [10] for more details).

4.3. Now we return to the general case. For each ideal $\mathfrak{a} \lhd A$ different from (0)

$$G_{\mathfrak{a}} := \bigcap_{a\in\mathfrak{a}} \mathrm{Ker}(\phi_a)$$

defines a scheme over S – which by Theorem 2.5. is a scheme of $A/\mathfrak{a}$-modules – called the *scheme of* $\mathfrak{a}$*-torsion points* of (G,ϕ). The restriction of $G_{\mathfrak{a}}$ to $S\setminus\vartheta^{-1}(V(\mathfrak{a}))$, where $V(\mathfrak{a})$ denotes the closed subscheme of Spec(A) defined by $\mathfrak{a}$, is an étale sheaf of $A/\mathfrak{a}$-modules. Locally, for the étale topology, this sheaf is isomorphic to the constant étale sheaf with stalk $(\mathfrak{a}^{-1}/A)^r$. In other words, the restriction of $G_{\mathfrak{a}}$ to $S\setminus\vartheta^{-1}(V(\mathfrak{a}))$ becomes isomorphic to the constant étale sheaf $(\mathfrak{a}^{-1}/A)^r$, after a finite étale extension of $S\setminus\vartheta^{-1}(V(\mathfrak{a}))$.

In the case that $\vartheta^{-1}(V(\mathfrak{a}))=\emptyset$, one defines a *level-*$\mathfrak{a}$ *structure on* (G,ϕ) as an isomorphism of (étale) sheaves of $A/\mathfrak{a}$-modules over S

$$\iota_{\mathfrak{a}} : (\mathfrak{a}^{-1}/A)^r \longrightarrow G_{\mathfrak{a}}$$

In more down to earth language, a level $\mathfrak{a}$ *structure on* (G,ϕ) is an A-module isomorphism between the A-module $(\mathfrak{a}^{-1}/A)^r$ and the A module of the global sections of $G_{\mathfrak{a}}$.

4.4. *Remark.* In the paper [4] of Drinfeld, the definition of a level-$\mathfrak{a}$ structure is extended to the case where the characteristic of (G,ϕ) divides $\mathfrak{a}$, i.e. $\vartheta^{-1}(V(\mathfrak{a}))\neq\emptyset$. In lecture 2 this definition will be given.

4.5 *The standard example.*
We continue and extend here the example 2.7. A Drinfeld module of rank r over an A-scheme S is given by a line bundle L above S and a homomorphism $\phi : A \to End(L)$. The scheme S is the union of open affine schemes $Spec(R)$, such that the restriction of L to each open piece is isomorphic to the trivial line bundle associated with the rank one module R over R. Above $Spec(R)$ the Drinfeld structure ϕ is equivalent to a homomorphism $\phi : A \to R$. Let the A-algebra structure of R be given by a homomorphism $\gamma : A \to R$.
Suppose again that $A=\mathbf{F}_q[T]$. Then this restriction of ϕ is determined by $\phi_T=\gamma(T)+c_1\tau+\cdots+c_r\tau^r$ with $c_1,\ldots,c_r\in R$ and moreover $c_r\in R^*$.
Let $f\in A$ be a polynomial of positive degree, such that $\gamma(f)\neq 0$. Then a level f-structure is given by an isomorphism of A-modules $(A/fA)^r\to Ker(\phi_f)(R)$.

References

[1] A. Blum, U. Stuhler: *Drinfeld modules and elliptic sheaves*, Springer-Verlag, Berlin (to appear).

[2] N. Bourbaki: *Commutative Algebra*, Springer-Verlag, Berlin 1989.

[3] P. Deligne, D. Husemöller: *Survey of Drinfeld modules*, Contemp. Math. **67**, 25-91 (1987).

[4] V.G. Drinfeld: *Elliptic modules*, Math. USSR-Sbornik **23**, 561-592 (1976).

[5] E.-U. Gekeler: *Drinfeld modular curves*, Springer-Verlag, 1986.

[6] D. Goss: *Basic structures of function field arithmetic*, Springer-Verlag, Berlin 1996.

[7] D.R. Hayes: *A brief introduction to Drinfeld modules*, in D.Goss(eds.) The arithmetic of function fields, W.d.Gruyter, Berlin 1992.

[8] G. Laumon: *Cohomology of Drinfeld modular varieties I*, Cambridge University Press 1996.

[9] B. Lopez: *Analytic theory of Drinfeld modules*, this volume.

[10] M. Saïdi: *Moduli schemes of Drinfeld modules*, this volume.

Lecture 2

Moduli schemes of Drinfeld modules

Mohamed Saïdi
Mathematisches Institut, Universität Heidelberg
Im Neuenheimer Feld 288, D-69120 Heidelberg Germany
email SAIDI@harmless.mathi.uni-heidelberg.DE

Introduction. In his paper [Dr] Drinfeld introduced the notion of *elliptic module* (Drinfeld module) of rank r over an A-scheme S, where A is the Dedekind ring of the functions on a proper, smooth, and geometrically irreducible curve, over a finite field $\mathbb{F}_q$, which are regular outside one point. He proves that elliptic modules endowed with level n-structures, for admissible ideals n of A, have a *fine moduli scheme*. This means that there exists an $\mathbb{F}_q$-scheme $M^r(n)$, and a one to one functorial correspondence between isomorphism classes of elliptic modules with level n-structure over S, and points of $M^r(n)$ with value in S, for any A-scheme S. Moreover the scheme $M^r(n)$ is affine, of finite type, and smooth over $\mathbb{F}_q$ of dimension r.

The fact that the scheme $M^r(n)$ exists, is affine, and of finite type, is not difficult to prove. The difficult part is to prove that $M^r(n)$ is smooth of dimension r over $\mathbb{F}_q$. Drinfeld's method is the following. Consider an elliptic module (i.e., a Drinfeld module) of rank r and with level n-structure over an algebraically closed field. This corresponds to a geometric point of the scheme $M^r(n)$. The space of all *deformations* of this elliptic module can be identified with the spectrum of the completion of the local ring of $M^r(n)$ at the geometric point. Drinfeld defines the notion of *divisible module* endowed with a level structure, and associated in a functorial way to every elliptic module with level structure a divisible module with level structure. He then proves that the spaces of deformations of the two objects are canonically isomorphic. This is an analogue of the Serre-Tate Theorem concerning deformations of an *abelian variety* in characteristic $p > 0$ and deformations of its associated *p-divisible group*. Drinfeld proves that the deformations space of the divisible module over $\overline{\mathbb{F}}_q$ associated to an elliptic module is the spectrum of a regular ring of dimension r over $\overline{\mathbb{F}}_q$. This implies that $M^r(n)$ is smooth. We refer to [Dr] for more details.

In this note we give the definitions of the objects introduced by Drinfeld, including Drinfeld modules over schemes, level structures and moduli functors. We discuss the existence of the moduli scheme $M^r(n)$, for admissible ideals n, and give a proof, due to Van der Put, of the fact that these schemes are smooth of dimension r over $\mathbb{F}_q$, in the case where $A = \mathbb{F}_q[T]$ is a polynomial ring. For other references concerning the existence and properties of the moduli schemes of Drinfeld modules, apart from [Dr], the reader can also consult [Ge], [Ge-1], [La], and [Bl-St].

I would like to thank M. Van der Put for very useful and interesting discussions, and also for communicating me his private notes on the subject. His comments and suggestions helped a lot during the preparation of this manuscript.

0. Notations.

Throughout this paper we adopt the following notations

1. $\mathbb{F}_q$ will denote the finite field of q elements and characteristic p.

2. C is a projective smooth and geometrically irreducible algebraic curve over $\mathbb{F}_q$.

3. ∞ is a fixed closed point of C, $\deg(\infty) := [\mathbb{F}_q(\infty) : \mathbb{F}_q]$ its degree, and ord_∞ is the corresponding discrete valuation of the function field $\mathbb{F}_q(C)$ of C.

4. $A := \Gamma(C - \infty, \mathcal{O}_C)$ is the ring of functions on C (these are elements of $\mathbb{F}_q(C)$) which are regular (i.e. have no poles) outside ∞.

The basic example is $C = \mathbb{P}^1$ the projective line over $\mathbb{F}_q$, and ∞ is the point "at infinity" with local parameter $1/T$. In this case A is the polynomial ring $\mathbb{F}_q[T]$.

1. Drinfeld modules over A-schemes.

1.1. Let $\gamma^* : S \to \operatorname{Spec} A$ be an A-scheme, and let $\gamma : A \to \Gamma(S, \mathcal{O}_S)$ be the canonical ring homomorphism, where $\Gamma(S, \mathcal{O}_S)$ denotes the ring of global sections of the structural sheaf $\mathcal{O}_S$. Consider a *line bundle* L over S (cf. [Gr-Di], 9.4). If L is trivial on an open $S' \subset S$, then the restriction of L to S' is isomorphic to the trivial S'-line bundle $\mathbb{A}^1_{S'} := \operatorname{Spec} S'[X] := S' \times_{\operatorname{Spec}\mathbb{Z}} \operatorname{Spec}\mathbb{Z}[X]$, and on $\mathbb{A}^1_{S'}$ one has the *additive* group structure $\mathbb{G}_{a,S'} := S' \times_{\operatorname{Spec}\mathbb{Z}} \mathbb{G}_{a,\mathbb{Z}}$ (cf. the paper of Shatz in [Co-Si] for the functorial definition of $\mathbb{G}_a$). As L

is locally trivial, there exists a unique structure of S-group scheme on L, which is, locally for the Zariski topology, isomorphic to the additive group scheme $\mathbb{G}_a$. Moreover there is an equivalence of categories between line bundles L over S, and invertible (locally free of rank 1) sheaves $\mathcal{L}$ of $\mathcal{O}_S$-modules on S (cf. loc. cit).

For a line bundle L over S and for a positive integer n, we will denote by $\tau^n : L \to L^{q^n}$ the n-th power of the Frobenius endomorphism, which is a morphism of group schemes. For an open affine $S' := \operatorname{Spec} R \subset S$ over which L is trivial, $\tau^n : \mathbb{G}_{a,S'} \to \mathbb{G}_{a,S'}$ corresponds to the ring homomorphism $R[X] \to R[X]$ given by $X \to X^{q^n}$. For an affine scheme $S = \operatorname{Spec} R$, we will denote by $R\{\tau\}$ the skew polynomial ring over R of endomorphisms of the additive group scheme $\mathbb{G}_{a,S}$, where τ is the Frobenius endomorphism, and the commutation rule is given by $\tau.b = b^q.\tau$ for $b \in R$ (cf. [Ma] 1).

1.2. Definition. A *(standard) Drinfeld A-module* $(L, \varphi)_S$ of rank $r \in \mathbb{N}$ over S consists of a line bundle L over S and a ring homomorphism

$$\varphi : A \to \operatorname{End}_{S-\text{groups}}(L)$$

from A to the ring of S-endomorphisms of the group scheme underlying L, which satisfies the following conditions:

(i) for $a \in A - \{0\}$, $\varphi(a) = \sum_{n=0}^{-r\deg(\infty)\operatorname{ord}_\infty(a)} \alpha_n(a).\tau^n$, where $\tau^n : L \to L^{q^n}$ is the n-th power of the Frobenius endomorphism.

(ii) $\alpha_n(a) \in \Gamma(S, \mathcal{L}^{1-q^n})$. In the expression above for $\varphi(a)$, the term $\alpha_n(a)$ can be seen as the morphism $\mathcal{L}^{q^n} \to \mathcal{L}^{1-q^n} \otimes \mathcal{L}^{q^n} \simeq \mathcal{L}$ given by the multiplication with the global section $\alpha_n(a)$ (where $\mathcal{L}$ is the invertible sheaf on S associated to L). Then $\alpha_n(a).\tau^n$ is an endomorphism of $\mathcal{L}$ since τ^n is a morphism of $\mathcal{L}$ to $\mathcal{L}^{q^n}$. (See also [Ma], Remark 4.2).

(iii) $\alpha_0(a) = \gamma(a)$, and $\alpha_n(a)$ vanishes nowhere on S (i.e., it is a global generator of the line bundle $\mathcal{L}^{1-q^n}$) for $n = -r\deg(\infty)\operatorname{ord}_\infty(a)$. It is the leading coefficient of $\varphi(a)$.

1.3. Remark. If $S = \operatorname{Spec} R$ is affine, where R is given a structure of A-algebra via $\gamma : A \to R$, and if L is the trivial line bundle over S, then the Drinfeld module $(L, \varphi)_S$ is given by a ring homomorphism

$$\varphi : A \to \operatorname{End}_{S-\text{groups}} \mathbb{G}_{a,S} = R\{\tau\}$$

such that for $a \in A - \{0\}$, $\varphi(a) = \sum_{n=0}^{-r\deg(\infty)\operatorname{ord}_\infty(a)} a_n\tau^n$, where $a_i \in R$, a_n is invertible in R if $n = -r\deg(\infty)\operatorname{ord}_\infty(a)$, and $a_0 = \gamma(a)$.

1.4. Remark. In his paper [Dr] Drinfeld defined a slightly different notion of *elliptic A-module* over S (cf. [Dr] §5, and [Ma] §4). He also defined the notion of a standard elliptic A-module as presented above. The latter definition is better adapted when dealing with the existence of moduli schemes. Moreover a standard Drinfeld A-module over S gives rise (by forgetting the $\mathcal{O}_S$-linear structure on the line bundle) to an elliptic A-module over S as in [Dr], and such an elliptic A-module is isomorphic (cf. 1.6) by [Dr], prop. 5.2. (cf. also [Ma] 1.4) to a standard one of the same rank.

1.5. Let $S' \to S$ be a morphism of schemes. If $(L, \varphi)_S$ is a Drinfeld A-module over S, one deduces a Drinfeld A-module $(L', \varphi')_{S'}$ over S' of the same rank, where $L' := L \times_S S'$, and φ' is deduced from φ. We will say that $(L', \varphi')_{S'}$ is the *pull-back* of $(L, \varphi)_S$ via the morphism $S' \to S$. In particular $(L, \varphi)_S$ induces by pull-back a Drinfeld A-module over the residue class fields at each point $s \in S$. In that sense $(L, \varphi)_S$ can be thought as a "*continuous family*" of Drinfeld A-modules over the residue class fields of S.

1.6. A *morphism* $\psi : (L', \varphi')_S \to (L, \varphi)_S$ between Drinfeld A-modules over S, is a morphism of S-group schemes $\psi : L' \to L$ which is compatible with φ and φ', i.e. $\psi \circ \varphi'(a) = \varphi(a) \circ \psi$ for each $a \in A$, and ψ is an isomorphism if $\psi : L' \to L$ is an isomorphism. Non zero morphisms can exist only between Drinfeld A-modules of the same rank (cf. [Ma]). Note that each invertible element of A defines an automorphism of the Drinfeld A-module $(L, \varphi)_S$ (cf. [Ma] 2.3).

1.7. The moduli problem. Consider the following contravariant functor $\mathcal{M}^r$ from the category of A-schemes to the category of sets

$$\mathcal{M}^r : (\text{Schemes}\,/A) \to (\text{Sets}), \quad S \mapsto \mathcal{M}^r(S)$$

which to an A-scheme S associate the set $\mathcal{M}^r(S)$ of isomorphism classes of Drinfeld A-modules of rank r over S.

For any A-scheme X we denote by $\mathcal{F}_X$ its functor of points

$$\mathcal{F}_X : (\text{Schemes}\,/A) \to (\text{Sets})$$

$$S \mapsto \mathcal{F}_X(S) := X(S) := \mathrm{Mor}_{A-\text{Schemes}}(S, X).$$

The functor $\mathcal{M}^r$ is said to be *representable*, if there exists an A-scheme M^r such that $\mathcal{M}^r$ is isomorphic to the functor $\mathcal{F}_{M^r}$, i.e., if there exist a one to one functorial bijection between the sets $\mathcal{M}^r(S)$ and $\mathcal{F}_{M^r}(S)$, for any A-scheme S.

Suppose that the functor $\mathcal{M}^r$ is representable by the A-scheme M^r. To the identity morphism id : $M^r \to M^r$ will then correspond a "*universal*" Drinfeld A-module $(\tilde{L}, \tilde{\varphi})^{\text{univ}}_{M^r}$ of rank r over M^r, and any Drinfeld A-module $(L, \varphi)_S$ of rank r, over an A-scheme S, is isomorphic to the Drinfeld module $(\tilde{L}, \tilde{\varphi})^{\text{univ}}_{M^r}$, obtained by pull-back via a unique morphism $S \to M^r$. This implies that

$(L, \varphi)_S$ has no automorphism different from the identity. But this is not the case, as is mentioned in 1.6. Hence we deduce that the functor $\mathcal{M}^r$ is not representable.

1.8. Example. We will give an example which shows another reason why the functor $\mathcal{M}^2$ is not representable. Assume that $A = \mathbb{F}_q[T]$ is a polynomial ring, and suppose that the functor $\mathcal{M}^2$ is representable, by an A-scheme X. Then for any field K which is an A-algebra via $\gamma : A \to K$, the set of points of X with values in K

$$X(K) := \mathrm{Mor}_{A-\mathrm{Schemes}}(\mathrm{Spec}\, K, X) = \{(c_1, c_2) |\, c_1, c_2 \in K,\ c_2 \neq 0\} / \sim_K$$

is equal to the set of pairs (c_1, c_2) of elements of K, with c_2 unequal to 0, up to the equivalence relation given by $(c_1, c_2) \sim_K (c_1', c_2')$ if there exists $u \in K^\times$ such that $(c_1', c_2') = (u^{q-1} c_1, u^{q^2-1} c_2)$. Indeed, any rank two Drinfeld module over $\mathrm{Spec}\, K$ is given by a ring homomorphism $\varphi : \mathbb{F}_q[T] \to K\{\tau\}$, which is determined by the image

$$\varphi(T) = \gamma(T) + c_1 \tau + c_2 \tau$$

of T, and $\varphi(T)$ is isomorphic to $\varphi'(T) = \gamma(T) + c_1' \tau + c_2' \tau$ if and only if $(c_1, c_2) \sim_K (c_1', c_2')$. The natural map $X(K) \to X(\overline{K})$, where $\overline{K}$ is an algebraic closure of K, is the inclusion and therefore injective. However, the map

$$\{(c_1, c_2) |\, c_1, c_2 \in K,\ c_2 \neq 0\} / \sim_K \to \{(c_1', c_2') |\, c_1', c_2' \in \overline{K}, c_2' \neq 0\} / \sim_{\overline{K}}$$

is not injective. This contradiction shows that the functor $\mathcal{M}^2$ is not representable.

In order to obtain representability results for functors associated to Drinfeld modules, Drinfeld rigidified the situation by introducing the notion of *level structure* on a Drinfeld A-module, with respect to ideals of A. In particular a Drinfeld A-module with *suitable* level structure will have only the identity as automorphism.

2. Level structures.

Drinfeld introduced in [Dr] the notion of *level structures* on Drinfeld A-modules, with respect to ideals of A. This notion has been extended, and intensively discussed by Katz and Mazur for *elliptic curves* (cf. [Ka-Ma]).

2.1. Let n be a non-zero ideal of A, and let $V(n)$ be the closed subscheme of $\mathrm{Spec}\, A$ defined by n (as a set, $V(n)$ consists of those prime ideals of A containing n). Let $\gamma^* : S \to \mathrm{Spec}\, A$ be an A-scheme, and consider a Drinfeld A-module $(L, \varphi)_S$ of rank r over S. The morphism γ^* is called the *characteristic* of $(L,\varphi)_S$ over S, (or more precisely the characteristic of the A-scheme S). We will say that the ideal n is *prime to the characteristic* of $(L,\varphi)_S$, if $\gamma^*(S)\cap V(n)=\emptyset$.

2.2. Definition. The subscheme of *n-division points* of L is defined as the subgroup scheme L_n of L given by

$$L_n := \cap_{a\in n}\operatorname{Ker}(L \xrightarrow{\varphi(a)} L)$$

where the intersection is the intersection (i.e., fiber product) of subschemes.

Note that the group scheme L_n has a natural structure of scheme of A/n-modules, which is induced by the homomorphism φ. Moreover if n is generated by $a_1,\ldots,a_m$ then

$$L_n = L_{a_1}\cap\ldots\cap L_{a_m} \text{ where } L_a := \operatorname{Ker}(L \xrightarrow{\varphi(a)} L).$$

The following proposition is a consequence of [Dr], prop. 2.2. (cf. also [Ma]). For the analogous result concerning structure of division points for elliptic curves we refer the reader to [Ka-Ma].

2.3. Proposition. *The subgroup scheme L_n of L is finite and flat over S, of rank* $\operatorname{Card}(A/n)^r$. *If the ideal n is prime to the characteristic of S (cf. 2.1), then L_n is étale over S, and locally for the étale topology isomorphic to the constant scheme of A/n-modules $(n^{-1}/A)^r_S$ with stalk $(n^{-1}/A)^r$.*

2.4. Definition. A *level n-structure* on the Drinfeld A-module $(L,\varphi)_S$ is a homomorphism of A-modules

$$l : (n^{-1}/A)^r \to L(S) := \operatorname{Mor}_{S-\text{Schemes}}(S,L)$$

where $L(S)$ is endowed with the A-module structure induced by φ, and which induces an equality of effective Cartier divisors in L

$$L_n = \sum_{\alpha\in(n^{-1}/A)^r} [l(\alpha)].$$

Here $[l(\alpha)]$ denotes the *Cartier divisor* in L, image of the section $l(\alpha): S\to L$.

2.5. Example. Assume $A=\mathbb{F}_q[T]$. Let $\gamma : A\to K$ be a ring homomorphism, where K is an algebraically closed field. Consider a Drinfeld module of rank r over $\operatorname{Spec} K$, given by the ring homomorphism $\varphi:\mathbb{F}_q[T]\to K\{\tau\}$, with $\varphi(T)=\sum_{i=0}^r c_i\tau^i$. Let $n=(f)$ be a non zero ideal of $\mathbb{F}_q[T]$. If $\varphi(f)=\sum_{j=0}^{r\,\mathrm{ord}_\infty(f)} b_j\tau^j$, we denote by $\varphi_f(Z)$ the polynomial $\sum_{j=0}^{r\,\mathrm{ord}_\infty(f)} b_jZ^{q^j}\in K[Z^q]$. A level n-structure on the Drinfeld module φ is a homomorphism of A-modules

$l : (f^{-1}A/A)^r \to K$, where K is given the structure of A-module induced by φ, and such that the polynomial $\varphi_f(Z)$ is equal to $b\prod_{\alpha\in(f^{-1}A/A)^r}(Z - l(\alpha))$ for some element $b \in K$.

More specifically, let $f = T$ and $r = 2$. The Drinfeld module (L, φ) is given by $\varphi(T) = \gamma(T) + c_1\tau + c_2\tau^2$ with c_2 invertible. Let us distinguish the following three cases:
(a) $\mathrm{char}(K) \neq (T)$,
(b) $\mathrm{char}(K) = (T)$ and $c_1 \neq 0$,
(c) $\mathrm{char}(K) = (T)$ and $c_1 = 0$.
The Cartier divisor $L_{(T)}$ of (T)-division points equals $\mathrm{Spec}(K[Z]/(\varphi_T(Z)))$. Its points over the algebraic closure $\bar{K}$ of K therefore form an $A/(T)$-module. In case (a), it is a vector space of dimension two over $\mathbb{F}_q$ and the points have multiplicity one. In case (b), it is a vector space of dimension one over $\mathbb{F}_q$ and the points have multiplicity q. In case (c), it consists of a single point 0 with multiplicity q^2.
A level (T)-structure on (L, φ) over $\bar{K}$ turns out to be a surjective $A/(T)$-homomorphism l from $(T^{-1}A/A)^2$ to the $\bar{K}$-points $L_{(T)}(\bar{K})$. In case (a) this simply means that l is an isomorphism.

2.6. Remark. If the ideal n is prime to the characteristic of S, then a level n-structure on $(L, \varphi)_S$ is an isomorphism

$$l : (n^{-1}/A)^r{}_S \simeq L_n$$

of schemes of A/n-modules. And it follows in this case from the structure of n-division points of the geometric fibers of L over S that such level n-structures exist locally for the étale topology on S. This means that, if n is prime to $\mathrm{Char}(S)$, there exists a finite étale covering $S' \to S$ such that the pull-back of $(L, \varphi)_S$ to S' carries a level (n)-structure over S'.

We will denote by $(L, \varphi, l)_S$ the Drinfeld A-module $(L, \varphi)_S$ with level n-structure l. A *homomorphism* between Drinfeld A-modules over S with level n-structures is a homomorphism of Drinfeld A-module over S (cf. 1.5) which respects the given level n-structures.

2.7. Consider the following contravariant functor

$$\mathcal{M}^r(n) : (\text{Schemes}/A) \to (\text{Sets})$$

$$S \mapsto \mathcal{M}^r(n)(S)$$

from the category of A-schemes to the category of sets which to an A-scheme S associates the set $\mathcal{M}^r(n)(S)$ of isomorphism classes $(L, \varphi, l)_S$ of Drinfeld A-modules $(L, \varphi)_S$ of rank r over S with a level n-structure l.

Note that if $0 \neq m \subset n$ is an inclusion of ideals of A, then one has a natural transformation of functors $\mathcal{M}^r(m) \to \mathcal{M}^r(n)$, which is induced by the canonical homomorphism $n^{-1}/A \to m^{-1}/A$ of A-modules.

3. Existence and properties of moduli schemes.

The following result of existence of moduli schemes for Drinfeld A-modules of rank r with level n-structures, for "*admissible*" ideals n of A, is due to Drinfeld.

3.1. Theorem. ([Dr], 5) *Let $0 \neq n \subset A$ be an admissible ideal of A, by which we mean that* Card $V(n) \geq 2$. *Then*

(i) The functor $\mathcal{M}^r(n)$ is representable by an affine A-scheme $M^r(n)$ of finite type, which has dimension $r-1$ over Spec A.
(ii) $M^r(n)$ is a smooth $\mathbb{F}_q$-scheme of dimension r, and the canonical morphism $M^r(n) \to$ Spec A is smooth over Spec $A - V(n)$.
(iii) If $0 \neq m \subset n$ is an inclusion of ideals, then the canonical morphism $M^r(m) \to M^r(n)$ is finite and flat.

3.2. Existence of the scheme $M^r(n)$.

3.2.1. We will first consider the special case where $A = \mathbb{F}_q[T]$ is a polynomial ring. For a non zero ideal $n = (f)$ of A, we will denote by $\mathcal{M}^r(f) := \mathcal{M}^r(n)$ the corresponding functor.

3.2.2. Lemma. *Let $n = (f)$ be a non-zero ideal of A and let $f_1 \in A$ be a monic irreducible divisor of f. Then the functor $\mathcal{M}^r(f)$, restricted to the category of $A[1/f_1]$-schemes, is representable by an affine scheme of dimension r over $\mathbb{F}_q$.*

Proof. We first introduce a ring

$$B := A[1/f_1][c_1, ..., c_r, c_r^{-1}, \{x_\alpha\}]_{\alpha \in (f^{-1}A/A)^r}$$

generated over $A[1/f_1]$ by $c_1, ..., c_r, c_r^{-1}$ $(c_r c_r^{-1} = 1)$, and the x_α for $\alpha \in (f^{-1}A/A)^r$. The relations between the generators will be specified later. In order to formulate them we introduce an $\mathbb{F}_q$-homomorphism $\tilde{\varphi} : A \to B\{\tau\}$ given by

$$\tilde{\varphi}_T = T + c_1\tau + ... + c_r\tau^r.$$

The relations are:

(i) $x_\alpha + x_\beta = x_{\alpha+\beta}$, for α, $\beta \in (f^{-1}A/A)^r$, and $x_{b\alpha} = bx_\alpha$, for $b \in \mathbb{F}_q$.

(ii) $x_0 = 0$, $\tilde{\varphi}_T(x_\alpha) := Tx_\alpha + c_1 x_\alpha^q + \ldots + c_r x_\alpha^{q^r} = x_{T\alpha}$, for every non-zero element $\alpha \in (f^{-1}A/A)^r$.

(iii) Let $\tilde{\varphi}_f = \sum_{j=0}^{r\,\mathrm{ord}_\infty(f)} b_j \tau^j$. Then the polynomial $\tilde{\varphi}_f(Z) := \sum_{j=0}^{r\,\mathrm{ord}_\infty(f)} b_j Z^{q^j}$ is equal to $b\prod_{\alpha\in(f^{-1}A/A)^r}(Z - x_\alpha)$, for a suitable $b \in B$.

(iv) Finally, in order to rigidify the situation, we choose an element $\alpha_0 \in (f^{-1}A/A)^r$ with $\alpha_0 \neq 0$, and such that $f_1\alpha_0 = 0$, and we add the relation $x_{\alpha_0} = 1$.

We note that the above data define a Drinfeld module $(\tilde{L}, \tilde{\varphi}, \tilde{l})$ over Spec B, of rank r, and with level (f)-structure $\tilde{l}$, where $\tilde{L}$ is the trivial line bundle over Spec B, $\tilde{\varphi}$ is as above, and $\tilde{l} : (f^{-1}A/A)^r \to B$ is defined by $\tilde{l}(\alpha) = x_\alpha$. We claim that Spec B with its additional Drinfeld structure represents the functor $\mathcal{M}^r(f)$ restricted to the category of $A[1/f_1]$-schemes. In order to prove the claim, we consider a ring R, which is an $A[1/f_1]$-algebra via $\gamma : A[1/f_1] \to R$, provided with the additional data:

(a) A line bundle L over Spec R.

(b) A Drinfeld A-module $\varphi : A \to \mathrm{End}(L)$ of rank r.

(c) A homomorphism of A-modules $l : (f^{-1}A/A)^r \to L(\mathrm{Spec}\, R)$ such that $\ker \varphi(f)$ coincides as Cartier divisor with $\sum_{\alpha\in(f^{-1}A/A)^r} l[(\alpha)]$.

Consider the section $l(\alpha_0)$ of L. It satisfies $\varphi(f_1)(l(\alpha_o)) = 0$. Using property (c) and the fact that $\varphi(f_1)(Z)$ is a separable polynomial, one sees that the section $l(\alpha_0)$ has no zeros. Thus we can uniquely identify the line bundle L with the ring R such that $l(\alpha_0) = 1 \in R$. In particular the homomorphism $\varphi : A \to R\{\tau\}$ is given by $\varphi_T = d_0 + d_1\tau + \ldots. + d_r\tau^r$.

All the $l(\alpha)$'s are now elements of R, and condition (c) can be translated into : the polynomials $\varphi_f(Z)$ and $d\prod_{\alpha\in(f^{-1}A/A)^r}(Z - l(\alpha))$ are identical for some unit d of R. Consider the $A[1/f_1]$-homomorphism $h : B \to R$ given by $h(c_i) = d_i$ for $i \in \{1, \ldots, r\}$, and $h(x_\alpha) = l(\alpha)$ for $\alpha \in (f^{-1}A/A)^r$. This is well defined since the $\{d_i\}$ and the $\{l(\alpha)\}$ satisfy the relations of the $\{c_i\}$ and the $\{x_\alpha\}$. Clearly h transports the Drinfeld structure $(\tilde{L}, \tilde{\varphi}, \tilde{l})$ above Spec B to (L, φ, l) over Spec R, and h is uniquely determined by this property. Thus we proved that Spec B represents the functor $\mathcal{M}^r(f)$ restricted to $A[1/f_1]$-schemes. It is easy to see that Spec B maps via a finite map to an open subscheme of the affine space $\mathbb{A}^r$ of dimension r over $\mathbb{F}_q$. This map is obtained by forgetting the level structure in the above considerations. Thus Spec B is also of dimension r.

3.2.3. Corollary. *Suppose that the non-zero ideal $n = (f)$ is admissible,* *i.e., f has at least two distinct monic irreducible divisors f_1 and f_2. Then the functor $\mathcal{M}^r(f)$ is representable by an affine scheme over A.*

Proof. The restriction of $\mathcal{M}^r(f)$ to the category of $A[1/f_i]$-scheme, for $i \in \{1,2\}$, is representable by some affine $A[1/f_i]$-scheme X_i. The restriction of the functor $\mathcal{M}^r(f)$ to $A[1/f_1, 1/f_2]$ is represented by $X_1 \otimes A[1/f_1, 1/f_2]$ and also by $X_2 \otimes A[1/f_1, 1/f_2]$. The two affine schemes are isomorphic. This defines an isomorphism of an open part of X_1 with an open part of X_2. Let $M^r(f)$ be obtained by gluing X_1 and X_2 over these open parts. It is clear that $M^r(f)$ represents the functor $\mathcal{M}^r(f)$ on the category of A-schemes. By construction, $M^r(f) \times_{\operatorname{Spec} A} \operatorname{Spec} A[1/f_i] \simeq X_i$, for $i \in \{1,2\}$. Using [Ha] page 81, exercise 2.17 (b), one concludes that $M^r(f)$ is affine.

3.2.4. Example. We will show that the functor $\mathcal{M}^2(T)$ is not representable (same notation as above). To see this, it suffices to produce a Drinfeld A-module of rank 2 and level T-structure, having non trivial automorphisms. Consider the homomorphism $\gamma : A = \mathbb{F}_q[T] \to \mathbb{F}_q = A/TA$. Let $\varphi : \mathbb{F}_q[T] \to \mathbb{F}_q\{\tau\}$ be the Drinfeld A-module of rank 2 over $\operatorname{Spec} \mathbb{F}_q$, given by $\varphi(T) = \tau^2$, and endowed with the level T-structure defined by $l : (T^{-1}A/A)^2 \to \mathbb{F}_q$ is the zero map. This Drinfeld module with the given level T-structure has $(\mathbb{F}_{q^2})^*$ as group of automorphisms.

3.2.5. The proof of theorem 3.1 part (i) for the general case, i.e., where A is not necessarily a polynomial ring, follows from the next proposition and the arguments used in the proof of 3.2.3.

3.2.6 Proposition. *Let $n \subset A$ be a non-zero ideal of A. Then the functor $\mathcal{M}^r(n)$, restricted to the category of A-schemes S such that the image of $S \to \operatorname{Spec} A$ is contained in $\operatorname{Spec} A - V(n)$, is representable by an affine A-scheme of finite type.*

Proof. (cf. [La] p. 7, 8) The ring A has a presentation $A = \mathbb{F}_q[a_1, ..., a_t]/(f_1, ..., f_m)$ such that the ideal n is generated by $a_1, ..., a_s$, for $s \leq t$. We will construct the affine A-scheme $\operatorname{Spec}(B)$, required in the proposition, by defining B by generators $X_{\lambda,n}, L(\alpha)$ and relations. Thus $B = A[X_{\lambda,n}, L(\alpha)]/(R) = A[x_{\lambda,n}, l(\alpha)]$. The line bundle above $\operatorname{Spec}(B)$ is taken to be the trivial one. The homomorphism $\varphi : A \to B\{\tau\}$ will be induced by a homomorphism $\Phi : \mathbb{F}_q[a_1, \ldots, a_t] \to A[X_{\lambda,n}, L(\alpha)]_{\lambda,n,\alpha}/(R)\{\tau\}$ The level structure l will be induced by L. The generators of B are specified by:

(a) $X_{\lambda,n}$ for $\lambda \in \{1, \ldots, t\}$, and $n = 0, ..., N_\lambda$, where $N_\lambda := -r \deg(\infty) \operatorname{ord}_\infty(a_\lambda)$.

(b) $L(\alpha)$, for $\alpha \in (n^{-1}A/A)^r$.

The homomorphism Φ is given by $\Phi(a_i) = \sum_{i=0}^{N_\lambda} X_{\lambda,n}\tau^n$ modulo (R) for $\lambda = 1, ..., t$.

(i) The first set of relations is obtained by requiring that the expressions $\Phi(a_{\lambda'})\Phi(a_{\lambda''})) - \Phi(a_{\lambda''})$ are identical zero. Thus R contains the coefficients of those expressions w.r.t. the variable τ. This guarantees that Φ is indeed a homomorphism.

(ii) The second set of relations are the coefficients of the expressions $f_i(\Phi(a_1), ..., \Phi(a_m))$ for $i = 1, ..., m$. By adding them to R, we have that Φ induces a homomorphism $\varphi : A \to B\{\tau\}$.

(iii) The next set of relations ensures that l is a homomorphism of A-modules with a certain normalization. They are given the following:
$L(0)$, and $L(\alpha_0) - 1$ for a suitable fixed $\alpha_0 \in (n^{-1}/A)^r - \{0\}$,
$L(\alpha + \beta) - L(\alpha) - L(\beta)$ for $\alpha, \beta \in (n^{-1}/A)^r$.
$L(\lambda\alpha) - \lambda L(\alpha)$ for $\lambda \in \mathbb{F}_q$.
$\Phi(a_\lambda)(L(\alpha)) - L(a_\lambda\alpha)$, for all α, and $\lambda = 1, ..., t$.

(iv) The last set of relations is given by requiring that the polynomial $\prod_{\alpha\in(n^{-1}/A)^r}(Z - L(\alpha))$ is, modulo R, the g.c.d. of the polynomials $\sum_{n=0}^{N_\lambda} X_{\lambda,n}Z^{q^n} = \Phi(a_\lambda)(Z)$, for $\lambda = 1, ..., a$.

Thus we have found B and a Drinfeld structure above Spec(B) with level n. The arguments, used in the proof of 3.2.2, show that Spec B with its additional Drinfeld structure represents the functor $\mathcal{M}^r(n)$, restricted to the category of A-schemes S such that the image of $S \to$ Spec A is contained in Spec $A - V(n)$.

3.3. Smoothness of the scheme $M^r(n)$.

3.3.1. In the following we present a proof of the fact that the moduli scheme $M^r(n)$, for admissible ideals n, is smooth of dimension r over $\mathbb{F}_q$, in the case where $A = \mathbb{F}_q[T]$ is a polynomial ring.

3.3.2. Proposition. *Assume that $A = \mathbb{F}_q[T]$ is a polynomial ring, and let $f \in A$ be a non-zero polynomial which has at least two different monic irreducible divisors. Then the moduli scheme $M^r(f)$ of Drinfeld A-modules of rank r, with level (f)-structure, is smooth of dimension r over $\mathbb{F}_q$.*

Proof. (Van der Put) We suppose, as we may, that f is monic. According to corollary 3.2.3, the functor $\mathcal{M}^r(f)$ on A-schemes is representable by the affine scheme $M^r(f)$ over $\mathbf{F}_q$. There is a finite map from $M^r(f)$ to an open subspace of $\mathbf{A}^r_{\mathbf{F}_q}$. This map is obtained by forgetting the level structure from the Drinfeld data (or more precisely by dividing $M^r(f)$ by the automorphism group $Gl(r, A/fA)/\mathbb{F}_q^*$). The dimension of $M^r(f)$ is therefore r. Thus we have to show that the tangent space at each closed point x of $M^r(f)$ has dimension r. It suffices in fact to show that the dimension of T_x is $\leq r$. Let k denote the residue field of the closed point x of $M^r(f)$. One can translate the point x and its tangent space T_x in terms of the functor $\mathcal{M}^r(f)$, as follows:

The point x corresponds to (the equivalence class of) a Drinfeld module, which is denoted by (γ, φ, l), (rank r and level (f)) over the field k, given by:

1. A homomorphism $\gamma : A \to k$.

2. A homomorphism $\phi : A \to k\{\tau\}$ determined by $\varphi_T = \gamma(T) + c_1\tau + \ldots + c_{r-1}\tau + c_r\tau$. The φ is normalized by requiring that $c_r = 1$. We note that φ induces a new structure of A-module on k.

3. A morphism $l : (f^{-1}A/A)^r \to k$ of A-modules (where k is given the new structure of A-module).

4. One requires that $\varphi_f(z) = \prod_{\alpha \in (f^{-1}A/A)^r}(z - l(\alpha))$.

We note that the line bundle, appearing in the general definition of a Drinfeld module, is here of course the trivial line bundle. We will also consider Drinfeld modules over the algebra $k[\epsilon]$, defined by $\epsilon^2 = 0$. The line bundles over $k[\epsilon]$ are also trivial. In the sequel we may and will suppress the line bundles in our considerations. The tangent space T_x is equal to the k-vector space of the equivalence classes of the lifts $(\tilde{\gamma}, \tilde{\varphi}, \tilde{l})$ to $k[\epsilon]$ of (γ, φ, l).

We make two observations. The first one is that it suffices to consider lifts $(\tilde{\gamma}, \tilde{\varphi}, \tilde{l})$, such that $\tilde{\varphi}_T = \tilde{\gamma}(T) + \tilde{c}_1\tau + \ldots + \tilde{c}_r\tau^r$ and $\tilde{c}_r = 1$.
The second one is the following. Suppose that $f = f_1 f_2$ with monic f_1, f_2 such that g.c.d.$(f_1, f_2) = 1$. Then one can write $(f^{-1}A/A)^r = M_1 \oplus M_2$ with $M_i = (f_i^{-1}A/A)^r$. Let l_i denote the restriction of l to M_i. Then it suffices to consider lifts $\tilde{\varphi}, \tilde{l}_1, \tilde{l}_2$ with $\tilde{\varphi}_T$ as above and such that $\tilde{\varphi}_{f_i}(z) = \prod_{\alpha \in M_i}(z - \tilde{l}_i(\alpha))$ for $i = 1, 2$. We have to consider two cases.

Case 1: *f does not lie in the kernel of $\gamma : A \to k$ (i.e., the characteristic)*

$$\text{Then } \tilde{\gamma}(T) = \gamma(T) + \epsilon d_0,\ \tilde{c}_1 = c_1 + \epsilon d_1, \ldots, \tilde{c}_{r-1} = c_{r-1} + \epsilon d_{r-1},$$

with $(d_0, \ldots, d_{r-1}) \in k^r$. We claim that for given $(d_0, \ldots, d_{r-1}) \in k^r$ there is a unique lift $\tilde{l} : (f^{-1}A/A)^r \to k[\epsilon]$ with the required properties. Using this claim one obtains a surjective linear map from k^r to T_x, the k-vector space of all equivalence classes of lifts. This proves the statement in this situation.
For proving the claim we observe that $\varphi_f(z)$ is a separable polynomial since

$\gamma(f)$, which is the coefficient of z in this polynomial, is invertible. The derivative of $\varphi_f(z)$ with respect to z is in fact $\gamma(f)$. This implies that the map

$$\{\alpha+\epsilon\beta \in k[\epsilon] | \tilde{\varphi}_f(\alpha+\epsilon\beta)=0\} \to \{\alpha \in k | \varphi_f(\alpha)=0\},$$

given by $\alpha+\epsilon\beta \mapsto \alpha$ is bijective. Indeed, let α be a zero of $\varphi(z)$ and consider $\alpha+\epsilon\beta \in k[\epsilon]$. Then $\tilde{\varphi}_f(\alpha+\epsilon\beta)=\tilde{\varphi}_f(\alpha)+\epsilon\beta\gamma(f)$. The term $\tilde{\varphi}_f(\alpha)$ lies in ϵk and $\gamma(f)\in k^*$. Thus there is a unique $\beta \in k$ such that $\alpha+\epsilon\beta$ is a zero of $\tilde{\varphi}_f(z)$. Hence there is a unique lift $\tilde{l}$ with the required properties.

Case 2: *$f=f_1f_2$, $f_1=g^s$, the g.c.d.$(f_1,f_2)=1$ and the monic irreducible polynomial g generates the kernel of γ.*

We use the notations above, i.e., $(f^{-1}A/A)^r = M_1 \oplus M_2$ and l_1, l_2. The lift $\tilde{\varphi}_T$ is already described above. The idea of the following calculation is that one views (γ,φ,l) as a pair of data (γ_i,φ_i,l_i), $i=1,2$ for the levels f_1 and f_2, such that $(\gamma_1,\varphi_1)=(\gamma_2,\varphi_2)$. Each of the data is lifted to $(\tilde{\gamma}_i,\tilde{\varphi}_i,\tilde{l}_i)$, $i=1,2$ with the requirement that $(\tilde{\gamma}_1,\tilde{\varphi}_1)=(\tilde{\gamma}_2,\tilde{\varphi}_2)$. One first describes the lifts $(\tilde{\gamma}_1,\tilde{\varphi}_1,\tilde{l}_1)$ of (γ_1,φ_1,l_1). It turns out that the only possible lifts are $\tilde{\gamma}_1=\gamma_1$, $\tilde{\varphi}_1=\varphi_1$ and $\tilde{l}_1$ described by a vector space k^r. Then one combines this with the lifts $(\tilde{\gamma}_2,\tilde{\varphi}_2,\tilde{l}_2)$ of (γ_2,φ_2,l_2). Now $\tilde{\gamma}_2=\tilde{\gamma}_1=\gamma$ and $\tilde{\varphi}_2=\tilde{\varphi}_1=\varphi$. Using case 1 above, one finds $\tilde{l}_2=l_2$. In other words, the "level f_2 part" does not contribute to the dimension of the space of the equivalence classes of the lifts of (γ,φ,l). Thus we find a surjective linear map $k^r\to T_x$ and the dimension of T_x is $\le r$.

Consider a lift $(\tilde{\gamma},\tilde{\varphi},\tilde{l}_1)$ of (γ,φ,l_1). Let $e_1,\dots e_r$ be a free basis of M_1 as A/f_1A-module. Then $\tilde{l}_1$ is determined by a $(d_1,\dots d_r)\in k^r$ with $\tilde{l}_1(e_i)=l(e_i)+\epsilon d_i$ for $i=1,\dots,r$.
It follows that also for the divisor g of f the relation $\tilde{\varphi}_g(z)=\prod_{\alpha\in g^{s-1}M_1}(z-\tilde{l}_1(\alpha))$ holds. We note that $\varphi_g(z)$ is an inseparable polynomial and that $l: g^{s-1}M_1\to k$ has a non trivial kernel. Using the formula $\prod_{\lambda\in\mathbf{F}_q}(z-\epsilon\lambda v)=z^q$ (for any $v\in k$) one sees that $\prod_{\alpha\in g^{s-1}M_1}(z-\tilde{l}_1(\alpha))$ is equal to $\prod_{\alpha\in g^{s-1}M_1}(z-l_1(\alpha))$ and hence to $\varphi_g(z)$. Thus $\varphi_g=\tilde{\varphi}_g$.
We will show that this implies that $\tilde{\varphi}_T=\varphi_T$ (and thus also $\tilde{\gamma}(T)=\gamma(T)$). Suppose that $\tilde{\varphi}_T\neq\varphi_T$. Then one can write $\tilde{\varphi}_T=\varphi_T+\epsilon(m_i\tau^i+\dots+m_{r-1}\tau^{r-1})$ with all $m_j\in k$ and $m_i\neq 0$. An easy calculation yields the formula

$$0=\tilde{\varphi}_g-\varphi_g=\epsilon(\overline{\frac{dg}{dT}}m_i\tau^i+*\tau^{i+1}+\dots+*\tau^{r-1}),$$

where $\overline{\frac{dg}{dT}}$ denotes the image of $\frac{dg}{dT}$ in k. Since the irreducible polynomial g is separable, one has $\overline{\frac{dg}{dT}} \neq 0$. This is a contradiction.
On the other hand, it is easily seen that any $(\gamma, \varphi, \tilde{l}_1)$ with $\tilde{l}_1$ described by $(d_1, \ldots, d_r) \in k^r$ is a lift of (γ, φ, l_1). This ends the proof of case 2.

3.3.3. Proposition. *Assume that $A = \mathbb{F}_q[T]$ is a polynomial ring, and let $n = (f)$ be a non-zero admissible ideal of A. Let $M^r(f)$ be the moduli scheme of Drinfeld A-modules of rank r, with level (f)-structure, and let $M^r(f) \to$ Spec A be the canonical morphism. Then $M^r(f)$ is smooth over Spec $A - V(n)$, and has relative dimension $r-1$ over Spec A.*

Proof. Since Spec A has dimension 1, and $M^r(f)$ has dimension r over $\mathbb{F}_q$, then $M^r(f)$ has relative dimension $r-1$ over Spec A. The proof of the relative smoothness of $M^r(f)$ above Spec $A[1/f]$ is similar to the proof of 3.3.2, case 1. In this case the lift $\tilde{\gamma}$ of γ is already fixed, and there are only parameters $(d_1, ..., d_{r-1}) \in k^r$ which are needed to uniquely determine the lifts $(\tilde{\varphi}, \tilde{l})$ of (φ, l).

In the general case (i.e., A is not isomorphic to $\mathbb{F}_q[T]$) one can prove that $M^r(n)$ is a smooth $\mathbb{F}_q$-scheme of dimension r by proving that its geometric fiber $M^r(n) \times_{\mathrm{Spec}\,\mathbb{F}_q} \mathrm{Spec}\,\overline{\mathbb{F}}_q$, where $\overline{\mathbb{F}}_q$ is the algebraic closure of $\mathbb{F}_q$, is a regular scheme of dimension r. It turns out that the completed local ring at a closed point of this scheme is isomorphic to the *deformation space* of the isomorphy class of the Drinfeld A-module with level structure corresponding to this point. This leads to the study of deformations of Drinfeld modules, and of some related objects. We refer to §4, §5 of [Dr] for details.

3.4.4. The morphism $M^r(m) \to M^r(n)$.

As a consequence of the smoothness of $M^r(n)$ we have the following concluding remark. Let $0 \neq m \subset n$ be an inclusion of ideals of A, where n is admissible. The natural transformation of functors $\mathcal{M}^r(m) \to \mathcal{M}^r(n)$ induces a canonical morphism $M^r(m) \to M^r(n)$. This morphism is finite of degree the cardinality of the kernel of the natural homomorphism $\mathbb{GL}_r(A/m) \to \mathbb{GL}_r(A/n)$, and since both $M^r(n)$ and $M^r(m)$ are regular $\mathbb{F}_q$-schemes this morphism is flat (cf. [Mat]).

References.

[Bl-St] A.Blum and U.Stuhler - *Drinfeld Modules and Elliptic Sheaves* - Springer-Verlag, (1996), to appear.

[Dr] V.G.Drinfeld - *Elliptic Modules* - Math. Sbornik 94, 594-627, (1974), english translation in Math. USSR-Sbornik 23, 561-592, (1976).

[Ge] E.-U.Gekeler - *Drinfeld Modular Curves* - Lecture Notes in Math., 1231, Springer, (1986).

[Ge-1] E.-U.Gekeler - *Moduli for Drinfeld Modules* - in The Arithmetic of Function Fields, Proc. Workshop at the Ohio State University, edited by D.Goss, D.R.Hayes and M.I.Rosen, Walter de Gruyter (1992).

[Gr-Di] A.Grothendieck and J.Dieudonné, - *Éléments de Géometrie Algébrique* - Chap. 2, Pub. Math. IHES, 8.

[Ha] R. Hartshorne - *Algebraic Geometry* - Graduate Texts in Math., 52, Springer-Verlag, (1977).

[Ka-Ma] N.Katz and B.Mazur - *Arithmetic moduli of elliptic curves* - Annals of Math. Studies 108, Princeton University Press, (1985).

[La] G.Laumon - *Cohomology of Drinfeld Modular Varieties, part I* - Cambridge University Press, (1996).

[Lu-Ta] J.Lubin and J.Tate - *Formal moduli for one parameter Lie Groups* - Bull. Soc. Math. France 94, 49-59, (1966).

[Mat] H.Matsumura - *Commutative Algebra* - Mathematics Lecture Notes Series, Second Edit., (1980).

[Ma] B.H.Matzat - *An introduction to Drinfeld modules* - This volume.

Lecture 3

Analytic theory of Drinfeld modules

Bartolomé López
Dpto. Algebra
Universidad Complutense de Madrid
28040 Madrid, Spain
bartolo@sunal1.mat.ucm.es

We present in this work the construction of Drinfeld modules over $\mathbf{C}$ from lattices in $\mathbf{C}$. This construction establishes a bijection between lattices and Drinfeld modules.

1 Non-archimedean valued fields

We recall some notions and results on non-archimedean valued fields.

1. *A non-archimedean valued field* L is a field with a valuation (or norm) $| \ | : L \to \mathbf{R}$ defined by the properties

$$|x| \geq 0,\ |x| = 0 \Leftrightarrow x = 0,\ |xy| = |x||y| \text{ and } |x+y| \leq \max\{|x|, |y|\}$$

We note that for $x, y \in L$ with $|x| \neq |y|$ one has $|x - y| = \max\{|x|, |y|\}$. The *value group* of L is $|L^*|$. The expression $d(x, y) := |x - y|$ is a *metric* on L. *Open and closed disk* are defined as usual, they are denoted by $D(a, s)$ and $\overline{D}(a, s)$ respectively. Any point in the disk can be considered as its center. Indeed, for $b \in D(a, s)$ (resp. $b \in \overline{D}(a, s)$), one has $D(b, s) = D(a, s)$ (resp. $\overline{D}(b, s) = \overline{D}(a, s)$).

2. The *completion* $\hat{L}$ *of* L is the completion of the metric space L. The elements of $\hat{L}$ are equivalence classes of Cauchy sequences in L with respect to the metric. $\hat{L}$ is again a non-archimedean field whose valuation is given by

$$|(a_n)_{n \in \mathbf{N}}| = \lim |a_n|$$

If the Cauchy sequence $(a_n)_{n \in \mathbf{N}}$ is not equivalent to the zero sequence, then there exists $n_0 \in \mathbf{N}$ such that if $n \geq n_0$, then $|a_n| = |a_{n_0}|$. As a consequence L and $\hat{L}$ have the same value group.

Example 1 (a) The p-adic valuation on $\mathbf{Q}$ is determined by the properties: $|p|_p = p^{-1}$ and $|m|_p = 1$ if $m \in \mathbf{Z}$ and $(p, m) = 1$. Indeed, any $x \in \mathbf{Q}^*$ can

uniquely be written as $p^k \frac{a}{b}$ with $a, b, k \in \mathbf{Z}$, $b > 0$ and $(p, a) = (p, b) = (a, b) = 1$. Then $|x| = p^{-k}$.
The completion $\mathbf{Q}_p$ of $\mathbf{Q}$ with respect to this norm is *the field of p-adic numbers* whose elements can be written as

$$a_n p^n + a_{n+1} p^{n+1} + \cdots, \qquad n \in \mathbf{Z},\ a_i \in \{0, 1, 2, \ldots, p-1\}.$$

(b) Similarly, we define on the field $\mathbf{F}_q(T)$ of characteristic p ($\mathbf{F}_q$ is a finite field and q a power of p) a valuation by

$$|a/b|_\infty = q^{\deg a - \deg b}$$

The completion of $\mathbf{F}_q(T)$ is $\mathbf{F}_q((1/T))$. This is the field of the *formal Laurent series* over $\mathbf{F}_q$ in the variable $1/T$. The elements of this field are of the form

$$a_n T^{-n} + a_{n+1} T^{-n-1} + \ldots, \qquad n \in \mathbf{Z},\ a_i \in \mathbf{F}_q.$$

(c) More generally, let $C/\mathbf{F}_q$ be a smooth, absolutely irreducible and projective curve with a chosen point $\infty \in C$. Let $K = \mathbf{F}_q(C)$ be its function field and let $\mathrm{ord}_\infty = v_\infty$ denote the (additive) valuation associated to ∞. Put $\delta = \deg \infty$ (i.e. the degree of the residue field at ∞ over $\mathbf{F}_q$). The valuation on K is defined by: For $f \in K^*$ one has $|f| = q^{-\delta \, \mathrm{ord}_\infty(f)}$. The completion K_∞ of K is isomorphic to $\mathbf{F}_{q^\delta}((\pi))$, where π is a uniformizer at ∞.

3. For non-archimedean complete fields we have the following results, which can be proved by using the non-archimedean property (i.e. $|x + y| \le \max(|x|, |y|)$)
(a) $\sum_n a_n$ converges $\Leftrightarrow \lim a_n = 0$. In this case, $|\sum_n a_n| \le \max_n\{|a_n|\}$.
(b) If $a_n \neq 0\ \forall n$, then $\prod_n a_n$ converges to an element $\neq 0 \Leftrightarrow \lim a_n = 1$

4. Let L be a complete non-archimedean field. If $L \subset M$ is a finite field extension of degree n, then there exists a unique extension of $|\ |$ to M given by

$$|\alpha|^n = N_{M/L}(\alpha), \qquad \alpha \in M$$

where $N_{M/L}(\alpha)$ is the norm of α with respect to the extension $L \subset M$. The non-archimedean valued field M is again complete.
A *(non-archimedean) local field* can be defined as a finite extension of either the field $\mathbf{Q}_p$ or the field $\mathbf{F}_p((X))$.
If $L \subset M$ is an infinite algebraic extension and L is complete, then there is again a unique extension of the valuation to M. However M is not complete. In particular if the algebraic closure L^{alg} of the complete field L is an infinite extension then L^{alg} is not complete. However, by Krasner's lemma, $\widehat{L^{alg}}$ is complete and algebraically closed.

5. From now on, we will use the following *standard notations*:
$C/\mathbf{F}_q$ curve, $K = \mathbf{F}_q(C)$, $A = \mathcal{O}(C - \{\infty\})$, $K_\infty = \hat{K}$ and $\mathbf{C} = \widehat{K_\infty^{alg}}$.
We note that $K_\infty \subset K_\infty^{alg}$ is an infinite extension and thus $K_\infty^{alg} \neq \mathbf{C}$.

2 Function theory on C

Let L be a complete non-archimedean valued field. In this section, we present first a brief introduction to the analytic functions in L, and later on we will restrict to the case $L = \mathbf{C}$. For a more general theory, see Lecture 6 of these Proceedings.

Notation 1 *Let the expression $\sum_{n\geq 0} a_n X^n$ denote an element in the ring of formal power series $L[[X]]$.*

The *radius of convergence* ρ of $\sum_n a_n X^n$ is defined as

$$\rho = \frac{1}{\overline{\lim}\sqrt[n]{|a_n|}} \tag{1}$$

Let $z \in L$. If $|z| < \rho$, then the series $\sum_n a_n z^n$ converges. Thus $\sum_n a_n X^n$ induces a map $f : D(0,\rho) \to L$ given by $f(z) = \sum_n a_n z^n$. Let $b \in D(0,\rho)$. Then there exists $\sum_n b_n X^n \in L[[X]]$ such that $f(z) = \sum_n b_n (z-b)^n$ in $D(b,\rho) = D(0,\rho)$. The radius of convergence of the formal power series $\sum_n b_n X^n$ is again ρ.

A function $f : D(a,s) \to L$ is *holomorphic* if it is given by a power series in $D(a,s)$. A function $f : L \to L$ is called *entire* if there exists $\sum_n a_n X^n$ (with $\rho = \infty$ such that $f(z) = \sum_n a_n z^n$, for all $z \in L$.

In the sequel we will suppose $L = \mathbf{C}$. Entire functions can be expressed by means of their zeroes as a (possibly infinite) product of polynomials of degree 1 with coefficients in $\mathbf{C}$. This result is based on the Weierstrass preparation theorem (cf. Lect. 6, Sect. 1).
The *Tate algebra* T_1 is defined by $T_1 = \{\sum_n a_n X^n : \lim a_n = 0\}$. The elements of T_1 are called the holomorphic functions on the closed unit disk $\overline{D}(0,1)$. The *norm* on T_1 is defined by:

$$||\sum_{n\geq 0} a_n X^n|| = \max_n |a_n| \tag{2}$$

This is actually the *spectral norm*, i.e. $||f|| = \max_{z\in\mathbf{C},\,|z|\leq 1}(|f(z)|)$. The norm is multiplicative, i.e. $||fg|| = ||f||||g||$. We have the following result which is a restatement of the Weierstrass preparation theorem for dimension one.

Theorem 1 *Let $f(X) \in T_1$ have norm 1. Then there exist a polynomial $g(X)$ and a $h(X) \in T_1$ such that*

$f(X) = g(X)h(X)$ where $g(X)$ and $h(X)$ satisfy

(a) $g(X) = X^m + \ldots + c_1 X + c_0$ and $||g|| = 1$.
(b) $h(X) = h_0 + h_1 X + \ldots$, $|h_0| = 1$ and $|h_n| < 1$ if $n > 0$.

In the sequel we will identify a convergent power series with the function it defines on a disk (or on the whole of **C**). The g, h of Theorem 1 satisfy the additional properties

Corollary 1 (a) *All the roots of g are in $\overline{D}(0,1)$.*
(b) *The function h has constant absolute value 1 on $\overline{D}(0,1)$.*

Definition 1 *Let f be a holomorphic function in the disk $\overline{D}(0,s)$ and $a \in \overline{D}(0,s)$. The order $ord_a(f)$ of f at a is $m \in \mathbf{N}$ if*

$$f(z) = a_m(z-a)^m + a_{m+1}(z-a)^{m+1} + \ldots, \quad a_m \neq 0 \text{ for } z \in \overline{D}(0,s).$$

The divisor of f on the unit disk is $\sum_{a,\, |a| \leq 1} ord_a(f)[a]$.

According to the corollary the divisor of any $f \in T_1$, $f \neq 0$ coincides with the divisor of a polynomial and is therefore a *finite* formal sum.
Let $f \neq 0$ be an entire function in **C**. The divisor of f on **C** is again defined as $\sum_{a \in \mathbf{C}} ord_a(f)[a]$. The *support* of this divisor is defined as $\{a \in \mathbf{C} |\ ord_a(F) \neq 0\}$. The restriction of the divisor of an entire function to any closed disk is finite. In general, the divisor of an entire function is infinite but has a *discrete support*, in the sense that this set has a finite intersection with any closed disk.

On the other hand, let a discrete positive divisor $D := \sum_{a \in \mathbf{C}} n_a[a]$ (i.e. the support is discrete and all $n_a \geq 0$) be given. Then one considers the following expression

$$f_D := X^{n_0} \prod_{a \neq 0} (1 - \frac{X}{a})^{n_a}.$$

One can, formally, rewrite this product as a formal power series in $\mathbf{C}[[X]]$. It is easily verified that this formal power series has infinite radius of convergence. Thus f_D is an entire function. It is an exercise to show that the divisor of f_D is D.

Let the entire function $f \neq 0$ be given. Let D denote the divisor of f and let f_D denote the entire function constructed above. Using Theorem 1 and its corollary one sees that the quotient $h := \frac{f}{f_D}$ represents an entire function on **C** which has no zeroes. Unlike the complex case, this implies that h is constant. Indeed, by the corollary of Theorem 1 the expansion of h as element of T_1 has the form $h_0 + h_1 X + \ldots$, where $|h_n| < |h_0| = |h(0)|$ for all $n > 0$. Consider h as holomorphic function on the closed disk $\overline{D}(0,r)$ then one finds similar inequalities $|h_n| r^n < |h(0)|$ for all $n > 0$. Since r can be any element in $|\mathbf{C}^*|$ this implies that $h_n = 0$ for $n > 0$. Hence h is constant. Let f be a non constant entire function, then $f - c$, for any $c \in \mathbf{C}$, is also a non constant entire function and has therefore a zero in **C**. Thus we have proved the following result.

Theorem 2 (1) *Every entire function $f \neq 0$ on $\mathbf{C}$ has uniquely the form*

$$f = cX^{n_0} \prod_{a \neq 0} (1 - X/a)^{n_a},$$

where $\sum_a n_a[a]$ is a discrete and positive divisor on $\mathbf{C}$ and with $c \in \mathbf{C}^*$.
(2) *Any non constant entire function $f : \mathbf{C} \to \mathbf{C}$ is surjective.*

We finish this study of entire functions in $\mathbf{C}$ with the expression for those functions which are $\mathbf{F}_q$-linear.

Proposition 1 *Let f be an entire function. Then f is $\mathbf{F}_q$-linear if and only if the formal series attached to f has the form $\sum_{n \geq 0} a_n z^{q^n}$.*

Proof. Let $f(z) = \sum_{n \geq 0} a_i z^n$. Then

$$f(z_1 + z_2) = \sum_{i \geq 0} a_n (z_1 + z_2)^n = \sum_{i,j \geq 0} a_{i+j} \binom{i+j}{i} z_1^i z_2^j.$$

This expression is equal to $f(z_1) + f(z_2)$ if and only if all the terms $\binom{i+j}{i} a_{i+j} z_1^i z_2^j$ with $i \neq 0 \neq j$ are zero. This amounts to $a_i = 0$ if i is not a power of p. Further for $\zeta \in \mathbf{F}_q$ one has $f(\zeta z) = \sum_n a_n \zeta^n z^n$. Thus $f(\zeta z) = f(z)$ for all $\zeta \in \mathbf{F}_q$ if and only if $a_n = 0$ if $n - 1$ is not a multiple of $q - 1$. This proves the statement. □

3 Lattices and Drinfeld modules

In this section we will use the standard notation $A, K, K_\infty, \mathbf{C}$. Further the results on modules over Dedekind domains used here can be found in ([1], Chapter 13).

Definition 2 *An A-lattice in $\mathbf{C}$ of rank r is a discrete, finitely generated A-module $\Lambda \subset \mathbf{C}$ such that* $\dim_K K\Lambda = r$.

In the complex number field case, there exist only lattices of rank 1 and 2. In our case, $\mathbf{C}$ is an infinite extension of K_∞ and there exists A-lattices of any rank.

Example 2 Let $K_\infty \subset M$ be a field extension of degree m, and $\{\alpha_1, \ldots, \alpha_m\}$ a basis for M over K_∞. Take an $r \leq m$ and choose non zero ideals $I_1, \ldots, I_r$ of A. Then $\Lambda = \alpha_1 I_1 \oplus \ldots \oplus \alpha_r I_r \subset M \subset \mathbf{C}$ is an A-lattice of rank r. Indeed, Λ is clearly an A-module of rank r. Since M is complete the topology of M coincides with the product topology of $M \cong K_\infty^m$. Since A is discrete in K_∞, one has that Λ is discrete in $\mathbf{C}$.

Remark 1 An A-lattice of rank r is torsion free. Since A is a Dedekind ring it is a projective module and isomorphic to a direct sum $A^{r-1} \oplus I$, where I is a non zero ideal of A (cf. [1], pag. 462-463).

Definition 3 *Let Λ_1, Λ_2 be A-lattices of rank r. A morphism from Λ_1 to Λ_2 is an element $c \in \mathbf{C}$ such that $c\Lambda_1 \subset \Lambda_2$. If $c \in \mathbf{C}$ satisfies that $c\Lambda_1 = \Lambda_2$, then c is an isomorphism from Λ_1 to Λ_2.*

In this way, the set of lattices of rank r in $\mathbf{C}$ forms a category.

Proposition 2 *Let Λ be an A-lattice of rank r. Then,* End(Λ) *is a commutative A-algebra.*

Proof. As Λ is an A-module, $A \subset \operatorname{End}(\Lambda)$. On the other hand, it is clear that $\operatorname{End}(\Lambda)$ is a subring of $\mathbf{C}$. □

We have seen that every element $a \in A$ is in $\operatorname{End}(\Lambda)$. Next, we give a non-trivial example of an endomorphism of an A-lattice of rank 2.

Example 3 Let $A = \mathbf{F}_q[T]$ and suppose that the characteristic is not 2. Then any quadratic extension M of K_∞ can be given as $M = K_\infty(\alpha)$, where $\alpha^2 = c^a T^b$ with $a, b \in \{0, 1\}$ not both 0 and where $c \in \mathbf{F}_q^*$ is an element which is not a square.
Then $\Lambda = A\alpha + A$ is an A-lattice which has as ring of endomorphism the integral closure of A in the field $K(\alpha)$.

We can associate to an A-lattice Λ of rank r an A-Drinfeld module of rank r over $\mathbf{C}$ as follows. Λ is discrete, i.e. the points of Λ form a sequence whose limit is ∞. The divisor $\sum_{\lambda \in \Lambda} [\lambda]$ is positive and discrete and defines an entire function e_Λ given by

$$e_\Lambda(z) = z \prod_{\lambda \in \Lambda - \{0\}} (1 - z/\lambda). \tag{3}$$

This function which is called the *Carlitz exponential function* associated to the A-lattice Λ.

Proposition 3 *The function $e_\Lambda(z)$ is $\mathbf{F}_q$-linear.*

Proof. Let $\Lambda_l = \Lambda \cap \overline{D}(0, q^{\delta l})$. It is easy to check that Λ_l is a finite $\mathbf{F}_q$-vector space. Then,

$$e_{\Lambda_l}(X) = X \prod_{\lambda \in \Lambda_l - \{0\}} (1 - X/\lambda)$$

is a $\mathbf{F}_q$-linear polynomial (cf. [5], pag. 295). As $e_\Lambda(z)$ is the limit of $e_{\Lambda_l}(z)$ and $e_{\Lambda_l}(z)$ is $\mathbf{F}_q$-linear, so is e_Λ. □

The function $e_\Lambda(z)$ is surjective by Theorem 2 and its zeroes are the points of Λ. In summary, we have

Proposition 4 *The function* $e_\Lambda : \mathbf{C} \to \mathbf{C}$ *is surjective,* $\mathbf{F}_q$*-linear and* $\ker e_\Lambda = \Lambda$. *Thus* $\mathbf{C}/\Lambda \simeq \mathbf{C}$ *as* $\mathbf{F}_q$*-vector spaces.*

One can show that the isomorphism $\mathbf{C}/\Lambda \simeq \mathbf{C}$ is an isomorphism of analytic rigid spaces (cf. Lecture 6). Unlike the complex case (e.g. an elliptic curve as a quotient $\mathbf{C}/\Lambda$), the quotient $\mathbf{C}/\Lambda$ is just the additive group over $\mathbf{C}$. The new thing however is the A-structure induced by the isomorphism of Proposition 4. For $a \in A$ and $z \in \mathbf{C}$ we choose a $\tilde{z} \in \mathbf{C}$ with $e_\Lambda(\tilde{z}) = z$ and we define this new A-module structure by $a * z = e_\Lambda(a\tilde{z})$. This does not depend on the choice of $\tilde{z}$. Indeed, any other choice would be $\tilde{z} + b$ with $b \in \Lambda$ and

$$e_\Lambda(a(\tilde{z} + b)) = e_\Lambda(a\tilde{z}) + e_\Lambda(ab) = e_\Lambda(a\tilde{z}).$$

In order to make this structure more explicit we consider the polynomial

$$\phi_a^\Lambda(X) = aX \prod_{\substack{\mu \in a^{-1}\Lambda/\Lambda \\ \mu \neq 0}} \left(1 - \frac{X}{e_\Lambda(\mu)}\right).$$

This is well defined, since Λ is the kernel of e_Λ. One easily verifies the formula $e_\Lambda(a\tilde{z}) = \phi_a^\Lambda(e_\Lambda(\tilde{z}))$. In other words $a * z = \phi_a^\Lambda(z)$.

Now we study the polynomial $\phi_a^\Lambda(X)$ more closely. Since $\{e_\Lambda(\mu)\}_{\mu \in a^{-1}\Lambda/\Lambda}$ is a finite $\mathbf{F}_q$-vector space, ϕ_a^Λ is a $\mathbf{F}_q$-linear polynomial. Let $\tau : \mathbf{C} \to \mathbf{C}$ be the endomorphism $z \to z^q$ of the additive group over $\mathbf{C}$. The ring of all $\mathbf{F}_q$-linear endomorphisms of the additive group over $\mathbf{C}$ is the skew polynomial ring $\mathbf{C}\{\tau\}$, where the product structure is defined by the rule $\tau c = c^q \tau$ for $c \in \mathbf{C}$. Clearly, ϕ_a^Λ considered as operator on $\mathbf{C}$ belongs to $\mathbf{C}\{\tau\}$. Moreover ϕ_a^Λ has the form $a + *\tau + \ldots + *\tau^n$ (the last $*$ is not zero), where n is the dimension of the $\mathbf{F}_q$-vector space $a^{-1}\Lambda/\Lambda$.

We claim that $a^{-1}\Lambda/\Lambda$ is isomorphic to $(A/aA)^r$. This would imply that $n = r \cdot \mathit{degree}(a)$.
Let $S \subset A$ denote the multiplicative set of the elements which become invertible in the ring A/aA. Both A-modules above are "killed" by a, i.e. they are A/aA-modules. Hence we may localize at S and the two modules are isomorphic to $a^{-1}\Lambda_S/\Lambda_S$ and $(A_S/aA_S)^r$. The Dedekind ring A_S has only finitely many prime

ideals and is therefore a principal ideal domain. Then Λ_S is isomorphic to $(A_S)^r$ and the claim follows from this. In total we have proved that the map

$$\phi^\Lambda : A \to \mathbf{C}\{\tau\} \text{ given by } a \mapsto \phi_a^\Lambda$$

is a Drinfeld A-module of rank r over $\mathbf{C}$.

We have associated to an A-lattice Λ of rank r a Drinfeld module ϕ^Λ over $\mathbf{C}$ by means of the function e_Λ. Let Λ_1, Λ_2 be lattices in $\mathbf{C}$ of rank r and let $c \in \mathbf{C}$ be a morphism from Λ_1 to Λ_2. We want to associate with c an homomorphism $\phi^{\Lambda_1} \to \phi^{\Lambda_1}$ of Drinfeld modules. It suffices to consider $c \neq 0$. Since $c\Lambda_1 \subset \Lambda_2$ there is a map $u : \mathbf{C} \to \mathbf{C}$ such that $u(e_{\Lambda_1}(z)) = e_{\Lambda_2}(cz)$. This map is $\mathbf{F}_q$-linear since both e_{Λ_1} and e_{Λ_2} are $\mathbf{F}_q$-linear. In fact one can produce the following explicit formula for u:

$$u = cX \prod_{\substack{\mu \in c^{-1}\Lambda_2/\Lambda_1 \\ \mu \neq 0}} \left(1 - \frac{X}{e_{\Lambda_1}(\mu)}\right) \tag{4}$$

Observe that $c^{-1}\Lambda_2/\Lambda_1$ is a $\mathbf{F}_q$-vector space of finite dimension and that u can be considered as an element of $\mathbf{C}\{\tau\}$. The equality $u(e_{\Lambda_1}(z)) = e_{\Lambda_2}(cz)$ holds since both expressions are $\mathbf{F}_q$-linear, have kernel $c^{-1}\Lambda_2$ and have c as coefficient of z. After replacing z by az with $a \in A$, the formula gives the following identity in the skew polynomial ring $\mathbf{C}\{\tau\}$: $u\phi_a^{\Lambda_1} = \phi_a^{\Lambda_1} u$.
Hence u defines an element of $\mathrm{Hom}(\phi^{\Lambda_1}, \phi^{\Lambda_2})$. Thus we found a functor from the category of A-lattices of rank r to the category of Drinfeld A-modules of rank r over $\mathbf{C}$. Our goal is to show that this functor is an isomorphism of categories. First, we give a lemma.

Lemma 1 *Let Λ be a discrete A-module in $\mathbf{C}$, lying in a finite dimensional K_∞-vector space $M \subset \mathbf{C}$. Then Λ is finitely generated.*

Proof. As a topological vector space M is isomorphic to a product K_∞^m. We may suppose that Λ contains a basis $e_1, \ldots, e_m$ of M. Hence Λ contains the A-lattice $\Lambda' := Ae_1 + \ldots + Ae_m$. Further Λ/Λ' is a discrete subset of $(K_\infty/A)^m$. We will show that K_∞/A is compact. This will imply that $(K_\infty/A)^m$ is compact, that Λ/Λ' is finite and that Λ is a finitely generated A-module.

Let π be a local parameter at ∞, then K_∞ is isomorphic to $\mathbf{F}_{q^\delta}((\pi))$. We note that for any $d \in \mathbf{Z}$, $\pi^d \mathbf{F}_{q^\delta}[[\pi]]$ is a compact subgroup of K_∞. By Riemann-Roch there exists an integer n_0 such that for $n \geq n_0$ the space $H^0(C, O(n \cdot \infty))$ has dimension $1 - g + n\delta$ over $\mathbf{F}_q$. This implies that for suitable d the map

$$A \to K_\infty \to K_\infty/(\pi^d \mathbf{F}_{q^\delta}[[\pi]])$$

is surjective. Then also $\pi^d \mathbf{F}_{q^\delta}[[\pi]] \to K_\infty/A$ is surjective and this proves that K_∞/A is compact. □

Theorem 3 *The category of lattices of rank r is isomorphic to the category of Drinfeld modules of rank r.*

Proof. We are going to prove that the map $\Lambda \mapsto \phi^\Lambda$ is a bijection between the set of A-lattices and the set of A-Drinfeld modules. Let a A-Drinfeld module ϕ be given.

Instead of searching for the A-lattice Λ we will first try to find its associated Carlitz exponential function, which will be denoted by $e(z) = z + \sum_n c_n z^{q^n}$, and regard this as a formal power series over $\mathbf{C}$. Fix an $a \in A$ which is not a constant, i.e. $|a| > 1$. Then e is supposed to satisfy the equation

$$\phi_a(e(z)) = e(az) \tag{5}$$

Put $\phi_a = a\tau^0 + \alpha_1\tau + \ldots + \alpha_d\tau^d$. We first remark that there is a unique formal power series e such that Equation (5) holds. Indeed, this equation is equivalent to the following set of relations for the coefficients c_i

$$c_0 = 1 \text{ and for } n \geq 1,\ c_n(a^{q^n} - a) = \sum_{i=1}^{d} \alpha_i c_{n-i}^{q^i} \tag{6}$$

Now we will show that $e(z)$ is in fact an entire function by estimating the values $|c_n|$.
From Equation (6) we get that

$$|c_n|^{q^n}|a| \leq \max_{1\leq i\leq d}\{|\alpha_i|^{q^{-n}}|c_{n-i}|^{q^{i-n}}\}$$

We have that $\lim|\alpha_i|^{-q^n} \leq 1$. Then, given $\epsilon > 0$, there exists $n_0 \in \mathbf{N}$ such that if $n \geq n_0$, then $|\alpha_i|^{q^{-n}} < 1 + \epsilon$. As $|a| > 1$, we may choose an ϵ and a $\theta \in \mathbf{R}$ which satisfies $\frac{1+\epsilon}{|a|} < \theta < 1$ and thus for all $n \geq n_0$,

$$|c_n|^{q^{-n}} \leq \theta \max_{1\leq i\leq d}\{|c_{n-i}|^{q^{i-n}}\}$$

Hence, there exists $\beta \in \mathbf{R}$ such that for all n one has $|c_n|^{q^{-n}} \leq \theta^n\beta$. This shows that $\lim|c_n|^{q^{-n}} = 0$ and that $e(z)$ is an entire function.

The next thing to show is that $\phi_b(e(z)) = e(bz)$ holds for every $b \in A$. It suffices to take $b \neq 0$. We see ϕ_a, ϕ_b, e, b et cetera as operators on $\mathbf{C}$. As we know, the formula $\phi_a \circ e = e \circ a$ holds with a uniquely determined e of the form $z + \sum_{n\geq 1} c_n z^{q^n}$. Now consider

$$\phi_b \circ e \circ b^{-1} = \phi_b \circ \phi_a \circ e \circ a^{-1} \circ b^{-1} = \phi_a \circ (\phi_b \circ e \circ b^{-1}) \circ a^{-1}$$

As $(\phi_b \circ e \circ b^{-1})(z) = z + \sum_{n\geq 1} \gamma_n z^{q^n}$ we conclude that $\phi_b \circ e \circ b^{-1} = e$. This proves $\phi_b(e(z)) = e(bz)$.

The derivative of e is equal to 1 and therefore every zero of e has order 1. Let Λ denote the kernel of e. Then e is the Carlitz exponential function associated the discrete $\mathbf{F}_q$-module Λ. From $\phi_b(e(z)) = e(bz)$ for $b \in A$ it follows that Λ is a (discrete) A-module. We will check now that Λ is finitely generated. Suppose that $x_1, \ldots, x_s \in \Lambda$ are linearly independent over K_∞. Let W be the K_∞-vector space spanned by $x_1, \ldots, x_s$. By Lemma 1, $\Lambda' = \Lambda \cap W$ is an A-lattice of rank s. Choose $a \in A$ with $|a| > 1$. Then $a^{-1}\Lambda'/\Lambda' \subset a^{-1}\Lambda/\Lambda$ and the last A-module can be identified with $\ker(\phi_a, \mathbf{C})$. This implies that $s \leq$ the rank of ϕ. We conclude that the K_∞-vector space spanned by Λ has dimension $\leq r$. Again using Lemma 1 one finds that Λ is an A-lattice. By construction $\phi^\Lambda = \phi$.

Finally we will show that the map from $\mathrm{Hom}(\Lambda_1, \Lambda_2)$ to $\mathrm{Hom}(\phi^{\Lambda_1}, \phi^{\Lambda_2})$ is bijective. Let $u \in \mathrm{Hom}(\phi^{\Lambda_1}, \phi^{\Lambda_2})$ and write $u = c\tau^0 + \ldots + \beta_d \tau^d$. By definition $\phi_a^{\Lambda_2} \circ u = u \circ \phi_a^{\Lambda_1}$ holds for all $a \in A$. Take some $a \in A$ with $|a| > 1$. Then

$$\phi_a^{\Lambda_2} \circ (u \circ e_{\Lambda_1}) = u \circ \phi_a^{\Lambda_1} e_{\Lambda_1} = (u \circ e_{\Lambda_1}) \circ a.$$

Using the uniqueness, proved in the first part of the proof, one finds that $u \circ e_{\Lambda_1}$ is equal to $e_{\Lambda_2} \circ d$ for a unique $d \in \mathbf{C}$. Clearly $d = c$ and so $u(e_{\Lambda_1}(z)) = e_{\Lambda_2}(cz)$. This implies that $c\Lambda_1 \subset \Lambda_2$ and again by uniqueness

$$u = cX \prod_{\substack{\mu \in c^{-1}\Lambda_2/\Lambda_1 \\ \mu \neq 0}} \left(1 - \frac{X}{e_{\Lambda_1}(\mu)}\right).$$

□

Example 4 Let $A = \mathbf{F}_q[T]$ and consider the rank one A-lattice $cA \subset \mathbf{C}$ with $c \in \mathbf{C}^*$. We note that $e_{cA} = c \circ e_A \circ c^{-1}$. As a consequence $\phi^{cA} = c \circ \phi^A \circ c^{-1}$. The Drinfeld module ϕ^{cA} is determined by ϕ_T^{cA}. Put $\phi_T^A = T + \alpha\tau$. Then $\phi_T^{cA} = T + c^{1-q}\alpha\tau$.

We will do some calculations on e_A and ϕ_T^A. By definition $e_A(z) = z\prod_{a\neq 0}(1 - \frac{z}{a})$, The derivative $e_A(z)'$ equals 1. Thus the logarithmic derivative of e_A yields

$$e_A^{-1} = z^{-1}(1 - \sum_{a \neq 0} \frac{-z/a}{1 - z/a}).$$

Using that $\sum_{a\neq 0} \frac{1}{a^n}$ is zero if n is not a multiple of $q-1$ and using the definition $s_m = \sum_{a\neq 0} \frac{1}{a^{m(q-1)}}$ one obtains the formula

$$e_A = z(1 - \sum_{m \geq 1} s_m z^{m(q-1)})^{-1} = z + s_1 z^q + *z^{q^2} + \ldots.$$

The element s_1 lies in K_∞ and satisfies $|s_1 + 1| < 1$. It can be shown that s_1 is transcendental over K. From $\phi_T^A \circ e_A = e_A \circ T$ it follows that $\alpha = s_1(T^q - T)$. This is no t very satisfactory as expression for the Drinfeld structure. There is a $c \in K_\infty$ with $c^{1-q}\alpha = -T$. Indeed, we can take $c = T \sqrt[q-1]{-s_1(1 - T^{-q+1})}$. The expression $\sqrt[q-1]{-s_1(1 - T^{-q+1})}$ is well defined and lies in K_∞, since the element $-s_1(1 - T^{-q+1})$ lies in $1 + T^{-1}\mathbf{F}_q[[T^{-1}]]$.
We conclude that for the $c \in K_\infty^*$, defined above, one has

$$\phi_T^{cA} = T - T\tau.$$

This is quite acceptable as Drinfeld structure. In particular, the kernel of ϕ_f^{cA} lies in K_∞ for all $f \in A$.
One can make a further transformation with a d such that $d^{q-1} = -T$. This leads to the nice formula

$$\phi_T^{dcA} = T + \tau.$$

However d does not lie in K_∞. This has the disadvantage that the kernel of ϕ_f^{dcA} lies in $K_\infty((-T)^{1/(q-1)})$ but not in K_∞.

Level structures play an important role in the theory of Drinfeld modules. We introduce them now as well as the notion of level structure on a lattice in $\mathbf{C}$. First we notice that if $\underline{n} \subset A$ is a nonzero ideal and Λ is a lattice, then the function e_Λ induces an isomorphism of the A-modules $\underline{n}^{-1}\Lambda/\Lambda$ and $\bigcap_{a\in\underline{n}} \ker \phi_a^\Lambda$.

Definition 4 *Let $\underline{n} \subset A$ be a nonzero ideal, Λ an A-lattice of rank r and ϕ an A-Drinfeld module over $\mathbf{C}$ of rank r. A level-$\underline{n}$ structure on Λ is an isomorphism of A-modules $\alpha : (A/\underline{n})^r \to \underline{n}^{-1}\Lambda/\Lambda$. A level-$\underline{n}$ structure on ϕ is an isomorphism of A-modules $\alpha : (A/\underline{n})^r \to \bigcap_{a\in\underline{n}} \ker \phi_a$.*

Let (Λ_1, α_1), (Λ_2, α_2) be pairs such that Λ_1, Λ_2 are A-lattices in $\mathbf{C}$ of rank r and α_1, α_2 are level-$\underline{n}$ structures on Λ_1, Λ_2 respectively. The pairs (Λ_1, α_1), (Λ_2, α_2) are *isomorphic* if there exists an isomorphism $c \in \mathbf{C}$ from Λ_1 to Λ_2 such that $\alpha_2 = c \circ \alpha_1$.

Similarly, let (ϕ_1, α_1), (ϕ_2, α_2) be pairs such that ϕ_1, ϕ_2 are A-Drinfeld modules over $\mathbf{C}$ of rank r and α_1, α_2 are level-$\underline{n}$ structures on ϕ_1, ϕ_2 respectively. The pairs (ϕ_1, α_1), (ϕ_2, α_2) are *isomorphic* if there exists an isomorphism $c \in \mathbf{C}^*$ from ϕ_1 to ϕ_2 such that $\alpha_2 = c \circ \alpha_1$.

Finally, we have the following result which is an immediate consequence of Theorem 3.

Theorem 4 *There is a one-to-one correspondence between the set of isomorphism classes of A-lattices in $\mathbf{C}$ of rank r with level-$\underline{n}$ structure and the set of isomorphism classes of Drinfeld modules over $\mathbf{C}$ of rank r with level-$\underline{n}$ structure.*

vvv

References

[1] M. Auslander and D. Buchsbaum - *Groups, Rings, Modules.* - Harper's Series in Modern Math., Harper & Row (1974).

[2] P. Deligne and D. Husemöller - *Survey of Drinfeld modules.* - Contem. Math. 67, (1987), 25-91.

[3] V. Drinfeld - *Elliptic Modules* - Math. USSR-Sbornik, 23, N. 4, (1976), 561-592.

[4] E. Gekeler - *Drinfeld Modular Curves.* - Lecture Notes in Math., 1231, Springer (1986).

[5] L. Gerritzen and M. van der Put - *Schottky Groups and Mumford Curves.* - Lecture Notes in Math., 817, Springer (1980).

[6] D. Hayes - *A Brief Introduction to Drinfeld Modules.* - The Arithmetic of Function Fields (D. Goss, D. Hayes and M. Rosen Eds.), 1-32, de Gruyter, 1992.

Lecture 4

A guide to explicit class field theory in global function fields

Imke Rust and Ortwin Scheja
Fachbereich Mathematik, Universität des Saarlandes,
Postfach 15 11 50, 66041 Saarbrücken, Germany.
e-mail: turtle@math.uni-sb.de, and scheja@math.uni-sb.de

0 Introduction

The aim of class field theory is to classify abelian extensions of global fields. The norm $N_K^L : C_L \mapsto C_K$ on the idèle class group induces a bijection between finite abelian extensions L/K and open subgroups W_L of finite index in C_K. Here, the group $\mathrm{Gal}(L/K)$ is isomorphic to $J_K/\widetilde{W_L}$, where $\widetilde{W_L}$ is the inverse image of W_L in the idèle group J_K. We will study the case where K is a function field of degree one with $\mathbf{F}_q$ as field of constants and ∞ an arbitrarily chosen place of K. Important examples of subgroups are given by the idèle groups $K^* \cdot (U_{\mathfrak{m}} \times K_\infty^*)$, where $\mathfrak{m} = \prod_{\mathfrak{p}} \mathfrak{p}^{m_{\mathfrak{p}}}$ is an ideal of the ring A of elements of K which have non-negative valuation at each finite place, and $U_{\mathfrak{m}} := \Pi_{\mathfrak{p} \not| \infty} U^{(\mathfrak{m}_{\mathfrak{p}})}$ is the product of $\mathfrak{m}_{\mathfrak{p}}$-units. The corresponding class fields are called *ray class fields* modulo $\mathfrak{m}$ and we denote them by $H(\mathfrak{m})$. Note that by admitting *all* of K_∞^* at the infinite place, we ensure that the corresponding subgroup is of finite index in J_K. As a consequence the place ∞ is totally split in $H(\mathfrak{m})$, which you have to consider in the formulation of the theorem of Kronecker-Weber: For any finite abelian extension L/K that is totally split at ∞ there exists an ideal $\mathfrak{m}$, such that L is contained in $H(\mathfrak{m})$. As a special ray class field you obtain the Hilbert class field $H := H(1)$, which is the maximal abelian extension of K where all finite primes are unramified and ∞ is totally split. Moreover, $\mathrm{Gal}(H/K)$ is isomorphic to the ideal class group $Pic(A)$.

The next scope is to describe the fields $H(\mathfrak{m})$ explicitly in terms of the field K. In the classical case $K = \mathbf{Q}$ we admit at the infinite place of the idèle group either the open subgroup $\mathbf{R}_{>0}$ or $\mathbf{R}^*$. For an ideal $\mathfrak{m} = (m)$ in $\mathbf{Z}$, these choices yield the ray class fields $\mathbf{Q}(\zeta)$ resp. the totally real subfield $\mathbf{Q}^+(\zeta)$. Here ζ denotes a primitive mth-root of unity. Since we impose for $\mathbf{Q}(\zeta)$ conditions at the real place ∞, we call it the ray class field of the divisor $\mathfrak{m}\infty$. So we get the ray class field modulo the *ideal* $\mathfrak{m}$ by adjoining the torsion of the operation of $\mathbf{Z}$ on $\mathbf{C}^*$ by $\zeta \mapsto \zeta^m$ and taking the fixed field of the decomposition group of ∞.

Hayes showed that the same procedure works in the case of function fields K, if you adjoin the elements of the torsion of a Drinfeld module ϕ of rank one to a certain finite extension of K, over which ϕ is defined. Analogously to the classical case you have to take the fixed field of the decomposition group of ∞ to get the ray class field modulo $\mathfrak{m}$. The aim of this paper is not to prove this program in all detail, what is done in [H2, H4]. Instead we will start by giving a shortened proof for the case of the rational function field $K = \mathbf{F}_q(T)$ with the infinite place $\frac{1}{T}$ and consequently $A = \mathbf{F}_q[T]$. This ring has already been used by Carlitz [C] in the 1930s to construct abelian extensions, and the construction has been put in the frame of class field theory by Hayes [H1]. In this easiest situation it becomes clear how the Drinfeld modules come in, and it gives furthermore a flavour of the method, undisturbed by technical difficulties. Afterwards we will provide a guide to the general case along the lines of the work of Hayes. We will concentrate on motivations and discuss the methods used. Furthermore, we will give some details left out in the original papers, while at the same time we refer the reader to the work of Hayes for a significant part of the proofs. Finally, we will illustrate the method by computing non-trivial examples.

We always regard Drinfeld modules to be modules over the ring A defined indirectly by the field K and the infinite place examined. Drinfeld modules are supposed to be of generic characteristic, unless we speak explicitly of reduction.

1 The rational function field

In this section, K denotes the rational function field $\mathbf{F}_q(T)$, the infinite place is the one belonging to $\frac{1}{T}$ and consequently A is the polynomial ring in T over $\mathbf{F}_q$. Any $\mathbf{F}_q$-algebra homomorphism is then determined by the image of the indeterminate T, which has to commute with the coefficients from $\mathbf{F}_q$. We define the Carlitz module to be the homomorphism $\phi : A \to \mathcal{C}\{\tau\}$ that takes T to $\tau + T$. Since $\phi_\alpha \phi_T = \phi_T \phi_\alpha$ for all $\alpha \in \mathbf{F}_q$, this is a Drinfeld module of rank one. In general, the fact that τ is the Frobenius endomorphism can be expressed by the isomorphism $(L\{\tau\}, \cdot) \cong (\{\sum_{i=0}^{l} \alpha_i Z^{q^i} : l \in \mathbf{N}, \alpha_i \in L\}, \circ)$, where polynomials in the set on the right are multiplied by composition. We will write $\phi_f(Z)$, if we look at the image ϕ_f of f as a polynomial in Z. Further, for an ideal $\mathfrak{m}(= (m))$ of A, we abbreviate the evaluation ${}_\mathfrak{m}\phi(\mathcal{C})$ in the field $\mathcal{C}$ by ${}_\mathfrak{m}\phi$, since we basically only need this evaluation in the course of this lecture. Thus ${}_\mathfrak{m}\phi$ stands for the set {zeroes of $\phi_m(Z)$ in $\mathcal{C}$}. Obviously, this definition is independent of the choice of the generator m of $\mathfrak{m}$. The ring A operates on the torsion via the Drinfeld module ϕ, and by looking at the primary decomposition we obtain

Proposition 1.1 *The torsion ${}_\mathfrak{m}\phi$ is a cyclic A-module, isomorphic to $A/\mathfrak{m}$, whence $Aut_A({}_\mathfrak{m}\phi) \cong (A/\mathfrak{m})^*$.*

Now we let $\widetilde{H}(\mathfrak{m}) := K({}_{\mathfrak{m}}\phi)$, and since $\widetilde{H}(\mathfrak{m})$ is the splitting field of the separable polynomial ϕ_m, it is a galois extension with galois group $G_{\mathfrak{m}} := \mathrm{Gal}(\widetilde{H}(\mathfrak{m})/K)$. In fact, if you fix one generator λ of ${}_{\mathfrak{m}}\phi$, any other generator is of the form $\phi_a(\lambda)$ for suitable $a \in A$, which is a polynomial in λ over A. Thus we have $\widetilde{H}(\mathfrak{m}) = K(\lambda)$. Moreover we can define a homomorphism

$$G_{\mathfrak{m}} \to (A/\mathfrak{m})^*$$

by taking an automorphism σ to a mod $\mathfrak{m}$, where $\phi_a(\lambda) = \sigma(\lambda)$. This is a well defined injective homomorphism, which is independent of the choice of the generator λ, because the Galois action and the action via the Drinfeld module commute. Indeed, this homomorphism is bijective as we will show by computing ramification numbers of primes. Because of the close analogy to the extensions $\mathbf{Q}(\zeta)/\mathbf{Q}$ in the classical case, we will call the fields $\widetilde{H}(\mathfrak{m})$ *cyclotomic extensions*.

Lemma 1.2 *Let a be any monic prime in A. Then $\phi_{a^{s-1}}(Z)$ divides $\phi_{a^s}(Z)$ in $A[Z]$ and the quotient is Eisenstein at (a) over A.*

Proof. First let $s = 1$, where we have to examine the polynomial $g(Z) := \phi_a(Z)/Z$. The leading coefficient of g is one and the constant just a. Therefore, the morphism $\pi \circ \phi$ obtained by reducing ϕ mod (a) is a Drinfeld module of rank 1 and has non-trivial height, which implies $height(\pi \circ \phi) = 1$. Therefore g is Eisenstein at (a).

Now let $s > 1$. Since $\phi_{a^s}(Z) = \phi_a(\phi_{a^{s-1}}(Z))$ we have $\phi_{a^s}(Z) = \tilde{g}(Z) \cdot \phi_{a^{s-1}}(Z)$ where $\tilde{g}(Z) := g(\phi_{a^{s-1}}(Z))$ is a polynomial over A that inherits the Eisenstein property from $g(Z)$. □

We now assume $\mathfrak{m}$ to be primary, i.e. $\mathfrak{m} = (a^s)$, where a is a monic prime. Then the Eisenstein polynomial $\tilde{g}$ of the above proposition is the minimal polynomial of $\widetilde{H}(\mathfrak{m})/K$ and is of degree $\#(A/\mathfrak{m})^*$. On the other hand, the ring $A[\lambda]$ is contained in the integral closure of A in $\widetilde{H}(\mathfrak{m})$, which implies via an argument of discriminants that finite primes coprime to $\mathfrak{m}$ are unramified in this extension. So we have proved

Theorem 1.3 *i) The ideal (a) is totally ramified in $\widetilde{H}(\mathfrak{m})/K$ of degree $\#(A/\mathfrak{m})^*$*
ii) If b is a prime in A, coprime to $\mathfrak{m}$, then b is unramified in $\widetilde{H}(\mathfrak{m})/K$.□

This theorem yields directly

$$G_{\mathfrak{m}} \xrightarrow{\cong} (A/\mathfrak{m})^*$$

for primary $\mathfrak{m}$. Furthermore, the ramification numbers just computed show that $\widetilde{H}(\mathfrak{m})$ and $\widetilde{H}(\mathfrak{n})$ are linearly disjoint over K for $\mathfrak{m}$ and $\mathfrak{n}$ coprime, wherefrom the isomorphy follows for general $\mathfrak{m}$.

We have established $\widetilde{H}(\mathfrak{m})$ to be an abelian extension of K. Now we have to relate idèles resp. ideals with $G_{\mathfrak{m}}$ (again only for primary $\mathfrak{m}$). This is done via the Artin map $(\ \cdot\ , \widetilde{H}(\mathfrak{m})/K)$, that carries elements of the group $I(\mathfrak{m})$ of fractional ideals coprime to $\mathfrak{m}$ into $G_{\mathfrak{m}}$. If you pick a monic prime b, the proof of the lemma shows $\phi_b(\lambda) \equiv \lambda^{q^{\deg b}} \bmod \mathcal{B}$, where $\mathcal{B}$ is a prime lying over b. But this property characterizes the Frobenius automorphism for (b), so we conclude

Theorem 1.4 *Let b be a monic prime in A coprime to $\mathfrak{m}$. Then we have*

$$\Big((b), \widetilde{H}(\mathfrak{m})/K\Big)(\lambda) = \phi_b(\lambda).$$

□

We have given the Artin map $I(\mathfrak{m}) \to G_{\mathfrak{m}}$ explicitly by $((x), \widetilde{H}(\mathfrak{m})/K)(\lambda) = \phi_x(\lambda)$ on integral ideals (x), where x is monic. This map extends to $I(\mathfrak{m})$. What is its kernel? Take any monic $x \in A$, coprime to $\mathfrak{m}$. Then $\phi_x(\lambda) = \lambda$ is equivalent to $\phi_{x-1}(\lambda) = 0$, which means $x \equiv 1 \bmod \mathfrak{m}$, since the isomorphism between $A/\mathfrak{m}$ and ${}_{\mathfrak{m}}\phi$ is induced by ϕ. If we set $\widetilde{\mathcal{P}}_{\mathfrak{m}} := \{(x) \subseteq A : x \equiv 1 \bmod \mathfrak{m}, x \text{ monic}\}$, we obtain

$$I(\mathfrak{m})/\widetilde{\mathcal{P}}_{\mathfrak{m}} \cong G_{\mathfrak{m}}.$$

But from the study of this ideal group we cannot derive the behaviour of the place ∞ in $\widetilde{H}(\mathfrak{m})$. Therefore we construct a homomorphism

$$\psi : J_K/K^* \xrightarrow{\cong} J_{\mathfrak{m}}/K^*_{\mathfrak{m}} \longrightarrow I(\mathfrak{m})/\widetilde{\mathcal{P}}_{\mathfrak{m}}\,(\cong G_{\mathfrak{m}}),$$

where $J_{\mathfrak{m}} = \{(x_{\mathfrak{p}})_{\mathfrak{p}} \in J_K : x_{\mathfrak{p}} \in U^{(\mathfrak{m}_{\mathfrak{p}})} \text{ for } \mathfrak{p}|\mathfrak{m}\}$, and $K^*_{\mathfrak{m}}$ is $J_{\mathfrak{m}} \cap K^*$. We take a representative $(a_{\mathfrak{p}})_{\mathfrak{p}}$ of an idèle class and choose it to be in $J_{\mathfrak{m}}$. The value a_∞ at the place ∞ is in $\mathbf{F}_q((\frac{1}{T}))$ and is thus of the form $\sum_{i \geq n} \alpha_i (\frac{1}{T})^i$, $\alpha_n \neq 0$. Then we define $\psi((a_{\mathfrak{p}})_{\mathfrak{p}}) := \prod_{\mathfrak{p} \neq \infty} \mathfrak{p}^{v_{\mathfrak{p}}(\alpha_n^{-1} \cdot a_{\mathfrak{p}})} \bmod \widetilde{\mathcal{P}}_{\mathfrak{m}}$.

The kernel of ψ, i.e., the set of elements of J_K/K^* that stabilize $\widetilde{H}(\mathfrak{m})$ is $K^*(U_{\mathfrak{m}} \times (\frac{1}{T})^{\mathbf{Z}} \times U_\infty^{(1)})$. Local class field theory tells us that the fixed field of $K^*(U_{\mathfrak{m}} \times K^*_\infty)$ is the desired ray class field $H(\mathfrak{m})$. The corresponding fixing ideal group is $\mathcal{P}_{\mathfrak{m}} := \{Ax : x \equiv 1 \bmod \mathfrak{m}\}$. We collect the information in the following diagram and refer to section 6, where the map ψ will be discussed in a broader context.

$$\begin{array}{ccc}
\widetilde{H}(\mathfrak{m}) & \widetilde{\mathcal{P}}_{\mathfrak{m}} & K^*(U_{\mathfrak{m}} \times (\frac{1}{T})^{\mathbf{Z}} \times U_\infty^{(1)}) \\
| & | & | \\
H(\mathfrak{m}) & \mathcal{P}_{\mathfrak{m}} & K^*(U_{\mathfrak{m}} \times K_\infty^*) \\
| & | & | \\
K & I(\mathfrak{m}) & J_K
\end{array}$$

We will compute generating polynomials for $H(\mathfrak{m})$ explicitly for more general examples.

2 The smallest field of definition

The first remark one ought to make when trying do describe a general procedure to construct ray class fields is the following: When dealing with the Carlitz module over the ring $\mathbf{F}_q[T]$ the coefficients of the torsion polynomial of an ideal automatically lie in the field K, because for any element $x \in A$ the polynomial ϕ_x has coefficients in K. This observation gives rise to the first definition:

Definition 2.1 *Let $\phi : A \to C\{\tau\}$ be a Drinfeld module. Then we say that ϕ is* defined over *the field $L \subseteq C$, whenever we have $\phi_x \in L\{\tau\}$ for all $x \in A$. A subfield $L \subseteq C$ is called a* field of definition *for ϕ, if there is a Drinfeld module ϕ' which is defined over L and isomorphic to ϕ over C.*

Hence we see that K is a field of definition for the Carlitz module. In general no Drinfeld module of rank 1 might be defined over the base field K, and one has to look for another appropriate field. The following two propositions (for proofs cf. [H4, section 10]) say that there is a "smallest" field of definition H_ϕ for a given Drinfeld module ϕ, and that this field is totally split at the infinite place. Later we will see that this field does not depend on the choice of ϕ and that it is indeed the Hilbert class field as explained in the introduction.

Proposition 2.2 *Let ϕ be a Drinfeld module of rank one. Then the field K_∞ is a field of definition for ϕ.*

Proposition 2.3 *For any Drinfeld module ϕ there is an extension H_ϕ of K which is a field of definition for ϕ, and which is contained in any other field of definition.*

In particular, if ϕ is of rank one, H_ϕ is contained in K_∞ and therefore it is separable. (To see this, combine e.g. [BAC, chap. 6, §8 n.5 Cor. 3 à théorème 2; chap. 6, §8 n.3 Prop. 2b] and [BA, §7 n.3 théorème 1].)

3 The torsion polynomial $\phi_{\mathfrak{m}}$ and the action of ideals on Drinfeld modules

By passing from the field K to the field H_ϕ we have overcome the first difficulty. The next task is to find the torsion polynomials of ideals $\mathfrak{m} \subseteq A$. In the case of the Carlitz module this was easy: Any ideal is a principal ideal, and the zeroes of the polynomial $\phi_x(Z)$ in C are just the elements of the (x)-torsion ${}_{(x)}\phi$. For a general ring A we make use of the fact (proven in lecture 1) that for any subfield $L \subseteq C$ the ring $L\{\tau\}$ is a left principal ideal ring. We can therefore define the polynomial $\phi_{\mathfrak{m}}$ for a Drinfeld module ϕ and an ideal $\mathfrak{m}$ of A to be the monic generator of the left ideal $I_{\phi,\mathfrak{m}}$ generated by the set $\{\phi_x : x \in \mathfrak{m}\}$. We then have:

Proposition 3.1 *Let ϕ be a Drinfeld module and $\mathfrak{m} \subseteq A$ an ideal. Then the kernel of $\phi_{\mathfrak{m}}$ as an endomorphism of $L \subseteq C$ (i.e., the set of zeroes of $\phi_{\mathfrak{m}}(Z)$ in L) consists precisely of the elements $z \in L$ with $\phi_a(z) = 0$ for all $a \in \mathfrak{m}$.*

Proof. By definition of $\phi_{\mathfrak{m}}$, the inclusion

$$\{z \in L : \phi_{\mathfrak{m}}(z) = 0\} \subseteq \{z \in L : \phi_x(z) = 0 \ \forall x \in \mathfrak{m}\}$$

is obvious. On the other hand, if some $z \in L$ is a zero of $\phi_x(Z)$ for all $x \in \mathfrak{m}$ then it is a zero of all elements of $I_{\phi,\mathfrak{m}}$, hence in particular of the element $\phi_{\mathfrak{m}}(Z)$, and the reverse inclusion follows. □

Corollary 3.2 *Let ϕ be a Drinfeld module of rank r_ϕ and $\mathfrak{m}$ an ideal of A. If $\deg \mathfrak{m}$ denotes the dimension of $A/\mathfrak{m}$ over $\mathbf{F}_q$, then we have*

$$\deg \phi_{\mathfrak{m}} = r_\phi \deg \mathfrak{m}.$$

Proof. From the previous lecture we know that ${}_{\mathfrak{m}}\phi(C) \cong (A/\mathfrak{m})^{r_\phi}$, so in particular the cardinality of ${}_{\mathfrak{m}}\phi(C)$ is $q^{\deg \mathfrak{m} \cdot r_\phi}$. Further, as for any $0 \neq x \in \mathfrak{m}$ the constant coefficient of ϕ_x is x, hence non-zero, also $\phi_{\mathfrak{m}}$ has a non-trivial constant coefficient and gives therefore rise to a separable polynomial in Z. By the proposition, the degree of this polynomial in Z is $q^{\deg \mathfrak{m} \cdot r_\phi}$ and consequently the degree of $\phi_{\mathfrak{m}}$ in τ is $\deg \mathfrak{m} r_\phi$. □

Now we have defined torsion polynomials of ideals, which we can use to define field extensions of H_ϕ. However, we have not yet exploited all the use we can make of these polynomials. Indeed, we observe that by definition of $I_{\phi,\mathfrak{m}}$ for any $x \in \mathfrak{m}$, the polynomial $\phi_{\mathfrak{m}} \cdot \phi_x$ lies in $I_{\phi,\mathfrak{m}}$. Hence there is a polynomial ϕ'_x such that

$$\phi_{\mathfrak{m}} \cdot \phi_x = \phi'_x \cdot \phi_{\mathfrak{m}}.$$

One can show (cf. [H4, section 4]) that by this commutation rule a Drinfeld module ϕ' is given, and that ϕ' is defined over the same field as ϕ is. We denote the resulting Drinfeld module by $\mathfrak{m} * \phi$, and we can easily prove:

Proposition 3.3 *(i) If $\mathfrak{m} = (m)$ is a non-zero principal ideal and μ is the leading coefficient of ϕ_m, then $\phi_{\mathfrak{m}} = \mu^{-1}\phi_m$, and $(\mathfrak{m} * \phi)_x = \mu^{-1}\phi_x\mu$ for any $x \in A$.*

(ii) If $\mathfrak{a}$ and $\mathfrak{b}$ are non-zero ideals of A then

(a) $\phi_{\mathfrak{ab}} = (\mathfrak{b} * \phi)_{\mathfrak{a}} \cdot \phi_{\mathfrak{b}}$, and

(b) $\mathfrak{a} * (\mathfrak{b} * \phi) = (\mathfrak{ab}) * \phi$.

This shows in particular that the ideals of A act on the set of Drinfeld modules, and the group $Pic(A)$ acts on the set of isomorphism classes of Drinfeld modules.

Note also that from the analytic theory we have a natural action of ideals on Drinfeld modules, simply by multiplying the lattice corresponding to the Drinfeld module with the inverse of an ideal. The connection between these two actions is described by the following proposition:

Proposition 3.4 *If Λ is a lattice in C corresponding to the Drinfeld module ϕ^Λ, and $\mathfrak{m} \subseteq A$ is an ideal, then the lattice Λ' corresponding to the Drinfeld module $\mathfrak{m} * \phi^\Lambda$ fulfils the relation*

$$\Lambda' = \delta(\phi^\Lambda_{\mathfrak{m}}) \cdot \mathfrak{m}^{-1} \cdot \Lambda,$$

where $\delta(\phi^\Lambda_{\mathfrak{m}})$ is the constant coefficient of the polynomial $\phi^\Lambda_{\mathfrak{m}}$.

At this point of the discussion it is very tempting to choose the following strategy: Pick a Drinfeld module ϕ of rank one defined over H_ϕ, construct fields $\widetilde{H}_\phi(\mathfrak{m})$ by adjoining torsion points of ideals $\mathfrak{m}$ and consider the subextension $H_\phi(\mathfrak{m})$ of this extension fixed by the decomposition group at infinity. This is a feasable way to go, but it has some disadvantages, the most notable being the fact that the field $\widetilde{H}_\phi(\mathfrak{m})$ depends on the choice of the Drinfeld module ϕ. This was not the case for the Carlitz module: The extension $\widetilde{H}(\mathfrak{m})$ was characterized by the action of ideals, and the ideals acting trivially on $\widetilde{H}(\mathfrak{m})$ had a description which did not seem to contain any data of the module. In order to create a similar situation in the general case, we make the following observation: In the above case we worked with a Drinfeld module, which had the property that any monic polynomial $x \in A$ was mapped by ϕ to a monic polynomial ϕ_x in τ. The idea to generalize this fact, i.e., to work with Drinfeld modules satisfying some similar condition, is due to Hayes, and we will see now, how to proceed.

4 Sign-functions, the leading coefficient function and sgn-normalized Drinfeld modules

The first step we have to make in our generalization is to find a suitable substitute for the notion "monic" in a general ring A. This is done by introducing so-called sign-functions, which essentially correspond to choices of uniformizers at the infinite place modulo 1-units.

We consider the residue class field $\mathbf{F}_\infty$ as embedded into the field K_∞ and denote by d the degree of the infinite place.

Definition 4.1 *A* sign-function $sgn : K_\infty^* \to \mathbf{F}_\infty^*$ *is a homomorphism such that* sgn *is trivial on* $U_\infty^{(1)}$ *and* $sgn|_{\mathbf{F}_\infty^*} = id_{\mathbf{F}_\infty^*}$. *We extend it to all of* K_∞ *by letting* $sgn(0) := 0$.

A twisted sign-function *is a map of the form* $\rho \circ sgn$, *where* sgn *is a sign-function and* $\rho \in \mathrm{Gal}(\mathbf{F}_\infty/\mathbf{F}_q)$.

Sign-functions can be given by choosing a uniformizer π at ∞ and using the decomposition

$$K_\infty^* = \mathbf{F}_\infty^* \times \pi^{\mathbf{Z}} \times U_\infty^{(1)};$$

the function sgn will then simply be the projection onto $\mathbf{F}_\infty^*$. Note that in the case $A = \mathbf{F}_q[T]$ we obtain $sgn(x) = 1$ if and only if $x \in A$ is monic by choosing the "natural" parameter $\frac{1}{T}$ at infinity.

Proposition 4.2 *If* sgn *and* sgn' *are sign-functions on* K_∞ *then there exists* $\zeta \in \mathbf{F}_\infty^*$ *such that*

$$sgn(x) = sgn'(x)\zeta^{-v_\infty(x)}$$

for all $x \in K_\infty$.

Proof. The function $\frac{sgn}{sgn'}$ is a homomorphism of K_∞^* which vanishes on the subgroup of units of K_∞^*. Therefore it factorizes over the function v_∞. The proposition follows. □

This proposition tells us that any sign-function can be described by choosing a parameter: If sgn is given by π and $\zeta \in \mathbf{F}_\infty^*$ is as in the proposition, then sgn' can be given by $\zeta\pi$. On the other hand, if $\frac{\pi}{\pi'} \in U^{(1)}$, then π and π' yield the same sign-function.

Corollary 4.3 *The number of sign-functions on K_∞ is $\frac{q^d-1}{q-1}$.*

Having a sign-function at hand to generalize the notion "monic", we come back to our problem. We need to get a good grasp of the connection between an element $x \in A$ and the leading coefficient of ϕ_x, or as will turn out later, between the sign of an element $x \in K_\infty$ and some extension of the leading coefficient function.

Definition 4.4 *If ϕ is a Drinfeld module defined over some field L, we let $\mu_\phi : A \to L$ be the leading coefficient map given by*

$$\mu_\phi(x) := \textit{leading coefficient of } \phi_x.$$

We will extend μ_ϕ to K_∞ in the sequel.

The function is not multiplicative, but for any $x, y \in A$ it satisfies the rule

$$\mu_\phi(xy) = \mu_\phi(x)\mu_\phi(y)^{N(x)^{r_\phi}} = \mu_\phi(y)\mu_\phi(x)^{N(y)^{r_\phi}},$$

where $N(z) = \#(A/(z))$ for $z \in A$, as one sees easily applying the commutation rules in $L\{\tau\}$.

Let for the moment $\mathfrak{m}_\infty$ denote the maximal ideal of K_∞. For $l \ll 0$ the theorem of Riemann-Roch tells us that

$$\dim_{\mathbf{F}_\infty} \mathfrak{m}_\infty^l \cap A/\mathfrak{m}_\infty^{l+1} \cap A = d = \dim_{\mathbf{F}_\infty} \mathfrak{m}_\infty^l/\mathfrak{m}_\infty^{l+1},$$

hence for any $x \in K_\infty$ with $v_\infty(x) = l \ll 0$ there is an $a \in A$ with $x \equiv a \bmod \mathfrak{m}_\infty^{l+1}$. If x and a are such, we let $\mu_\phi(x) := \mu_\phi(a)$. Then $\mu_\phi(x)$ is well defined, because if both $a \in A$ and $b \in A$ represent $x \bmod \mathfrak{m}_\infty^{l+1}$, then $\deg a = \deg b = -ld$ and $\deg\,(a-b) < ld$, and therefore

$$\mu_\phi(b) = \mu_\phi(b + (a-b)) = \mu_\phi(a).$$

Consider now the norm map N extended to K_∞ by $N(x) = q^{-d \cdot v_\infty(x)}$ for $x \in K_\infty$. From the corresponding facts in A one checks easily that for any $x, y \in K_\infty$ with $v_\infty(x), v_\infty(y)$ small enough, we have

$$\begin{aligned} \mu_\phi(x+y) &= \mu_\phi(x) + \mu_\phi(y), \text{ if } v_\infty(x) = v_\infty(y) \text{ and} \\ \mu_\phi(xy) &= \mu_\phi(x)\mu_\phi(y)^{N(x)^{r_\phi}}. \end{aligned}$$

The second equation will now be used to define $\mu_\phi(x)$ for the remainder of K_∞: If x is any element of K_∞, we choose $y \in K_\infty^*$ such that both $\mu_\phi(y)$ and $\mu_\phi(xy)$ are defined (i.e., such that $v_\infty(y)$ and $v_\infty(xy)$ are small enough). Then we let

$$\mu_\phi(x) := \mu_\phi(xy)\mu_\phi(y)^{-N(x)^{r_\phi}}.$$

Observe that if we choose y' instead of y, with both $v_\infty(y')$ and $v_\infty(xy')$ small enough, we have in particular that $q^d - 1$ divides $N(y')^{r_\phi} - 1$ and $N(y)^{r_\phi} - 1$, and thus we obtain

$$\begin{aligned} \mu_\phi(xy')\mu_\phi(y)^{N(x)^{r_\phi}} &= \mu_\phi(xy')\mu_\phi(y)^{N(xy')^{r_\phi}}\mu_\phi(y)^{N(x)^{r_\phi}(1-N(y')^{r_\phi})} \\ = \ \mu_\phi(xyy')\cdot 1 &= \mu_\phi(xy)\mu_\phi(y')^{N(xy)^{r_\phi}}\mu_\phi(y')^{N(x)^{r_\phi}(1-N(y)^{r_\phi})} \\ = \ \mu_\phi(xy)\mu_\phi(y')^{N(x)^{r_\phi}}. \end{aligned}$$

Hence μ_ϕ is well-defined as a function on K_∞, and by construction its values fulfil the multiplicative relations given above for all $x, y \in K_\infty$. We further observe that by restricting μ_ϕ to $\mathbf{F}_\infty$ we obtain a non-trivial endomorphism of $\mathbf{F}_\infty$, hence an isomorphism. This shows that for any Drinfeld module ϕ defined over a field L we have $\mathbf{F}_\infty \subseteq L$, because the image of μ_ϕ is generated as a field by the elements $\mu_\phi(x)$, $x \in A$.

We are now in a position to describe the type of Drinfeld modules we want to work with:

Definition 4.5 *A Drinfeld module ϕ of rank one is called* normalized *if for all $x \in A$ we have $\mu_\phi(x) \in \mathbf{F}_\infty$. It is called sgn-*nomalized *for some sign-function sgn, if the map $\mu_\phi : K_\infty \to \mathbf{F}_\infty$ is a twist of sgn.*

From now on, we will rigidify our ground situation by working with a given triple (K, ∞, sgn) rather than with the pair (K, ∞), in a similar way in which we always think of the real numbers $\mathbf{R}$ as naturally provided with a sign. In analogy with that situation, we call an element $x \in K_\infty$ *positive* if $sgn(x) = 1$. Finally, in what follows, all Drinfeld modules will have rank one unless otherwise stated.

The first fact we ought to state is that we do not lose anything essential when restricting our devotion to sgn-normalized Drinfeld modules.

Theorem 4.6 *Every Drinfeld module ϕ' is isomorphic over C to a sgn-normalized A-Drinfeld module ϕ.*

Proof. [H4, section 12] □

We can also count the precise number of sgn-normalized Drinfeld A-modules in each isomorphism class by using the following proposition:

Proposition 4.7 *If ϕ is sgn-normalized and $\xi \in C$ such that $\phi' := \xi\phi\xi^{-1}$ is also sgn-normalized, then $\xi \in \mathbf{F}_\infty^*$ and the functions μ_ϕ and $\mu_{\phi'}$ coincide.*

Proof. [H4, section 13] □

From these theorems we can derive easily:

Corollary 4.8 *Each isomorphism class of rank one Drinfeld modules contains*

$$\kappa := \frac{q^d - 1}{q - 1}$$

sgn-normalized elements.

Corollary 4.9 *There are exactly $\kappa \cdot h(A)$ sgn-normalized Drinfeld modules.*

From now on, we let $\mathbf{X}$ denote the set of *sgn*-normalized Drinfeld modules. Using the definition of the $*$-action of ideals on Drinfeld modules and the fact that for a *sgn*-normalized Drinfeld module the coefficient $\mu_\phi(x)$ lies in $\mathbf{F}_\infty^*$ for all $x \in A$, we can show that $\mathbf{X}$ is stable under the operation of ideals. Applying further the fact that isomorphism classes of Drinfeld modules are fixed precisely by the principal ideals, and then examining the action of a principal ideal on $\phi \in \mathbf{X}$, we see:

Proposition 4.10 *Let $\phi \in \mathbf{X}$ and $\mathfrak{a} \subseteq A$ an ideal. Then $\mathfrak{a} * \phi = \phi$ if and only if $\mathfrak{a} = (x)$ for some $x \in A$ with $sgn(x) = 1$.*

By letting $\mathfrak{b} * \phi := ((a)\mathfrak{b}) * \phi$ for $\mathfrak{b}$ a fractional ideal of A and a positive such that $(a)\mathfrak{b}$ is an integral ideal of A, we extend the $*$-action on $\mathbf{X}$ to the whole ideal group of A; the kernel will then be the subgroup

$$\widetilde{\mathcal{P}} := \{(x) : x \in K^*, x \text{ positive}\}.$$

Definition 4.11 *We define $\widetilde{Pic}(A)$ to be the quotient group*

$$\widetilde{Pic}(A) := I(A)/\widetilde{\mathcal{P}}.$$

Due to the equality $\tilde{h}(A) := \#(\widetilde{Pic}(A)) = \kappa h(A) = \#(\mathbf{X})$ and the fact that $\widetilde{Pic}(A)$ acts faithfully on $\mathbf{X}$, we conclude that the operation of $\widetilde{Pic}(A)$ on $\mathbf{X}$ is transitive.

We will interpret this operation later on as the action of the Galois group of a suitable extension of K by using the Artin map. More precisely, we will construct a finite abelian extension H of K such that

(i) any $\phi \in \mathbf{X}$ is defined over $\widetilde{H}$,

(ii) all finite primes of K are unramified in $\widetilde{H}$,

(iii) the action of $(\mathfrak{p}, \widetilde{H}/K)$ on $\mathbf{X}$ coincides with the $*$-action of $\mathfrak{p}$ for any prime $\mathfrak{p}$ of K,

(iv) whenever $\mathfrak{a}$ operates trivially on $\mathbf{X}$ then $(\mathfrak{a}, \widetilde{H}/K)$ is trivial on $\widetilde{H}$.

Using then the above fact, we will conclude that there is a natural isomorphism $\mathrm{Gal}(\widetilde{H}/K) \cong \widetilde{Pic}(A)$.

Before we construct the field $\widetilde{H}$, we note that not only the ideals of A operate on the set $\mathbf{X}$, but we also have $\sigma\phi \in \mathbf{X}$ whenever ϕ is an element of $\mathbf{X}$ and σ is an element of the automorphism group $\mathrm{Aut}(\mathcal{C}/K)$; this is true because σ will just change the Galois twist involved in the definition of *sgn*-normalized Drinfeld modules, and this is permitted. Finally note that for an ideal $\mathfrak{a}$ and a Drinfeld module ϕ, the module $\mathfrak{a} * \phi$ is simply defined by a polynomial division, which entails that the $*$-action of ideals commutes with the action of $\mathrm{Aut}(\mathcal{C}/K)$ on Drinfeld modules. Now we will construct the field $\widetilde{H}$; that it has the properties stated above will be shown in the sequel.

Proposition 4.12 *Let ϕ be an element of $\mathbf{X}$ and $x \in A$ be transcendental over $\mathbf{F}_q$. Let further $\widetilde{H}$ be the field obtained by adjoining the coefficients of ϕ_x to K. Then $\widetilde{H}$ is independent of x and ϕ.*

Proof. [H4, section 14] □

Definition 4.13 *Let $\widetilde{H}$ be defined as in the proposition. It will be called the* normalizing field for rank one Drinfeld modules *over* (K, ∞, sgn).

Example 1. We will now have a close look at the first non-trivial example, that you can also find in [H2]: We let K be the rational function field $\mathbf{F}_3(T)$, which we can also write as $\mathbf{F}_3(x, y)$ with

$$x = \frac{1}{T^2+1}, \; y = \frac{T}{T^2+1}.$$

We have the relation

$$y^2 = -x^2 + x,$$

and if we choose the infinite place to be the one belonging to (T^2+1), then we can calculate the ring A of elements integral at every finite place to be $A = \mathbf{F}_3[x, y]$. The residue field $\mathbf{F}_\infty$ at ∞ has degree 2 over $\mathbf{F}_3$, hence it is the field $\mathbf{F}_3(c)$, where c fulfils the relation $c^2 = -1$. To define the function sgn, we choose the parameter $\pi = T^2 + 1$. With respect to the decomposition $K_\infty^* = \mathbf{F}_\infty^* \times U_\infty^{(1)} \times \pi^{\mathbf{Z}}$, we have

$$x = 1 \cdot 1 \cdot \pi^{-1} \quad \text{and} \quad y = \pm c \cdot u \cdot \pi^{-1},$$

for some $u \in U_\infty^{(1)}$, as $y = T \cdot \pi^{-1}$ and T fulfills $T^2 \equiv -1 \bmod (\pi)$. Therefore we have $sgn(x) = 1$ and $sgn(y) \in \{\pm c\}$. The degree of both x and y is two, so a sgn-normalized rank one Drinfeld module of generic characteristic is given by two polynomials in $C\{\tau\}$ of the form $\phi_x = \tau^2 + a_1\tau + x$ and without loss of generality $\phi_y = c\tau^2 + b_1\tau + y$. (The leading coefficient of ϕ_y could of course also be $-c$, but we cannot really distinguish c and $-c$ anyway.) The polynomials ϕ_x and ϕ_y have to commute in $C\{\tau\}$, and this commutation rule is actually a sufficient condition for the result being a Drinfeld module. (This follows from the analytical theory; for a proof cf. [DH]). Hence by comparing the coefficients of $\phi_x\phi_y$ and $\phi_y\phi_x$ we obtain equations determining the coefficients a_1 and b_1. The result is:

$$\begin{aligned}\phi_x &= \tau^2 + y\eta\xi\tau + x\\ \phi_y &= c\tau^2 + (1+x)\eta\xi\tau + y\end{aligned}$$

where $\eta = 1 + x + cy$ is a fundamental unit in the ring of integers of $K(c)$, and $\xi^4 = \eta$. Hence in our example the field $\widetilde{H}$ is generated over K by c and ξ. Note that you obtain 8 different sgn-normalized Drinfeld modules this way: You have two choices for the leading coefficient of ϕ_y (namely c and $-c$) and then you can choose among four different elements ξ which satisfy $\xi^4 = \eta$. Knowing that in the example the number κ is four, we see that the class number of A must be two - which is fortunately true, cf. page 16.

Proposition 4.14 *The extension $\widetilde{H}$ is galois, and the Galois group is isomorphic to a subgroup of $\widetilde{Pic}(A)$.*

Proof. The normality of the extension follows from the fact that $\mathrm{Aut}(C, K)$ acts on the set $\mathbf{X}$, and that on the other hand $\widetilde{H}$ is independent on the element of $\mathbf{X}$ used to construct it. Separability is shown in two steps: First we recall that for any Drinfeld module ϕ of rank one the field H_ϕ is separable over K (cf. the end of §2). For the second step we observe that for $\xi \in C$ such that $\xi^{-1}\phi\xi$ is defined over H_ϕ, in particular $\xi^{q^{dv_\infty(x)}-1} \in H_\phi$. But then $H_\phi(\xi)$, and therefore also $\widetilde{H}$, being contained in $H_\phi(\xi)$, is separable over H_ϕ.

Further, for any element σ of the galois group, we know by prop. 4.10 and the fact that $\widetilde{Pic}(A)$ acts transitively on $\mathbf{X}$, that there is a unique element $\mathfrak{a} \bmod \widetilde{\mathcal{P}}$ of $\widetilde{Pic}(A)$ with $\sigma\phi = \mathfrak{a}*\phi$. We can check that this delivers an injective homomorphism of $\mathrm{Gal}(\widetilde{H}/K)$ into $\widetilde{Pic}(A)$, as stated in the proposition. □

5 Reduction

The method of reduction was already used in lemma 1.2 to prove ramification properties for the case $A = \mathbf{F}_q[T]$. There we saw by reducing the Drinfeld module ϕ mod (a) that the polynomial $\phi_{a^s}(Z)/\phi_{a^{s-1}}(Z)$ is Eisenstein at (a)

over A. Here, since K is in general no field of definition for ϕ_J we will study reduction in a fixed finite extension K'/K (thinking of K' as $\widetilde{H}$). Afterwards, we will use the technical results, which we cite from [H4, sections 11 , 14], to prove missing properties of $\widetilde{H}$ itself, before applying the method to cyclotomic extensions. In this section, let $\mathfrak{P}$ be a place in K' lying over some finite place $\mathfrak{p} \subseteq A$, let $\mathfrak{O}_{\mathfrak{P}}$ be the valuation ring of $\mathfrak{P}$ and ϕ be any Drinfeld module of rank one. We remind the reader that $\mu_\phi(x)$ stands for the leading coefficient of ϕ_x.

Definition 5.1 *Let B be any subring of K'. We say that ϕ is* defined over B *if for all $x \in A$, $\phi_x \in B\{\tau\}$ and moreover $\mu_\phi(x)$ is a unit in B.*

Suppose ϕ is defined over $\mathfrak{O}_{\mathfrak{P}}$. Then we can show easily that for every non-zero ideal $\mathfrak{a} \subseteq A$, $\phi_{\mathfrak{a}} \in \mathfrak{O}_{\mathfrak{P}}\{\tau\}$, and $\mathfrak{a} * \phi$ is defined over $\mathfrak{O}_{\mathfrak{P}}$. Before we will examine the extension $\widetilde{H}/K$, we will prove the generalization of lemma 1.2:

Lemma 5.2 *Suppose ϕ is defined over $\mathfrak{O}_{\mathfrak{P}}$ and $\mathfrak{p}$ unramified in K'. Then $\phi_{\mathfrak{p}^{s-1}}(Z)$ divides $\phi_{\mathfrak{p}^s}(Z)$ in $\mathfrak{O}_{\mathfrak{P}}[Z]$ and the quotient is Eisenstein at $\mathfrak{P}$.*

Proof. Let $s = 1$. As in lemma 1.2 we see that the height of ϕ reduced at $\mathfrak{P}$ is one, since the constant coefficient $\delta(\phi_a)$ of ϕ_a is equal to a for all $a \in A$. To prove that $g(Z) := \phi_{\mathfrak{p}}(Z)/Z$ is Eisenstein, it therefore suffices to show that $v_{\mathfrak{P}}(\delta(\phi_{\mathfrak{p}})) \leq 1$.

Choose $x \in A$ with $v_{\mathfrak{P}}(x) = v_{\mathfrak{p}}(x) = 1$, thus $Ax = \mathfrak{b}\mathfrak{p}$, with $\mathfrak{b}$ an ideal of A coprime to $\mathfrak{b}$. Then we have $\phi_x = \mu_\phi(x)(\mathfrak{p} * \phi)_{\mathfrak{b}} \cdot \phi_{\mathfrak{p}}$ and

$$1 = v_{\mathfrak{P}}(\delta(\phi_x)) = v_{\mathfrak{P}}(\delta(\mathfrak{p} * \phi)_{\mathfrak{b}}) + v_{\mathfrak{P}}(\delta(\phi_{\mathfrak{p}})) \geq v_{\mathfrak{P}}(\delta(\phi_{\mathfrak{p}}))$$

using the remark above.

For $s > 1$ we note that $\phi_{\mathfrak{p}^s}(Z) = f(\phi_{\mathfrak{p}^{s-1}}(Z)) \cdot \phi_{\mathfrak{p}^{s-1}}(Z)$ in analogy to the proof of lemma 1.2, with f the polynomial $f(Z) = (\mathfrak{p}^{s-1} * \phi)_{\mathfrak{p}}(Z)/Z$, as one sees taking into account the action of the ideals. Applying the case $s = 1$ to the Drinfeld module $\mathfrak{p}^{s-1} * \phi$, we derive that $f(\phi_{\mathfrak{p}^{s-1}}(Z))$ is Eisenstein. □

We are looking for a good criterion for a given ϕ to be defined over $\mathfrak{O}_{\mathfrak{P}}$. We can always show that at least a Drinfeld module isomorphic to ϕ is defined over a valuation ring of a finite extension of K' sitting over $\mathfrak{O}_{\mathfrak{P}}$. If we know furthermore that $\mu_\phi(x)$ is a unit in $\mathfrak{O}_{\mathfrak{P}}$, a valuation argument shows

Proposition 5.3 *Suppose $\mu_\phi(x)$ is a unit of the valuation ring $\mathfrak{O}_{\mathfrak{P}}$ for some non-constant $x \in A$. If ϕ has rank one, then ϕ is defined over $\mathfrak{O}_{\mathfrak{P}}$.*

For the rest of the section (and the paper) we let $K' := \widetilde{H}$, ϕ be a *sgn*-normalized Drinfeld module and $\mathfrak{P}$ a place in $\widetilde{H}$, lying over some finite prime $\mathfrak{p}$ in A. Then we obtain by the definition of *sgn*-normalized Drinfeld modules that for all $x \in A$, $\mu_\phi(x)$ is a unit in the integral closure $\mathfrak{O}_{\widetilde{H}}$ of A in $\widetilde{H}$. Thus the above proposition allows us to reduce ϕ modulo $\mathfrak{P}$. We cite from [H4, section 14]

Lemma 5.4 *The reduction map $\phi \mapsto \mathrm{red}_{\mathfrak{P}} \circ \phi$ is injective on* $\mathbf{X}$.

This yields

Proposition 5.5 *$\widetilde{H}/K$ is unramified at all finite primes.*

Proof. Let σ be an element of the inertia group of the extension $\widetilde{H}/K$ at a finite place $\mathfrak{p}$. By definition of the inertia group, the congruence $\sigma\phi \equiv \phi \bmod \mathfrak{P}$ holds for any $\phi \in \mathbf{X}$. But reduction modulo $\mathfrak{P}$ is injective on $\mathbf{X}$, so we have $\sigma\phi = \phi$. Since $\widetilde{H}$ is generated by the coefficients of ϕ_y for any transcendental $y \in A$, σ must be the identity. □

Moreover, we can give the Artin map effectively: Comparing the coefficients of the Drinfeld modules ϕ and $\mathfrak{p} * \phi$, you obtain

$$(\mathfrak{p}, \widetilde{H}/K)\phi \equiv \mathfrak{p} * \phi \bmod \mathfrak{P}.$$

Since reduction is injective, we deduce as before $(\mathfrak{p}, \widetilde{H}/K)\phi = \mathfrak{p} * \phi$. Proceeding inductively on the number of prime ideals dividing $\mathfrak{a}$, we see

Theorem 5.6 *For an ideal $\mathfrak{a} \neq 0$ in A we have $(\mathfrak{a}, \widetilde{H}/K)\phi = \mathfrak{a} * \phi$.* □

We can combine this with Prop. 4.14 and the fact that $\widetilde{Pic}(A)$ acts transitively on $\mathbf{X}$ to conclude

Corollary 5.7 *The Galois group $\mathrm{Gal}(\widetilde{H}/K)$ is isomorphic to the group $\widetilde{Pic}(A)$, and this isomorphism is given by the Artin map.* □

Knowing that in $\widetilde{H}$ all finite primes are unramified, it is clear by definition that the Hilbert class field of A as defined in the introduction must contain the decomposition field of the infinite place in this extension. Using the idèlic description of the Galois group, which will be found in section 6, we can identify the decomposition group at ∞ to be the subgroup $\mathcal{P}/\widetilde{\mathcal{P}}$ of $\widetilde{Pic}(A)$. But then the Galois group of this decomposition field over K is isomorphic to $Pic(A)$, and consequently we obtain:

Proposition 5.8 *The fixed field of $\mathcal{P}/\widetilde{\mathcal{P}}$ in $\widetilde{H}$ is the Hilbert class field H of A.*

So we have several subfields of $\widetilde{H}$ which seem significant: We have the fields of definition H_ϕ of rank one Drinfeld modules, and we have on the other hand the Hilbert class field. Indeed, by examining the elements of $\mathcal{C}$ which by conjugation carry an element ϕ of $\mathbf{X}$ to a Drinfeld module defined over the field H_ϕ, one can show:

Theorem 5.9 *For any rank one Drinfeld module ϕ the field H_ϕ is the Hilbert class field H. In particular, there is a common smallest field of definition for all rank one Drinfeld modules.*

Proof. [H4, section 15] □

Example 1. If we look at the example from the preceding section, the smallest field of definition for any rank one Drinfeld module over A is easy to determine: The class number of the ring A is two, since it is the product of the divisor class number of the field $K = \mathbf{F}_q(T)$, which has class number one, and the degree of the infinite place, which is two. Therefore the degree of H over K is two. On the other hand, we already know from a remark after the definition of the function μ_ϕ that any field of definition of a Drinfeld module must contain the residue field at the infinite place - hence H must contain the field $K(c)$, where $c^2 = 1$, which has degree two over K; we obtain $H = K(c)$. In [H2] you can find a Drinfeld module with coefficients in H: It is obtained from the sgn-normalized Drinfeld module introduced in the previous section module by conjugating with a square root of ξ.

6 Cyclotomic extensions

We are now ready to construct the cyclotomic extension $\widetilde{H}(\mathfrak{m}) := \widetilde{H}({}_{\mathfrak{m}}\phi)$ via a sgn-normalized Drinfeld module ϕ. As before, the torsion ${}_{\mathfrak{m}}\phi$ is the set of zeroes of $\phi_{\mathfrak{m}}$ in $\mathcal{C}$. Since proposition 1.1 holds in the general case too, we obtain an embedding

$$\mathrm{Gal}(\widetilde{H}(\mathfrak{m})/\widetilde{H}) \hookrightarrow (A/\mathfrak{m})^*$$

by replacing K by $\widetilde{H}$. Again, the computation of ramification numbers will show that this is an isomorphism.

At first, we restrict ourselves to the primary case $\mathfrak{m} = \mathfrak{p}^s$. Since $\mathfrak{p}$ is unramified in $\widetilde{H}$, we can apply the lemma 5.2 and get in the same way as in theorem 1.3

Theorem 6.1 *$\widetilde{H}(\mathfrak{m})/\widetilde{H}$ is totally ramified at $\mathfrak{p}$ of degree $\#(A/\mathfrak{m})^*$ and unramified at all other finite primes of A.*

Proof. Because of lemma 5.2 there only remains to prove the second part of the theorem. Let h be an integer such that $\mathfrak{p}^h = Ax$ is a principal ideal. As in theorem 1.3 we can show that a finite prime $\mathfrak{q} \neq \mathfrak{p}$ is unramified even in $\widetilde{H}({}_{(x)}\phi)/\widetilde{H}$. □

We have established $\widetilde{H}(\mathfrak{m})/\widetilde{H}$ to be an abelian extension with galois group isomorphic to $(A/\mathfrak{m})^*$. Moreover, the theorem allows to restrict consideration to primary $\mathfrak{m}$ further on, since the fields $\widetilde{H}(\mathfrak{m})$ and $\widetilde{H}(\mathfrak{n})$ are linearly disjoint for coprime $\mathfrak{m}$ and $\mathfrak{n}$.

In the next step we want to show that $\widetilde{H}(\mathfrak{m})$ is galois even over K, and give the Artin map from $I(\mathfrak{m})$ to $\mathrm{Gal}(\widetilde{H}(\mathfrak{m})/K)$ explicitly. To control the extension $\widetilde{H}(\mathfrak{m})/K$, we let ideals of A coprime to $\mathfrak{m}$ operate on a set $\mathbf{X}_{\mathfrak{m}}$. The operation on this space should reflect the galois operation of ideals on $\widetilde{H}$, which was described by the set $\mathbf{X}$ and the operation of ideals on the extension $\widetilde{H}(\mathfrak{m})$ of $\widetilde{H}$, which is generated by torsion points. Therefore we let $\mathbf{X}_{\mathfrak{m}}$ be the set of pairs (ϕ, λ), where ϕ is a *sgn*-normalized Drinfeld module and λ a generator of ${}_{\mathfrak{m}}\phi$. We proceed as before by letting $I(\mathfrak{m})$ operate on $\mathbf{X}_{\mathfrak{m}}$ via

$$\mathfrak{a} * (\phi, \lambda) := (\mathfrak{a} * \phi, \phi_{\mathfrak{a}}(\lambda)).$$

Using the definition of $\phi_{\mathfrak{a}}$ we can show that $\phi_{\mathfrak{a}}(\lambda)$ generates ${}_{\mathfrak{m}}(\mathfrak{a} * \phi)$. We proved in 4.10 that $\widetilde{\mathcal{P}}$ is the stabilizer of the operation of ideals on *sgn*-normalized Drinfeld module. But for $(x) \in \widetilde{\mathcal{P}}$, x positive, $\phi_x(\lambda) = \lambda$ is equivalent to $x \equiv 1 \bmod \mathfrak{m}$ for any pair (ϕ, λ). Thus, counting the cardinalities of both sides, we see that $I(\mathfrak{m})/\widetilde{\mathcal{P}}_{\mathfrak{m}}$ operates not only faithfully but also transitively on $\mathbf{X}_{\mathfrak{m}}$.

Lemma 6.2 *For any class of $A/\mathfrak{m}$ there is a positive representative in A.*

Proof. Let $a \in A$ be arbitrarily chosen, and let y be a positive element of $\mathfrak{m}$ (such a y exists: take any element of $\mathfrak{m}$ and consider its $(q^d - 1)$th power). In particular, $v_\infty(y) < 0$. By replacing y by a power of itself, we can obtain $v_\infty(y) < v_\infty(a)$. But then $sgn(a + y) = sgn(y) = 1$ and $a + y$ represents a in $A/\mathfrak{m}$. □

Theorem 6.3 *i) The extension $\widetilde{H}(\mathfrak{m})/K$ is abelian with* $\mathrm{Gal}(\widetilde{H}(\mathfrak{m})/K)$ *isomorphic to $I(\mathfrak{m})/\widetilde{\mathcal{P}}_{\mathfrak{m}}$.*

ii) If $\mathfrak{a}$ is an ideal of A coprime to $\mathfrak{m}$, then $\left(\mathfrak{a}, \widetilde{H}(\mathfrak{m})/K\right)(\lambda) = \phi_{\mathfrak{a}}(\lambda)$.

Proof. Let σ be any element of $\mathrm{Aut}(\mathcal{C}/K)$ and $\xi \in \widetilde{H}$, $\lambda \in {}_{\mathfrak{m}}\phi$ elements of $\mathcal{C}$, such that $\widetilde{H} = H(\xi)$ and $\widetilde{H}(\mathfrak{m}) = \widetilde{H}(\lambda)$. We already know that $\sigma\xi$ is an element of $\widetilde{H}$. Thus, to prove $\widetilde{H}(\mathfrak{m})$ to be galois over K, it suffices to show that $\sigma\lambda$ lies in $\widetilde{H}(\mathfrak{m})$. The conjugate $\sigma\lambda$ is a generator of ${}_{\mathfrak{m}}\sigma\phi = {}_{\mathfrak{m}}(\mathfrak{a} * \phi)$ for some ideal $\mathfrak{a}$. We can assume $\mathfrak{a}$ to be coprime to $\mathfrak{m}$, as we can multiply it by a fractional principal ideal generated by $x \in K$ with $x \equiv 1(\text{mod } \infty)$ and $v_{\mathfrak{P}}(x) = -v_{\mathfrak{P}}(\mathfrak{a})$. Such an element exists by the weak approximation theorem. Further, the above lemma allows us to find an element of $\widetilde{\mathcal{P}}$ which maps (via the Drinfeld module $\mathfrak{a} * \phi$) the element $\phi_{\mathfrak{a}}(\lambda)$ to $\sigma\lambda$. Multiplying $\mathfrak{a}$ by this ideal gives us an ideal $\mathfrak{b}$ with $\sigma\phi = \mathfrak{a} * \phi = \mathfrak{b} * \phi$ and $\sigma\lambda = \phi_{\mathfrak{b}}(\lambda)$, which is a polynomial in λ with coefficients from $\widetilde{H}$, thus an element of $\widetilde{H}(\mathfrak{m})$. Moreover, the operation of $I(\mathfrak{m})$ on $\mathbf{X}_{\mathfrak{m}}$ allows us to embed $\text{Gal}(\widetilde{H}(\mathfrak{m})/K) \hookrightarrow I(\mathfrak{m})/\widetilde{\mathcal{P}}_{\mathfrak{m}}$, which is in fact an isomorphism, because of the equality $\#(I_{\mathfrak{m}}/\widetilde{\mathcal{P}}_{\mathfrak{m}}) = [\widetilde{H}(\mathfrak{m}) : K]$.

For the second part of the theorem look at some prime ideal $\mathfrak{q}$ coprime to $\mathfrak{m}$. Lemma 5.2 allows us to apply the argument of theorem 1.4. The general case is done via induction on the numbers of prime ideals dividing $\mathfrak{a}$. □

Note that the theorem implies in particular that $\widetilde{H}(\mathfrak{m})$ is independent of the choice of the sgn-normalized Drinfeld module we used to define it, because we have shown that the generators of the $\mathfrak{m}$-torsion with respect to any sgn-normalized Drinfeld module lie in $\widetilde{H}(\mathfrak{m})$.

The above theorem defines the Artin map on the group $I(\mathfrak{m})$, extending it to all fractional ideals. As for the case $A = \mathbf{F}_q[T]$, we will obtain the ray class field $H(\mathfrak{m})$ by inspecting the corresponding idèle groups. But how does an idèle operate on $\widetilde{H}(\mathfrak{m})$? We know that the elements $a_{\mathfrak{p}}$ from $K_{\mathfrak{p}}^*$ operate like the ideals $\mathfrak{p}^{v_{\mathfrak{p}}(a_{\mathfrak{p}})}$, and principal idèles operate trivially. Hence an element of the form

$$(1, \ldots, 1, x) = x \cdot (x^{-1}, \ldots, x^{-1}, 1) \text{ with } x \in K_{\mathfrak{m}}^* \text{ positive,}$$

acts like the ideal $(x^{-1}) \in \widetilde{\mathcal{P}}_{\mathfrak{m}}$, that is, trivially. But this determines the action of K_∞^*: Since by the weak approximation theorem $\{x \in K_{\mathfrak{m}}^* : sgn(x) = 1\}$ is dense in $\{x \in K_\infty^* : sgn(x) = 1\}$ and the Artin map is continuous, the action of any element x of K_∞^* with $sgn(x) = 1$ is trivial. Therefore, we can give the Artin map on the idèles via the mapping

$$J_K/K^* \xrightarrow{\cong} J_{\mathfrak{m}}/K_{\mathfrak{m}}^* \longrightarrow I(\mathfrak{m})/\widetilde{\mathcal{P}}_{\mathfrak{m}}\, (\cong G_{\mathfrak{m}}),$$

where an idèle $(a_{\mathfrak{p}})_{\mathfrak{p}}$ in $J_{\mathfrak{m}}$ is carried to the class of $\prod_{\mathfrak{p} \neq \infty} \mathfrak{p}^{v_{\mathfrak{p}}(a_{\mathfrak{p}} x)} \bmod \widetilde{\mathcal{P}}_{\mathfrak{m}}$. Here, x is an element of $K_{\mathfrak{m}}$ with $sgn(x) = sgn(a_\infty)^{-1}$. This function is well defined on $J_{\mathfrak{m}}/K_{\mathfrak{m}}^*$ and lifted to J_K/K^*.

Knowing the groups associated to H, $\widetilde{H}$, and $\widetilde{H}(\mathfrak{m})$ on the ideal side, we are able to compute the corresponding idèle groups as seen in the diagram below. But this idèlic interpretation gives immediately

Proposition 6.4 *The decomposition and inertia group of $\widetilde{H}(\mathfrak{m})/K$ at ∞ coincide and are isomorphic to $\mathbf{F}_\infty^*$.*

With the help of the homomorphism ψ, we can calculate the ideal group associated to the decomposition field $H(\mathfrak{m})$ of $\widetilde{H}(\mathfrak{m})$ to be $\mathcal{P}_\mathfrak{m} := \{(x) : x \in K_\mathfrak{m}^*\}$. We collect all this information in the following diagram.

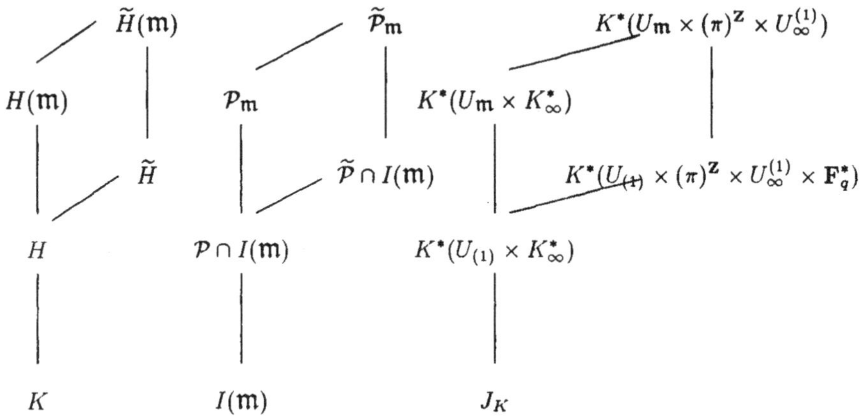

A last theorem allows us to give an explicit generator of $H(\mathfrak{m})$ over H. In the examples we shall see that the minimal polynomial of this generator can easily be computed from a generating polynomial for $\widetilde{H}(\mathfrak{m})$ over $\widetilde{H}$. The theorem is due to Hayes and can also be found in [G, chapter 7].

Theorem 6.5 *We have $\widetilde{H}(\mathfrak{m}) = K(\lambda)$, where λ is a generator of the torsion ${}_\mathfrak{m}\phi$.*

Proof. Take any ideal $\mathfrak{a} \in I(\mathfrak{m})$ with $\phi_\mathfrak{a}(\lambda) = \lambda$. If we show that $\mathfrak{a} \in \widetilde{\mathcal{P}}_\mathfrak{m}$, we are done, because $\widetilde{\mathcal{P}}_\mathfrak{m}$ is the fixing group of $\widetilde{H}(\mathfrak{m})$. But the elements of the torsion ${}_\mathfrak{m}\phi$ are polynomials in λ over A, so $\phi_\mathfrak{a}(\lambda) = \lambda$ implies $\phi_\mathfrak{a}(\alpha) = \alpha$ for any element of the $\mathfrak{m}$-torsion with respect to ϕ. But then by definition of the operation $*$, we obtain ${}_\mathfrak{m}\phi \subseteq {}_\mathfrak{m}(\mathfrak{a} * \phi)$ and because both these sets have the same cardinality, equality holds. We conclude $\phi_\mathfrak{m} = (\mathfrak{a} * \phi)_\mathfrak{m}$. This equation extends to $\sigma\phi$ for any Galois automorphism $\sigma \in \mathrm{Gal}(\widetilde{H}/K)$, since σ operates on the coefficients of the polynomial $\phi_\mathfrak{m}$. Especially, if e is some integer with $\mathfrak{m}^e = (\alpha)$ principal and α positive, we obtain

$$\phi_\alpha = \phi_{\mathfrak{m}^e} = (\mathfrak{m}^{e-1} * \phi)_\mathfrak{m} \cdot \ldots \cdot \phi_\mathfrak{m} = (\mathfrak{a}\mathfrak{m}^{e-1} * \phi)_\mathfrak{m} \cdot \ldots \cdot (\mathfrak{a} * \phi)_\mathfrak{m} = (\mathfrak{a} * \phi)_\alpha.$$

Since Drinfeld modules are determined by their value on any non-constant element, herefrom follows $\phi = \mathfrak{a} * \phi$, which means $\mathfrak{a} \in \widetilde{\mathcal{P}}_\mathfrak{m}$. □

Corollary 6.6 *We have $H(\mathfrak{m}) = K(\lambda^{q^d-1})$, where d is the degree of the infinite place.*

Proof. Any element of $\mathrm{Gal}(\widetilde{H}(\mathfrak{m})/H(\mathfrak{m}))$ can be represented by (a) where $a \in K^*_{\mathfrak{m}}$. It maps λ to $\phi_{(a)}(\lambda)$, which equals $\mu_\phi(a)^{-1} \cdot \lambda$, since $a \equiv 1 \bmod \mathfrak{m}$. As ϕ is sgn-normalized, there is a $\tau \in \mathrm{Gal}(\mathbf{F}_\infty/\mathbf{F}_q)$ such that (a) carries λ to $\tau(sgn(a))^{-1} \cdot \lambda$.

A system of representatives of $\mathcal{P}_{\mathfrak{m}}/\widetilde{\mathcal{P}}_{\mathfrak{m}}$ is given by a set of ideals $\{(a)\}$ where $a \in K^*_{\mathfrak{m}}$ and $sgn(a)$ runs through $\mathbf{F}^*_\infty$. The norm of λ is therefore

$$N(\lambda) = \prod_{\alpha \in \mathbf{F}^*_\infty} \alpha \cdot \lambda = (-1)^{q-1} \cdot \lambda^{q^d-1}.$$

This shows that λ^{q^d-1} is contained in $H(\mathfrak{m})$. On the other hand, the degree $[\widetilde{H}(\mathfrak{m}) : K(\lambda^{q^d-1})]$ is evidently less than or equal to $q^d - 1$ by the theorem, so the fields $K(\lambda^{q^d-1})$ and $H(\mathfrak{m})$ coincide. □

Example 1. We take up the example from section 4. The ideal $\mathfrak{p} := (x, y)$ is prime and $\mathfrak{p}^2 = (x)$, since $y^2 = -x^2 + x$. We want to give polynomials for the ray class fields $H(\mathfrak{p})$ and $H(\mathfrak{p}^2)$. We already know that $\widetilde{H}$ equals $H(\xi)$. The extension $\widetilde{H}(\mathfrak{p})/\widetilde{H}$ is generated by a zero of the polynomial

$$\phi_{\mathfrak{p}}(Z)/Z = rgcd(\phi_x, \phi_y)/Z = -\frac{1}{c}\xi Z^2 + (-\frac{1}{c}y + x),$$

which yields $Z^2 + \frac{1}{\xi}y - c\frac{x}{\xi}$ as minimal polynomial. By the corollary this implies $H(\mathfrak{p}) = H$, which is also clear from the fact that $\mathfrak{p}$ is an ideal of degree one. We obtain the minimal polynomial for $\widetilde{H}(\mathfrak{p}^2)/\widetilde{H}$ dividing $\phi_{(x)}(Z) = \phi_x(Z)$ by $\phi_{\mathfrak{p}}(Z)$, which results to

$$Z^6 - \xi^3(cx + y)Z^4 + \xi^2 x Z^2 + (cx + y)\xi.$$

Now take some generator λ of ${}_{\mathfrak{p}^2}\phi$. Again by the corollary we know that λ^8 generates the ray class field $H(\mathfrak{p}^2)$. But λ^8 fullfills the relation

$$\lambda^8 = \xi\Big(c(x^2 - x) - y(1 + x)\Big)\lambda^2 - x$$

over $\widetilde{H}$, which implies $\widetilde{H}(\lambda^8) = \widetilde{H}(\lambda^2)$. The minimal polynomial of λ^2 over $\widetilde{H}$ is just

$$Z^3 - \xi^3(cx + y)Z^2 + \xi^2 x Z + (cx + y)\xi.$$

This leads via the corresponding affine transformation to the minimal polynomial of λ^8 over $\widetilde{H}$

$$Z^3 + (x^2\eta + x)Z^2 - (x^3(x^2 + x - 1)\eta + x^4)Z - x^2\overline{\eta},$$

which has coefficients already in H. Thus it is the polynomial we looked for.

Example 2. The second example we examine is the function field of the elliptic curve given by

$$y^2 = x^3 - x^2 - x$$

over the field $\mathbf{F}_3$. The infinite place is the one which lies over the place belonging to $\frac{1}{x}$ of $\mathbf{F}_3(x)$. There is only one such place as $(\frac{1}{x})$ ramifies in the extension $\mathbf{F}_3(x,y)/\mathbf{F}_3(x)$. For the same reason the degree of the infinite place is one. With this choice of the infinite place, the ring A can be computed to be $\mathbf{F}_3[x,y]$. We will work with the sgn-function given by the parameter $\pi = \frac{x}{y}$. As $x = \pi^{-2} \cdot \frac{x^3}{y^2}$ and $y = \pi^{-3} \cdot \frac{x^3}{y^2}$, we know that $sgn(x) = sgn(y) = sgn(\frac{x^3}{y^2})$. Further, $\frac{x^3}{y^2}$ is a 1-unit, hence $sgn(x) = sgn(y) = 1$. Either by computing the coefficients using the commutation rules of the Drinfeld module, or by consulting [H3], one obtains the formulae:

$$\phi_x = \tau^2 + \xi\tau + x, \qquad \phi_y = \tau^3 + c_1\tau^2 + c_2\tau + y,$$

where ξ is a zero of the polynomial

$$W^2 - (x^3 - x^2 - x + 1)yW + (x^6 + x^5 + x^4 + x^3 + x^2 + x)$$

and c_1, c_2 are given by

$$\begin{aligned} c_1 &= \xi\frac{y^3 - y}{x^3 - x} &= \xi\frac{(x^3 - x^2 - x - 1)y}{x(x^2-1)} \quad \text{and} \\ c_2 &= \frac{y^9 - y + \xi c_1^3 - c_1\xi^3}{x^9 - x} &= \frac{(x^{13}-1)\xi + (-x^{13} + x^{11} + x^9 - x^5)y}{x^4 - 1}. \end{aligned}$$

We now know that $\widetilde{H}$ is generated over K by ξ, and so is H, because of the equality $[\widetilde{H} : H] = \frac{q^d - 1}{q-1} = 1$.

We examine the prime ideal $\mathfrak{p} = (x,y)$, and by computing the right gcd of ϕ_x and ϕ_y we obtain:

$$\phi_{\mathfrak{p}} = Z^3 + \frac{(-x^3 + x)y + (x^3 - x^2)\xi}{\xi y - x^4 + x^2} Z.$$

Its quotient by Z is then the generating polynomial of $\widetilde{H}(\mathfrak{p})$. Since $\deg(\mathfrak{p})$ is one, we have $H(\mathfrak{p}) = H = \widetilde{H} = K(\xi)$.

If we let $\mathfrak{m} := \mathfrak{p}^2$, we observe first that $\mathfrak{m} = (x)$ due to the relation between x and y. As ϕ_x is monic this entails $\phi_{\mathfrak{m}} = \phi_x$, and the extension $\widetilde{H}(\mathfrak{m})$ can be generated by a zero λ of the polynomial

$$\phi_x(Z)/\phi_{\mathfrak{p}}(Z) = Z^6 + \frac{(-x+1)y - \xi}{(x-1)(x^2+1)} Z^4 + \frac{\xi y - x(x-1)^2}{(x-1)(x^2+1)} Z^2 + \frac{x^2(x-1)y - \xi}{(x-1)(x^2+1)}.$$

Applying once more the corollary, we deduce that $H(\mathfrak{m})$ is generated over H by λ^2, which is evidently a zero of the irreducible polynomial

$$Z^3 + \frac{(-x+1)y-\xi}{(x-1)(x^2+1)}Z^2 + \frac{\xi y - x(x-1)^2}{(x-1)(x^2+1)}Z + \frac{x^2(x-1)y-\xi}{(x-1)(x^2+1)},$$

obtained from $\phi_x(Z)/\phi_{\mathfrak{p}}(Z)$ by substituting Z^2 by Z.

References

[AFF] The arithmetic of function fields, D.Goss et al. (eds), Proceedings of the workshop at the Ohio State University, June 17-26, Columbus, Ohio (USA). Berlin: Walter de Gruyter (1991).

[BAC] Bourbaki, N., Eléments de mathématique, Fasc. XXX, Algebre commutative. Chap. 5: Entiers. Chap. 6: Valuations. Hermann, Paris (1964).

[BA] Bourbaki, N., Eléments de mathématique, Fasc. XXIII, Livre 2, Algèbre, Chap. 8, Hermann, Paris (1958).

[C] L. Carlitz, A class of polynomials, *Trans. AMS* **43** (1938), 167-182.

[DH] Dummit, D.S., Hayes, David, Rank-one Drinfeld modules on elliptic curves. *Math. Comput.* **62**, No.206, (1994) 875-883 (3 microfiche supplements).

[G] D. Goss, Basic structures of Function Filed Arithmetic, Springer (1996).

[H1] D. Hayes, Explicit class field theory of rational function fields, *Trans. AMS* **189** (1974), 77-91.

[H2] D. Hayes, Explicit class field theory in global function fields, G.C. Rota (ed.), *Studies in Algebra and Number Theory*, New York, Academic Press 1979, 173-217.

[H3] D. Hayes, On the reduction of rank-one Drinfeld modules, *Math. Comp.* **57** (1991), 339-349.

[H4] D. Hayes, A brief introduction to Drinfeld modules, in [AFF], 1-32.

Lecture 5

Drinfeld modules over finite fields

Ernst-Ulrich Gekeler and Brian A. Snyder
Fachbereich 9 Mathematik,Universität des Saarlandes
Postfach 15 11 50, D-66041 Saarbrücken, Germany, gekeler@math.uni-sb.de and
Department of Mathematics, 100 Mathematics Tower, 231 W. 18th Avenue, The Ohio State University, Columbus, OH 43210-1174 USA, snyder@math.ohio-state.edu

0. Introduction.

Drinfeld modules over finite A-fields are important for several reasons. They give rise to algebraic curves with many rational points, which are effectively computable. The determination of their isomorphism and isogeny classes yields a method to calculate zeta functions and cohomology groups of moduli schemes. Their endomorphism rings are orders in simple K-algebras, hence they may be used for investigating the ideal theory of such algebras. In certain cases, the combinatorics of the finite set of supersingular isomorphism classes determines, via some *Eichler-Shimura isomorphism*, the action of Hecke operators on spaces of automorphic forms.

Of course, we will not pursue these advanced topics here, but we rather lay the foundations and supply the prerequisites for such a deeper study.

We start with an overview on central division algebras and their ideal theory. In section 2, we collect and prove some properties (projectivity, rank estimates) of End(ϕ), where ϕ is a Drinfeld module defined over an arbitrary A-field. In section 3, the main results on the structure of End(ϕ) are proved (Thm. 3.8, 3.13) or discussed (Thm. 3.15), where now ϕ is defined over a finite field. These results had been anticipated by Drinfeld in [3], but without giving full proofs. We next show, in section 4, how a Z-function $Z_\phi(X)$ for ϕ can be defined that enjoys the formal properties of the complex Z-function of a variety over a finite field. In section 5, we discuss supersingularity and its relation to orders in division algebras. We conclude with a section that contains numerous examples for all the concepts introduced so far.

The exposition is largely based on the papers [7] and [20], to which we often refer for details.

1. Central division algebras.

We recall some basic facts on central simple and division algebras over local and global fields.

(1.1) A *central simple algebra* (CSA) M over a field L is a matrix ring $M_n(D)$ over a central division algebra D/L, which is uniquely determined through M. Two such, M and M', are *equivalent* if their Brauer kernels D and D' are isomorphic. The set $\{[M]\}$ of equivalence classes forms a group, the *Brauer group* Br(L) of L, through

$$[M] \cdot [M'] = [M \otimes_L M'].$$

(The neutral element is $[L]$, the inverse of $[M]$ is the class of the opposite algebra M^{op}.)

We will need the following facts on the CSA M over L (see [2]).

(1.2) After a suitable finite extension L' of L, $M_{L'} := M \otimes L'$ becomes isomorphic with a full matrix algebra M_n over L'. In particular, the dimension $\dim_L M$ is a perfect square. The composite mapping $M \hookrightarrow M_{L'} \xrightarrow{\det} L'$ takes values in L and is independent of the splitting field L'. It is called the *reduced norm* $Nrd = Nrd_L^M$.

(1.3) Let S be a simple subalgebra of M. Then its centralizer $\mathcal{C}(S)$ is simple, we have $(\dim_L S)(\dim_L \mathcal{C}(S)) = \dim_L M$, and $\mathcal{C}(\mathcal{C}(S)) = S$.

(1.4) Let $\alpha : S \longrightarrow S'$ be an isomorphism of simple subalgebras of M. Then there exists an invertible $x \in M$ such that $\alpha(y) = xyx^{-1}$ for all $y \in S$ (Theorem of Skolem-Noether).

(1.5) Let now L be a local field, i.e., a complete discretely valued field with finite residue class field $\mathbb{F}_q$, and let L'/L be a finite extension of degree n. It is a basic fact that

$$inv_L : Br(L) \xrightarrow{\cong} \mathbb{Q}/\mathbb{Z}$$

with a canonical isomorphism inv_L that fits into a commutative diagram

$$\begin{array}{ccc} Br(L) & \xrightarrow{inv_L} & \mathbb{Q}/\mathbb{Z} \\ \downarrow & & \downarrow n \\ Br(L') & \xrightarrow{inv_{L'}} & \mathbb{Q}/\mathbb{Z}. \end{array}$$

Here the left hand arrow is induced from base extension $M \longmapsto M_{L'}$, while the right hand map is multiplication by n. The invariant map inv_L may be defined by cohomological means, or through the description of a CSA as a crossed product ([17] Ch. X).

(1.6) The principal example for our purposes is as follows. Let $L = \mathbb{F}_q((T))$ be the field of formal Laurent series over $\mathbb{F}_q$ with ring of integers $O_L = \mathbb{F}_q[[T]]$, let

$\mathbb{F}_{q^r}$ be the extension of degree r of $\mathbb{F}_q$, and let $O_D = \mathbb{F}_{q^r}\{\{\tau\}\}$ be the ring of non-commutative power series over $\mathbb{F}_{q^r}$ in the variable τ with the commutation rule $\tau a = a^q\tau$, $a \in \mathbb{F}_{q^r}$. We embed the $\mathbb{F}_q$-algebra O_L into O_D through $T \longmapsto \tau^r$, which identifies O_L with the center of O_D. Hence the tensor product $D := O_D \otimes_{O_L} L$ is well-defined and agrees with the quotient skew-field $\mathbb{F}_{q^r}((\tau))$ of formal Laurent series in τ. It is central over L with invariant $inv_L([D]) = \frac{1}{r}$.

(1.7) Let now L be a global field, i.e., a function field in one variable over a finite field $\mathbb{F}_q$, or a finite extension of $\mathbb{Q}$. From the embedding $L \hookrightarrow L_v$ (v a place of L) we get a map $Br(L) \longrightarrow Br(L_v)$. One of the main results of class field theory can be stated as follows: The canonical map from $Br(L)$ to the direct product of the $Br(L_v)$, where v runs through the places of L, has values in the direct sum, and the sequence

$$0 \longrightarrow Br(L) \longrightarrow \oplus_v Br(L_v) \longrightarrow \mathbb{Q}/\mathbb{Z} \longrightarrow 0$$

is exact. Here the right hand map is the sum of local invariants, and $Br(L_v) = \frac{1}{2}\mathbb{Z}/\mathbb{Z}$ or 0, if v is a real or complex archimedean place, respectively.

(1.8) We can easily globalize the example from (1.6). Let now $L = \mathbb{F}_q(T)$ be the field of rational functions with ring of integers $O_L = \mathbb{F}_q[T]$, and let $O_D = \mathbb{F}_{q^r}\{\tau\}$ be the *polynomials* in τ. Again $T \longmapsto \tau^r$ identifies O_L with the center of O_D, and $D := O_D \otimes_{O_L} L$ is the skew-field $\mathbb{F}_{q^r}(\tau)$ of *rational functions* in τ.

Taking T-adic completions, i.e., applying $\otimes_{O_L} \hat{O}_L$, where $\hat{O}_L = \varprojlim_n (O_L/T^n O_L) = \mathbb{F}_q[[T]]$, yields the situation of (1.6). For a place v of L, we write inv_v for inv_{L_v}. Then $inv_v(D) = \frac{1}{r}, -\frac{1}{r}, 0$ if $v = (T)$, $v = \infty$, $v \neq (T), \infty$, respectively.

(1.9) Let again L be local and D/L a central division algebra of dimension r^2. The valuation $v : L^* \longrightarrow \mathbb{Z}$ always extends to a valuation $v_D : D^* \longrightarrow \frac{1}{r}\mathbb{Z}$ formally ramified of index r. The ring $O_D = \{x \in D \mid v_D(x) \geq 0\}$ consists of the elements of D that are integral over the integers O_L of L. The notation used here is compatible to that of (1.6), where $v_D(f) = \frac{1}{r}\mathrm{ord}_\tau(f)$ for $f \in O_D$.

Quite generally, a *lattice* Λ in D is a finitely generated O_L-submodule (necessarily free of rank r^2) that generates the L-vector space D. An *order* in D is a subring that is also a lattice. The *left* and *right order*, respectively, of a lattice Λ is defined as the set of $x \in D$ such that $x\Lambda \subset \Lambda$ or $\Lambda x \subset \Lambda$. Each order is contained in the maximal order O_D.

(1.10) Let now L be global, quotient field of the Dedekind ring B, and D/L a central simple algebra of dimension r^2. Still the notion of B-integral elements of D makes sense, but contrary to the commutative or local case, the integral elements do not form a subring of D. As in (1.9), we define a *B-lattice* in D as a finitely generated B-submodule Λ (necessarily projective of rank r^2) with $L\Lambda = D$, a *B-order* as a subring O that is also a B-lattice, and the left and right order of a B-lattice Λ. Each B-order embeds into a maximal B-order. The theory of lattices and orders *localizes*, i.e., a lattice Λ in D is fully determined by its completions Λ_v in $D_v := D \otimes_L L_v$, all v, and similarly for orders.

Two left ideals $\mathfrak{a}, \mathfrak{b}$ of the B-order O are *equivalent* if $\mathfrak{a} = \mathfrak{b} \cdot f$ with some $f \in D^*$. The *class number* of O is the number of equivalence classes of proper left O-ideals, where $\mathfrak{a}$ is a *proper left O-ideal* if O equals the left order of $\mathfrak{a}$. The following is crucial: The class number of a maximal B-order O in D depends only on D and B but not on the choice of O (and it agrees with its right class number defined analogously); it is called the *class number $h(D,B)$ of (D,B)*. As long as B is fixed, we write $h(D)$ for $h(D,B)$. We then have $h(D) \geq t(D)$, where the *type number* $t(D) = t(D,B)$ is the number of conjugacy classes of maximal B-orders in D. Standard arguments show that under our assumption *L global*, $h(D)$ is always finite. In the example (1.8), $h(D) = t(D) = 1$, since $\mathbb{F}_q\{\tau\}$ is left and right principal. Proofs and more details for all the assertions made can be found e.g. in [14].

2. The endomorphism ring of a Drinfeld module.

Let now A, K, ∞, K_∞ be as usual in this series, i.e., K a function field in one variable over $\mathbb{F}_q$ (which by definition means that $\mathbb{F}_q$ is algebraically closed in K), ∞ a fixed place of K of degree d_∞, say, $A = \{a \in K \mid v(a) \geq 0 \text{ for all places } v \neq \infty \text{ of } K\}$, K_∞ the completion of K at ∞, provided with its normalized valuation v_∞. We use the terms *primes* and *places* as synonyms and denote them by $\mathfrak{p}, \mathfrak{q} \ldots$ or $v, w \ldots$, where the valuation $v = v_\mathfrak{p}$ attached to $\mathfrak{p}$ is always assumed normalized, if not stated otherwise.

We fix an A-field $\gamma : A \longrightarrow L$ with algebraic closure $\overline{L}$ and A-characteristic $\mathfrak{p}$, where either $\mathfrak{p} = \{0\}$ or $\mathfrak{p}$ is a (maximal) prime of A of degree d. The ring $L\{\tau\}$ of non-commutative polynomials in τ with the rule $\tau x = x^q \tau$ $(x \in L)$ is as usual identified with the ring of $\mathbb{F}_q$-endomorphisms of the additive group over L, where τ corresponds to the q-th power map $(x \longmapsto x^q)$. Each $0 \neq f \in L\{\tau\}$ can uniquely be written as $f = f_s \circ \tau^h$, where f_s is separable, i.e., has non-zero constant coefficient. The number h is the *height* $ht(f)$ of f.

Next, we fix a Drinfeld A-module $\phi : A \longrightarrow L\{\tau\}$, $a \longmapsto \phi_a$, over L, of rank

$r \in \mathbb{N}$. Each ϕ_a has degree $\deg_\tau(\phi_a) = -r \cdot d_\infty \cdot v_\infty(a)$ in τ. For an element a or an ideal $\mathfrak{a} \subset A$, we let ${}_a\phi$ or ${}_\mathfrak{a}\phi$ be the scheme of a- or $\mathfrak{a}$-division points of ϕ, regarded as a scheme in $\mathbb{F}_q$-vector spaces.

2.1 Proposition. (i) *Let* $\mathfrak{a}$, $\mathfrak{b}$ *be coprime ideals of* A. *Then*

$${}_{\mathfrak{ab}}\phi(\overline{L}) = {}_\mathfrak{a}\phi(\overline{L}) \ \oplus \ {}_\mathfrak{b}\phi(\overline{L}).$$

(ii) *If* $\mathfrak{a}$, $\mathfrak{p}$ *are coprime then* ${}_\mathfrak{a}\phi(\overline{L}) \cong (A/\mathfrak{a})^r$.
(iii) *If* $\mathfrak{a} = \mathfrak{p}^k \neq \{0\}$ *then* ${}_\mathfrak{a}\phi(\overline{L}) \cong (A/\mathfrak{a})^{r-ht(\phi)}$, *where the integer* $ht(\phi) := \frac{1}{d}\min\{ht(\phi_a) \mid 0 \neq a \in \mathfrak{p}\}$ *satisfies* $1 \leq ht(\phi) \leq r$.

Proof. (i) is obvious. For (ii), note that [1]

$$(*) \qquad \sharp\ {}_\mathfrak{a}\phi(\overline{L}) = \sharp(A/\mathfrak{a})^r$$

at least if $\mathfrak{a} = (a)$ is principal, as follows from the structure of $\phi_a \in L\{\tau\}$. Together with (i), this implies the validity of $(*)$ for all $\mathfrak{a}$ coprime with $\mathfrak{p}$, since for each such $\mathfrak{a}$ there exists an ideal $\mathfrak{b}$ such that $\mathfrak{a}\cdot\mathfrak{b}$ is principal and $\mathfrak{a}$, $\mathfrak{b}$, $\mathfrak{p}$ are mutually coprime. Next, let $A_{(\mathfrak{a})}$ be the ring A localized in $\mathfrak{a}$ (i.e., we invert the elements of $A - \cup\mathfrak{q}$, where $\mathfrak{q}$ runs through the prime divisors of $\mathfrak{a}$). Then $A_{(\mathfrak{a})}$ is semi-local, thus principal. Since both ${}_\mathfrak{a}\phi(\overline{L})$ and $(A/\mathfrak{a})^r$ are $A_{(\mathfrak{a})}$-modules, the assertion follows from $(*)$ and the structure theorem for finitely generated modules over principal ideal domains.
(iii) The function $a \longmapsto ht(\phi_a)$ defines a valuation on A equivalent with the $\mathfrak{p}$-adic valuation $v_\mathfrak{p}$. Let $\pi \in \mathfrak{p}$ be a uniformizer, $(\pi) = \mathfrak{p}\cdot\mathfrak{b}$ with $\mathfrak{b}$, $\mathfrak{p}$ coprime. From (i) and (ii) we have

$$\sharp\ ({}_\mathfrak{p}\phi(\overline{L})) = \frac{\sharp(A/(\pi))^r}{q^{ht(\phi_\pi)}\sharp(A/\mathfrak{b})^r} = \frac{\sharp(A/\mathfrak{p})^r}{q^{ht(\phi_\pi)}},$$

i.e., $\log_q \sharp({}_\mathfrak{p}\phi(\overline{L})) = r\cdot d - ht(\phi_\pi)$. Since ${}_\mathfrak{p}\phi(\overline{L})$ is an $A/\mathfrak{p}$-module, the left hand side is divisible through $d = \deg\ \mathfrak{p}$, hence $ht(\phi) = d^{-1}ht(\phi_\pi) \in \mathbb{N}$. We thus have for $\mathfrak{a} = \mathfrak{p}$ and similarly for powers $\mathfrak{a} = \mathfrak{p}^k$ of $\mathfrak{p}$: $\sharp({}_\mathfrak{a}\phi(\overline{L})) = \sharp(A/\mathfrak{a})^{r-ht(\phi)}$. The assertion now follows as in (ii). □

(2.2) For a place $v = v_\mathfrak{q}$ of A we let A_v, K_v be the completions, ${}_v\phi = \lim_{\rightarrow}\ {}_{\mathfrak{q}^k}\phi$, and

$$T_v(\phi) = \mathrm{Hom}_{A_v}(K_v/A_v, {}_v\phi(\overline{L}))$$

the *v-adic Tate module.* Since by the proposition, ${}_{\mathfrak{q}^k}\phi(\overline{L}) \cong (\mathfrak{q}^{-k}/A)^{r'}$, where $r' = r$ if $\mathfrak{q} \neq \mathfrak{p}$ and $r' = r - ht(\phi)$ if $\mathfrak{q} = \mathfrak{p}$, we get

2.3 Corollary. $T_v(\phi) \cong A_v^{r'}$.

2.4 Remark. If $\mathfrak{q} = (a)$ is principal, the modules ${}_{\mathfrak{q}^k}\phi(\overline{L})$ form a projective system with transition maps $\phi_{a^l} : {}_{\mathfrak{q}^{k+l}}\phi(\overline{L}) \longrightarrow {}_{\mathfrak{q}^k}\phi(\overline{L})$, and $T_v(\phi) \xrightarrow{\cong} \lim_{\leftarrow} {}_{\mathfrak{q}^k}\phi(\overline{L})$. The advantage of the definition given above is that it doesn't refer to the choice of the generator a, and that it also works for non-principal prime ideals $\mathfrak{q}$. We also note that $T_v(\phi)$ is provided with mutually commuting actions of the Galois group $\mathrm{Gal}(\overline{L}/L)$ and the endomorphism ring $\mathrm{End}(\phi)$ of ϕ.

Recall that an *isogeny* u from ϕ to another Drinfeld module ψ over L is some $0 \neq u \in L\{\tau\}$ such that for all $a \in A$ the equation $\psi_a \circ u = u \circ \phi_a$ in $L\{\tau\}$ holds.

2.5 Proposition. (i) *Let $u : \phi \longrightarrow \psi$ be an isogeny. There exist $a \in A$ and $u' : \psi \longrightarrow \phi$ such that $u' \circ u = \phi_a$ and $u \circ u' = \psi_a$.*
(ii) *Let ϕ be defined over some L of A-characteristic $\mathfrak{p}$, and let $H \subset \mathbb{G}_a/L$ be a finite subscheme of $\mathbb{F}_q$-vector spaces, given as the kernel of $u \in L\{\tau\}$. Then u is an isogeny if and only if $H(\overline{L})$ is an A-submodule of $(\overline{L}, \phi)$ and $ht(u) = 0$ (if $\mathfrak{p} = \{0\}$) or $ht(u) \equiv 0(d)$ (if $\mathfrak{p}$ is a prime of degree d).*

Proof. (i) Let $H = L[X]/(u(X))$ be the scheme-theoretic kernel of u, where $u(X) = u_0X + u_1X^q + \cdots$ is the q-additive polynomial corresponding to u. Since H is finite, there exists $0 \neq a \in A$ such that ϕ_a is the zero morphism on H. In other words: ϕ_a factors $\phi_a = u' \circ u$ with some $u' \in L\{\tau\}$. Then $\psi_a \circ u = u \circ \phi_a = u \circ u' \circ u$, i.e., $\psi_a = u \circ u'$ since $L\{\tau\}$ has no zero divisors. We claim that u' is an isogeny from ψ to ϕ. For let $b \in A$. Then

$$\begin{aligned} u' \circ \psi_b \circ \psi_a \circ u &= u' \circ \psi_{ab} \circ u = u' \circ u \circ \phi_{ab} = \phi_a \circ \phi_{ab} \\ &= \phi_b \circ \phi_a \circ \phi_a = \phi_b \circ u' \circ u \circ \phi_a = \phi_b \circ u' \circ \psi_a \circ u, \end{aligned}$$

hence $u' \circ \psi_b = \phi_b \circ u'$.
(ii) The conditions are necessary: for $\mathfrak{p} = \{0\}$, this follows from (i) and the fact that all the ϕ_a are separable. For $\mathfrak{p}$ a prime of degree d, let $h = ht(u)$, $u = u_h\tau^h + \cdots$ with $u_h \neq 0$, and a an element of $A - \mathfrak{p}$. Looking at the first non-vanishing terms of $u \circ \phi_a = \psi_a \circ u$, we have $u_h \cdot \gamma(a)^{q^h} = \gamma(a) \cdot u_h$. Hence $x^{q^h} = x$ for all x in the prime field $A/\mathfrak{p} \hookrightarrow L$ of degree d over $\mathbb{F}_q$, and therefore $h \equiv 0(d)$. Conversely, suppose the conditions are fulfilled. Then we find $a \in A$ and $u' \in L\{\tau\}$ such that $\phi_a = u' \circ u$. In view of the conditions, $u \circ \phi_b$ induces for each $b \in A$ the zero morphism on H, which means that $u \circ \phi_b = \psi_b \circ u$ with some $\psi_b \in L\{\tau\}$. It is now easily verified that $b \longmapsto \psi_b$ is a Drinfeld module.
□

Recall that $\mathrm{End}(\phi)$ is the centralizer of $A \xrightarrow{\cong} \phi(A)$ in $L\{\tau\}$. In order to prove its projectivity as an A-module, we need the next two lemmata.

2.6 Lemma. *Let V be a K_∞-vector of dimension n.*
(i) *There exists a unique topology on V such that each one-dimensional subspace is topologically isomorphic with K_∞.*
(ii) *Let $H \subset V$ be an A-submodule which is discrete with respect to the above topology. Then H is projective of rank $\leq n$.*

Proof. (i) Obvious. (ii) Straightforward exercise, using induction on n and the fact that A is discrete in K_∞ with compact quotient. □

2.7 Lemma. *Let $a \in A - \mathfrak{p}$. Then the canonical representation* $i_a : \mathrm{End}(\phi) \otimes A/(a) \longrightarrow \mathrm{End}_A({}_a\phi(\overline{L}))$ *is injective.*

Proof. Let $u \in \mathrm{End}(\phi)$ vanish on ${}_a\phi(\overline{L})$. Then ϕ_a is a right factor of u in $L\{\tau\} : u = u' \circ \phi_a$. Since u commutes with ϕ, so does u', which therefore is an endomorphism. Hence $u \in a \cdot \mathrm{End}(\phi)$. □

2.8 Theorem. *The endomorphism ring End(ϕ) is a projective A-module of rank less or equal to r^2 (r = rank of ϕ), and End(ϕ)$\otimes_A K_\infty$ is a division ring.*

Proof. The map $u \longmapsto -\deg_\tau u$ on $\mathrm{End}(\phi) - \{0\}$ extends uniquely to a discrete valuation w_∞ on $\mathrm{End}(\phi) \otimes_A K = \mathrm{Quot}(\mathrm{End}(\phi))$, whose restriction to K is equivalent with v_∞. The induced topology is discrete on $\mathrm{End}(\phi)$ and agrees with the ∞-adic topology on finite-dimensional K-subspaces of $\mathrm{End}(\phi) \otimes K$. Hence w_∞ even extends to $\mathrm{End}(\phi) \otimes K_\infty$, which already shows that $\mathrm{End}(\phi) \otimes K_\infty$ is a division ring. Now (2.6) implies that for each K_∞-subspace V of dimension n of $\mathrm{End}(\phi) \otimes K_\infty$, the A-module $\mathrm{End}(\phi) \cap V$ is projective of rank $\leq n$. If $\mathrm{End}(\phi) \otimes K_\infty$ had K_∞-dimension strictly larger than r^2, we would get a contradiction e.g. to (2.7). □

2.9 Corollary. *Let $v = v_\mathfrak{q}$, where $\mathfrak{q}$ is a prime of A different from $\mathfrak{p}$. The canonical representation* $i_v : \mathrm{End}(\phi) \otimes A_v \longrightarrow \mathrm{End}_{A_v}(T_v(\phi))$ *is injective.*

Proof. Let $\mathfrak{q}^k = (a)$ be principal. Then $A_v = \varprojlim_n A/(a^n)$ and, since $\mathrm{End}(\phi)$ is finitely generated, $\mathrm{End}(\phi) \otimes A_v = \varprojlim(\mathrm{End}(\phi) \otimes A/(a^n))$. Now i_v is the limit of the

$$i_{a^n} : \mathrm{End}(\phi) \otimes A/(a^n) \longrightarrow \mathrm{End}_A({}_{a^n}\phi(\overline{L})) = \mathrm{End}_A(T_v(\phi)/a^n T_v(\phi)),$$

which are injective by (2.7). □

For further use, we note two observations on the behavior of $\mathrm{End}(\phi) = \mathrm{End}_L(\phi)$ under separable algebraic extensions L'/L of L.

2.10 Proposition. $\mathrm{End}_L(\phi)$ *is a cotorsion-free submodule of* $\mathrm{End}_{L'}(\phi)$.

Proof. Without restriction, L'/L is galois with group G. Let $u \in \mathrm{End}_{L'}(\phi)$ and $a \in A$ be such that $\phi_a \circ u \in \mathrm{End}_L(\phi)$. Then, since ϕ_a and $\phi_a \circ u$ commute with G, so does u, and $u \in \mathrm{End}_L(\phi)$. □

2.11 Proposition. *The automorphism group* $\mathrm{Aut}(\phi) = \mathrm{End}(\phi)^*$ *is isomorphic with a subgroup* $\mathbb{F}_{q^s}^*$ *of* $\mathbb{F}_{q^r}^*$, *where* s *divides* r. *In particular,* $\mathrm{Aut}(\phi)$ *is abelian. Hence, for* L'/L *galois with group* G, *the* L'*-forms of* ϕ/L *are in canonical bijection with the cohomology group* $H^1(G, \mathrm{Aut}_{L'}(\phi))$.

Proof. If $u \in L^*$ commutes with $\phi_a = \Sigma a_i \tau^i$ and $a_i \neq 0$ then $u^{q^i-1} = 1$, which implies that u is a $q^r - 1$-th root of unity. The precise form of $\mathrm{Aut}(\phi)$ results since for any $u \in \mathrm{Aut}(\phi)$, $\mathbb{F}_q(u)^*$ embeds into $\mathrm{Aut}(\phi)$. The assertion on L'-forms is standard, see e.g. [17], Ch. X. □

3. Structure of End(ϕ).

From now on we assume that L is a finite A-field, extension of degree m of $A/\mathfrak{p}$, where $\mathfrak{p}$ has degree d. Then $\sharp L = q^n$ with $n = d \cdot m$. We let $F = \tau^n$ be the associated Frobenius element. For a fixed Drinfeld module ϕ of rank r over L, we regard A via ϕ as a subring of $L\{\tau\}$. Similarly, we have embeddings $K \hookrightarrow D := \mathrm{End}(\phi) \otimes K \hookrightarrow L(\tau) =$ quotient skew field of $L\{\tau\}$. Then $F \in \mathrm{End}(\phi) = \mathcal{C}_{L\{\tau\}}(A)$ and $D = \mathcal{C}_{L(\tau)}(K)$ ($\mathcal{C}$ = centralizer). Let E be the commutative subfield $K(F)$ generated by F. The picture is (explanations below):

(3.1)

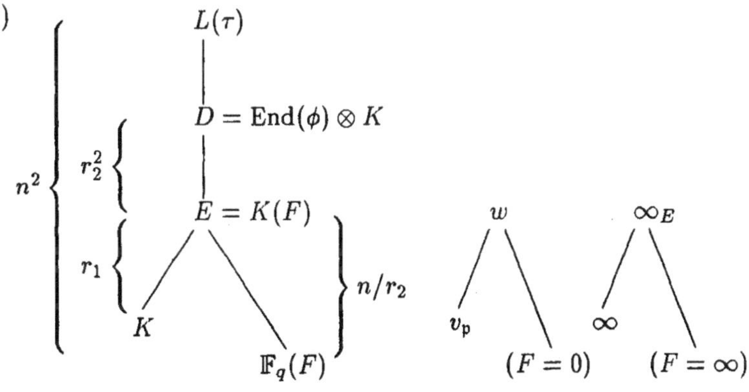

Here we consider $\mathbb{F}_q(F)$ as the field of rational functions in F. We collect some facts about these algebras.

(3.2) $L(\tau)$ is a central division algebra of dimension n^2 over $\mathbb{F}_q(F)$ with invariants $\frac{1}{n}$, $-\frac{1}{n}$, 0 at places v of $\mathbb{F}_q(F)$ corresponding to $(F=0)$, $(F=\infty)$, neither one, respectively. This is (1.8).

(3.3) The places $(F=0)$ and $(F=\infty)$ of $\mathbb{F}_q(F)$ have unique extensions to E, labelled by "w" and "∞_E". This follows since the completions of $L(\tau)$ at these places are divison algebras. Note that w lies above the place $v_{\mathfrak{p}}$ of K, and that ∞_E is the unique extension of the given place ∞ of K to E. We let $v_\infty : E^* \longrightarrow \mathbb{Q}$ be the extension of the normalized valuation $v_\infty : K^* \longrightarrow \mathbb{Z}$ to E.

(3.4) From $\deg_\tau(\phi_a) = -r \cdot d_\infty \cdot v_\infty(a)$ we see that

$$v_\infty(F) = -\frac{n}{r \cdot d_\infty}.$$

(3.5) We have $[E:K] = [E_\infty : K_\infty] = e \cdot f$ with the ramification index e and residue degree f of E_∞/K_∞. The normalized valuation associated to ∞_E is $e \cdot v_\infty : E^* \longrightarrow \mathbb{Z}$.

(3.6) Similarly, $[E : \mathbb{F}_q(F)] = [E_\infty : \mathbb{F}_q(F)_\infty] = e' \cdot f'$, where now $f' = d_\infty \cdot f$ and $e' = -$normalized valuation of F in $E_\infty = -e \cdot v_\infty(F)$.

Combining (3.5) and (3.6) yields

$$[E:\mathbb{F}_q(F)]/[E:K] = n/r. \tag{3.7}$$

We let $r_1 := [E:K]$. Since n equals the degree over $\mathbb{F}_q(F)$ of a maximal commutative subfield of $L(\tau)$, the number

$$r_2 := r/r_1 = n/[E:\mathbb{F}_q(F)]$$

is an integer.

"$\mathcal{C}$" denoting the centralizer in $L(\tau)$, we have $D = \mathcal{C}(K) = \mathcal{C}(E)$, since F centralizes D. Hence (1.3) implies that D is central over E of degree r_2^2. The invariants $inv_v([D])$ of D over E are obtained from those of $L(\tau)$ over $\mathbb{F}_q(F)$ by multiplication with $[E : \mathbb{F}_q(F)] = n/r_2$ (this follows from (1.5)), and are therefore given by $\frac{1}{r_2}$, $-\frac{1}{r_2}$, 0 for places v of E with $v = w$, $v = \infty_E$, $v \neq w, \infty_E$.

We have thus proved the following theorem stated already in [3].

3.8 Theorem. *Let E be the subfield of $D = \mathrm{End}(\phi) \otimes K$ generated over K by F and $r_1 = [E:K]$ its degree. Then r/r_1 is an integer r_2, and D is a central division algebra over E of degree r_2^2. There exists a unique place w of E that* *divides F, and w lies above the place $v_{\mathfrak{p}}$ of K. The place ∞ of K has a unique extension ∞_E to E. D splits at places v of E different from w and ∞_E, and has invariants $\frac{1}{r_2}$, $-\frac{1}{r_2}$ at $v = w$, $v = \infty_E$, respectively.*

Let now A' be the order $\mathrm{End}(\phi) \cap E$ in E. Each finite place v of K splits in E into a certain number of places w_i. Taking completions, we have $A' \otimes_A A_v = \oplus A'_{w_i}$ and $T_v(\phi) = \oplus T_{w_i}(\phi)$ (where we refrain from making the obvious definitions). In general, the A_v-order A'_{w_i} in E_{w_i} is not maximal, and $T_{w_i}(\phi)$ need not be A'_{w_i}-free. However, the A_v-ranks are as expected.

3.9 Corollary. *Let w be the distinguished place of E above $v_{\mathfrak{p}}$. Then $T_w(\phi) = 0$, and for finite places $w' \neq w$ of E, $T_{w'}(\phi)$ is a free A_v-module of rank $r_2 \cdot [E_{w'} : K_v]$.*

Proof. $T_w(\phi) = 0$ since $F : \overline{L} \longrightarrow \overline{L}$ is purely inseparable and so ${}_F\phi(\overline{L}) = 0 = {}_w\phi(\overline{L})$. On the other hand, for places $w' \neq w$ (whether or not $w|v_{\mathfrak{p}}$), the scheme ${}_{w'}\phi$ is reduced. Comparing degrees as in the proof of (3.8) yields the result. We leave details to the reader. □

3.10 Remark. The order A' in E is a ring of almost the same type as A, except that it need not be maximal, i.e., integrally closed. Instead of considering ϕ as a Drinfeld A-module over L, we can consider it as a "Drinfeld A'-module" $\phi' : A' \longrightarrow L\{\tau\}$, where the A'-characteristic of L is the distinguished prime w. As an A'-module, its "rank" is $r/[E : K] = r_2$, which by arguments similar to those used in the proof of (2.1) gives $\dim_{K_v} T_{w'}(\phi') \otimes E_{w'} = r_2$ ($w' \neq w$ a place of E lying above v). Since $T_{w'}(\phi') = T_{w'}(\phi)$ possibly fails to be free over $A'_{w'}$, the equality makes sense only after tensoring with $E_{w'}$. It is equivalent with the one in (3.9). We use the quotation marks "..." as a warning sign, since important parts of the theory doe not work for "Drinfeld A'-modules", if A' is not integrally closed.

Note finally that (3.9) implies

$$ht(\phi) = r_2 \cdot [E_w : K_{\mathfrak{p}}],$$

since both sides may be described as $r - \mathrm{rank}_{A_{\mathfrak{p}}} T_{\mathfrak{p}}(\phi)$. (Here the subscript $\mathfrak{p}$ refers to $v_{\mathfrak{p}}$-adic completions and Tate modules, respectively.)

3.11 Corollary. *With notations as above, the $A_{\mathfrak{p}}$-orders $A'_w = (\mathrm{End}(\phi) \cap E)_w$ in E_w and $(\mathrm{End}(\phi))_w$ in $D_w = D \otimes_E E_w$ are maximal.*

Proof. Since D_w is a division algebra, the respective maximal orders are characterized as the sets of $A_{\mathfrak{p}}$-integral elements or, what is the same, of elements with non-negative valuations (see (1.9)). Now by section 1,

$$\begin{array}{lcl} L(\tau)_{(F=0)} & = & L((\tau)), \\ D_w & = & \text{closure of } D \text{ in } L((\tau)) = \text{centralizer of } K \text{ or of } E \text{ in } L((\tau)) \\ (\mathrm{End}(\phi))_w & = & \text{closure of End}(\phi) \text{ in } L((\tau)) \\ & = & \text{closure (centralizer of } K \text{ or } E \text{ in } L\{\tau\}) \\ & = & \text{centralizer of } K \text{ or } E \text{ in } L\{\{\tau\}\} \\ & = & \text{elements with non-negative valuation in } D_w. \end{array}$$

Hence $(\mathrm{End}(\phi))_w$ is maximal, which also shows the maximality of A'_w. □

The results of (3.8) to (3.11) are the starting point of [1] for a more detailed study of endomorphism rings of Drinfeld modules over finite fields.

As an element of $\mathrm{End}(\phi)$, F has a minimal polynomial $M_\phi(X)$ of degree r_1 over A. On the other hand, for each place $v \neq v_\mathfrak{p}$ of A, $i_v(F) \in \mathrm{End}(T_v(\phi))$ has a characteristic polynomial $P_\phi(X)$ over A_v. We let $N : D \longrightarrow K$ be the composite of the reduced norm Nrd_E^D and the field norm N_K^E.

3.12 Lemma. *Let $u \in \mathrm{End}(\phi)$. We have:*

(i) $N(u) = \det(i_v(u))$

(ii) $-v_\infty(N(u)) = (rd_\infty)^{-1} \deg_\tau N(u) = d_\infty^{-1} \deg_\tau u$

(iii) $P_\phi(X) = M_\phi(X)^{r_2}$. *So it has coefficients in A independent of $v \neq v_\mathfrak{p}$.*

Proof. (i) First note that N agrees on maximal commutative subfields H of D (which have degree r over K) with the field norm N_K^H. Therefore $N|_H = (\det \circ i_v)|_H$ and $N = \det \circ i_v$.
(ii) $\deg_\tau \circ N$ defines a valuation on D equivalent with the ∞-adic valuation. Hence $\deg_\tau N(u)$ is proportional with $\deg_\tau u$. The proportionality constant comes out by evaluating on $u = \phi_a$. The first equality is (3.4).
(iii) Let $t \in E$. Then

$$P_\phi(t) = \det_{A_v}(i_v(t - F)) = N(t - F) = N_K^E \circ Nrd_E^D(t - F)$$
$$= N_K^E(t - F)^{r_2} = (N_K^E(t - F))^{r_2} = (M_\phi(t))^{r_2},$$

and the polynomials $P_\phi(X)$ and $M_\phi(X)^{r_2}$ agree. □

This now leads to a classification of the isogeny classes of Drinfeld modules over L.

3.13 Theorem. *For two Drinfeld modules ϕ and ψ of rank r over L, the following are equivalent:*

(a) *ϕ and ψ are isogeneous (i.e., L-isogeneous)*

(b) $\mathrm{End}(\phi) \otimes K$ *and $\mathrm{End}(\psi) \otimes K$ are F-isomorphic K-algebras (i.e., there exists a K-isomorphism compatible with the respective Frobenius actions)*

(c) *The subfields $E_\phi \hookrightarrow$* $\mathrm{End}(\phi) \otimes K$ *and $E_\psi \hookrightarrow$* $\mathrm{End}(\psi) \otimes K$ *are F-isomorphic*

(d) $P_\phi = P_\psi$

(e) $M_\phi = M_\psi$.

(Note that the condition on F in (b) and (c) was assumed implicitly in [7] Thm. 3.5.)

Proof. We have (d) $\Leftrightarrow$ (e) by the last lemma and (e) $\Leftrightarrow$ (c) $\Leftrightarrow$ (b) from (3.8). (b) $\Rightarrow$ (a): Let $\alpha : \mathrm{End}(\phi) \otimes K \longrightarrow \mathrm{End}(\psi) \otimes K$ be an F-isomorphism. By the Skolem-Noether theorem (1.4), $\alpha(f) = u \circ f \circ u^{-1}$ with some $u \in L(\tau)$, which we can actually choose in $L\{\tau\}$. Such an u defines an L-isogeny from ϕ to ψ, as is easily checked (cf. proof of (2.5)). (a) $\Rightarrow$ (c): Let $u : \phi \longrightarrow \psi$ be an isogeny and $M_\phi(X) = \Sigma a_i X^i$. Since $u \circ \phi_a = \psi_a \circ u$ for $a \in A$, we have in $L\{\tau\}$:

$$0 = u \circ \Sigma F^i \circ \phi_{a_i} = \Sigma F^i \circ \psi_{a_i} \circ u,$$

thus $\Sigma F^i \circ \psi_{a_i} = 0$. This means that M_ϕ divides M_ψ, and thus $M_\phi = M_\psi$. □

By Theorem 3.8 and the preceding discussion, the Frobenius element F of ϕ, regarded as an element of a field extension of K in its algebraic closure $\overline{K}$, enjoys the following properties:

(3.14) (a) F is integral, and $[K(F) : K]$ divides r.
(b) There is a unique place of $K(F)$ over ∞.
(c) There is only one place w of $K(F)$ which is a zero of F, and w lies above $v_\mathfrak{p}$.
(d) If v_∞ denotes the unique extension to $K(F)$ of the normalized ∞-adic valuation on K then $v_\infty(F) = - - (r \cdot d_\infty)^{-1} \cdot n$.

An element φ of $\overline{K}$ that satisfies these conditions is called a *Weil number of rank r for L* (cf. [20]). From the above we get a map

$$\left\{ \begin{array}{c} L\text{-isogeny classes of Drinfeld} \\ A\text{-modules of rank } r \text{ over } L \end{array} \right\} \longrightarrow \left\{ \begin{array}{c} \text{isomorphism classes of Weil} \\ \text{numbers of rank } r \text{ for } L \end{array} \right\},$$

with the obvious notion of isomorphism for Weil numbers.

3.15 Theorem. *The above map is bijective.*

This has been stated in [3]; a complete proof can be found in [20], to which we refer for details. Note that the injectivity is obvious from 3.13; hence the proof amounts to constructing Drinfeld modules with prescribed minimal equation for their Frobenius elements.

4. The Z-function of ϕ.

Let for the moment E be an elliptic curve defined over a finite field $\mathbb{F}_q$. Its Z-function is defined as the rational function

$$Z_{E/\mathbb{F}_q}(X) = \frac{qX^2 - cX + 1}{(1-X)(1-qX)},$$

where $c = q + 1 - \sharp\{\mathbb{F}_q - \text{rational points of } E\}$. Note that the numerator of Z is the *inverse characteristic polynomial* $\det(1 - XF)$ of the Frobenius endomorphism F of $E/\mathbb{F}_q$. The *raison d'être* for Z is the following formal power series identity:

$$X\frac{d}{dX} \log Z_{E/\mathbb{F}_q}(X) = \sum_{k \geq 1} a_k X^k,$$

where $a_k = \sharp E(\mathbb{F}_{q^k})$, $\mathbb{F}_{q^k}$ the extension of degree k of $\mathbb{F}_q$, and "$\frac{d}{dX}$ log" is the operator $f \longmapsto f'/f$ on power series. (See [18] Ch. V for details and generalizations.)

(4.2) We now come back to the situation and notations of the last section. Our aim is to associate a Z-function $Z_{\phi/L}(X)$ with similar properties to the Drinfeld module ϕ of rank r over L.

Which quantity corresponds, under the usual analogy of Drinfeld modules with elliptic curves, to the number a_k above?

Now since $E(\mathbb{F}_{q^k})$ is a finite abelian group, a_k is nothing else but its Euler-Poincaré characteristic, i.e., the value on $E(\mathbb{F}_{q^k})$ of the uniquely determined function on finite abelian groups that associates p to $\mathbb{Z}/p\mathbb{Z}$ (p a rational prime) and is multiplicative in short exact sequences. Hence we define the Euler-Poincaré characteristic χ for A-modules as the uniquely determined map from {isomorphism classes of finite A-modules} to {A-ideals} that satisfies

(a) $\chi(A/\mathfrak{q}) = \mathfrak{q}$ ($\mathfrak{q}$ prime) and

(b) $\chi(M) = \chi(M_1)\chi(M_2)$, if there exists a short exact sequence

$$0 \longrightarrow M_1 \longrightarrow M \longrightarrow M_2 \longrightarrow 0.$$

We further put $\chi(L_k, \phi)$ for $\chi(L_k)$, where L_k is the field extension of degree k of L, regarded as an A-module via ϕ.

(4.3) As in section 3, we let $P(X) = P_\phi(X)$ be the characteristic polynomial of the Frobenius endomorphism F of ϕ over L. We further put for $0 \leq i \leq r$

$$Q_i(X) = \det(1 - X\Lambda^i F),$$

which is the *inverse characteristic polynomial* of $\Lambda^i F$ acting on the i-th exterior power $\Lambda^i T_v(\phi)$ of $T_v(\phi)$, $v \neq v_\mathfrak{p}$ a place of A. Note that $Q_i(X)$ is determined through $P(X)$, similar to which it has coefficients in A independent of v. We finally put

$$Z_{\phi/L}(X) = \prod_{0\leq i\leq r} Q_i(X)^{(-1)^{i+1}},$$

a rational function over K. If e.g. $r = 1,2,3$ and $P(X) = X - a$, $X^2 - aX + b$, $X^3 - aX^2 + bX - c$ then

$$\begin{aligned} Z_{\phi/L}(X) &= \tfrac{1-aX}{1-X} && (r=1) \\ &= \tfrac{1-aX+bX^2}{(1-X)(1-bX)} && (r=2) \\ &= \tfrac{(1-aX+bX^2-cX^3)(1-cX)}{(1-X)(1-bX+acX^2-c^2X^3)} && (r=3). \end{aligned}$$

Recall that v_∞ is the extension of the normalized valuation v_∞ to the algebraic closure $\overline{K}_\infty$, $\mathfrak{p} = \mathrm{char}_A(L)$ has degree d, $m = [L : A/\mathfrak{p}]$ and $n = [L : \mathbb{F}_q] = m\cdot d$.

4.4 Theorem. (i) *The principal ideal* $(P(0))$ *of* A *equals* $\mathfrak{p}^m$.
(ii) $(P(1)) = \chi(L,\phi)$.
(iii) *The zeroes* x_i *of* $P(X)$ *in* $\overline{K}_\infty$ *satisfy* $v_\infty(x_i) = -\frac{n}{r\cdot d_\infty}$.

Proof. By (3.8) and (3.12) we see that $\mathfrak{p}$ is the only prime that divides $P(0)$. The exponent m results from (3.4) and the product formula in K, thus (i). Again by (3.12), it suffices to prove (iii) for the minimal polynomial $M(X)$. But $M(X)$ is also the minimal polynomial of F over the local field K_∞. Hence (iii) follows from (i), since the Newton polygon of $M(X)$ has only one slope. For (ii), first note that L = kernel of $F-1$ on $\overline{L}$ and therefore $\chi(L,\phi) = \chi(\ker(F-1))$. We calculate, for a prime $\mathfrak{q} \neq \mathfrak{p}$ and $v = v_\mathfrak{q}$ the $\mathfrak{q}$-primary part of the ideal $(P(1))$:

$$\begin{aligned} (P(1))_\mathfrak{q} &= (\det \circ i_v(F-1)) = \chi(T_v(\phi)/im(i_v(F-1))) \\ &= \chi(ker(i_v(F-1))) = \chi(ker(F-1))_\mathfrak{q}. \end{aligned}$$

Furthermore,

$$\begin{aligned} &\log_q \sharp A/(P(1)) = -d_\infty v_\infty(P(1)) \overset{(a)}{=} -d_\infty v_\infty(P(0)) \\ &\overset{(b)}{=} dm = \log_q \sharp L = \log_q \sharp A/\chi(ker(F-1)), \end{aligned}$$

where (a) comes from (iii) and (b) from the product formula in K. This means that the two ideals $(P(1))$ and $\chi(ker(F-1))$ agree in their $\mathfrak{q}$-components ($\mathfrak{q} \neq \mathfrak{p}$) and their size, and thus are equal. □

4.5 Corollary. *If there exists a Drinfeld module of arbitrary rank r over the field L of degree m over $A/\mathfrak{p}$, the ideal $\mathfrak{p}^m$ is principal.*

4.6 Remark. Conversely, if $\mathfrak{p}^m = (a)$ is principal, there always exists a rank-one Drinfeld module ϕ over the field L of degree m over $A/\mathfrak{p}$, as follows from the results of [12]. Whether or not the same assertion holds for an arbitrary given rank $r > 1$ seems to be an interesting question.

We are now in a position to prove the analogue of (4.1) for Drinfeld modules.

4.7 Theorem. *Let $Z_{\phi/L}(X)$ be the Z-function of the Drinfeld module ϕ over L. With notations as in (4.1) and (4.2), the coefficients a_k of*

$$X\frac{d}{dX} \log Z_{\phi/L}(X) = \sum_{k\geq 1} a_k X^k$$

lie in A and satisfy $(a_k) = \chi(L_k, \phi)$ for all k.

Proof. We briefly write F, F^k, $\Lambda^i(F)\ldots$ for the operators $i_v(F)$, $i_v(F)^k$, $\Lambda^i(i_v(F))$ etc. In view of (4.4) (ii), it suffices to show the following identity of formal power series

$$\sum_{k\geq 1} \det(1-F^k)X^k = X\frac{d}{dX} \log \prod_{0\leq i\leq r} Q_i(X)^{(-1)^{i+1}}. \tag{1}$$

But (1) is an easy consequence of

$$\frac{d}{dX} \log(f\cdot g) = \frac{d}{dX} \log(f) + \frac{d}{dX} \log(g) \tag{2}$$

(valid for arbitrary non-zero power series f, g) and the following two identities, which are known to hold for arbitrary endomorphisms F of a free module T of finite rank r:

$$\det(1-XF) = \sum_{0\leq i\leq r} (-1)^i Tr(\Lambda^i F)X^i \tag{3}$$

$$-X\frac{d}{dX} \log \det(1-XF) = \sum_{k\geq 1} Tr(F^k)X^k. \tag{4}$$

(Apply (3) to $F \longrightarrow F^k$ and put $X = 1$!) □

4.8 Remark. As is explained in [7] and, much more deeply, in [11], the *local* Z-function considered here may be used as the $\mathfrak{p}$-th Euler factor of a *global* zeta function $\zeta_{\phi/L}$, where now the Drinfeld module ϕ is defined over a finite extension L of K. From such ϕ, one gets Drinfeld modules over finite A-fields through reduction modulo places of L. It is commonly expected that $\zeta_{\phi/L}$ encodes meaningful information on ϕ, although not very much is known about such functions.

5. Supersingularity.

We keep notations of the preceding sections and consider in more detail the case where eventually $r_1 = 1$.

5.1 Proposition. *For a Drinfeld module ϕ over L the following are equivalent:*

(a) *There exists a finite extension L' of L such that* $[\mathrm{End}_{L'}(\phi)\otimes K : K] = r^2$.

(b) *Some power of F lies in A.*

(c) $({}_{\mathfrak{p}}\phi(\overline{L})) = 0$.

Proof. By (2.8), there exists a finite extension L' of L such that $\mathrm{End}_{\overline{L}}(\phi) = \mathrm{End}_{L'}(\phi)$. This together with Theorem 3.8 yields the equivalence of (a) and (b). If (b) is satisfied then $F^k = \phi_a$ for some $a \in A$ with $\mathfrak{p}$-primary divisor $(a) = \mathfrak{p}^l$. Hence $({}_{\mathfrak{p}^l}\phi)(\overline{L}) = 0$ and therefore $({}_{\mathfrak{p}}\phi)(\overline{L}) = 0$. Conversely, let $({}_{\mathfrak{p}}\phi)(\overline{L}) = 0$. Then also $({}_{\mathfrak{p}^l}\phi)(\overline{L}) = 0$ for $l \in \mathbb{N}$ and, if $(a) = \mathfrak{p}^l$ as above, some powers of F and ϕ_a agree in $L\{\tau\}$. □

Drinfeld modules ϕ over L subject to the conditions of Proposition 5.1 are called *supersingular*. Let m_0 be the order of $\mathfrak{p}$ in the class group of A and let $\tilde{L}$ be the extension of degree $m = m_0 \cdot r$ of $A/\mathfrak{p}$ (where r is the rank of all the Drinfeld modules that appear).

5.2 Proposition. (i) *Each supersingular Drinfeld module ϕ of rank r over $A/\mathfrak{p}$ is isomorphic to one which is defined over $\tilde{L}$.*
(ii) *If $\phi/\tilde{L}$ is supersingular,* $\mathrm{End}_{\tilde{L}}(\phi) = \mathrm{End}_{\overline{L}}(\phi)$. *Conversely, if ϕ/L is supersingular and* $\mathrm{End}_L(\phi) = \mathrm{End}_{\overline{L}}(\phi)$ *then* $\tilde{L} \subset L$.

Proof. (i) Let ϕ be defined over a finite extension L' of $\tilde{L}$ and $F = \tau^n$, $n = d\cdot m = d \cdot m_0 \cdot r$ the Frobenius relative to $\tilde{L}$. Let $a \in A$ with $(a) = \mathfrak{p}^{m_0}$, thus $\phi_a = \mathrm{const.}\ \tau^n$. Possibly replacing ϕ by an $A/\mathfrak{p}$-isomorphic Drinfeld module, we may assume $\phi_a = \tau^n$. If $b \in A$ and $\phi_b = \Sigma b_i\tau^i$, the commutation rule $\phi_a \circ \phi_b = \phi_b \circ \phi_a$ implies $b_i^{q^n} = b_i$, i.e., ϕ defined over L.
(ii) The Frobenius F of ϕ with respect to $\tilde{L}$ satisfies $F = \phi_a$ with some $a \in A$, hence $\mathrm{End}_{\tilde{L}}(\phi)$ has maximal rank r^2, and $\mathrm{End}_{\tilde{L}}(\phi) = \mathrm{End}_{\overline{L}}(\phi)$ follows from (2.10). Conversely, if $\mathrm{End}_L(\phi) = \mathrm{End}_{\overline{L}}(\phi)$ then its Frobenius $F = \tau^n$ equals ϕ_a for some $a \in A$, and $n/d\cdot m_0$ must be divisible by r. □

We next define $\Sigma(r, \mathfrak{p})$ as the set of $\overline{A/\mathfrak{p}}$-isomorphism classes of supersingular Drinfeld modules of rank r over $\overline{A/\mathfrak{p}}$.

5.3 Proposition. *$\Sigma(r, \mathfrak{p})$ is finite, and all its elements are $\overline{A/\mathfrak{p}}$-isogeneous.*

Proof. The first assertion is clear from (5.2) (i). Let ϕ, ϕ' represent two elements of $\Sigma(r, \mathfrak{p})$, defined over some finite extension L of $\tilde{L}$. Their Frobenius elements F, F' correspond to elements a, a' of A that generate the same ideal. Hence $a^{q-1} = (a')^{q-1}$, which by (3.13) implies that ϕ and ϕ' are isogeneous at least over the extension of degree $q-1$ of L. □

Let now $D(r, \mathfrak{p}, \infty)$ be the central division algebra over K of dimension r^2 and with local invariants $\frac{1}{r}$, $-\frac{1}{r}$, 0 at places $v = v_{\mathfrak{p}}$, $v = \infty$, $v \neq v_{\mathfrak{p}}, \infty$, respectively. It appears as $\mathrm{End}_L(\phi) \otimes K$ for every $\phi \in \Sigma(r, \mathfrak{p})$ defined over a finite field $L \supset \tilde{L}$. The next result relates the arithmetic of $D(r, \mathfrak{p}, \infty)$ with the theory of Drinfeld modules and their moduli schemes (see next section), and enables e.g. the calculation of its class and type numbers.

5.4 Theorem [7]. *Let ϕ be a supersingular Drinfeld module of rank r over a finite field L that contains $\tilde{L}$.*
(i) $\mathrm{End}(\phi) = \mathrm{End}_L(\phi)$ *is a maximal order in* $\mathrm{End}(\phi) \otimes K$.
(ii) *The left ideal classes of* $\mathrm{End}(\phi)$ *correspond bijectively to the elements of* $\Sigma(r, \mathfrak{p})$.

The proof (*loc. cit.*) uses a counting argument that requires some prerequisites from the theory of orders, and will therefore not be given here. We restrict to giving the construction of the map underlying (ii).

Let $\mathfrak{u} \subset \mathrm{End}(\phi)$ be a left ideal. The left ideal $L\{\tau\}\mathfrak{u}$ of $L\{\tau\}$ is principal, generated by a uniquely determined monic $u \in L\{\tau\}$. As $\phi(A)$ is central in $\mathrm{End}(\phi)$, we have $\mathfrak{u} = \mathfrak{u}\phi(A)$, which for each $a \in A$ implies the existence of $\psi_a \in L\{\tau\}$ with $u \circ \phi_a = \psi_a \circ u$. One then verifies that $a \longmapsto \psi_a$ is a Drinfeld module, u is an isogeny, and (class of $\mathfrak{u}$) $\longmapsto$ (class of ψ in $\Sigma(r, \mathfrak{p})$) is well-defined. Statement (ii) of the theorem asserts that it is in fact bijective.

6. Examples and applications.

In all the examples to follow except for (6.3), we assume that A is a polynomial ring $\mathbb{F}_q[T]$ (hence $d_\infty = 1$ and $q_\infty = q$) and $\mathfrak{p}$ a prime of A, generated by the monic polynomial f of degree d. We first consider Z-functions. Let $\nu : A/\mathfrak{p} \longrightarrow A$ be the map composed of the norm $A/\mathfrak{p} \longrightarrow \mathbb{F}_q$ and the canonical inclusion $\mathbb{F}_q \hookrightarrow A$. We consider Drinfeld modules of rank r over the *prime field* $A/\mathfrak{p}$.

6.1 Example. Let $r = 1$, ϕ given by $\phi_T = \gamma(T) + g\tau$, $g \in A/\mathfrak{p}$. Then an easy computation yields $P_\phi(X) = M_\phi(X) = X - \nu(g) \cdot f$, thus

$$Z_\phi(X) = \frac{1 - \nu(g)fX}{1 - X}.$$

We get

$$\chi(L_k, \phi) = \text{principal ideal generated by } (\nu(g)f)^k - 1,$$

and, since in this rank one case the A-module (L_k, ϕ) is cyclic,

$$(L_k, \phi) \cong A/((\nu(g)f)^k - 1).$$

6.2 Example. Let $r = 2$, ϕ given through $\phi_T = \gamma(T) + g\tau + \Delta\tau^2$ with $g \in A/\mathfrak{p}$, $\Delta \in (A/\mathfrak{p})^*$. Then $P_\phi(X) = X^2 - aX + cf$ with $a \in A$ of degree $\leq d/2$ and $c \in \mathbb{F}_q^*$. The precise values of a and c can be found by the methods of [6]. Suppose further that $\Delta = 1$, and let first

$\boxed{d = 1}$ Then $P_\phi(X) = X^2 + gX - f$, where $f = (T - f_0)$, f_0 and $g = \nu(g) \in \mathbb{F}_q$,

$$\begin{aligned} Z_\phi(X) &= \tfrac{1 + gX - fX^2}{(1-X)(1+fX)}, \\ X\tfrac{d}{dX} \log Z_\phi(X) &= (1 + g - f)X + (1 - g^2 + f^2 - 2f)X^2 + \cdots \end{aligned}$$

Let now

$\boxed{d = 2}$ Then $P_\phi(X) = X^2 - (\nu(g) + f')X + f \quad (f' = \frac{d}{dT}f)$,

$$\begin{aligned} Z_{\phi(X)} &= \tfrac{1 - (\nu(g) + f')X + fX^2}{(1-X)(1-fX)}, \\ X\tfrac{d}{dX} \log Z_\phi(X) &= (1 - (\nu(g) + f') + f)X + \\ &\quad ((1 + f)^2 - (\nu(g) + f')^2)X^2 + \cdots \end{aligned}$$

From these formulas, we can easily calculate e.g. the smallest field extension L_k of $L_1 = A/\mathfrak{p}$ over which ϕ acquires a $\mathfrak{q}$-division point ($\mathfrak{q}$ a prime of A).

Next, we consider the sets $\Sigma(r, \mathfrak{p})$ or, what amounts to the same, the sets of ideal classes of $D(r, \mathfrak{p}, \infty)$.

6.3 Example. Here we allow a generic A subject only to our general conditions as described in section two. Our knowledge of the geometry of Drinfeld modular curves suffices to determine $\Sigma(2, \mathfrak{p})$ and therefore the class number $h(D(2, \mathfrak{p}, \infty)) = \sharp\Sigma(2, \mathfrak{p})$ of the quaternion algebra $D(2, \mathfrak{p}, \infty)$. It depends only on (a) the zeta function ζ_K of the global field $K = \mathrm{Quot}(A)$ and (b) the degrees d and d_∞ of $\mathfrak{p}$ and ∞, respectively. The precise formula is Théorème 1 in [8]. Here we restrict to cite a non-trivial example from there. Others may be worked out following the same lines.

Let the characteristic p be unequal to two and $A = \mathbb{F}_q[X, Y]$, where X, Y satisfy $Y^2 = aX^2 + bX + c$ with $b^2 - 4ac \neq 0$. Hence K is again a rational function field, but class number of $A = d_\infty = 2$. The resulting class number is

$$h(D(2, \mathfrak{p}, \infty)) = 2(q^d - 1)/(q - 1).$$

If further $d = 1$, the type number $t(D(2, \mathfrak{p}, \infty))$ may be verified to be one.

The class number $h(D(r, \mathfrak{p}, \infty)) = \sharp\Sigma(r, \mathfrak{p})$ can be determined for arbitrary r at least in the case $A = \mathbb{F}_q[T]$, to which we restrict from now on.

Let r_1 be the largest divisor of r coprime with $d = \deg \mathfrak{p}$ and μ the Möbius function. Then ([9] Thm. 5.13):

$$\sharp\Sigma(r, \mathfrak{p}) = \frac{1}{q^r - 1} \sum_{s|r_1} (q^s - 1) \sum_{i|r_1/s} \mu(i) \prod_{\substack{0<j<r \\ j \equiv 0(is)}} \frac{q^{jd} - 1}{q^j - 1} \tag{6.4}$$

The intersection of cases (6.3) and (6.4), i.e., that of $\mathbb{F}_q[T]$-orders in the quaternion algebra $D(2, \mathfrak{p}, \infty)$ over $K = \mathbb{F}_q(T)$ or, equivalently, of supersingular isomorphism classes of rank-two Drinfeld $\mathbb{F}_q[T]$-modules, is described in the next example. Its result already appears in [5].

6.5 Example. For $A = \mathbb{F}_q[T]$ and a prime $\mathfrak{p} = (f)$ of degree d, we have

$$\begin{aligned} \sharp\Sigma(2, \mathfrak{p}) &= \tfrac{q^d - 1}{q^2 - 1} && (d \text{ even}) \\ &= \tfrac{q^d - q}{q^2 - 1} + 1 && (d \text{ odd}). \end{aligned}$$

If we define the j-invariant of a rank-two Drinfeld $\mathbb{F}_q[T]$-module ϕ given by $\phi_T = \gamma(T) + g\tau + \Delta\tau^2$ through $j = g^{q+1}/\Delta$, the following j-invariants in $A/\mathfrak{p}$ are supersingular:

$$\begin{aligned} d = 1 &: \quad j = 0 \\ d = 2 &: \quad j = -f' \\ d = 3 &: \quad j = 0,\ j = (T^q - T)(1 - (T^q - \alpha)^{q-1}), \quad \alpha \in \mathbb{F}_q, \end{aligned}$$

where the values have to be taken mod $\mathfrak{p}$. More information on supersingular j-invariants can be found in [16].

We briefly mention some applications of the above.

(6.6) First, we point out that $\sharp\Sigma(2, \mathfrak{p}) = g + 1$, where g is the genus of the Drinfeld modular curve $X_0(\mathfrak{p})/K$ of Hecke type. That equality is more than coincidence; it comes from an isomorphism as Hecke modules of $\mathbb{C}^{\Sigma(2,\mathfrak{p})}$ with a certain space of automorphic forms related to $X_0(\mathfrak{p})$ (see [15], [10]). That isomorphism allows to carry out calculations for $X_0(\mathfrak{p})$ or its Jacobian on the finite set $\Sigma(2,\mathfrak{p})$, provided with its Hecke action.

(6.7) Second, we may use supersingular invariants to construct series of (smooth, projective, geometrically irreducible) curves C_k over $\mathbb{F}_q$ with asymptotically many rational points. The basic idea is due to Tsfasman, Vladut and Zink [19]; its prototype is given as follows: For a prime $\mathfrak{p}$ of odd degree d, let $C(\mathfrak{p})$ be the curve $X_0(\mathfrak{p})$ reduced mod (T). For $\mathfrak{p} \neq (T)$, $X_0(\mathfrak{p})$ has good reduction at $\mathfrak{p}$, and $C(\mathfrak{p})$ is a smooth curve over $A/(T) = \mathbb{F}_q$, with genus $g(C((\mathfrak{p})) = \frac{q^d-q}{q^2-1}$. Above the supersingular point $j = 0$ there are $\frac{q^d+1}{q+1}$ points of $C(\mathfrak{p})$, which, similar to (5.2) (i), may be verified to be $\mathbb{F}_{q^2}$-rational. Together with its two cusps, which are always $\mathbb{F}_q$-rational, $C(\mathfrak{p})$ has $N(C(\mathfrak{p})),q^2) \geq \frac{q^d+1}{q+1} + 2$ rational points over $\mathbb{F}_{q^2}$. (In fact, one has equality in all but a finite number of exceptional cases.) Finally, the lim sup of the ratio $\frac{N(C(\mathfrak{p}),q^2)}{g(C(\mathfrak{p}))}$ for $d \longrightarrow \infty$ is $\geq q-1$, hence equals $q-1$ by the inequality of Drinfeld and Vladut [4]. The series $C(\mathfrak{p})$ is therefore asymptotically optimal.

(6.8) Next, for several purposes it is useful to know the precise number of L-isomorphism classes of rank-r Drinfeld modules ϕ over the field L of degree m over $A/\mathfrak{p}$. Since ϕ is given by $\phi_T = \gamma(T) + g_1\tau + \cdots g_{r-1}\tau^{r-1} + \Delta\tau^r$ with $g_i, \Delta \in L$, $\Delta \neq 0$, these classes are in 1-1 correspondence with the orbits of L^* on $L^{r-1} \times L^*$, upon which L^* acts through

$$a(g_1,\ldots,g_{r-1},\Delta) = (a^{q-1}g_1,\ldots,a^{q^{r-1}-1}g_{r-1},a^{q^r-1}\Delta).$$

It is elementary but rather complicated to work out a general formula (see [13] 4.1). For small r, the result is given in the following table, where $n = m \cdot d = [L : \mathbb{F}_q]$, $n_r = gcd(n,r)$, $\sharp$ = number L-isomorphism classes of rank-r Drinfeld modules over L:

r	1	2	2	3	3
n_r	1	1	2	1	3
$\sharp$	$q-1$	$(q-1)q^n$	$q^{n+1}-q^n+q^2-q$	$(q-1)q^{2n}$	$q^{2n+1}-q^{2n}+q^3-q$

(6.9) We finally describe the Weil numbers of rank two for the A-field L (notations as before). It is easy to see that their minimal polynomials $M(X)$ over

K are of one of the following types, which all occur:

(1)	$X^2 - aX + cf^m$	
(2)	$X^2 + cf^m$	if m is odd
(3)	$X^2 + c_0 f^{m/2} X + cf^m$	if m is even and d odd
(4)	$X + cf^{m/2}$	if m is even.

Here $a \in A$ with $(a, f) = 1$, $c \in \mathbb{F}_q^*$, $c_0 \in \mathbb{F}_q$ are such that the polynomial $M(X)$ remains irreducible over K_∞. This means that we have the additional conditions in the respective cases:

(1) $2 \deg a < d \cdot m = \deg f \cdot m$ or $2 \deg a = d \cdot m$ and $X^2 - a_0 X + c$ is irreducible over $\mathbb{F}_q$, $a_0 =$ leading coefficient of a

(2) d odd or $-c \notin (\mathbb{F}_q^*)^2$

(3) $X^2 + c_0 X + c$ irreducible over $\mathbb{F}_q$.

The associated isogeny classes of Drinfeld modules are supersingular in cases (2), (3) and (4) and ordinary = non-supersingular in case (1). The algebra $\mathrm{End}(\phi) \otimes K$ is an imaginary quadratic field extension of K in cases (1), (2), (3) and equals $D(2, \mathfrak{p}, \infty)$ in case (4). It is interesting to know how the $\approx q^{n+1}$ isomorphism classes of rank-two Drinfeld modules over L are distributed over the isogeny classes given by (1)..(4). The answer is given in [20] in terms of class numbers of imaginary quadratic orders over A.

References

[1] B. Anglès: Modules de Drinfeld sur les corps finis. Thesis, Toulouse 1994.

[2] N. Bourbaki: Algèbre, Ch. 8. Masson, Paris 1981.

[3] V.G. Drinfeld: Elliptic modules II. Math. USSR-Sbornik 31, 159-170 (1977).

[4] V.G. Drinfeld and S.G. Vladut: On the number points of an algebraic curve. Funct. Anal. Appl. 17 (1983).

[5] M. Eichler: Zur Zahlentheorie der Quaternionen-Algebren. J. reine angew. Math. **195**, 127-151 (1955).

[6] E.-U. Gekeler: On the coefficients of Drinfeld modular forms. Invent. Math. **93**, 667-700 (1988).

[7] E.-U. Gekeler: On finite Drinfeld modules. J. Algebra **141**, 187-203 (1991).

[8] E.-U. Gekeler: Sur la géométrie de certaines algèbres de quaternions. Sém. Th. des Nomb. Bordeaux 2, 143-152 (1990).

[9] E.-U. Gekeler: On the arithmetic of some division algebras. Comm. Math. Helv. **67**, 316-333 (1992).

[10] E.-U. Gekeler: Jacquet-Langlands theory ... This volume.

[11] D. Goss: Basic structures of function field arithmetic. Springer 1996.

[12] D. Hayes: Explicit class field theory in global function fields. Studies in Algebra and Number Theory. G.C. Rota (ed.) Academic Press, New York 1979.

[13] I. Potemine: Arithmétique des corps globaux de fonctions et géométrie des schémas modulaires de Drinfeld. Thesis, Grenoble 1996.

[14] I. Reiner: Maximal orders. Academic Press, New York 1975.

[15] M. Reversat: Automorphic forms, Drinfeld's reciprocity law. This volume.

[16] Andreas Schweizer: On singular and supersingular invariants of Drinfeld modules. Ann. Fac. Sci. Toulouse (to appear).

[17] J.-P. Serre: Corps locaux. Hermann, Paris 1968.

[18] J.H. Silverman: The arithmetic of elliptic curves. Springer 1986.

[19] M.A. Tsfasman, S.G. Vladut and T. Zink: Modular curves, Shimura curves, and Goppa codes better than the Varshamov-Gilbert bound. Math. Nachr. **109**, 21-28 (1982).

[20] J.-K. Yu: Isogenies of Drinfeld modules over finite fields. J. Numb. Th. **54**, 161-171 (1995).

Lecture 6

Some rigid geometry

Guido Van Steen
University of Antwerpen, R.U.C.A.
Department of Mathematics and Computer Science
Groenenborgerlaan 171, B-2020 Antwerpen, Belgium
vansteen@ruca.ua.ac.be

Almost all of the material in this text can be found in [F-vdP] and in [G-vdP]. Therefore there are no specific references to these books.

Let L be a field, complete with respect to a non-archimedean valuation. Let $O \subset L$ be the valuation ring with maximal ideal M and residue field $\bar{L} = O/M$.

1 The Tate algebra T_n.

Definition 1.1 *The Tate algebra $T_n = T_n(L) = L < X_1, \ldots, X_n >$ is the algebra of formal power series in $X_1, \ldots, X_n$ with coefficients in L with the following condition :*

$$\sum_{k \in \mathbf{N}^n} a_k X_1^{k_1} \ldots X_n^{k_n} \in T_n \iff \lim_{k_1 + \cdots + k_n \to \infty} |a_k| = 0$$

The algebra is provided with the following norm :

$$|| \sum_{k \in \mathbf{N}^n} a_k X_1^{k_1} \ldots X_n^{k_n} || = max \left\{ |a_k| \mid k \in \mathbf{N}^n \right\}$$

Proposition 1.2 *The norm on $T_n(L)$ is multiplicative and $T_n(L)$ is a Banach algebra with respect to this norm.*

The O-algebra T_n^o and its ideal T_n^{oo} are defined as follows :

$$T_n^o = \left\{ f \in T_n \mid ||f|| \leq 1 \right\}$$

$$T_n^{oo} = \left\{ f \in T_n \mid ||f|| < 1 \right\}$$

The quotient ring $\bar{T}_n = T_n^o / T_n^{oo}$ ist a $\bar{L}$-algebra. It is clear that

$$\bar{T}_n = \bar{L}[\bar{X}_1, \ldots, \bar{X}_n]$$

Many important properties of T_n follow from the Weierstrass preparation theorem.

Definition 1.3 *A power series*

$$g = \sum_{\nu \in \mathbf{N}} g_\nu(X_1, \ldots, X_{n-1}) X_n^\nu$$

(with $g_\nu \in L < X_1, \ldots, X_{n-1} >$) is X_n-distinguished of degree s if

1. *g_s is a unit in T_{n-1}*
2. *$||g_s|| = ||g||$ and $||g_\nu|| < ||g_s||$ for all $\nu > s$*

It is easy to show that a series $g \in T_n$ with $||g|| = 1$ is X_n-distinguished of degree s if and only if $\bar{g} \in \bar{T}_n$ is a unitary polynomial of degree s in the polynomial ring $\bar{L}[X_1, \ldots, X_{n-1}][X_n]$.

Theorem 1.4 (Weierstrass division and preparation)

1. *Let $g \in T_n$ be X_n-distinguished of degree s. For each $f \in T_n$ there exist unique elements $q \in T_n$ and $r \in T_{n-1}[X_n]$ such that*

   $$f = gq + r \text{ and such that } r = 0 \text{ or } deg(r) < s$$

   Furthermore

   $$||f|| = max\{||g||.||q||, ||r||\}$$

2. *Let $g \in T_n$ be X_n-distinguished of degree s. Then there exist a unique monic polynomial $\omega \in T_{n-1}[X_n]$ of degree s and a unique unit $e \in T_n$ such that $g = e.\omega$. Furthermore $||\omega|| = 1$. If $g \in T_{n-1}[X_n]$ then also $e \in T_{n-1}$.*

3. *Let $f \in T_n$ be an element with $||f|| = 1$. Then there exists a L-automorphism σ of T_n such that $\sigma(f)$ is X_n-distinguished.*

Proposition 1.5

1. *T_n is noetherian.*
2. *Every ideal in T_n is closed.*
3. *If $I \subset T_n$ is an ideal then there exist a finite injective morphism $T_d \to T_n/I$ with d the Krull dimension of T_n/I.*
4. *If $I \subset T_n$ is a maximal ideal then T_n/I is a finite extension of L.*

As an example we use 1.4 to prove 1.5.3.
Assume that $I \neq 0$. After applying an automorphism on T_n we may assume that I contains an element $f \neq 0$ which is X_n-distinguished of some degree t, (cf 1.4.3). It follows from 1.4.2 that I contains an element

$$g = X_n^t + \sum_{i=0}^{t-1} a_i X_n^i$$

with $a_i \in T_{n-1}$. Hence $T_n/(g)$ is a free T_{n-1}-module generated by $1, X_n, \ldots, X_n^{t-1}$.
Let $J = I \cap T_{n-1}$. We find that the canonical map $T_{n-1}/J \to T_n/I$ is a finite injective homomorphism. By induction on n there exists a finite injective homomorphism $T_d \to T_{n-1}/J$.
If $T_d \to T_n/I$ is a finite injective homomorphism then T_d and T_n/I have the same Krull dimension. It is not very difficult to prove that T_d has Krull dimension d.

Let

$$B_n(L) = \{x \in L^n \mid max|x_i| \leq 1\}$$

and let

$$D_n = D_n(L) = max(T_n(L)) = \{x \subset T_n(L) \mid x \text{ is a maximal ideal}\}$$

Elements of T_n can be regarded as functions on D_n. If $f \in T_n$ an if $x \in D_n$ then $f(x)$ is the residue class of f in the field T_n/x. This field can be embedded in the algebraic closure L_a of L. Since this embedding is not unique, $f(x) \in L_a$ is only defined up to conjugation over L. The valuation on L can be extended in a unique way to T_n/x
Let Γ be the Galois group of L_a over L. The elements of T_n can than be regarded as functions on D_n taking values in L_a/Γ.

Elements of T_n can also be regarded as functions on $B_n(L_a)$ with values in L_a. If $f = \sum_k a_k X_1^{k_1} \ldots X_n^{k_n} \in T_n$ and if $x \in B_n(L_a)$ then the series

$$f(x) = \sum_k a_k x_1^{k_1} \ldots x_n^{k_n}$$

converges and hence defines an element in L_a.
The canonical injection

$$L < X_1, \ldots, X_n > \to L_a < X_1, \ldots, X_n >$$

defines a map $\tau : B_n(L_a) \to D_n$ with

$$\tau(x_1, \ldots, x_n) = \{f \in T_n \mid f(x_1, \ldots, x_n) = 0\}$$

This map is surjective and has finite fibres. Furthermore $\tau(x) = \tau(y)$ if and only if there exists an element $\gamma \in \Gamma$ such that $\gamma(x) = y$. (In particular there is a canonical bijection between $D_n(L_a)$ and $B_n(L_a)$.) The folowing diagram is then commutative.

$$\begin{array}{ccc} B_n(L_a) & \xrightarrow{f} & L_a \\ \tau\downarrow & & \downarrow \\ D_n & \xrightarrow{f} & L_a/\Gamma \end{array}$$

The spectral norm on T_n is defined as follows :

$$||f||_{sp} = \sup\left\{|f(x)| \mid x \in D_n(L)\right\}$$

It follows from the previous commutative diagram that $|f(x)| = |f(\tau(x))|$ for all $x \in B_n(L_a)$ and $f \in T_n(L)$. If $f = \sum_\nu a_\nu X_1^{\nu_1} \ldots X_n^{\nu_n}$ and if $x \in B_n(L_a)$ then

$$|f(x)| \leq max\left\{|a_\nu x_1^{\nu_1} \ldots x_n^{\nu_n}| \mid \nu \in \mathbf{N}^n\right\} \leq max\left\{|a_\nu| \mid \nu \in \mathbf{N}^n\right\} = ||f||$$

It follows that $||f||_{sp} \leq ||f||$.

Proposition 1.6 Maximum modulus principle

$$\forall f \in T_n(L) \quad \exists x \in D_n(L) \text{ such that } ||f|| = |f(x)|$$

Corollary 1.7 $\forall f \in T_n(L) : ||f|| = ||f||_{sp}$

2 Affinoid algebras

Let $I \subset T_n$ be an ideal. The algebra $A = T_n/I$ is called an affinoid L-algebra. Since I is closed in T_n the algebra A is noetherian and is a Banach algebra with respect to the quotient norm on A, i.e.:

$$||\bar{f}|| = inf\left\{||f+g|| \mid g \in I\right\}$$

Let $Y \subset D_n(L)$ be the zero set of I. By restriction the elements of T_n can be regarded as functions on Y. Elements $f, g \in T_n$ define the same function on Y if and only if $f \equiv g \pmod{rad(I)}$. This follows from the fact that T_n is a Jacobson ring. We will always assume that A is reduced, i.e. $I = rad(I)$. The elements of A can be then be regarded as functions on Y. They are called holomorphic functions on Y.

The canonical surjection $\sigma : T_n \to A$ induces a map $\sigma^* : max(A) \to max(T_n)$ defined by $\sigma^*(x) = \sigma^{-1}(x)$. This map is a 1-1 correspondence between $max(A)$

and Y. It follows that the elements of A can be regarded as functions on $max(A)$.

In a similar way as for T_n the spectral norm on A is defined :

$$||f||_{sp} = \sup\{|f(x)| \mid x \in max(A)\}$$

It is easy to verify that $||.||_{sp}$ is a power-multiplicative seminorm on A, i.e.

$$\forall f \in A : ||f^n||_{sp} = ||f||_{sp}^n$$

Since A is assumed to be reduced $||.||_{sp}$ is in fact a norm. Furthermore

$$\forall f \in A : ||f||_{sp} \leq ||f||$$

Proposition 2.1 Maximum modulus principle

$$\forall f \in A \quad \exists x \in max(A) \textit{ such that } |f(x)| = ||f||_{sp}$$

The set $X = max(A)$ is provided with the topology generated by the subsets $U \subset X$ which satisfy the following condition.

There exists an affinoid algebra B and a morphism $\sigma : A \to B$ such that the induced map

$$\begin{array}{rccc} \sigma^* & : & max(B) & \longrightarrow & max(A) \\ & & x & \mapsto & \sigma^{-1}(x) \end{array}$$

has its image in U.
Furthermore if C is an affinoid algebra and if $\tau : A \to C$ is a homomorphism such that the induced map

$$\tau^* : max(C) \to max(A)$$

has its image in U, then there exists a unique morphism $\nu : B \to C$ such that $\tau = \nu \circ \sigma$.
It is easy to prove that σ^* is a bijection between $max(B)$ and U.
Subsets of X of this type are called open affinoid subsets. The property of being an open affinoid subset is transitive in the following sense. If $U \subset X$ is an open affinoid, corresponding to the morphism $\sigma : A \to B$ and if $V \subset max(B)$ is an open affinoid corresponding to the morphism $\tau : B \to C$, then $\sigma^*(V) \subset U$ is an open affinoid in X corresponding to the morphism $\tau \circ \sigma : A \to C$.
A special class of open affinoid subsets are the rational subsets of X.

Definition 2.2 *Let $f_0, \ldots, f_n \in A$ be elements such that $A = f_0A + \cdots + f_nA$. This means that $f_0, \ldots, f_n$ have no common zeroes on $X = max(A)$. The set*

$$X(f_0, \ldots, f_n) = \{x \in X \mid |f_i(x)| \leq |f_0(x)|, \quad i = 1, \ldots, n\}$$

is a rational subset of X.

Proposition 2.3 *$X(f_0,\ldots,f_n)$ is an open affinoid subset of $X = max(A)$. The corresponding affinoid algebra is*

$$A(f_0,\ldots,f_n) = A < T_1,\ldots,T_n > / \, (f_0T_i - f_i | i = 1,\ldots,n)$$

Proposition 2.4 *If R_1 and R_2 are rational subsets of $X = max(A)$ then $R_1 \cap R_2$ is also a rational subset of X.*
If $R_1 \subset X$ is rational and if $R_2 \subset R_1$ is rational then R_2 is a rational subset of X.

Proof. 1) Let

$$R_1 = X(f_0,\ldots,f_n) = \{x \in X \mid |f_0(x)| \geq |f_i(x)|, \quad i = 1,\ldots,n\}$$

and

$$R_2 = X(g_0,\ldots,g_m) = \{x \in X \mid |g_0(x)| \geq |g_i(x)|, \quad i = 1,\ldots,m\}$$

Then

$$R_1 \cap R_2 = \{x \in X \mid |f_0g_0(x)| \geq |f_ig_j(x)|, \; i = 0,\ldots,n \text{ and } j = 0,\ldots,m\}$$

2) Let

$$R_1 = X(f_0,\ldots,f_n) = \{x \in X \mid |f_0(x)| \geq |f_i(x)|, \quad i = 1,\ldots,n\}$$

and

$$R_2 = X(g_0,\ldots,g_m) = \{x \in R_1 \mid |g_0(x)| \geq |g_i(x)|, \quad i = 1,\ldots,m\}$$

then $g_i \in A(f_0,\ldots,f_n) = A < T_1,\ldots,T_n > /I$ where I is generated by $f_0T_1 - f_1,\ldots,f_0T_n - f_n$.
If $g_i' \in A(f_0,\ldots,f_n)$ such that $\|g_i' - g_i\|$ is small enough then $g_0',\ldots,g_m'$ determine the same rational subset R_2. So we may assume that $g_0,\ldots,g_m$ are images of elements in $A[T_1,\ldots,T_n]$ and that the total degree of g_i is less than N for some $N \in \mathbf{N}$. It follows that $f_0^N g_0,\ldots,f_0^N g_m \in A$. These elements still determine the same rational subset R_2. So we may assume that $g_0,\ldots,f_m \in A$. For suitable constants $\lambda_1,\ldots,\lambda_n \in L^*$ and for all $x \in R_2$ we have

$$|g_0(x)| \geq |\lambda_i f_i(x)|, \quad i = 1,\ldots,n$$

Hence $R_2 = R_1 \cap R_3$ where R_3 is defined by the inequalities $|g_0(x)| \geq |g_i(x)|$, $i = 1,\ldots,m$ and $|g_0(x)| \geq \lambda_i f_i(x)|$, $\quad i = 1,\ldots,n$. It follows that R_2 is rational in X. □

The canonical topology on $X = max(A)$ is generated by the rational subsets of X. This follows from the following theorem by L. Gerritzen and H. Grauert, (see e.g [B] p. 309).

Proposition 2.5 *Every open afinoid subset of X is a finite union of rational subsets of X.*

In order to complete the strucure on $max(A)$ we need the notion of a Grothendieck topology.

Definition 2.6 *A Grothendieck topology on a topological space X consists of the following data :*

1. *a family $\mathcal{G}$ of open subsets of X satisfying the following conditions :*
   (a) $\emptyset, X \in \mathcal{G}$
   (b) $\forall U, V \in \mathcal{G} : U \cap V \in \mathcal{G}$
2. *for all $U \in \mathcal{G}$ a set $cov(U)$ of coverings of U such that :*
   (a) *for all $\mathcal{U} \in cov(U)$ and for all $V \in \mathcal{U}$: $V \in \mathcal{G}$*
   (b) $\{U\} \in cov(U)$
   (c) *if $\mathcal{U} \in cov(U)$ and if $V \subset U$ with $V \in \mathcal{G}$ then*
   $$\mathcal{U} \cap V = \{U' \cap V \mid U' \in \mathcal{U}\} \in cov(V)$$
   (d) *if $(U_i)_{i \in I} \in cov(U)$ and if $\mathcal{U}_i \in cov(U_i)$ then*
   $$\cup_{i \in I} \mathcal{U}_i = \{U' \mid \exists i \in I : U' \in \mathcal{U}_i\} \in cov(U)$$

Elements of $\mathcal{G}$ are called admissible subsets of X. Elements in $cov(U)$ are called admissible coverings of U.

It follows from Proposition 2.5 that a Grothendieck topology can be defined on $X = max(A)$ for which the rational subsets are admissible and for which the finite coverings by rational subsets are admissible.

If X is a topological space provided with a Grothendieck topology $\mathcal{G}$ then a (pre)sheaf $\mathcal{F}$ of abelian groups (rings ...) on X is defined by the giving of the following data :

1. for each $U \in \mathcal{G}$ an abelian group (ring...) $\mathcal{F}(U)$ with $\mathcal{F}(\emptyset) = 0$
2. for all $U, V \in \mathcal{G}$ with $U \subset V$ a (restriction) morphism
   $$\phi_{VU} : \mathcal{F}(V) \to \mathcal{F}(U)$$
   such that
   (a) $\forall U \in \phi_{UU} = Id_U$

(b) if $U \subset V \subset W$ with $U, V, W \in \mathcal{G}$ then $\phi_{WU} = \phi_{WV} \circ \phi_{VU}$

The presheaf $\mathcal{F}$ is a sheaf if for all $U \in \mathcal{G}$ and for all $(U_i)_{i \in I} \in cov(U)$ the sequence

$$\mathcal{F}(U) \to \prod_{i \in I} \mathcal{F}(U_i) \overset{\to}{\underset{\to}{}} \prod_{i,j \in I} \mathcal{F}(U_i \cap U_j)$$

is exact.
The Čech cohomology of a presheaf $\mathcal{F}$ with respect to a covering $\mathcal{U} \in cov(X)$ is defined in the usual way.

Let A be an affinoid algebra and let $X = max(A)$ be provided with the Grothendieck topology of rational subsets. Denote $\mathcal{O}(U)$ the affinoid algebra of a rational subset $U \subset X$. The L-algebras $\mathcal{O}(U)$, together with the canonical restriction maps define the presheaf $\mathcal{O}$ on X.
The following theorem is fundamental.

Theorem 2.7 *If $\mathcal{U} \in cov(X)$ then $H^0(\mathcal{U}, \mathcal{O}) = A$ and $H^n(\mathcal{U}, \mathcal{O}) = 0$ for $n > 0$.*

A much more general version of the theorem can be found in [T] or in [B].
As a consequence of the theorem we find that $\mathcal{O}$ is in fact a sheaf.

Definition 2.8 *An affinoid space is a topological space $X = max(A)$ with A an affinoid algebra, provided with the Grothendieck topology of rational subsets and the structure sheaf $\mathcal{O}$.*

Affinoid spaces can be glued in a natural way. General rigid analytic paces are obtained in this way.

Definition 2.9 *A rigid analytic space is a triple $(X, \mathcal{G}, \mathcal{O})$ where X is a topological space, provided with a Grothendieck topology $\mathcal{G}$ and a sheaf of L-algebras $\mathcal{O}$, such that there exists a covering $(X_i)_{i \in I} \in cov(X)$ and such that each triple $(X_i, \mathcal{O}|_{X_i}, \mathcal{O}|_{X_i})$ is an affinoid space.*

One of the main techniques in rigid analytic geometry is the use of reductions. If A is an affinoid algebra then we put

$$A^o = \{f \in A \mid ||f||_{sp} \leq 1\}$$

and

$$A^{oo} = \{f \in A \mid ||f||_{sp} < 1\}$$

So A^o is an L^o-algebra and A^{oo} is an ideal. The $\overline{L}$-algebra $\overline{A} = A^o/A^{oo}$ is called the reduction of A. In particular $\overline{T}_n = \overline{L}[X_1, \ldots, X_n]$.

If $\phi : A \to B$ is a homomorphism of affinoid algebras then $\phi(A^o) \subset B^o$ and $\phi(A^{oo}) \subset B^{oo}$. Hence ϕ induces a homomorphism $\overline{\phi} : \overline{A} \to \overline{B}$. Furthermore $\overline{\phi}$ is finite if and only if ϕ is finite. It follows that $\overline{A}$ is a finitely generated $\overline{L}$-algebra.

A maximal ideal $x \subset A$ induces a surjective homomorphism $\phi : A \to L'$ where $L' \equiv A/x$ is a finite field extension of L. It follows that the kernel of the induced morphism $\overline{\phi} : \overline{A} \to \overline{L'}$ is a maximal ideal $\overline{x} \subset \overline{A}$.
The map $R : max(A) \to max(\overline{A})$ with $R(x) = \overline{x}$ is called the reduction of $max(A)$. This map is surjective.
If X is an analytic space and if $\mathcal{U} = (U_i)_{i \in I} \in cov(X)$ has certain good properties then the canonical reductions $\overline{U}_i$ can be glued to an affine $\overline{L}$-space $\overline{X}$ with a surjective map $R : X \to \overline{X}$. This is called the reduction of X with respect to $\mathcal{U}$.

Examples (we assume that L is algebraically closed)

1) $A = T_n$ and $X = \{(x_1, \ldots, x_n) \mid |x_i| \le 1\}$. In this case $\overline{A} = \overline{L}[X_1, \ldots, X_n]$ and $\overline{X} = \mathbf{A}^n(\overline{L})$. The reduction maps an element $(x_1, \ldots, x_n)$ onto $(\overline{x}_1, \ldots, \overline{x}_n)$.

2) $X = \{x \in L \mid |q| \le |x| \le 1\}$ with $q \in L$ and $|q| < 1$. We have

$$A = L < X, Y > /(X.Y - q) \text{ and } \overline{A} = \overline{L}[X, Y]/(X.Y)$$

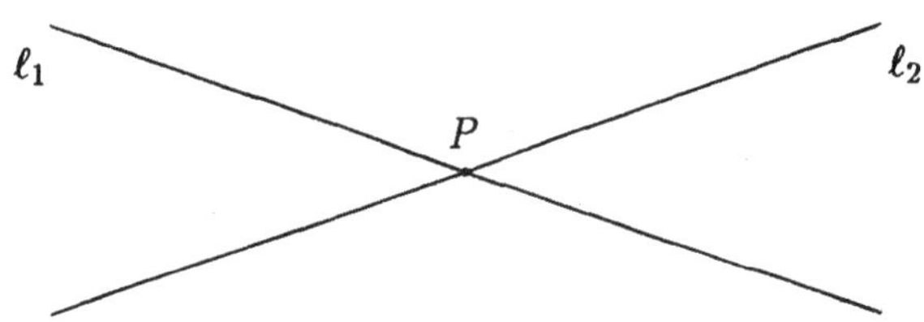

The reduction $\overline{X}$ consists of two lines l_1 and l_2 intersecting in a point P. The points in X which satisfy $|x| = 1$ are mapped onto $l_1 \setminus \{P\}$, the points which satisfy $|x| = |q|$ are mapped onto $l_2 \setminus \{P\}$ and the points which satisfy $|q| < |x| < 1$ are mapped onto P.

3) X is the complement in $\mathbf{P}^1$ of some disjoint open discs $D_1, \ldots, D_r$.
Assume that $\infty \in X$. X is then given by inequalities $|x - a_i| \ge |\pi_i|$ $(i = 1, \ldots, r)$ with $a_i, \pi_i \in L$. Assume that the closed discs $D_i^+ = \{x \mid |x - a_i| \le |\pi_i|\}$ are also disjoint. The corresponding affinoid algebra A is then isomorphic with

$$L < X_1, \ldots, X_r > /(\frac{\pi_i}{a_i - a_j} X_j - \frac{\pi_j}{a_j - a_i} X_i - X_i X_j)_{i \ne j}$$

and the reduction $\overline{A}$ is isomorphic with

$$\overline{L}[X_1, \ldots, X_r]/(X_iX_j)_{i \neq j}$$

$\overline{X}$ consists of r lines which intersect in one point.

4) $X = \mathbf{P}^1(L) = L \cup \{\infty\}$ is covered by

$$U_1 = \{x \mid |x| \leq 1\} \text{ and } U_2 = \{x \mid |x \geq 1\} \cup \{\infty\}.$$

Each $\overline{U}_i$ is an affine line over $\overline{L}$. The gluing of $\overline{U}_1$ and $\overline{U}_2$ results in a projective line $\mathbf{P}^1(\overline{L}) = \overline{L} \cup \{\infty\}$. The reduction maps a point x with $|x| \leq 1$ onto $\overline{x}$. Points with $|x| > 1$ and $\infty \in \mathbf{P}^1(L)$ are mapped onto $\infty \in \mathbf{P}^1(\overline{L})$.

3 Analytic subspaces of $\mathbf{P}^1$

The projective line $\mathbf{P}^1_L$ over L can be defined as an analytic space in the following way. Let $V_0 = max(L < X_0 >)$ and $V_1 = max(L < X_1 >)$ be two copies of de unit ball. Let

$$V_i^* = \{x \in V_i \mid |X_i(x)| = 1\} \subset V_i, \quad i = 0, 1$$

Then V_i^* is a rational subset of V_i corresponding with the affinoid algebra $T_i^* = L < X_i, X_i^{-1} >$.
The homomorphism $\psi : T_0^* \to T_1^*$ with $\psi(X_0) = X_1^{-1}$ is an isomorphism and induces an isomorphism $\psi^* : V_1^* \to V_0^*$. The projective line $\mathbf{P}^1_L$ is obtained by gluing the spaces V_0 and V_1 over the open subspaces $V_0^* \subset V_0$ and $V_1^* \subset V_1$ by means of the isomorphism ψ^*.

For the moment we suppose that the field L is algebraically closed. We can then identify $B_1(L)$ with $D_1(L)$ and similarly also $\mathbf{P}^1_L$ with $\mathbf{P}^1(L) = L \cup \{\infty\}$. The affinoid subsets of $\mathbf{P}^1_L$ can be described in a simple way.

Definition 3.1 *A standard rational subset in $D_1(L)$ is a subset of the form*

$$\{x \in D_1(L) \mid |x - a| \leq \rho, \quad |x - a_i| \geq \rho_i, \quad i = 0, \ldots, s\}$$

with $\rho, \rho_i \in |L^*|$.

Thus a standard rational subset is a closed disc minus a finite subset of open discs.
It is clear that this set is a rational subset of $D_1(L)$. It is easy to verify the following statements :

1. if U_1 and U_2 are standard rational and if $U_1 \cap U_2 \neq \emptyset$ then $U_1 \cap U_2$ and $U_1 \cup U_2$ are standard rational.

2. a finite union of standard rationals can uniquely be written as a disjoint union of finitely many standard rationals.

The main result for rational subsets of $D_1(L)$ is then :

Proposition 3.2 *Every rational subset of $D_1(L)$ is a finite union of standard rational subsets.*

Since every open affinoid subset is a finite union of rational subsets we find that every open affinoid subset of $D_1(L)$ is a disjoint union of standard rationals. These results for $D_1(L)$ can be generalized for $\mathbf{P}^1(L)$. An open disk in $\mathbf{P}^1(L)$ is by definition a set of the form

$$\{x \in L \mid |x-a| < \rho\}$$

or

$$\mathbf{P}^1(L) \setminus \{x \in L \mid |x-a| \leq \rho\} = \{x \in L \mid |x-a| > \rho\} \cup \{\infty\}$$

with $a \in L$ and $\rho \in |L|^*$.
A *standard rational subset in* $\mathbf{P}^1(L)$ is defined as the complement of finitely many open disks. A subset $U \subset \mathbf{P}^1(L)$ is admissible for the analytic structure on $\mathbf{P}^1_L$ if $U \cap V_i$ is admissible for the analytic structure on V_i,$(i = 0, 1)$.
It follows from the previous results that an admissible subset of $\mathbf{P}^1_L$ is a finite union of standard rational subsets.

In the general case, i.e. L is not necessarily algebraically closed, the description of the affinoid subsets of $\mathbf{P}_L$ is somewhat more complicated. The standard rational subsets defined above are still affinoid subspaces. There are however more affinoid subspaces, namely the subspaces which become, after a finite extension of the field L, a finite disjoint union of standard rational subsets. In the sequel we will *not* suppose that L is algebraically closed.

Analytic subspaces of $\mathbf{P}^1_L$ can be obtained in the following way. Let $\mathcal{A}$ be a family of affinoid subsets of $\mathbf{P}^1_L$ satisfying the following conditions :

1. if U_1 and U_2 are elements in $\mathcal{A}$ such that $U_1 \cup U_2$ is affinoid then $U_1 \cup U_2$ is an element in $\mathcal{A}$

2. if $U_1 \subset U_2$ are affinoids with $U_2 \in \mathcal{A}$ then $U_1 \in \mathcal{A}$

3. $\mathcal{A} \neq \emptyset$.

Let $\Omega \subset \mathbf{P}^1_L$ be the union of the U in $\mathcal{A}$. Define a Grothendieck topology $\mathcal{G}$ on Ω by :

1. $U \subset \Omega$ is admissible if for every $V \in \mathcal{A}$ one has $U \cap V \in \mathcal{A}$.

2. a covering $(U_i)_{i\in I}$ of a subset $U \subset \Omega$ is admissible if the following conditions are satisfied :

   (a) each U_i is admissible

   (b) if $V \subset U$ is an element of $\mathcal{A}$ then there exist finitely many $i_1, \ldots, i_n$ such that $V \subset U_{i_1} \cup \ldots \cup U_{i_n}$

The structure sheaf $\mathcal{O}$ on Ω is defined by

1. $\mathcal{O}(U)$ is the affinoid algebra of U if $U \in \mathcal{A}$

2. if $U \notin \mathcal{A}$ then $\mathcal{O}(U) = \lim_{\leftarrow} \mathcal{O}(V)$, where the projective limit is taken over all elements $V \in \mathcal{A}$ with $V \subset U$

Examples

(1) $\Omega = \mathbf{P}^1_L \setminus \{0, \infty\}$. This space is defined by the family $\mathcal{A}$ consisting of the affinoid subsets $U \subset \mathbf{P}^1_L$ with $0, \infty \notin U$. It is a rigid space over the field L.

$$\text{Put } \Omega_n := \left\{x \in \mathbf{P}^1_L \mid |\pi|^n \leq |x| \leq |\pi|^{-n}\right\}, \quad n \in \mathbf{N}$$

where $\pi \in L$ is an element with $0 < |\pi| < 1$. Then $\{\Omega_n\}_{n\in\mathbf{N}}$ is an admissible covering of Ω. Indeed, on can easily show that any $U \in \mathcal{A}$ is contained in some Ω_n.

(2) Let S be a compact subset of $\mathbf{P}^1(L)$. Consider the family $\mathcal{A}$ consisting of the affinoid subsets U of $\mathbf{P}^1_L$ such that $U \cap S = \emptyset$. This family has the required properties and defines a subspace (which is a rigid space over L) which we will denote by $\Omega := \mathbf{P}^1_L \setminus S$.

As before we choose a $\pi \in L$ with $0 < |\pi| < 1$. Let n be a positive integer. Then S is covered by the open disks $\{x \in \mathbf{P}^1_L \mid |x - a| < |\pi|^n\}$ with $a \in S \setminus \{\infty\}$ (and $\{x \in \mathbf{P}^1_L \mid |x| > |\pi|^{-n}\}$ in case $\infty \in S$). Since S is compact the union of those open disks is in fact equal to a finite disjoint union of open disks with radii $|\pi|^n$. The complement Ω_n of those finitely many open disks is a standard rational subspace of $\mathbf{P}^1_L$. The family $\{\Omega_n\}_{n\in\mathbf{N}}$ is an admissible covering of Ω. Indeed, one can show that every $U \in \mathcal{A}$ is contained in some Ω_n. The analytic structure of Ω is thus given by:

(a) A subset $U \subset \Omega$ is admissible if $U \cap \Omega_n$ is an affinoid subset of Ω_n for every $n \geq 0$.

(b) A covering $\{U_i\}_{i\in I}$ of an admissible U is called admissible if each U_i is admissble and if for each n the set $U \cap \Omega_n$ is covered by finitely many of the $U_i \cap \Omega_n$. We note further that a holomorphic (or meromorphic) function on Ω is a function whose restriction to every affinoid (or to every Ω_n) is holomorphic (resp. meromorphic).

4 Classification of local fields

In the sequel we will suppose that L is a local field. We recall the classification of local fields. There are two types of local fields, namely:
Type one: The field of the formal Laurent series in the parameter u over a finite field $\mathbf{F}_q$. This field is written as $\mathbf{F}_q((u))$. The valuation is given by the formula $|\sum_{n>>-\infty} a_n u^n| = q^{-m}$ if m is the smallest integer with $a_m \neq 0$ (and $|0| = 0$).
Type two: The finite extensions of the field of p-adic numbers $\mathbf{Q}_p$. Any element of the latter field can uniquely be written as an expression $\sum_{n>>-\infty} a_n p^n$ with all $a_n \in \{0, 1, \ldots, p-1\}$. The valuation on $\mathbf{Q}_p$ is given by the formula $|\sum_{n>>-\infty} a_n p^n| = p^{-m}$ if m is the smallest integer with $a_m \neq 0$ (and $|0| = 0$). For any finite field extension L of $\mathbf{Q}_p$ this valuation extends in a unique way to a valuation of L.
For a local field L we denote by π a *uniformizing parameter*, i.e. an element such that $0 < |\pi| < 1$ and $|\pi|$ generates the value group $|L^*|$ of L. Further $O = O_L$ will denote the *valuation ring* $\{a \in L|\ |a| \leq 1\}$ of L. The element π generates the maximal ideal of O and the *residue field* $O/\pi O$ is a finite field. Let $\mathbf{C}$ denote the completion of the algebraic closure of L.

The situation which interest us here is $S = \mathbf{P}^1(L)$ and $\Omega := \mathbf{P}^1_L \setminus \mathbf{P}^1(L)$. The aim is to describe a rather special admissible covering of Ω which reflects the combinatorial structure of Ω. This combinatorial structure is given by a certain tree, the Bruhat-Tits building of $PGl(2, L)$. We will first introduce this tree.

5 The Bruhat-Tits building of $PGl(2, L)$

There are various ways to define this building $\mathcal{T}$. In fact this building has dimension 1 and is moreover a tree (as we will see). Let again O denote the valuation ring of L. An *O-lattice* $M \subset L^2$ is an O-submodule, which is free of rank two. In other words $M = Of_1 + Of_2$ where f_1, f_2 is a basis of L^2 over L. Two submodules M_1, M_2 are called equivalent if $M_1 = aM_2$ for some $a \in L^*$. The equivalence class of M will be denoted by $[M]$. The set of vertices of $\mathcal{T}$ is the set of equivalence classes of O-lattices. The edges of $\mathcal{T}$ are given as the pairs $\{[M_0], [M_1]\}$ such that $M_1 \subset M_0$ and M_0/M_1 is a vector space over the residue field (which is $O/\pi O = \mathbf{F}_q$) of dimension 1. This condition on M_0, M_1 can also be written as: there is a basis f_1, f_2 of L^2 over L such that $M_0 = Of_1 + Of_2$ and $M_1 = Of_1 + O\pi f_2$. One observes that this condition is symmetric in the M_0, M_1. The edges with end point $[M_0]$ are the pairs $\{[M_0], [M_1]\}$ such that the inclusions $\pi M_0 \subset M_1 \subset M_0$ are strict. Each M_1 corresponds to a line in

the two dimensional vector space $M_0/\pi M_0$ over the residue field $\mathbf{F}_q$ of O. Thus the number of edges with end point $[M_0]$ is $q+1$. This describes the graph $\mathcal{T}$ locally.
It is easily seen that $\mathcal{T}$ is a connected graph. Consider a path $\{[M_0],[M_1]\},\{[M_1],[M_2]\},\ldots,\{[M_{n-1}],[M_n]\}$ in $\mathcal{T}$ which does not contain an "aller-retour". By an "aller-retour" we mean a path $\{[M],[N]\},\{[N],[M]\}$. A small calculation shows that there is a basis f_1,f_2 of L^2 over L such that the path above can be represented by the modules $M_i=Of_1+O\pi^i f_2$ for $i=0,\ldots,n$. In particular, there are no loops in $\mathcal{T}$ and so $\mathcal{T}$ is a tree.
An *end* of a graph is a subgraph of the form:
Two ends are called equivalent if they differ only by finitely many elements. An end in $\mathcal{T}$ is seen to be of the form $\{[M_0],[M_1],\ldots\}$ with each $M_i=Of_1+O\pi^i f_2 O$. To this end we can associate the line Lf_1 in L^2. This is well defined since $Of_1=\cap_{i=0}^{\infty}M_i$. This induces a bijection between the ends of $\mathcal{T}$ and the lines in L^2 (or the points of $\mathbf{P}^1(L)$). The following is a picture for the Bruhat-Tits tree with $q=3$.

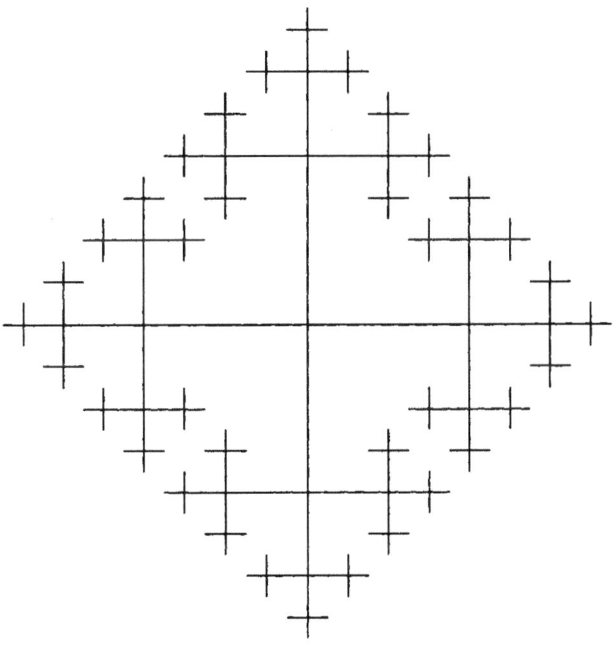

The group $Gl(2,L)$ acts on L^2, on the O-lattices in L^2 et cetera. Thus $PGl(2,L)$ acts on the Bruhat-Tits tree $\mathcal{T}$. This action is transitive on the vertices of $\mathcal{T}$. *The standard vertex* v is $O\oplus O\subset L^2$. The stabilizer of v

is equal to the open compact subgroup $PGl(2,O)$ of $PGl(2,L)$. In this way the set of vertices of $\mathcal{T}$ can be identified with (the discrete and countable set) $PGl(2,L)/PGl(2,O)$. The group $PGl(2,L)$ also acts transitively on the edges. *The standard edge* e is $\{[O \oplus O],[\pi O \oplus O]\}$. Its stabilizer, i.e. the elements in the group which leave the *two* vertices invariant, is equal to the subgroup I of $PGl(2,O)$ consisting of the matrices $\begin{pmatrix} a & b \\ c & d \end{pmatrix} \in PGl(2,O)$ such that $b \in \pi O$. This group is called the *Iwahori subgroup of* $PGl(2,L)$.
Thus we can identify the edges of $\mathcal{T}$ with $PGl(2,L)/I$. The two vertices of the edge gI are $gPGL(2,O)$ and $g\delta PGl(2,O)$, where $\delta = \begin{pmatrix} \pi & 0 \\ 0 & 1 \end{pmatrix}$.

References

[B] : S. Bosch, U. Güntzer and R. Remmert - *Non-Archimedean Analysis* - Grundlehren der math. Wissenschaften, 1984, Springer Verlag.

[G-vdP] : L. Gerritzen and M. van der Put - *Schottky Groups and Mumford Curves* - Lect. Notes in Math. 817, 1980, Springer Verlag.

[F-vdP] : J. Fresnel et M. van der Put - *Géométrie Analytique Rigide et Applications* - Progress in Math. 18, 1981, Birkhäuser.

[T] : J. Tate - *Rigid analytic spaces* - Invent. Math 12, 257-289 ,1971.

Lecture 7

The structure of Ω and its quotients $\Gamma\backslash\Omega$

Marius van der Put
Department of Mathematics, University of Groningen
P.O.Box 800, 9700 AV Groningen, The Netherlands
mvdput@mat.rug.nl

1 The special covering of Ω

The notations of lecture 6 will be used here. We choose representatives $0 = r_0, \ldots, r_{q-1}$ of the residue field of L (i.e. elements in O which map to the q elements of the residue field $O/\pi O$). For the field of formal Laurent series, we will take $\{r_0, \ldots r_{q-1}\} = \mathbf{F}_q$. Every element of L can be uniquely written as $\sum_{n>>\infty} a_n \pi^n$ with all $a_n \in \{r_0, \ldots, r_{q-1}\}$.

The basis for this construction are the two affinoid subsets V and E of Ω given by:

$$V = \{z \in \mathbf{P}^1_L |\ |z| = 1 \text{ and } |z - r_i| = 1 \text{ for } i = 1, \ldots, q-1\} \text{ and}$$

$$E = \{z \in \mathbf{P}^1_L |\ |\pi| \leq |z| \leq 1,\ |z - r_i| \geq 1 \text{ for } i = 1, \ldots, q-1$$

$$\text{and } |z - \pi r_i| \geq |\pi| \text{ for } i = 1, \ldots, q-1\}.$$

The first affinoid $V \subset \mathbf{P}^1_L$ is the complement of the union of the open disks of radii 1 around the points $0 = r_0, r_1, \ldots, r_{q-1}, \infty$. And E, the second one, is the complement of the union of the open disk around the points $r_1, \ldots, r_{q-1}, \infty$ with radii 1 and the open disks around the points $\pi r_0, \pi r_1, \ldots, \pi r_{q-1}$ with radii $|\pi|$.

At this point it is useful to introduce the real valued function $d(\ )$ on Ω given by the formula

$$d(\omega) = \min\{|\omega + \lambda| \mid \lambda \in L\}.$$

For any $\begin{pmatrix} a & b \\ c & d \end{pmatrix} \in Gl(2, L)$, one has the formula

$$d(\frac{a\omega + b}{c\omega + d}) = \frac{|ad - bc|}{|c\omega + d|^2} d(\omega).$$

The function $d(\ )$ can be seen as an equivalent of the function "imaginary part" on the complex upper half plane.
A proof of the formula can be given as follows. If the formula is valid for matrices A_1 and A_2, then it is also valid for the product A_1A_2. This reduces the verification to the case where A is one of the following matrices

$$\begin{pmatrix} 1 & b \\ 0 & 1 \end{pmatrix}, \begin{pmatrix} a & 0 \\ 0 & 1 \end{pmatrix}, \begin{pmatrix} 0 & 1 \\ 1 & 0 \end{pmatrix}.$$

For the first two matrices the verification is trivial. For the transformation $z \mapsto \frac{1}{z}$ one has to verify that $d(\frac{1}{\omega}) = \frac{d(\omega)}{|\omega|^2}$. Let $\lambda_0 \in L$ be an element with $d(\omega) = |\omega - \lambda_0|$. Then $|\omega|^2 d(\frac{1}{\omega}) = |\omega| \min\{|1 + \lambda\omega| \mid \lambda \in L\}$. Further

$$|1 + \lambda\omega| = |(1 + \lambda\lambda_0) + \lambda(\omega - \lambda_0)| = \max\{|1 + \lambda\lambda_0|, |\lambda| d(\omega)\}.$$

If $|\lambda_0| < |\omega|$ then $d(\omega) = |\omega|$ and one may take $\lambda_0 = 0$. Then clearly $|\omega|^2 d(\frac{1}{\omega}) = |\omega| = d(\omega)$.
If $|\omega - \lambda_0| < \max\{|\omega|, |\lambda_0|\}$ then $|\omega| = |\lambda_0|$ and the substitution $\lambda = \frac{-1}{\lambda_0}$ gives the minimal value. Hence also $|\omega|^2 d(\frac{1}{\omega}) = d(\omega)$.

The affinoid E is characterized by: $E(\mathbf{C})$ consists of the points ω such that $|\pi| \leq |\omega| \leq 1$ and $d(\omega) = |\omega|$.

The group $PGl(2, L)$ acts on $\mathbf{P}^1_L$ as a group of automorphisms by

$$\begin{pmatrix} a & b \\ c & d \end{pmatrix} z = \frac{az + b}{cz + d}.$$

Clearly $PGl(2, L)$ acts also on Ω. The stabilizer of V, i.e. $\{g \in PGl(2, L) |\ gV = V\}$, is equal to $PGl(2, O)$.
The stabilizer of the pair (E, V) is the subgroup of $PGl(2, O)$ consisting of the matrices such that $b \in \pi O$. This is the Iwahori group I introduced above. The stabilizer of E is generated by I and the transformation $z \mapsto \frac{\pi}{z}$. Hence I is a subgroup of index two in the stabilizer of E.

For every $g \in PGl(2, L)$ the set gE is also an affinoid subset of Ω. Moreover $g_1E = g_2E$ if $g_1I = g_2I$. The set gE depends only on gI and we can denote it as gE with $g \in PGl(2, L)/I$. We note that $PGl(2, L)$ is locally compact and I is an open compact subgroup. Thus $Gl(2, L)/I$ is a countable discrete set.
Similarly, $g_1V = g_2V$ if and only if $g_1PGl(2, O) = g_2PGl(2, O)$. The translates of V under $PGl(2, L)$ can be denoted by gV with $g \in PGl(2, L)/PGl(2, O)$. The last set is again a discrete and countable set.

Lemma 1.1
(1) *The collection* $\{gE |\ g \in Gl(2, L)/I\}$ *is an admissible covering of* Ω.

(2) *Two affinoids $g_1E \neq g_2E$ have a non empty intersection if and only if the edges g_1e and g_2e in $\mathcal{T}$ have a common vertex. If the common vertex of the edges g_1e and g_2e is g_3v then $g_1E \cap g_2E = g_3V$.*

Proof. (1) As before, we denote by Ω_n the complement in $\mathbf{P}^1_L$ of the union of the open disks with radii $|\pi|^n$ around the points of $\mathbf{P}^1(L)$. It suffices to show that each Ω_n is covered by finitely many translates gE of E.
Take a point $\omega \in \Omega_n(\mathbf{C})$. Then $|\omega| \leq |\pi|^{-n}$ and $|\omega + a| \geq |\pi|^n$ for all $a \in L$. The set $\{a \in L | 0 \leq |a| \leq |\pi|^{-n}\}$ is the union of s disjoint open disks of radii $|\pi|^n$. Choose a point c_i in each disk. Then for some i one has $|\omega + c_i| \leq |\omega + a|$ for all $a \in L$. This implies that $|a(\omega + c_i) + b| = \max(|a(\omega + c_i)|, |b|)$ for all $a, b \in L$. Take an integer m such that $|\pi| < |\pi^m(\omega + c_i)| \leq 1$. This integer m lies in $\{-n, -n+1, \dots, n-1, n\}$. The element $\omega' = \pi^m(\omega + c_i)$ satisfies again $|a\omega' + b| = \max(|a\omega'|, |b|)$ for all $a, b \in L$. It follows from this that $\omega' \in E$. Thus we have found a finite subset S of $PGl(2, L)$ such that for every $\omega \in \Omega_n$ there is a $g \in S$ with $g\omega \in E$. This proves that Ω_n is contained in the union of the gE with $g^{-1} \in S$.
(2) It suffices to study the intersections $E \cap gE$. Then one is reduced to proving the following:
(a) If the edges e and ge, with $e \neq ge$, have a common vertex which may be supposed to be v, then $E \cap gE = V$.
(b) If $E \cap gE \neq \emptyset$ then $e \cap ge \neq \emptyset$.

One proves (a) by writing down the q possibilities for the edge ge. They are given by the transformations g of the form $\omega \mapsto \omega + r_i$, with $i = 1, \dots q-1$ and the transformation $\omega \mapsto \omega^{-1}$. The verification is straightforward.
The proof of (b) is a more complicated combinatorial exercise. Using the element $z \mapsto \frac{\pi}{z}$ in the stabilizer of E one may suppose that there is a $\omega \in E$ with $|\pi|^{1/2} \leq |\omega| \leq 1$ such that $g(\omega) \in E$ and also $|\pi|^{1/2} \leq |g(\omega)| \leq 1$.
Suppose that $|\omega| < 1$. Then using

$$\frac{|a\omega + b|}{|c\omega + d|} = |g\omega| = d(g\omega) = \frac{|ad - bc|}{|c\omega + d|^2}|\omega|$$

one finds that $gE = E$. The same conclusion holds if $|g(\omega)| < 1$.
Suppose that $|\omega| = |g(\omega)| = 1$. Then using

$$1 = d(\omega) = |\omega| = |\frac{a\omega + b}{c\omega + d}| = \frac{|ad - bc|}{|c\omega + d|^2}$$

one finds that $g \in PGl(2, O)$. From this it follows that ge and e have a common vertex. □

The covering of the lemma will be called *the special covering of* Ω. The lemma states that the intersection graph of this covering coincides with the tree

$\mathcal{T}$.

1.1 The canonical reduction of Ω

Every affinoid space S has a canonical reduction $\bar{S}$. The reductions of the affinoid spaces E and V can be described as follows:
(a) The one dimensional space $\bar{E}$ can be compactified by adding smooth points. The result is a space $\bar{\bar{E}}$ which has two irreducible components L_1, L_0. Each L_i is a projective line over $\mathbf{F}_q$. They intersect normally in one point. The point of intersection is the point ∞ of L_1 and the point 0 of L_0. Further $\bar{E}$ is obtained from $\bar{\bar{E}}$ by deleting the $\mathbf{F}_q$-rational points of L_1, different from ∞ and the $\mathbf{F}_q$-rational points of L_0, different from 0.
(b) The reduction $\bar{V}$ of V is the open subset of $\bar{E}$, obtained by deleting the component L_1. Thus $\bar{V}$ is the open subset of $L_0 \cong \mathbf{P}^1_{\mathbf{F}_q}$, obtained by deleting the $q+1$ points $\mathbf{P}^1(\mathbf{F}_q)$.

Each of the affinoids gE and hV has a similar description. The collection of the reductions $\overline{gE}$ is glued over the Zariski open subsets $\overline{hV}$, according to the intersection graph of the gE, i.e. the tree $\mathcal{T}$. The result $\overline{\Omega}$ will be called the canonical reduction of Ω. The irreducible components of $\overline{\Omega}$ are projective lines over $\mathbf{F}_q$. They correspond to the vertices of $\mathcal{T}$. The double points (ordinary double points) correspond to the edges of $\mathcal{T}$. In other words $\overline{\Omega}$ is a tree of projective lines over $\mathbf{F}_q$. The group $PGL(2, L)$ acts on this tree in a way compatible with the actions on Ω and $\mathcal{T}$.

2 Discrete subgroups Γ and the quotients $\Gamma\backslash\Omega$

The group $PGl(2, L)$ acts on $\Omega, \mathcal{T}$ and the special covering $\{gE | \ g \in Gl(2, L)/I\}$ in a compatible way. Let the subgroup Γ of $PGl(2, L)$ be *discrete* for the topology induced by $PGl(2, L)$. Then for any affinoid subset B of Ω the number of elements $g \in \Gamma$ with $gB \cap B \neq \emptyset$ is finite.

Indeed, one may take for B the union of finitely many translates of E. From the action of $PGl(2, L)$ on the special covering (or from its action on $\mathcal{T}$) it is easily seen that the subset of the $g \in PGl(2, L)$ such that $gB \cap B \neq \emptyset$ is compact and open. The discreteness of Γ in $PGl(2, L)$ implies the statement.
This has as consequence that one can define the quotient $\Gamma\backslash\Omega$ as a rigid analytic space over L for any discrete Γ.
We note that the quotient $\Gamma\backslash\mathcal{T}$ is not always a graph.

Indeed, an element $\gamma \in \Gamma$ is called an *inversion* if for some edge $\{v_1, v_2\}$ of $\mathcal{T}$, one has $\gamma(v_1) = v_2$ and $\gamma(v_2) = v_1$. Then one has to identify this edge with its inverse. This will not result in a graph structure on $\Gamma\backslash\mathcal{T}$. One says that Γ *acts without inversions on* $\mathcal{T}$ if the situation above does not occur. If Γ acts without inversion then $\Gamma\backslash\mathcal{T}$ is indeed a graph.
The main results about the structure of the quotient spaces are summarized in the next theorem.

Theorem 2.1 *Let Γ denote a discrete subgroup of $PGl(2, L)$. Then the quotient $\Gamma\backslash\Omega$ has the structure of a rigid space over L. Suppose (for convenience) that Γ acts without inversion on $\mathcal{T}$. Then:*
(1) *$\Gamma\backslash\Omega$ is the analytification of a non singular connected projective curve if and only if $\Gamma\backslash\mathcal{T}$ is a finite graph. The genus of this curve is equal to the Betti number of the graph. The "stable reduction" of this projective curve is isomorphic to $\Gamma\backslash\overline{\Omega}$.*
(2) *$\Gamma\backslash\Omega$ is the analytification of a non singular connected affine curve X if and only if $\Gamma\backslash\mathcal{T}$ is the union of a finite graph with finitely many (equivalence classes of) ends.*
(3) *Suppose that Γ satisfies the equivalent conditions of (2). Let $\hat{X}$ denote the compactification of the curve X. The genus of $\hat{X}$ is equal to the Betti number of the graph $\Gamma\backslash\mathcal{T}$. The genus is also equal to the dimension over $\mathbf{Q}$ of the vector space $\mathbf{Q} \otimes \Gamma/[\Gamma, \Gamma]$.*
The points of $\hat{X} \setminus X$ are in a one to one correspondence with the (equivalence classes of the) ends of $\Gamma\backslash\mathcal{T}$.

Proof. We refer here to some literature [G-vdP,F-vdP,vdP,Ge] for a proof. The idea of the proof of the second part of the theorem is as follows. One makes an analytic compactification of the rigid space $\Gamma\backslash\Omega$ by adding a finite number of points $\{c_1, \ldots, c_s\}$. (More details are provided in Proposition 3.2 and in lecture 8). The points c_i are called the *cusps*. For each equivalence class of an end of the graph $\Gamma\backslash\mathcal{T}$ one has to add one point. Then from "GAGA" one obtains that this "compact rigid curve" is in fact the analytification of a smooth projective curve $\hat{X}$ over L. The analytification of the affine curve $X = \hat{X} \setminus \{c_1, \ldots, c_s\}$ is then isomorphic to $\Gamma\backslash\Omega$. □

Remark

Consider a discontinuous subgroup Γ of $PSl(2, \mathbf{R})$, acting on the complex upper half plane $\mathcal{H} := \{z \in \mathbf{C} \mid im(z) > 0\}$, such that the quotient $\Gamma\backslash\mathcal{H}$ is equal to $X = \hat{X} \setminus S$, where $\hat{X}$ is a projective, non singular curve over $\mathbf{C}$ of genus g and where S is a finite subset with cardinality s. The dimension of the $\mathbf{Q}$-vector space $\mathbf{Q} \otimes \Gamma/[\Gamma, \Gamma]$ turns out to be $2g$ if S is empty and $2g + s - 1$ if S is not empty (and maybe under the assumption that the group Γ is a free non abelian group). This is rather different from the formula given in part (3) of the last theorem.

The main reason for this is that in the category of rigid spaces "étale coverings" do not coincide with "analytic coverings". One can define an analytic covering $f : Y \to X$ of rigid spaces as follows:
There exists an admissible covering $\{X_i\}$ of X by admissible sets (say affinoid subsets) such that for every i the restriction of f to $f : f^{-1}X_i \to X_i$ is equivalent to the projection map $pr_1 : X_i \times A \to X_i$, where A denotes some non empty discrete set.
A (finite) étale covering $f : Y \to X$ can be defined as:
There exists an admissible covering $\{X_i\}$ of X by admissible sets (say affinoids subsets) such that for every i the preimage $f^{-1}X_i$ is again affinoid and the map $O(X_i) \to O(f^{-1}X_i)$ is an étale map of affinoid algebras (i.e. finite, flat and unramified).
A finite étale covering is in general not an analytic covering. Consider as an example the map $f : Z := \{z \in L \mid |z| = 1\} \to Z$ with L a non-archimedean valued complete field and $f(z) = z^n$ with $n > 1$ not divisible by the characteristic of L. This is certainly a finite étale covering. For every point $z \in Z$ there exists a small disk $D \subset Z$ around z such that the covering $f^{-1}D \to D$ is equivalent with the projection $pr_1 : D \times A \to D$, where A is a discrete set with cardinality n. However the covering of Z by this set of disks is not an admissible covering. One can show that the map f above is not an analytic covering and more generally that Z has only trivial analytic coverings. The same statement holds for the space $\{z \in L \mid 0 < |z| \leq 1\}$. This explains why the set S does not contribute to the dimension of the $\mathbf{Q}$-vector space $\mathbf{Q} \otimes \Gamma/[\Gamma, \Gamma]$.

For a curve Z (projective, non singular) over L of genus g there does exists an analytic uniformization $u : U \to Z$. There are two extreme cases. The first one is that u is an isomorphism (this is equivalent with Z having a good reduction). In the other extreme case the universal covering has a covering group Γ which is isomorphic to the free non abelian group on g generators. The curve has such a universal covering if and only if Z is a Mumford curve (i.e. all the components of the stable reduction of Z are rational curves over the residue field).

In the case considered in the last theorem, the curve $\hat{X}$ is in fact a Mumford curve. This explains the g in the formula instead of the expected $2g$.

3 The standard example

Notations: $A = \mathbf{F}_q[T]$, K is the field of fractions of A, $K_\infty = \mathbf{F}_q((T^{-1}))$ is the completion of K at ∞. The uniformizing parameter is $\pi = T^{-1}$.

(1) $\Gamma = PGl(2, A) \subset PGl(2, K_\infty)$.

It is rather obvious that Γ is a discrete subgroup. It acts without inversions on $\mathcal{T}$ since the determinant of any element of Γ lies in $\mathbf{F}_q^*$ and has therefore absolute value 1.
Consider the subtree S with vertices $\{[M_0], [M_1], [M_2], \ldots\}$ of $\mathcal{T}$, defined by $M_i = T^i O + O$ for $i \geq 0$.

Lemma 3.1 (After [S], p.118,119)
S maps isomorphically to the quotient $\Gamma \backslash \mathcal{T}$.

Proof. We have to prove two things:
(a) Every vertex (and also every edge) of $\mathcal{T}$ is equivalent modulo Γ to a vertex (or an edge) in this subtree.
(b) The vertices (and then also the edges) of this subtree are inequivalent.

For (a) one observes that the $q+1$ edges with end point $[M_0]$ are transitively permuted under the subgroup $PGl(2, \mathbf{F}_q)$ of Γ. Thus all those edges are equivalent to the edge $\{[M_0], [M_1]\}$. Also every vertex with "distance 1 " to $[M_0]$ is equivalent to $[M_1]$. The subgroup of Γ which stabilizes the edge $\{[M_0], [M_1]\}$ contains the subgroup

$$\{\begin{pmatrix} 1 & aT \\ 0 & 1 \end{pmatrix} \mid a \in \mathbf{F}_q\}.$$

This group acts transitively on the q edges with endpoint $[M_1]$, which are different from the edge $\{[M_0], [M_1]\}$. Hence every vertex with distance 2 to $[M_0]$ is equivalent to $[M_2]$. A similar reasoning shows that the subgroup of Γ which stabilizes $\{[M_0], \ldots, [M_n]\}$ permutes the remaining q edges of $[M_n]$ transitively. This proves statement (a).

For (b) one observes that an element

$$\gamma := \begin{pmatrix} a & b \\ c & d \end{pmatrix} \in \Gamma$$

with $\gamma[M_n] = [M_{n+m}]$ with $m \geq 0$ must satisfy $T^h \gamma M_n = M_{n+m}$ for some integer h. Taking determinants one finds that $m = 2h$. Moreover the equality $T^h \gamma M_n = M_{n+m}$ implies some restrictions on the degrees of a, b, c, d, namely $\leq h, n+h, n-h, -h$. There is no element γ satisfying those restrictions. □

(2) **The group $\Gamma(n)$.**

Let n denote a polynomial of degree ≥ 1. Define $\Gamma(n)$ as the kernel of the homomorphism $Gl(2,A) \to Gl(2,A/An)$. This group has a finite index in Γ. It follows that $\Gamma(n)\backslash\mathcal{T}$ is the union of a finite graph with finitely many ends. The (smooth, connected) projective curve associated to $\Gamma(n)\backslash\Omega$ by the theorem is denoted by $\hat{M}_{\Gamma(n)}$. As we will see later this is a connected component of the Drinfeld modular curve corresponding to $A = \mathbf{F}_q[T]$ and level (n), considered as a curve over the field K_∞. A calculation of the number of cusps and the genus of $\hat{M}_{\Gamma(n)}$ is given in lecture 8.

For the special case $n = T$ (or any polynomial of degree 1), the method of the lemma above can be used to show that the graph $\Gamma(T)\backslash\mathcal{T}$ has the form:

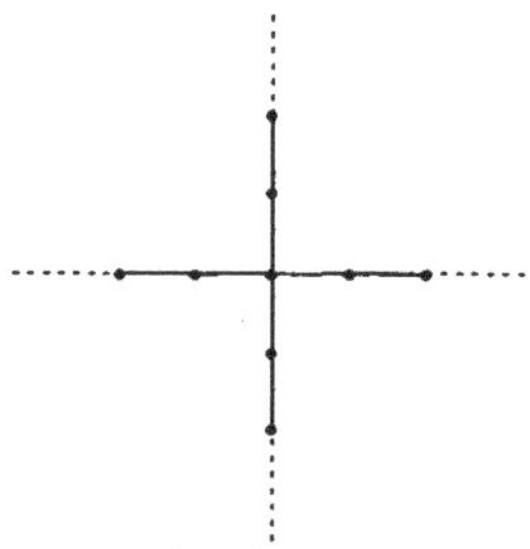

In words, the graph consists of $q+1$ ends starting from the image of the vertex $[M_0]$. The genus of the curve is 0 and there are $q+1$ cusps.

(3) $\Gamma = PGl(2,A)$ **and the construction of $\Gamma\backslash\Omega$.**

For the action of the discontinuous group $\Gamma := PSl(2,\mathbf{Z})$ on the upper half plane $\mathcal{H} = \{z \in \mathbf{C} \mid im(z) > 0\}$ one has the classical result $\Gamma\backslash\mathcal{H}$ is isomorphic to (the analytification of) the affine line over $\mathbf{C}$. One way to see this uses the fundamental domain $\{z \in \mathbf{C} \mid |z| \geq 1 \text{ and } -1/2 \leq re(z) \leq 1/2\}$ with its identification on the boundary for the group Γ. In the next proposition we give an analogue of this fundamental domain and prove that $PGl(2,A)\backslash\Omega$ is isomorphic to (the analytification of) the affine line over K_∞. The "fundamental domain" that we produce is maybe strictly speaking not a fundamental domain in the way one defines this for the complex case. There is a nice analogue $d(\ )$ for the "imaginary part" in the complex case. However, there is no analogue for the "real part" in the complex case. In our construction we allow in the fundamental domain infinite orbits under the action of Γ.

Proposition 3.2 *Let $\mathbf{A}^{1,an}_{K_\infty}$ denote the affine line over K_∞, considered as a* *rigid space over* K_∞. *Then* $PGl(2,A)\backslash\Omega \to \mathbf{A}^{1,an}_{K_\infty}$ *is an isomorphism.*

Proof. On Ω one considers the equivalence relation $\sim$ given by the action of the group $\Gamma = PGl(2,A)$. A "fundamental domain" is introduced, which is the union of two admissible sets U and V. This means that every point of Ω is equivalent with a point of $U \cup V$. The equivalence relation on U and V separately is transparent enough in order to construct the spaces $U/\sim$ and $V/\sim$. They turn out to be isomorphic to $\{z \in \mathbf{C} | \, |z| \geq 1\}$ and $\{z \in \mathbf{C} | \, |z| \leq 1\}$. The two spaces $U/\sim$ and $V/\sim$ are glued together over the open subset $U \cap V/\sim$ which is in both spaces identified with $\{z \in \mathbf{C} | \, |z| = 1\}$. The result is obviously equal to the analytification of the affine line (over K_∞). We give now more details.

The two admissible subsets U and V of Ω, are given by:

$$V = \{z \in \mathbf{C} | \, |z| \leq \sqrt{q}\} \setminus (\{z \in \mathbf{C} | \, |z| < \sqrt{q}^{-1}\} + \mathbf{F}_q)$$

$$U = \{z \in \mathbf{C} | \, d(z) \geq \sqrt{q}\}.$$

It has been shown that the half line in $\mathcal{T}$ with vertices $\{[M_0], [M_1], [M_2], \ldots\}$ is a fundamental domain for the action of Γ on the tree $\mathcal{T}$. The subspace of Ω corresponding to this half line is the admissible set $\{\omega \in \mathbf{C} | \, d(\omega) \geq 1\}$. Hence every element of Ω is equivalent to a ω with $d(\omega) \geq 1$. If $d(\omega) \geq q^{1/2}$ then ω belongs to U. Let $1 \leq d(\omega) \leq q^{1/2}$ then there is an $a \in A$ such that $d(\omega) = |\omega - a|$ and $\omega - a$ lies in V. This shows that $U \cup V$ is a fundamental domain for Γ.

V is the complement of $q+1$ open disk of radii $q^{-1/2}$ around the points of $\mathbf{P}(\mathbf{F}_q)$. For $g \in PGl(2,A)$ one has $gV \cap V \neq \emptyset$ if and only if $g \in PGl(2,\mathbf{F}_q)$. The quotient $V/\sim$ of V by this finite group of automorphisms is an affinoid subset of the curve $PGl(2,\mathbf{F}_q)\backslash\mathbf{P}^1$. This curve is again the projective line over K_∞ and the the image of V in this curve is easily seen to be a closed disk, i.e. isomorphic to $\{z \in \mathbf{C} | \, |z| \leq 1\}$. The subset $U \cap V$ of V is mapped isomorphically to $\{z \in \mathbf{C} | \, |z| = 1\}$.

U is an admissble subset of Ω and has the property:
Let $g \in PGl(2,A)$. Then $gU \cap U \neq \emptyset$ if and only if g has the form $\begin{pmatrix} a & b \\ 0 & d \end{pmatrix}$.
Let H denote the subgroup of $PGl(2,A)$ consisting of those matrices. Thus the image $U/\sim$ of U in $PGl(2,A)\backslash\Omega$ is isomorphic to $H\backslash U$. The last space is a subspace of $H\backslash\mathbf{C} \cong \mathbf{C}$ and can be seen to be isomorphic to $\{z \in \mathbf{C} | \, |z| \geq 1\}$. The subset $U \cap V$ is mapped under this isomorphism to $\{z \in \mathbf{C} \mid |z| = 1\}$. □

Remark We note that $U/\sim$ is isomorphic to a closed disk minus one point. Let $\overline{U/\sim}$ denote this closed disk. Then one can glue $\overline{U/\sim}$ to $V/\sim$ over $\{z \in \mathbf{C} | \, |z| = 1\}$. The result is the projective line over K_∞ seen as a rigid space. This is the proces of "adding one cusp" to $\Gamma\backslash\Omega$.

References

[F-vdP] J. Fresnel et M. van der Put - *Géométrie Analytique Rigide et Applications* - Progress in Math. 18, 1981, Birkhäuser.

[G-vdP] L. Gerritzen and M. van der Put - *Schottky Groups and Mumford Curves* - Lect. Notes in Math. 817, 1980, Springer Verlag.

[Ge] E.U. Gekeler - *Drinfeld modular curves* - Lect. Notes in Math. 1231, 1986, Spriger verlag.

[vdP] M. van der Put - *Discrete groups, Mumford curves and Theta functions* - Annales de la Faculté des Sciences de Toulouse, Vol 1 , n0 3, 1992, p. 399-438.

[S] J.-P. Serre - *Arbres, Amalgames, Sl_2* - Astérisque 46, 1977.

Lecture 8

Analytic compactification and modular forms

Marius van der Put and Jaap Top
Department of Mathematics, University of Groningen
P.O.Box 800, 9700 AV Groningen, The Netherlands
mvdput@math.rug.nl and top@math.rug.nl

1 The classical modular curves

As a motivation for the analytic construction of the Drinfeld modular curves we review the analytic construction of the classical modular curve corresponding to full level n with $n \in \mathbb{Z}$, $n \geq 3$. This curve, considered as an affine curve over the field of complex numbers $\mathbb{C}$, will be denoted by $Y(n)$.

The complex valued points of $Y(n)$ are in one to one correspondence with the isomorphy classes of the pairs (E, λ), where E denotes an elliptic curve over $\mathbb{C}$; $E[n] := \{e \in E \mid ne = 0\}$ is the group of the n-torsion points and $\lambda : n^{-1}\mathbb{Z}^2/\mathbb{Z}^2 \to E[n]$ is an isomorphism of groups.

Every elliptic curve E over $\mathbb{C}$ has the form $\mathbb{C}/\Lambda$ where $\Lambda \subset \mathbb{C}$ is a lattice of rank two. Two lattices Λ_1 and Λ_2 define isomorphic curves if and only if there is a constant $c \in \mathbb{C}^*$ with $c\Lambda_1 = \Lambda_2$. Moreover the group of the n-torsion points of E can be identified with $(\frac{1}{n}\Lambda)/\Lambda \subset \mathbb{C}/\Lambda$. Consider now the set

$$Z := (\mathbb{C} \setminus \mathbb{R}) \times Gl(n^{-1}\mathbb{Z}^2/\mathbb{Z}^2).$$

This can be interpreted as the disjoint union of $2 \cdot \#Gl(n^{-1}\mathbb{Z}^2/\mathbb{Z}^2)$ copies of the upper half plane $\mathcal{H} := \{z \in \mathbb{C} \mid im(z) > 0\}$. The group $Gl(2, \mathbb{Z})$ acts on this space by

$$\sigma := \begin{pmatrix} a & b \\ c & d \end{pmatrix} \in Gl(2, \mathbb{Z}) \text{ acts on } (\tau, \alpha) \in Z \text{ as } \sigma(\tau, \alpha) = (\frac{a\tau + b}{c\tau + d}, \tilde{\sigma}\alpha),$$

where $\tilde{\sigma}$ denotes the image of $(\sigma^t)^{-1}$ (i.e. the inverse of the transposed matrix) in $Gl(n^{-1}\mathbb{Z}^2/\mathbb{Z}^2)$.

To a point $(\tau, \alpha) \in Z$ we associate a pair (E, λ). Let $f_\tau : \mathbb{Z}^2 \to \mathbb{C}$ denote the map given by $f_\tau(a, b) = a\tau + b$. This map extends to a $\mathbb{Q}$-linear map $\mathbb{Q}^2 \to \mathbb{C}$. Then $E = \mathbb{C}/f_\tau(\mathbb{Z}^2)$ and

$$\lambda := n^{-1}\mathbb{Z}^2/\mathbb{Z}^2 \xrightarrow{\alpha} n^{-1}\mathbb{Z}^2/\mathbb{Z}^2 \xrightarrow{can} E[n],$$

where *can* denotes the canonical map induced by f_τ. It can be seen that the corresponding map $Z \to Y(n)$ is surjective and that two elements (τ_1, λ_1) and (τ_2, λ_2) have the same image in $Y(n)$ if and only if there exists a $\sigma \in Gl(2,\mathbb{Z})$ with $\sigma(\tau_1, \alpha_1) = (\tau_2, \alpha_2)$. In other words we have found a bijection

$$Gl(2,\mathbb{Z})\backslash Z \to Y(n).$$

This construction induces on $Y(n)$ the structure of a Riemann surface. Let $\Gamma(n)$ denote the subgroup of $Gl(2,\mathbb{Z})$ consisting of the matrices σ such that σ is congruent to the identity matrix modulo n. Clearly $\Gamma(n)$ is a normal subgroup of $Sl(2,\mathbb{Z})$. The action of $\Gamma(n)$ on Z leaves every component of Z (setwise) invariant. Let us write $\mathbb{C} \setminus \mathbb{R}$ as $\mathcal{H} \times \{\pm 1\}$. Then $\Gamma(n)\backslash Z$ can be identified with $(\Gamma(n)\backslash\mathcal{H}) \times \{\pm 1\} \times Gl(n^{-1}\mathbb{Z}^2/\mathbb{Z}^2)$. This space has to be divided out by the action of the subgroup $\Gamma(n)\backslash Gl(2,\mathbb{Z})$ of $Gl(2,\mathbb{Z}/n\mathbb{Z}) \cong Gl(n^{-1}\mathbb{Z}^2/\mathbb{Z}^2)$, consisting of the matrices with determinant ± 1. The action of this finite group on the collection of the components, i.e. on the set $\{\pm 1\} \times Gl(n^{-1}\mathbb{Z}^2/\mathbb{Z}^2)$, is faithful. The action of an element σ of this finite group on the factor $\{\pm 1\}$ is multiplication by the determinant of σ and the action on the second factor is left multiplication by $\tilde{\sigma}$. Let Ξ denote the finite set $Sl(2,\mathbb{Z})\backslash Gl(2,\mathbb{Z}/n\mathbb{Z})$. One easily sees that Ξ is isomorphic to the group $(\mathbb{Z}/n\mathbb{Z})^*$, which has cardinality $\phi(n)$. From the above one derives an isomorphism

$$(\Gamma(n)\backslash\mathcal{H}) \times \Xi \to Y(n).$$

The Riemann surface $Y(n)$ has therefore $\phi(n)$ connected components.

In the classical literature the (affine) modular curve with full level n is often defined as the quotient $\Gamma(n)\backslash\mathcal{H}$. This quotient is known to be the analytification of an affine curve over $\mathbb{C}$. The (complex valued) points on this (more restricted) curve correspond to the isomorphy classes of the pairs (E, λ) with λ a "symplective isomorphism". Our definition of $Y(n)$ is a finite disjoint union of $\phi(n)$ copies of an affine irreducible curve.

2 An analytic moduli space

We recall the standard notations $p, q, A, K, K_\infty, \mathbf{C}$ which are used here:
p is a prime number and q is a power of it; K is a function field in one variable over $\mathbf{F}_q$; a place ∞ of K, called the place at infinity, is chosen; A is the subring of K consisting of the elements which have no poles outside ∞; the completion of K with respect to the place ∞ is written as K_∞; the completion of the algebraic closure of K_∞ is denoted by $\mathbf{C}$. The degree $deg(a)$ of a nonzero element $a \in A$ is chosen such that $a \mapsto -deg(a)$ is the normalized additive valuation

associated with the place ∞.

We remark that $A, K, K_\infty, \mathbf{C}$ are analogues of $\mathbb{Z}, \mathbb{Q}, \mathbb{R}, \mathbb{C}$. The analogue of the complex upper half plane (more precisely, of $\mathbb{C}\backslash\mathbb{R}$) is the rigid analytic space Ω, which is defined over K_∞. This is Drinfeld's upper half plane introduced in lecture 6. By abuse of notation Ω will also denote the set $\Omega(\mathbf{C}) = \mathbf{C} \setminus K_\infty$.

As we have seen, any Drinfeld module of rank two over $\mathbf{C}$ can be obtained from an A-lattice $\Lambda \subset \mathbf{C}$ of rank two. We recall the definitions and the construction (see also lecture 3):

1. An *A-lattice Λ of rank two* is a finitely generated A-submodule of $\mathbf{C}$ such that Λ is discrete as subset of $\mathbf{C}$ and such that $K \otimes_A \Lambda$ has dimension two over K. For two A-lattices Λ_1, Λ_2 of rank two we define $\mathrm{Hom}(\Lambda_1, \Lambda_2)$ to be the additive group of the $c \in \mathbf{C}$ such that $c\Lambda_1 \subset \Lambda_2$. This makes the set of A-lattices of rank two in $\mathbf{C}$ into a category.

2. Let Λ be an A-lattice of rank two. One associates to this lattice the function $e_\Lambda : \mathbf{C} \to \mathbf{C}$, given by the formula

$$e_\Lambda(z) = z \prod_{\lambda \in \Lambda, \lambda \neq 0} (1 - \frac{z}{\lambda}).$$

e_Λ is an entire function on $\mathbf{C}$ and has an expansion $\sum_{n \geq 0} c(\Lambda)_n z^{q^n}$. This function is a surjective $\mathbf{F}_q$-linear map : $\mathbf{C} \to \mathbf{C}$ and its kernel is Λ. The induced isomorphism of groups $\mathbf{C}/\Lambda \to \mathbf{C}$ is actually an isomorphism of rigid analytic spaces.

3. The homomorphism

$$\phi : A \to \mathbf{C} \to \mathbf{C}/\Lambda \overset{\cong}{\to} \mathbf{C}$$

induces a new A-module structure on $\mathbf{C}$ for which $\mathbf{C}$ is a Drinfeld module of rank two over $\mathbf{C}$. The definition of ϕ implies that for any $a \in A$; $a \neq 0$ the map $\phi_a(z)$ has the form $\sum_{n=0}^{2\,deg(a)} c(\Lambda, a)_n z^{q^n}$. By construction $\phi_a(e_\Lambda(z)) = e_\Lambda(az)$. Let $D(\Lambda)$ denote this Drinfeld module. For a homomorphism $c\mathrm{Hom}(\Lambda_1, \Lambda_2)$ one has an induced map $c : \mathbf{C}/\Lambda_1 \to \mathbf{C}/\Lambda_2$ and as a consequence a homomorphism $D(c) : D(\Lambda_1) \to D(\Lambda_2)$ of Drinfeld modules.

4. The functor D is an equivalence of the category of the A-lattices of rank two in $\mathbf{C}$ with the category of the Drinfeld A-modules over $\mathbf{C}$ of rank two.

5. Since A is a Dedekind ring and Λ has no torsion it follows that Λ is a projective module of rank two. Such a module is isomorphic to $Y = A \oplus N$

where N is a projective module of rank one. The class group of A (or Picard group $Pic(A)$) is finite and can be represented by a finite set of nonzero ideals $\{N_1, \ldots, N_s\}$ of A. In the sequel we will suppose that Y has the form $A \oplus N$ for some $N \in \{N_1, \ldots, N_s\}$.

6. An A-lattice Λ will be called of type Y if Λ is isomorphic to Y as an A-module. The corresponding Drinfeld module $D(\Lambda)$ is also called of type Y. Let $\underline{n} \neq 0$ be an ideal of A. A (full) level $\underline{n}$-structure on an A-lattice Λ of rank two and of type Y is given by an A-linear isomorphism $\underline{n}^{-1}Y/Y \to \underline{n}^{-1}\Lambda/\Lambda$. A (full) level $\underline{n}$-structure on a Drinfeld module $\phi : A \to \mathbf{C}$ over $\mathbf{C}$ and of type Y is defined to be an A-linear isomorphism

$$\underline{n}^{-1}Y/Y \to \{c \in \mathbf{C} | \phi_a(c) = 0 \text{ for all } a \in \underline{n}\}.$$

The last set is easily seen to be the image of $\underline{n}^{-1}\Lambda \subset \mathbf{C}$ under the map e_Λ. One obtains in this way a 1-1 correspondence between the level $\underline{n}$-structures on Λ and on $D(\Lambda)$. Hence the operation $\Lambda \mapsto e_\Lambda$ induces an equivalence between the category of the A-lattices of rank two in $\mathbf{C}$ with level structure $\underline{n}$ and the category of the Drinfeld A-modules of rank two over $\mathbf{C}$ with level $\underline{n}$-structure.

Some comments

We note some differences with the classical situation. First of all $\mathbf{C}$ contains A-lattices of any rank because $\mathbf{C}$ is an infinite extension of K_∞. Moreover rank two lattices are in general not free since $Pic(A)$ can be nontrivial. As a consequence we will have to consider a Drinfeld modular curve for each element in $Pic(A)$. Finally, the rigid analytic space Ω is connected and $\mathbb{C} \setminus \mathbb{R}$ has two components.

Our aim is to describe the set of equivalence classes of A-lattices in $\mathbf{C}$ with level structure $\underline{n}$. Fix a choice for $Y = A \oplus N$ and a level $\underline{n}$. For $\omega \in \Omega$ one considers the map $f_\omega : Y \to \mathbf{C}$, given by $f_\omega(a + n) = a\omega + n$. This map extends to a K-linear map $f_\omega : K \otimes Y \to \mathbf{C}$. Then $\Lambda = f_\omega(Y)$ and the map $\underline{n}^{-1}Y/Y \to \underline{n}^{-1}\Lambda/\Lambda$, induced by f_ω, define an A-lattice of type Y with level structure $\underline{n}$.

In this way, every equivalence class of A-lattices of type Y in $\mathbf{C}$ is obtained. The obvious homomorphism $Gl(Y) \to Gl(\underline{n}^{-1}Y/Y)$ is in general not surjective. This has the consequence that the map from Ω to the equivalence classes of A-lattices of type Y and with level $\underline{n}$ is not surjective. We consider the map

$$\Omega \times Gl(\underline{n}^{-1}Y/Y) \to A\text{-lattices of type } Y \text{ and with level } \underline{n}$$

given by $(\omega, \alpha) \mapsto \Lambda = A\omega + N$ and with the level structure given by

$$\underline{n}^{-1}Y/Y \stackrel{\alpha}{\to} \underline{n}^{-1}Y/Y \stackrel{f_\omega}{\to} \underline{n}^{-1}\Lambda/\Lambda.$$

The last isomorphism is the one induced by f_ω defined above.

The action of the group $Gl(Y)$ on $\Omega \times Gl(\underline{n}^{-1}Y/Y)$ is defined by the formula $g(\omega, \alpha) = (g(\omega), \tilde{g}\alpha)$, where $\tilde{g}$ denotes the image of $(g^t)^{-1}$ (i.e. the inverse of the transposed map) in $Gl(\underline{n}^{-1}Y/Y)$. It is easy to verify that two elements (ω_1, α_1) and (ω_2, α_2) define equivalent A-lattices of type Y and with level $\underline{n}$, if and only if there is a $g \in Gl(Y)$ with $g(\omega_1, \alpha_1) = (\omega_2, \alpha_2)$.
Let $Gl(Y, \underline{n})$ denote the kernel of the map $Gl(Y) \to Gl(\underline{n}^{-1}Y/Y)$. One can verify that $Gl(Y, \underline{n})$ consists of the σ such that $\sigma - 1$ maps Y into $\underline{n}Y$. Put $\Xi := Gl(Y)\backslash Gl(\underline{n}^{-1}Y/Y)$. Then the quotient $Gl(Y)\backslash(\Omega \times Gl(\underline{n}^{-1}Y/Y))$ can also be written as a finite disjoint union over the $\xi \in \Xi$ of the quotients $(Gl(Y, \underline{n})\backslash\Omega) \times \xi$. Thus we have shown:

Proposition 2.1 *The set of Drinfeld A-modules of rank two with type Y and level $\underline{n}$ over the field* $\mathbf{C}$ *is parametrized by*

$$Gl(Y)\backslash(\Omega \times Gl(\underline{n}^{-1}Y/Y)).$$

This rigid analytic space is isomorphic to $(Gl(Y, \underline{n})\backslash\Omega) \times \Xi$ where Ξ denotes the finite set $Gl(Y)\backslash Gl(\underline{n}^{-1}Y/Y)$.

This result is the first step in showing that the rigid space

$$\mathcal{M}_Y(\underline{n}) := (Gl(Y, \underline{n})\backslash\, \Omega) \times \Xi$$

is a moduli space (in the rigid analytic category) for the Drinfeld modules of type Y and level $\underline{n}$.
We note further that the homomorphism $Sl(2, A) \to Sl(2, A/\underline{n})$ is surjective. It follows from this that the group $Gl(2, A)\backslash Gl(2, A/\underline{n})$ is isomorphic to $(A/\underline{n})^*/\mathbf{F}_q^*$. Thus the number of connected components of the space $Gl(2, A)\backslash(\Omega \times Gl(2, A/\underline{n})) = (Gl(A^2, \underline{n})\backslash\Omega) \times \Xi$ is equal to $\#\Xi = \#(A/\underline{n})^*/\mathbf{F}_q^*$.

To show that $\mathcal{M}_Y(\underline{n})$ is the desired moduli space, one has to construct a universal Drinfeld module (of type Y and level $\underline{n}$) above it. This will be obtained from the construction of a Drinfeld module above $\Omega \times Gl(\underline{n}^{-1}Y/Y)$ (of type Y and with level $\underline{n}$) and then dividing out the action of the group $Gl(Y)$.

Write again $Y = A \oplus N$. To a point $\omega \in \Omega$ we have associated the lattice $A\omega + N$ and the " Carlitz exponential" of the lattice

$$e(\omega, z) = z \prod_{\alpha \in A\omega + N,\ \alpha \neq 0} (1 - \frac{z}{\alpha}).$$

This function is holomorphic in the two variables ω and z. It has a convergent expansion $\sum_{n \geq 0} e(\omega)_n z^{q^n}$, where the $e(\omega)_n$ are holomorphic functions on

Ω. For $a \in A$, $a \neq 0$, the function $\phi_a(\omega, z)$, which is defined by the equality $\phi_a(\omega, e(\omega, z)) = e(\omega, az)$, is also holomorphic in the two variables and has the expansion $\phi_a(\omega, z) = \sum_{n=0}^{2\ deg(a)} e(\omega, a)_n z^{q^n}$ with respect to the variable z.
Line bundles (rigid analytic or algebraic) above a space X can be either defined as a sheaf on X which is locally isomorphic to the structure sheaf or as a space $L \to X$ which is locally isomorphic to a product $X \times \mathbf{C}$. The latter description will be used now since it is more convenient. The trivial analytic line bundle L above $\Omega \times Gl(\underline{n}^{-1}Y/Y)$ is $\Omega \times Gl(\underline{n}^{-1}Y/Y) \times \mathbf{C}$. (In fact any analytic line bundle above $\Omega \times Gl(\underline{n}^{-1}Y/Y)$ is trivial.) For any $a \in A$, $a \neq 0$ the map $\phi_a : (\omega, \alpha, z) \to (\omega, \alpha, \phi_a(\omega, z))$ is an endomorphism of this line bundle. This defines the Drinfeld module above $\Omega \times Gl(\underline{n}^{-1}Y/Y)$.

As usual the group $Gl(Y)$ operates on Ω by

$$\begin{pmatrix} a & b \\ c & d \end{pmatrix} \omega = \frac{a\omega + b}{c\omega + d}.$$

We extend this to an action on the trivial line bundle L by

$$\begin{pmatrix} a & b \\ c & d \end{pmatrix} (\omega, \alpha, z) = (\frac{a\omega + b}{c\omega + d}, \overline{\begin{pmatrix} a & c \\ b & d \end{pmatrix}}^{-1} \alpha, (c\omega + d)^{-1} z).$$

We note that $Gl(Y)$ is the subgroup of the matrices of $Gl(K)$ satisfying: $a, d \in A$, $b \in N^{-1}$, $c \in N$ and $ad - bc \in A^*$.

For $\gamma = \begin{pmatrix} a & b \\ c & d \end{pmatrix} \in Gl(Y)$ one has the following formulas:

$$(c\omega + d) e(\gamma(\omega), (c\omega + d)^{-1} z) = e(\omega, z) \text{ and}$$

$$(c\omega + d) \phi_a(\gamma(\omega), (c\omega + d)^{-1} z) = \phi_a(\omega, z).$$

The first formula is correct since the left hand side and the right hand side of the formula have expansions which start with z and the kernels of the two $\mathbf{F}_q$-linear maps are $A\omega + N$. The formula concerning $\phi_a(\omega, z)$ is a consequence of the first formula. One can now easily verify that the Drinfeld module $\phi : A \to End(L)$ above $\Omega \times Gl(\underline{n}^{-1}Y/Y)$ commutes with the action of $Gl(Y)$ on L.

The next step is the division of the Drinfeld module $\Omega \times Gl(\underline{n}^{-1}Y/Y), L, \phi$ by the group $Gl(Y)$. Here we have to make a general remark about moduli. For the existence of a moduli space for certain objects it is necessary that the objects have no automorphisms (different from the identity). In our situation this means that no A-lattice of type Y and level structure $\underline{n}$ has an automorphism different from the identity. The automorphisms (and endomorphisms) of an A-lattice of type Y (and of Drinfeld modules) are discussed in the lectures 1,3 and 5. One can show that for a level structure $\underline{n}$ such that the dimension of the $\mathbf{F}_q$-vector space $A/\underline{n}$ is large enough the A-lattices with level $\underline{n}$-structure have no automorphisms. This condition on the A-lattices of type Y with level $\underline{n}$ can be translated into:

The group $Gl(Y)$ has no fixed points on $\Omega \times Gl(\underline{n}^{-1}Y/Y)$

or into the equivalent statement:

The group $Gl(Y,\underline{n})$ has no fixed points for its action on Ω.

The standard example.
For $A = \mathbf{F}_q[T]$ and $\underline{n}$ generated by a (monic) polynomial of degree ≥ 1, the condition above is satisfied.

In the following we will assume that the group $Gl(Y,\underline{n})$ has no fixed points on Ω. Then $\Omega \times Gl(\underline{n}^{-1}Y/Y)$ and L can be divided by $Gl(Y)$ and one obtains a (rigid analytic) line bundle

$$Gl(Y)\backslash L \to Gl(Y)\backslash\Omega \times Gl(\underline{n}^{-1}Y/Y).$$

Since ϕ is equivariant for the action of $Gl(Y)$ one obtains a Drinfeld module with level $\underline{n}$-structure above the space $Gl(Y)\backslash(\Omega \times Gl(\underline{n}^{-1}Y/Y))$.

Theorem 2.2 *Suppose that the group $Gl(Y,\underline{n})$ has no fixed points for its action on Ω. Then the Drinfeld module above the rigid space $Gl(Y)\backslash(\Omega\times Gl(\underline{n}^{-1}Y/Y)$, constructed above, is the universal one for type Y and level $\underline{n}$ in the category of the rigid spaces over K_∞.*

Before indicating a proof we return to the "standard example".

The standard example. Put $A = \mathbf{F}_q[T]$ and $\Gamma = Gl(A^2)$. The function

$$e(\omega, z) = z\prod(1 - \frac{z}{a\omega + b}) = \sum_{n\geq 0} e(\omega)_n z^{q^n}$$

satisfies

$$e(\omega, z) = z(1 - \sum_{m\geq 1} s_m z^{m(q-1)})^{-1}, \text{ with } s_m = \sum_{(0,0)\neq(a,b)\in A^2} (a\omega + b)^{m(1-q)}.$$

One finds this formula by calculating $\frac{\frac{d}{dz}e(\omega,z)}{e(\omega,z)} = \frac{1}{e(\omega,z)}$ and by observing that $\sum_{(0,0)\neq(a,b)\in A^2}(a\omega + b)^{-k}$ is 0 for $k > 0$ and k not divisible by $q - 1$.

The formula implies that $e(\omega)_1 = s_1$ and $e(\omega)_2$ is some polynomial in $s_1, \ldots, s_{q+1}$.

Write $\phi_T(\omega, z) = Tz + c_1(\omega)z^q + c_2(\omega)z^{q^2}$. From $\phi_T(\omega, e(\omega, z)) = e(\omega, Tz)$ it follows that

$$c_1(\omega) = e(\omega)_1(T^q - T) \text{ and } c_2(\omega) = e(\omega)_2(T^{q^2} - T) - e(\omega)_1^{q+1}(T^q - T).$$

The $c_2(\omega)$ is invertible on Ω and we can form the expression $j(\omega) = \frac{c_1(\omega)^{q+1}}{c_2(\omega)}$. This expression is a holomorphic function on Ω and is invariant under the action of $Gl(A^2)$. Hence j induces a holomorphic function $j : \Gamma\backslash\Omega \to \mathbf{C}$. This map is a bijection.

Indeed, the set $\Gamma\backslash\Omega$ parametrizes all Drinfeld A-modules of rank two over $\mathbf{C}$. Such a Drinfeld module is characterized by the $\phi_T = T + c_1\tau + c_2\tau^2$. The polynomials $T+c_1\tau+c_2\tau^2$ and $T+d_1\tau+d_2\tau^2$ define isomorphic Drinfeld modules if and only if there exists a $c \in \mathbf{C}^*$ with $c^{-1}(T + c_1\tau + c_2\tau^2)c = T + d_1\tau + d_2\tau^2$. This last condition is equivalent with $\frac{c_1^{q+1}}{c_2} = \frac{d_1^{q+1}}{d_2}$. This proves the statement.

One can prove a stronger result, namely :

Proposition 2.3 *Let $\mathbf{A}_{K_\infty}^{1,an}$ denote the affine line over K_∞, considered as a rigid space over K_∞. Then $j : PGl(2, A)\backslash\Omega \to \mathbf{A}_{K_\infty}^{1,an}$ is an isomorphism.*

Proof. The space $PGl(2, A)\backslash\Omega$ has been identified with the affine line over K_∞ (see lecture 7). Any point $b \in \mathbf{F}_{q^2} \setminus \mathbf{F}_q$ is a zero of the function $c_1(\omega)$. It can be calculated that it is a zero of order one. The stabilizer in $PGl(2, A)$ of b is a cyclic group of order $q + 1$. Let $\xi \in PGl(2, A)\backslash\Omega$ denote the image of b. Using the expression $j = \frac{c_1^{q+1}}{c_2}$ one finds that j has a zero of order one at ξ. Thus j is a bijective analytic map on the rigid analytic affine line over K_∞ with a simple zero at ξ. Let z denote a coordinate of this affine line with $z(\xi) = 0$. Then j is equal to a convergent power series $\sum_{n>0} a_n z^n$ with $a_1 \neq 0$. Since j is bijective, this expression has only one zero and so $\sum_{n>0} a_n z^{n-1}$ has no zeroes on the affine line. It is known that the only invertible holomorphic functions on $\mathbf{A}_{K_\infty}^{1,an}$ are the constant functions (see lecture 3). Thus $\sum_{n>0} a_n z^{n-1}$ is constant and equals a_1. This shows that j is an isomorphism. □

Proof. **Sketch of the proof of theorem 2.2**
We first observe that the analytic construction is given over the field K_∞. Let a Drinfeld module (ψ, L, S) of type Y and level $\underline{n}$ be given in the category of the rigid analytic spaces over K_∞. This means that S is a rigid analytic space over K_∞, that L is an analytic line bundle over S and that there is given a

Drinfeld module of rank two on L, of type Y with a level $\underline{n}$- structure. It has to be shown that there exists a unique analytic map

$$f : S \to Gl(Y)\backslash(\Omega \times Gl(\underline{n}^{-1}Y/Y)$$

such that f transports the Drinfeld module above $Gl(Y)\backslash(\Omega \times Gl(\underline{n}^{-1}Y/Y)$ to one above S which is isomorphic to the given one (by a uniquely determined isomorphism).

We will restrict ourselves to the case $A = \mathbf{F}_q[T]$ and some level $\underline{n}$ generated by a polynomial f of degree ≥ 1. Write $\mathcal{M}(\underline{n})$ for the space

$$Gl(A^2)\backslash(\Omega \times Gl(\underline{n}^{-1}A^2/A^2)) = \cup_{\xi\in\Xi}(Gl(A^2,\underline{n})\backslash\Omega) \times \xi,$$

and $\mathcal{M}(1)$ for $PGl(A^2)\backslash\Omega$. On $\mathcal{M}(\underline{n})$ the group $Gl(2, A/\underline{n})/\mathbf{F}_q^*$ acts in the obvious way and the quotient by this action is $\mathcal{M}(1)$.
Let a Drinfeld module ψ with level $\underline{n}$-structure above a rigid space S (over K_∞) be given. We have to construct an analytic map $S \to \mathcal{M}(\underline{n})$ with the required properties. Using the j-invariant of this Drinfeld module and proposition 2.4 one finds an analytic map $f : S \to \mathcal{M}(1)$. We note further that the line bundle L is trivial since the level $\underline{n}$-structure provides a nowhere vanishing section of L.

Let S' be the fibered product of S and $\mathcal{M}(\underline{n})$ over $\mathcal{M}(1)$. The projection $pr_1 : S' \to S$ is finite and S is in fact the quotient of S' under the action of $Gl(2, A/\underline{n})/\mathbf{F}_q^*$ on S'. Further pr_1 induces a Drinfeld structure $pr_1^*\psi$ on S' (with respect to the trivial line bundle). The second projection $pr_2 : S' \to \mathcal{M}(\underline{n})$ induces a Drinfeld structure $pr_2^*\phi$ on S' (again with respect to the trivial line bundle). For each point of S' the two Drinfeld structures are isomorphic. This implies that $\ker(pr_1^*\psi_f, O_{S'})$=$\ker(pr_2^*\phi_f, O_{S'})$. Let us denote this constant subsheaf of $O_{S'}$ by $\mathcal{L}$.
From the level structure on S we obtain an isomorphism of constant sheaves $I_1 : \underline{n}^{-1}A^2/A^2 \to \mathcal{L}$ and from the level structure on $\mathcal{M}(\underline{n})$ one obtains another isomorphism of constant sheaves $I_2 : \underline{n}^{-1}A^2/A^2 \to \mathcal{L}$. This induces an automorphism of the constant sheaf $\underline{n}^{-1}A^2/A^2$ on S'. This automorphism is a section σ of the constant sheaf $Gl(2, A/\underline{n})$ on S'. Then S' can be written as a disjoint union $\cup S'_g$, over the elements g of the group $Gl(2, A/\underline{n})$. The subspace S'_g is the subset of S' where σ is equal to g. The group $Gl(2, A/\underline{n})$, acting on S', permutes the S'_g as follows: $h(S'_g) = S'_{hg}$. It follows from this that each $pr_1 : S'_g \to S$ is an isomorphism. In particular, the isomorphism $pr_1 : S'_1 \to S$ induces a morphism $S \to \mathcal{M}(\underline{n})$ which has the required properties as one easily verifies. □

Corollary 2.4 *Assume that the group $Gl(Y, \underline{n})$ has no fixed points for its action on Ω. Let $M_Y^2(\underline{n})$ denote the moduli space for Drinfeld A-modules of rank two with level structure $\underline{n}$ and type Y.*

Then the analytic space over K_∞ associated to the affine algebraic curve $M_Y^2(\underline{n})\otimes_A K_\infty$ together with its Drinfeld module and level $\underline{n}$-structure is isomorphic to $\mathcal{M}_Y(\underline{n})$ with its Drinfeld module and level $\underline{n}$-structure.

Proof. Generalities about representable functors imply that the analytification of $M_Y^2(\underline{n})\otimes_A K_\infty$ with its Drinfeld module and level structure is also universal in the rigid category. This yields the isomorphism of the corollary. □

3 Compactifying $Gl(Y,\underline{n})\backslash\Omega$

As a didactical excursion we recall first the compactification of the classical modular curve. The notation of section 1 will be used.

3.1 Compactifying $\Gamma(n)\backslash\mathcal{H}$

The integer n is supposed to be ≥ 3. Since all the components of the curve $Y(n)$ are isomorphic we may restrict our attention to the component $\Gamma(n)\backslash\mathcal{H}$. This object is an affine nonsingular irreducible curve over $\mathbb{C}$. Its compactification $\overline{\Gamma(n)\backslash\mathcal{H}}$ is simply the corresponding projective nonsingular curve. The new points are called the *cusps*.

Each rational point on the "boundary" of $\mathcal{H}$, i.e. an element of $\mathbf{P}^1(\mathbb{Q}) = \mathbb{Q}\cup\{\infty\}$, produces a cusp of $\Gamma(n)\backslash\mathcal{H}$. The simplest cusp to consider is ∞. A "horicycle neighbourhood" of ∞ is the set $U := \{z\in\mathbb{C} \mid im(z) > r\}$. For $r\geq 1$ one has:
Two points $u_1, u_2 \in U$ are equivalent under the action of $\Gamma(n)$ if and only if they are equivalent under the action of the subgroup

$$H_n = \{\begin{pmatrix} 1 & na \\ 0 & 1 \end{pmatrix} \mid a\in\mathbb{Z}\} \text{ of } \Gamma(n).$$

The image of U in $\Gamma(n)\backslash\mathcal{H}$ is therefore isomorphic to $H_n\backslash U$. The last space is isomorphic to $\{q\in\mathbb{C} \mid 0<|q|<r'\}$ (for some r') by means of the exponential map $z\in U\mapsto q=e^{2\pi iz/n}$. The complex analytic way to compactify $\Gamma(n)\backslash\mathcal{H}$ at the cusp ∞ consists of gluing $\{q\in\mathbb{C} \mid |q|<r'\}$ to $\Gamma(n)\backslash\mathcal{H}$ over the open subset $H_n\backslash U\cong\{q\in\mathbb{C} \mid 0<|q|<r'\}$.

The group $\Gamma(1)=PGl(2,\mathbb{Z})$ has only one cusp. Therefore the cusps of $\Gamma(n)$ are conjugated under the action of $\Gamma(1)$. This has the consequence that the set of cusps of $\Gamma(n)$ can be identified with $\Gamma(n)\backslash\mathbf{P}^1(\mathbb{Q})$.

Above the curve $Y(n)$ (or above the more restricted curve $\Gamma(n)\backslash\mathcal{H}$) there is a universal family of elliptic curves with full level n-structure. One can extend this family by placing a "generalized elliptic" curve above every cusp. In the case of the cusp ∞, considered above, this would mean that one has to find out what the natural extension of the elliptic family above $H_n \backslash U \cong \{q \in \mathbb{C} \mid 0 < |q| < r'\}$ to $q = 0$ is.

This family is essentially the "Tate elliptic curve" $\mathbb{C}((q))^*/\{q^{na} | \ a \in \mathbb{Z}\}$, which has the affine equation $y^2 + xy = x^3 + Bx + C$ where $B, C \in q^n\mathbb{C}[[q^n]]$. Using this equation one can see the Tate-curve as a closed subset of $\mathbf{P}^2_{\mathbb{C}[[q]]}$. Above the open subset $q \neq 0$ it is a honest elliptic curve. The fibre for $q = 0$ is equal to a rational curve with an ordinary double point (i.e. given by the affine equation $y^2 + xy = x^3$). After deleting the double point (i.e. the point $(0, 0)$) one finds a group structure and this group is the multiplicative group $\mathbf{G}_m$ over $\mathbb{C}$. This is not quite the correct object to put above the cusp, since the group of the n-torsion points $\mathbf{G}_m[n]$ of $\mathbf{G}_m$ is too small.

The remedy consists of replacing the Tate curve by its Néron model (or equivalently blowing up the double point a number of times). This Néron model is obtained by dividing out a canonical model for the multiplicative group over the field $\mathbb{C}((q))$, with respect to its valuation ring $\mathbb{C}[[q]]$, by the action of q^n. The fibre at $q = 0$ is the "generalized elliptic curve" $\tilde{E}$ that fits into this family of elliptic curves with level n.

We will now describe $\tilde{E}$. As an algebraic variety, $\tilde{E}$ is an n-gon of projective lines over $\mathbb{C}$. After deleting the n ordinary double points of $\tilde{E}$ one obtains a group $\tilde{E}^*$ which is isomorphic to the product $\mathbf{G}_m \times \mathbb{Z}/n\mathbb{Z}$. The group of the n-torsion points $\tilde{E}^*[n]$ of $\tilde{E}^*$ is isomorphic to $n^{-1}\mathbb{Z}^2/\mathbb{Z}^2$. A level n-structure on the generalized elliptic curve is an isomorphism $n^{-1}\mathbb{Z}^2/\mathbb{Z}^2 \to \tilde{E}^*[n]$. The generalized elliptic curve has automorphisms. An automorphism will be an isomorphism $f : \tilde{E} \to \tilde{E}$ which respects the group law on $\tilde{E}^*$. It is not difficult to see that the group of automorphisms is generated by the two elements:
(1) $i : a \mapsto a^{-1}$ on $\tilde{E}^*$, extended in the obvious way to $\tilde{E}$.
(2) $\sigma : \tilde{E}^* \to \tilde{E}^*$, extended in the obvious way to $\tilde{E}$, where $\sigma(a, i) = (a\zeta^i, i)$ with ζ a primitive nth root of unity, $a \in \mathbf{G}_m$, $i \in \mathbb{Z}/n\mathbb{Z}$.
Thus the group of automorphisms of the generalized elliptic curve has order $2n$.

Now one can count the number of cusps in two ways. The first method counts the number of level n structures on the generalized elliptic curve. This number is clearly $\frac{\#Gl(n^{-1}\mathbb{Z}^2/\mathbb{Z}^2)}{2n}$.
The other method is to count the number of cusps of $Y(n)$ as the number of the components of $Y(n)$ (i.e. $\phi(n)$) multiplied by the number of cusps of $\Gamma(n)\backslash\mathcal{H}$ (i.e. the cardinal of $\Gamma(n)\backslash\mathbf{P}^1(\mathbb{Q})$). Those two numbers are equal as they should be. We refer to [D-R] for more details. In the exposition above we have de-

scribed the situation over the field of complex numbers. In lecture 9 we will continue the description of the cusps of modular curves and Néron models of the Tate curve over a general base ring.

3.2 Describing the cusps of $Gl(Y, \underline{n})\backslash\Omega$

A consequence of 2.5 is that $Gl(Y, \underline{n})\backslash\Omega$ is the analytification of a (smooth, connected) affine curve over K_∞. This algebraic curve has a natural compactification to a projective curve. The analytification of the latter is of course equal to the analytic compactification $\overline{\text{of } Gl(Y, \underline{n})\backslash\Omega}$ as rigid analytic space. The new points are called the *cusps* of $\overline{Gl(Y, \underline{n})\backslash\Omega}$ (or of $Gl(Y, \underline{n})$). Thus the existence of this analytic compactification follows from the algebraic situation. However we will describe the analytic compactification of $Gl(Y, \underline{n})\backslash\Omega$ in more detail. The analytic compactification is also useful for the construction of the local parameters at the cusps.

An element $g \in PGl(2, K_\infty)$, $g \neq 1$, is called *parabolic* if g has only one fixed point in $\mathbf{P}^1(\mathbf{C})$. Such a g can be represented by a matrix having the single eigenvalue 1. As a consequence, the order of a parabolic element is p. Conversely, an element of order p is a parabolic element. Let Γ denote a discrete subgroup of $PGl(2, K_\infty)$. A point of $\mathbf{P}^1(\mathbf{C})$ will be called a *parabolic point of* Γ if it is the fixed point of infinitely many parabolic elements of Γ. Such a point belongs to $\mathbf{P}^1(K_\infty)$. We will now investigate the parabolic points of $PGl(Y)$.

Lemma 3.1
(1) *Let Γ be a subgroup of finite index in $PGl(Y)$. Then the set of parabolic points of Γ is equal to $\mathbf{P}^1(K)$.*
(2) *There is a natural bijection $PGl(Y)\backslash\mathbf{P}^1(K) \to Pic(A)$.*

Proof. (1) Two groups Γ and Γ' have the same set of parabolic points if $\Gamma \cap \Gamma'$ is a subgroup of finite index in both Γ and Γ'. It suffices therefore to consider $Y = A^2$ and the group $\Gamma := PGl(A^2)$. Any parabolic point lies in $\mathbf{P}^1(K)$ since it is the fixed point of a parabolic element in $PGl(2, K)$. It is further clear that 0 and ∞ are parabolic points. Let $\lambda \in K, \lambda \neq 0$. Choose any $c \in A$, $c \neq 0$ such that $c\lambda \in A$ and $c\lambda^2 \in A$. Then λ is the fixed point of the parabolic element $\begin{pmatrix} c\lambda - 1 & -c\lambda^2 \\ c & -c\lambda - 1 \end{pmatrix}$. This shows that λ is a parabolic point.
(2) Y is considered as a projective rank two A-submodule of K^2. Every point of $\mathbf{P}^1(K)$ is represented by a line $L \subset K^2$. We associate to L the class in $Pic(A)$ of the rank one projective module $L \cap Y$. For every $g \in PGl(Y)$, one has $gL \cap Y = g(L \cap Y)$. Hence L and gL have the same image in $Pic(A)$.

Let the two lines L_1, L_2 have the same image in $Pic(A)$. The projective submodules $L_i \cap Y$ are direct factors of Y. Thus we can write $Y = (L_1 \cap Y) \oplus M_1 = (L_2 \cap Y) \oplus M_2$. Since $L_1 \cap Y \cong L_2 \cap Y$, also $M_1 \cong M_2$. This leads to an automorphism g of Y with $g(L_1 \cap Y) = (L_2 \cap Y)$ and $gM_1 = M_2$. Thus we found an injective map $PGl(Y) \backslash \mathbf{P}^1(K) \to Pic(A)$.
Finally, let M be a rank one projective module representing an element of $Pic(A)$. Then $M \oplus (M^{-1} \otimes N) \cong Y = A \oplus N$, since the determinants of the two modules are isomorphic. Let $B \subset Y$ be the image of M under an isomorphism and put $L =: K \otimes B \subset K \otimes Y = K^2$. Then $L \in \mathbf{P}^1(K)$ is mapped to the class of M in $Pic(A)$. □

Proposition 3.2 *Let Γ be a subgroup of finite index in $PGl(Y)$. There is a bijective map from $\Gamma \backslash \mathbf{P}^1(K)$ to the set of cusps of Γ.*

Proof. We will sketch the proof. A full proof can be obtained by using [S]. Every point of $\mathbf{P}^1(K)$ is represented by a line $L = Ke \subset K^2$. This line induces an end, called $end(L)$ in the tree $\mathcal{T}$, which can be represented by the sequence of modules $Oe + Y \supset Oe + \pi Y \supset Oe + \pi^2 Y \supset \ldots$ (see the lectures 6 and 7).
Let the subgroup $H \subset \Gamma$ consist of the matrices h in $\Gamma \subset PGl(Y)$ with $h(e) = e$ and $det(h) = 1$. As we have seen this is an infinite group consisting of parabolic elements (indeed, L is a parabolic point). It is not difficult to show that $end(L)$ has the property:

For large n, m the modules $Oe + \pi^n Y$ and $Oe + \pi^m Y$ are Γ-equivalent only if $n = m$.

Thus the image of $end(L)$ in $\Gamma \backslash \mathcal{T}$ is also an end of $\Gamma \backslash \mathcal{T}$. This provides an injective map from $\Gamma \backslash \mathbf{P}^1(K)$ to the ends of $\Gamma \backslash \mathcal{T}$.

It is more difficult to see that this map is surjective. The structure of $PGl(Y) \backslash \mathcal{T}$ is completely given in [S]. In particular, he shows that the ends of $PGl(Y) \backslash \mathcal{T}$ are in one-to-one correspondence with the elements of $Pic(A)$. This implies the statement for the group $PGl(Y)$.

Let $\Gamma \subset PGl(Y)$ be a normal subgroup of finite index. The factor group $PGl(Y)/\Gamma$ acts on the graph $\Gamma \backslash \mathcal{T}$. The obvious map $pr : \Gamma \backslash \mathcal{T} \to Gl(Y) \backslash \mathcal{T}$, is just the quotient map under the action of $PGl(Y)/\Gamma$. The map from $\mathbf{P}^1(K)$ to the ends of $\Gamma \backslash \mathcal{T}$ is surjective, since the map

$$\mathbf{P}^1(K) \to \text{ the ends of } \Gamma \backslash \mathcal{T} \to \text{ the ends of } PGl(Y) \backslash \mathcal{T} \text{ is surjective.}$$

For any subgroup $\Gamma \subset PGl(Y)$ of finite index, there is a normal subgroup $\Gamma' \subset PGl(Y)$ of finite index which is contained in Γ. Then

$$\mathbf{P}^1(K) \to \text{ the ends of } \Gamma' \backslash \mathcal{T} \to \text{ the ends of } \Gamma \backslash \mathcal{T} \text{ is surjective.}$$

□

Remark
In the case of the standard example $A = \mathbf{F}_q[T]$, we have already shown that $PGl(2,A)\backslash\Omega$ has one cusp. The more complicated results of [S] are not needed in this case.

3.3 The analytic compactification of the curve $\Gamma\backslash\Omega$

Γ is supposed to be a subgroup of finite index in $PGl(Y)$. The aim is to compactify the rigid analytic space $\Gamma\backslash\Omega$ by "adding" a point for each end of the graph $\Gamma\backslash\mathcal{T}$. Thus we have to "add a point" for every element of $\Gamma\backslash\mathbf{P}^1(K)$. The idea is the following.
An element c of $\Gamma\backslash\mathbf{P}^1(K)$ is represented by a parabolic point $Ke \in \mathbf{P}^1(K)$. One considers the subgroup H of Γ consisting of the matrices h in $\Gamma \subset PGl(Y)$ with $h(Ke) = Ke$. One constructs a suitable subset U of Ω (called a "horicycle neighbourhood of the cusp") with the properties:

U is (setwise) invariant under H. Let $u_1, u_2 \in U$ and let $g \in \Gamma$ satisfy $g(u_1) = u_2$, then $g \in H$.

In other words, the image of U in $\Gamma\backslash\Omega$ is isomorphic to $H\backslash U$. It is shown that there is an isomorphism $f_c : \tilde{U}_c := H\backslash U \to \{z \in \mathbf{C} |\, 0 < |z| \leq 1\}$. The disk $\{z \in \mathbf{C} |\, |z| \leq 1\}$ is glued to $\Gamma\backslash\Omega$ using f_c. This is done for every $c \in \Gamma\backslash\mathbf{P}^1(K)$. The result is an analytic space $\overline{\Gamma\backslash\Omega}$ which is proper and smooth over K_∞. The rigid analytic version of GAGA states that $\overline{\Gamma\backslash\Omega}$ is the analytification of a smooth projective connected curve over K_∞. The local parameter of this curve at the cusp c is the function f_c.

We will now give some more information about this horicycle neighbourhood. For notational convenience, we suppose that Ke is the point at ∞. The group H consists of certain (classes of) matrices $\begin{pmatrix} a & b \\ 0 & d \end{pmatrix}$. The subgroup H_0 of H of the matrices with $a = d = 1$ is a normal subgroup and H/H_0 is a finite cyclic group. The group H_0 can be identified with an infinite discrete subgroup of $K \subset K_\infty$. It has the property that $H_0\backslash K_\infty$ is a compact set. For the construction of the admissible subset $U \subset \Omega$ we will use again the real valued function $d(\ )$ on Ω given by $d(z) = \min\{|z + \lambda| \mid \lambda \in K_\infty\}$. (See lecture 7)

Lemma 3.3 *There exists an element $r \in |K_\infty^*|$ such that the "horicycle neighbourhood of ∞", i.e. the admissible subset $U := \{z \in \Omega|\, d(z) \geq r\}$ of Ω, has the*

properties:
U is (setwise) invariant under H.
Let $u_1, u_2 \in U$ and $g \in \Gamma$ be such that $g(u_1) = u_2$, then $g \in H$.

Proof. This is a combinatorial exercise. See [G-vdP]. □

The quotient space $H_0\backslash U$ can be identified with a subset of the quotient space $H_0\backslash \mathbf{C}$. The latter is isomorphic with $\mathbf{C}$. Indeed, the function

$$e_{H_0}(z) = z \prod_{h\in H_0,\ h\neq 0} (1 - \frac{z}{h})$$

is known to be a surjective homomorphism of $\mathbf{C} \to \mathbf{C}$ with kernel H_0 (See lecture 3). Thus e_{H_0} provides an analytic isomorphism of $H_0\backslash\mathbf{C}$ with $\mathbf{C}$. Using that $H_0\backslash K_\infty$ is compact one can show that the image of U under e_{H_0} is equal to $\{z \in \mathbf{C} |\ r' \leq |z|\}$ for some $r' \in |K_\infty^*|$. This proves that $H_0\backslash U$ is isomorphic to $\{z \in \mathbf{C} |\ 0 < |z| \leq 1\}$. The action of the cyclic group H/H_0 on this space is multiplication of z by roots of unity. We conclude that also $H\backslash U$ is isomorphic to $\{z \in \mathbf{C} |\ 0 < |z| \leq 1\}$.

The function $e_{H_0}^{-m}$, where m is the order of H/H_0, is defined and analytic on $\overline{\Gamma\backslash\Omega}$ in some neighbourhood of the cusp ∞. It has a zero of order one at the cusp ∞ and can be taken as local parameter at the cusp ∞.

This finishes the construction of the analytic compactification of $\Gamma\backslash\Omega$.

4 The standard example

We consider here $A = \mathbf{F}_q[T]$ and the level is given by $f \in A$, a monic polynomial of degree n.

4.1 The number of cusps

The calculation of the number of cusps for the group $\Gamma(f) := Gl(A^2, (f)) \subset Gl(2, A)$ is somewhat delicate. First of all, in the case $f = 1$ there is only one cusp. Indeed, $PGl(2, A)$ and also $Sl(2, A)$ acts transitively on $\mathbf{P}^1(K)$.

Let f of degree ≥ 1 have the décomposition $f = f_1^{d_1} \dots f_s^{d_s}$, with $f_1, \dots, f_s$ distinct monic irreducible polynomials of degrees n_i. The degree of f is $n = \sum d_i n_i$.

Note that $\mathbf{P}^1(K)$ is the orbit of ∞ under $Sl(2,A)$. Hence we can identify $\mathbf{P}^1(K)$ with $Sl(2,A)/H$, where H is the group of the matrices of the form $\begin{pmatrix} a & b \\ 0 & a^{-1} \end{pmatrix}$, with $b \in A$ and $a \in A^*$. The set of cusps of the congruence subgroup $\Gamma(f)$ is then equal to the family of double cosets $\{g \in Sl(2,A) |\ g \equiv 1 \bmod f\}\backslash Sl(2,A)/H$.
The map $Sl(2,A) \to Sl(2,A/Af)$ is surjective. Hence our set of cusps is equal to $Sl(2,A/Af)/H_1$, where H_1 is the image of H in $Sl(2,A/Af)$. The group H_1 is equal to the group of matrices

$$\{\begin{pmatrix} a & b \\ 0 & a^{-1} \end{pmatrix} \mid a \in \mathbf{F}_q^*, b \in A/Af\}.$$

It is known that $Sl(2,A/Af)$ is isomorphic to the product $\prod_{i=1}^{s} Sl(2,A/Af_i^{d_i})$. The number of elements of each factor is equal to $q^{3n_i d_i}(1-q^{-2n_i})$. Counting leads to the following formula for the number of cusps of $\Gamma(f)$:

$$\frac{q^{2n}}{q-1}\prod_{i=1}^{s}(1-q^{-2n_i}).$$

The complete Drinfeld modular curve $Gl(A^2)\backslash(\Omega \times Gl(f^{-1}A^2/A^2)\ )$ has $\#(A/fA)^*/\mathbf{F}_q^*$ connected components, each one isomorphic to $\Gamma(f)\backslash\Omega$. And thus the number of cusps of this curve is

$$\frac{\prod_{i=1}^{s}(q^{n_i}-1)q^{n_i(d_i-1)}}{q-1}\cdot\frac{q^{2n}}{q-1}\prod_{i=1}^{s}(1-q^{-2n_i}).$$

4.2 Elliptic points and a formula for the genus

The covering

$$X(f) := \overline{Gl(A^2,(f))\backslash\Omega} \to X(1) := \overline{PGl(2,A)\backslash\Omega} \cong \mathbf{P}^1_{K_\infty}$$

can be used to find the formula for the genus of the curve $\overline{Gl(A^2,(f))\backslash\Omega}$. The degree d of the morphism is equal to the order of the group $Gl(A^2,(f))\backslash PGl(A^2)$, which is also equal to the order of the group $Sl(2,A/(f))$. This gives a formula for the degree

$$d = q^{3n}\prod_{i=1}^{s}(1-q^{-2n_i}).$$

Above the point ∞ of $X(1)$ there are the $\frac{d}{e}$ cusps, where e denotes the ramification of any cusp of $X(f)$ above $\infty \in X(1)$. The stabilizer of the cusp ∞ of $X(f)$ has the form

$$\{\begin{pmatrix} 1 & b \\ 0 & d \end{pmatrix} | d \in \mathbf{F}_q^*,\ b \in A/(f)\}$$

This implies that $e = (q-1)q^n$ and that the number of cusps of $X(f)$ is equal to $\frac{d}{(q-1)q^n}$. We have already found this number.

The contribution B of the cusp ∞ of $X(f)$ for the Riemann-Hurwitz-Zeuthen formula can be found by considering the morphism $\mathbf{P}^1_{K_\infty} \to \mathbf{P}^1_{K_\infty}$ obtained by dividing the first projective line by a group of the form

$$\{\begin{pmatrix} 1 & b \\ 0 & d \end{pmatrix} | d \in \mathbf{F}_q^*,\ b \in W\},$$

where $W \subset K_\infty$ is a linear subspace over $\mathbf{F}_q$ of dimension n. In this situation, ∞ is the only point above ∞ and gives a contribution B for the Riemann-Hurwitz-Zeuthen formula. The points of W are ramified points of order $q-1$. There is no further ramification. This leads to $-2 = -2(q-1)q^n + q^n(q-2) + B$ and thus $B = q^{n+1} - 2$.

The fixed points in Ω for the group $PGl(2, A)$ are called the "elliptic points" of Ω (we deviate here somewhat from the classical use of the expression "elliptic point"). Take a point $b \in \mathbf{F}_{q^2} \setminus \mathbf{F}_q$. This is an elliptic point. Its stabilizer is a cyclic group of order $q+1$. Any other elliptic point of Ω is $PGl(2, A)$-equivalent to b. Let ξ denote the image of b in the curve $X(1)$. Then the points of $X(f)$ which lie above $\xi \in X(1)$ are ramified with order $q+1$. Their number is clearly $\frac{d}{q+1}$ and their contribution in the Riemann-Hurwitz-Zeuthen formula is $\frac{dq}{q+1}$.

Let g denote the genus of $X(f)$. According to the Riemann-Hurwitz-Zeuthen formula one has $2g - 2 = -2d + \frac{dq}{q+1} + \frac{d(q^{n+1}-2)}{q^n(q-1)}$. This gives the formula

$$g = 1 + \frac{q^{2n}(q^n - q - 1)}{q^2 - 1} \prod_{i=1}^{s} (1 - q^{-2n_i}).$$

4.3 The Drinfeld module with level f at a cusp

The universal Drinfeld module with level f-structure above the space $Gl(A^2)\backslash(\Omega \times Gl(f^{-1}A^2/A^2))$, constructed in theorem 2.2, extends in a certain way to the cusps. In this section we study this extension. The cusps are equivalent under the automorphisms of this curve. Hence it suffices to consider the cusp at ∞ of the curve $X(f)$, which is the compactification of $Gl(A^2, (f))\backslash\Omega$.

Our first concern is to find out in what sense the Drinfeld structure ϕ extends at ∞. Let **cusps** denote the set of the cusps of the curve $X(f)$. The construction of theorem 2.2 produces a line bundle, which we will denote by L, on $X(f)\backslash$**cusps** and a homomorphism $\phi : A \to End(L)$. We want to extend this line bundle L and the homomorphism ϕ to all of $X(f)$. It suffices to describe this extension at the cusp ∞.

Let us first consider the case $f = 1$.

As we know, the cusp has the following description. Consider the horicycle neighbourhood at ∞, i.e. $U = \{\omega | \, d(\omega) \geq R\} \subset \Omega$ (for some $R > 1$), and put

$$H := \{\omega \mapsto \alpha\omega + \beta | \, \alpha \in \mathbf{F}_q^*, \, \beta \in A\}.$$

The space $H \backslash U \subset X(1)$ can be identified with $\{c \in \mathbf{C} | 0 < |c| \leq r\}$ for some $r > 0$ be means of the local parameter t at $\infty \in X(1)$. The local parameter t has the form

$$t(\omega) = \{\omega \prod_{a \in A, a \neq 0} (1 - \frac{\omega}{a})\}^{1-q}.$$

The action of H on the trivial line bundle $U \times \mathbf{C}$ is given as $h(\omega, z) = (h(\omega), z)$. Hence the restriction of L to $H \backslash U$ coincides with the structure sheaf. The extension of L to $\{c \in \mathbf{C} | \, |c| \leq r\}$ is chosen to be again the structure sheaf.

In order to find the extension of ϕ above $\{c \in \mathbf{C} | \, |c| \leq r\}$ we consider again the function

$$e(\omega, z) = z \prod_{(a,b) \neq (0,0)} (1 - \frac{z}{a\omega + b}) = z(1 - \sum_{m \geq 1} s_m z^{m(q-1)})^{-1},$$

where $s_m = \sum_{(a,b) \neq (0,0)} \frac{1}{(a\omega + b)^{m(q-1)}}$.

This function and the s_m are invariant under the group H. Hence s_m is a convergent Laurent series in t. In fact s_m is holomorphic at $t = 0$ and $s_m(t = 0) \neq 0$. Indeed, for $(a, b) \in A^2$, $(a, b) \neq (0, 0)$, one has $|a\omega + b| \geq |a|$ if $a \neq 0$ and $|a\omega + b| \geq |b|$ if $a = 0$. Hence, each term $\frac{1}{(a\omega+b)^{m(q-1)}}$ of s_m has absolute value ≤ 1. As a consequence s_m is bounded on U and has as function of t no pole at $t = 0$. The limit of $s_m(\omega)$ for $\omega \to \infty$, $\omega \in U$ is the expression $\sum_{b \in A, \, b \neq 0} \frac{1}{b^{m(q-1)}}$. This is the value of s_m at $t = 0$. One easily sees that $|s_m(t = 0) + 1| < 1$.

Also the function $e(\omega, z)$ can therefore be seen as a holomorphic function of the two variables t, z. The value of e for $t = 0$ is equal to $z \prod_{b \in A, b \neq 0} (1 - \frac{z}{b})$. This is the "Carlitz exponential function" of the rank one lattice $A \subset \mathbf{C}$.
The Drinfeld module of rank two above $H \backslash U$ is determined by $\phi_T = T + c_1\tau + c_2\tau^2$ with $c_1 = (T^q - T)s_1$ and some c_2. We know that the j-function $j = \frac{c_1^{q+1}}{c_2}$ has a pole of order 1 at $t = 0$. Thus c_2 has a zero of order 1 at $t = 0$. Further c_1

is not zero at $t = 0$. We conclude that $\phi : A \to End(L)$ has a natural extension at the cusp ∞ and that $\phi(t = 0)$ is the rank one Drinfeld module corresponding to the lattice $A \subset \mathbf{C}$.

The case where the level is given by a monic polynomial $f \in A$ of degree $n \geq 1$.

The group H_f is defined as $\{\omega \mapsto \omega + b | \, b \in Af\}$. The space $H_f \backslash U$ is also isomorphic to $\{c \in \mathbf{C} | \, 0 < |c| \leq r\}$ (for some $r > 0$). This isomorphism is given by the local parameter t_f of $X(f)$ at the cusp ∞. This local parameter is given by the formula

$$t_f = (\omega \prod_{b \in (f), b \neq 0} (1 - \frac{\omega}{b}))^{-1}.$$

Again c_1, c_2 are holomorphic functions of t_f, c_1 is not zero for $t_f = 0$. Further c_2 has a zero of order $(q-1)q^n$ at $t_f = 0$ since the ramfication index of $X(f) \to X(1)$ at the point $\infty \in X(f)$ is equal to $(q-1)q^n$. We note that this is also the index of H_f in H. We conclude that the Drinfeld module of rank two above $X(f) \setminus \{\mathbf{cusps}\}$ has an extension to all of $X(f)$ as a "stable Drinfeld module". The last expression means that at any cusp the rank of the Drinfeld module is equal to 1.

We now investigate the level f-structure. Above U the kernel of ϕ_f is isomorphic to the constant sheaf $f^{-1}A^2/A^2$. Its generators as $A/(f)$-module are the functions $e(\omega, \frac{1}{f})$ and $e(\omega, \frac{\omega}{f})$. Both functions are invariant under H_f. For the first function this is clear. For the second one, this follows from:

$$e(\omega + bf, \frac{\omega + bf}{f}) = e(\omega, \frac{\omega + bf}{f}) = e(\omega, \frac{\omega}{f}) + e(\omega, b) = e(\omega, \frac{\omega}{f}).$$

Hence both functions are defined on $H_f \backslash U$ and are Laurent series in t_f. The first function $e(\omega, \frac{1}{f})$ can be written in the form

$$\frac{1}{f}(1 - \sum_{m \geq 1} s_m f^{m(1-q)})^{-1}.$$

This is clearly holomorphic and nonzero at $t_f = 0$. This function generates a constant subsheaf (isomorphic to $f^{-1}A/A$ and moreover a direct summand) of the constant sheaf $ker\phi_f$ above $H_f \backslash U$. This subsheaf extends as a constant subsheaf (isomorphic to $f^{-1}A/A$) of the structure sheaf on $\{c \in \mathbf{C} | \, |c| \leq r\}$. (i.e. it extends at $t_f = 0$).

The second generator of the level structure $e(\omega, \frac{\omega}{f})$ can be written as

$$\frac{\omega}{f}\prod(1-\frac{\omega/f}{a\omega+b}).$$

On the subset of U given by $d(\omega)=|\omega|\geq R$, the absolute value of the expression above can be given as

$$\frac{|\omega|}{|f|}\prod_{b\in A,\, b\neq 0,\, |bf|<|\omega|}|\frac{\omega}{bf}|.$$

For ω as above, the function $\frac{1}{f}t_f(\omega)^{-1}=\frac{\omega}{f}\prod_{b\in A,\, b\neq 0}(1-\frac{\omega}{bf})$ has the same absolute value. It follows that $e(\omega,\frac{\omega}{f})$ has a pole of order 1 at $t_f=0$.

We investigate now the behaviour of the more general expression $e(\omega, \frac{c\omega+d}{f})$, with c, d polynomials of degree $< n$. (Again n denotes the degree of f). This expression is a $\mathbf{F}_q$-linear combination of the terms $v_i := e(\omega, \frac{T^i\omega}{f})$ and $w_j := e(\omega, \frac{T^j}{f})$ for $0 \leq i, j < n$. We have shown that the order of v_0 at the cusp is -1. Further we know that $\phi_T = T + c_1\tau + c_2\tau^2$ with c_1 invertible at the cusp and the order of c_2 at the cusp is $q^n(q-1)$. It follows from this that the order of v_i at the cusp is $-q^{i-1}$ and that the w_j have no poles at the cusp. We conclude that the order of $e(\omega, \frac{c\omega+d}{f})$ at the cusp is ≥ 0 if and only if $c = 0$ and that for $c \neq 0$ of degree i this order is $-q^{i-1}$.

What we have found is a rank two Drinfeld module over $K_\infty[[t_f]]$ given by $\phi_T = T + c_1\tau + c_2\tau^2$ with the following properties:

1. $c_1, c_2 \in K_\infty[[t_f]]$ have orders 0 and $q^n(q-1)$
2. $ker(\phi_f, K_\infty((t_f)))$ is isomorphic to the A-module $(f^{-1}A/A)^2$.
3. $ker(\phi_f, K_\infty[[t_f]])$ is isomorphic to the A-module $f^{-1}A/A$.

We will call this structure a *Tate-Drinfeld module* (rank two and level f) over the field K_∞. In lecture 9, a "universal" Tate-Drinfeld module (rank two and level f) will be constructed.

4.4 Geometric interpretation at a cusp

In the following the valuation of the field K_∞ will play no role. The field $L := K_\infty((t_f))$ is considered as a discrete valued field. Its valuation ring is $R := K_\infty[[t_f]]$.

The Tate-Drinfeld module above will be given a *geometric interpretation*. We will describe this by using the rigid analytic terminology. The algebraic counterpart will be explained in lecture 9.
The affine line over L is replaced by the projective line $\mathbf{P}^1_L$. The ring A acts, via ϕ, on the affine part of this projective line. We are interested in the set $F := ker(\phi_f, L) \cup \{\infty\} \subset \mathbf{P}^1_L$. The projective line $\mathbf{P}^1_L$ is considered as a rigid analytic space over the discrete valued field L. The obvious reduction (see lecture 6) of this space is the projective line over the residue field, i.e. K_∞, of L. On this reduction ϕ has rank one and many elements of F are mapped to the point ∞ of $\mathbf{P}^1_{K_\infty}$. We would like to find another reduction of the analytic space $\mathbf{P}^1_L$ such that the images of the points of F are distinct. We will briefly explain the concept of reductions.

4.4.1 Reductions

The canonical reduction of an affinoid space $X = max(B)$, where B is an affinoid algebra over a complete valued field L is the affine space $\bar{X} = max(\bar{B})$ over the residue field l of L. We recall that $\bar{B} = B^o/B^{oo}$ with $B^o = \{b \in B| \, ||b||_{sp} \leq 1\}$ and $B^{oo} = \{b \in B| \, ||b||_{sp} < 1\}$. Let $c_X : X \to \bar{X}$ denote the (canonical) reduction map.
For a general rigid space X we call a covering $\mathcal{U} = \{U_i\}_{i \in I}$ *pure (or formal)* if the following holds:

1. Each U_i is affinoid and the covering $\mathcal{U}$ is admissible.

2. For each pair U_i, U_j such that $U_i \cap U_j \neq \emptyset$ there is a Zariski open subset $V_{i,j} \subset \bar{U}_i$ such that $U_i \cap U_j$ is the preimage of $V_{i,j}$ under the canonical reduction map $c_{U_i} : U_i \to \bar{U}_i$.

The reduction of X with respect to this pure covering $\mathcal{U}$ is denoted by $\overline{(X,\mathcal{U})}$. It is defined as the gluing of the affine space $\bar{U}_i$ over the open subsets $V_{i,j}$. There is a reduction map $r_\mathcal{U} : X \to \overline{(X,\mathcal{U})}$, which is obtained by gluing the reduction maps c_{U_i}.

The "obvious" reduction of $\mathbf{P}^1_L$ is given by

$$\mathcal{U} = \{\{z \in \mathbf{P}^1_L| \, |z| \leq 1\}, \{z \in \mathbf{P}^1_L| \, |z| \geq 1\}\}.$$

The space $\overline{(\mathbf{P}^1_L,\mathcal{U})}$ is equal to the projective line over the residue field l. The reduction map $r_\mathcal{U} : \mathbf{P}^1_L \to \mathbf{P}^1_l$ is given by $r_\mathcal{U} : (x_0 : x_1) \mapsto (\bar{x}_0 : \bar{x}_1)$, where the x_0, x_1 are chosen such that $\max(|x_0|, |x_1|) = 1$.

For certain pure finite affinoid coverings $\mathcal{U}$ of $\mathbf{P}^1_L$ one has a corresponding analytic reduction $\overline{(\mathbf{P}^1_L, \mathcal{U})}$ which is a tree of projective lines over the residue field of L. By induction on the cardinality of F one can show the following statement:

Lemma 4.1 *For any finite set F (with cardinality ≥ 3) of points in $\mathbf{P}^1(L)$ there is a unique* reduction map $r_{\mathcal{U}}$ *(the covering $\mathcal{U}$ is not quite unique) such that:*
(1) The space $\overline{(\mathbf{P}^1_L, \mathcal{U})}$ is a tree of lines over the residue field of L.
(2) $r_{\mathcal{U}}$ is injective on F and $r_{\mathcal{U}}(F)$ contains no double point.
(3) On every irreducible component of $\overline{(\mathbf{P}^1_L, \mathcal{U})}$ there are at least three special points. Here special point means either a double point or an element in $r_{\mathcal{U}}(F)$.

In the terminology of [G-H-vdP], $\overline{(\mathbf{P}^1_L, \mathcal{U})}$ with the set of points $r_{\mathcal{U}}(F)$ is a stable $\#F$-pointed tree.

The algebraic version of this lemma reads as follows:
There is a model M over the valuation ring R of L such that

1. M is projective and flat over R.
2. $M \otimes L \cong \mathbf{P}^1_L$.
3. The special fibre $M_s = M \otimes l$ is a tree of lines over the residue field l of L.
4. The map $r : \mathbf{P}^1_L(L) = M(R) \to M_s(l)$ is the reduction map.
5. r is injective on F and $r(F)$ contains no double point of M_s.
6. On every irreducible component of M_s there are at least three special points (i.e. double points or points of $r(F)$).

The lemma is applied to the field $\mathbf{F}_q(x)$ with as valuation ring the local ring at $x = 0$, i.e. the local ring $\mathbf{F}_q[x]_{(x)}$. For the set F we choose

$$\{\mathbf{F}_q + \mathbf{F}_q x^{-1} + \mathbf{F}_q x^{-q} + \ldots + \mathbf{F}_q x^{-q^{n-1}}\} \cup \{\infty\}.$$

Let $RT = RT(q,n)$ denote the reduction given by the lemma. We will describe RT in more detail.

Each projective line in RT is identified with $\mathbf{P}^1_{\mathbf{F}_q}$ in a suitable way. The intersection graph $\mathcal{G}$ of RT is by definition the graph with vertices the irreducible components of RT and with edges the double points of RT.
There is one vertex of $\mathcal{G}$, the "root" R, which has q edges. This root is the irreducible component on which the image of ∞ lies. The image of $\infty \in S$

is the infinite point of R. The q edges correspond to the intersection of the line R with the other irreducible components of RT. The points of intersection on R are the points $\mathbf{F}_q \subset R$. A neighbour of R can be labelled as V_{a_1}, with $a_1 \in \mathbf{F}_q$, according to the point of intersection R. The point of intersection of V_{a_1} with R is, as point on V_{a_1}, the infinite point. Each V_{a_1} has q new neighbours corresponding to the new double points which form the set $\mathbf{F}_q \subset V_{a_1}$. The new neighbours of V_{a_1} can be labelled by V_{a_1,a_2} with $a_2 \in \mathbf{F}_q$, according to their point of intersection on V_{a_1}. This point, seen as a point of V_{a_1,a_2}, is the infinite point. And so on. We stop at the nth level. The following picture illustrates the situation for $q = 2$ and $n = 2$.

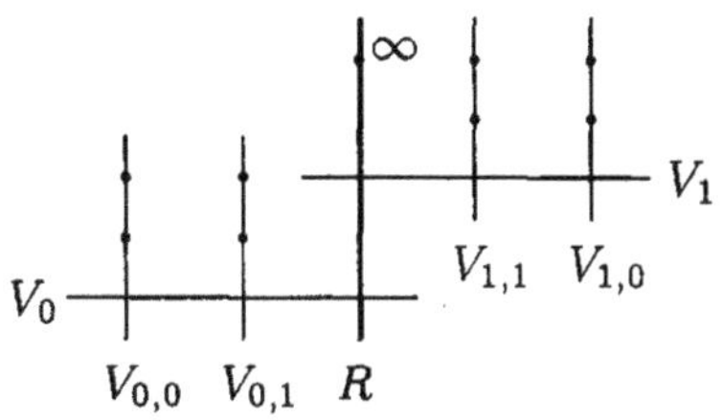

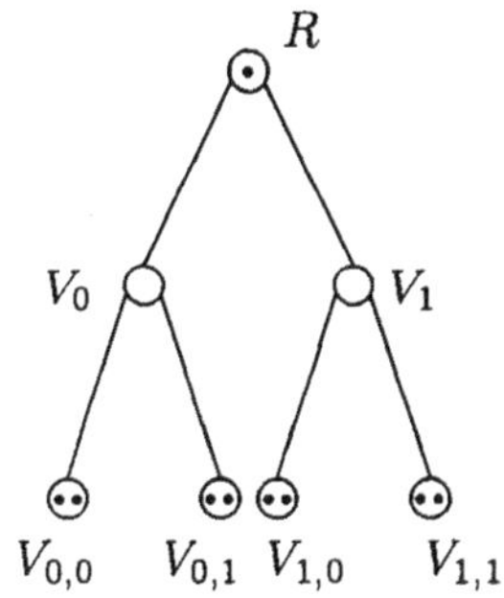

Now we introduce the following definition. Let k be any field containing $\mathbf{F}_q$. *The q, n-rooted tree over k* is the k-scheme $RT(q,n) \otimes_{\mathbf{F}_q} k$.

4.4.2 Structure on the reduction B

We apply lemma 4.1 to the subset $F = ker(\phi_f, L) \cup \{\infty\}$ of $\mathbf{P}^1_L(L)$. Let us write $r : \mathbf{P}^1_L \to B$ for the corresponding reduction. The set F is rather special, apart from ∞ it consists of

$$ker(\phi_f, R) + \mathbf{F}_q v + \mathbf{F}_q \phi_T(v) + \ldots + \mathbf{F}_q (\phi_T)^{n-1} v,$$

with $v = e(\omega, \frac{\omega}{f})$. The order of $(\phi_T)^i v$ is equal to $-q^i$ for $i = 0, \ldots, n-1$. It is a combinatorial exercise to show that the data of F imply that B is isomorphic to $RT(q,n)_s \otimes K_\infty$, the q, n-rooted tree over K_∞.

The point ∞ is mapped (under r) to (the projective line corresponding with) the root R. The other points of F are mapped to the end vertices $V_{a_1,\ldots,a_n}$. One can choose the labelling of the vertices of $\mathcal{G}$ such that the points

$$ker(\phi_f, R) + a_n v + a_{n-1}\phi_T(v) + \ldots + a_1 (\phi_T)^{n-1} v$$

are mapped to the (projective line corresponding to the) vertex $V_{a_1,\ldots,a_n}$.

The reduction B has much more structure. Let B^* denote the open subset of B obtained be deleting all the lines on which no points of $ker(\phi_f, L)$ lie. B^* is the disjoint union of affine lines over K_∞ (corresponding to the vertices $V_{a_1,\ldots,a_n}$). The line which contains the images of $ker(\phi_f, R)$ is denoted by B^{**} (corresponding to the vertex $V_{0,\ldots,0}$).

The preimage $r^{-1}(B^*)$ is the affinoid subset

$$\{z \in L|\ |z| \leq 1\} + \mathbf{F}_q v + \mathbf{F}_q \phi_T(v) + \ldots + \mathbf{F}_q (\phi_T)^{n-1} v$$

of $\mathbf{P}^1_L$. Further $r^{-1}(B^{**})$ is equal to $\{z \in L|\ |z| \leq 1\}$. In fact $r^{-1}(B^*)$ and $r^{-1}(B^{**})$ are affinoid subgroups of $\mathbf{A}^1_L$. Hence B^* and B^{**} are algebraic groups over the residue field K_∞. The second group is the additive group over K_∞ and has a Drinfeld structure of rank one induced by ϕ. The algebraic group B^* is isomorphic to $B^{**} \times f^{-1}A/A$ and has therefore an A-action induced by the Drinfeld structure on B^{**} and the obvious action on $f^{-1}A/A$.
We note that the A-module of the f-torsion points of B^* is equal to the image of $ker(\phi_f, L)$. We note further that the group law $B^* \times B^* \to B^*$ extends in a unique way to an algebraic group action $B^* \times B \to B$. Indeed, the group law $r^{-1}(B^*) \times r^{-1}(B^*) \to r^{-1}(B^*)$ extends to a group action $r^{-1}(B^*) \times \mathbf{P}^1_L \to \mathbf{P}^1_L$, which respects the pure covering and induces the required algebraic group action on the reduction.

4.4.3 Generalized Drinfeld modules over a field

The calculations above concerning the reduction B lead to a definition. Let k be any field which has an A-algebra structure. Fix a level (f) of degree n. A *generalized Drinfeld module* (of rank two and with level (f)) is a special q,n-rooted tree N over k with the additional structure:

(1) Let N^* be the open subset obtained by deleting all the lines of N which are not corresponding to end vertices. Then N^* is given a structure of an algebraic group over k for which it is isomorphic to the commutative algebraic group $\mathbf{G}_{a,k} \times f^{-1}A/A$.

(2) The action $N^* \times N^* \to N^*$ extends to an algebraic action $N^* \times N \to N$.

(3) The component N^{**} of the identity of N^* has the structure of a rank one Drinfeld module $\psi : A \to End_{\mathbf{F}_q}(N^{**})$. The kernel of ψ_f acting upon N^{**} is isomorphic to $f^{-1}A/A$.

Let N_1, N_2 be two generalized Drinfeld modules (of rank two and with level (f)). An *isomorphism* $h : N_1 \to N_2$ is an isomorphism of schemes over k such that $h : N_1^* \to N_2^*$ is an isomorphism of algebraic groups and commutes with the A-action.

Suppose that the field k is separably algebraically closed. Then any two generalized Drinfeld modules (rank two and level (f)) are isomorphic. Let N denote the generalized Drinfeld module. Then $ker(f\cdot, N^*)$ is isomorphic to the A-module $(f^{-1}A/A)^2$.

It can be seen that the automorphisms h of N are parametrized by the matrices $\begin{pmatrix} a & b \\ 0 & d \end{pmatrix}$ with $a, d \in \mathbf{F}_q^*$ and $b \in A/fA$. For such a matrix the corresponding h on $N^* \cong k \times f^{-1}A/A$ is given by the formula

$$h(z, \frac{g}{f} \bmod A) = (az + \lambda(\frac{bg}{f}), \frac{dg}{f} \bmod A),$$

where $\lambda : f^{-1}A/A \to ker(f\cdot, N^{**})$ is a chosen isomorphism of A-modules. This action h is extended to an isomorphism of the k-scheme N.

4.4.4 The universal generalized Drinfeld module

We will briefly describe this universal object which lives above the compactification $Z := \hat{M}^2(f) \otimes K_\infty$ of the affine modular curve $M^2(f) \otimes K_\infty$ over K_∞. We have already constructed the universal Drinfeld module of rank two and with level f. (See 2.2 and 2.5) The line bundle L above $M^2(f) \otimes K_\infty$ is extended to a line bundle, also called L, on the compactification $\hat{M}^2(f) \otimes K_\infty$. Using "GAGA" this line bundle is algebraic. We will continue the description in the category of algebraic varieties over K_∞.

The line bundle $L \to Z$ is made into a bundle of projective lines, say $PL \to Z$, by adding an infinite section. The fibre of PL above an ordinary point of Z is a projective line with a Drinfeld structure of rank two and level f on the corresponding affine line. The fibre of PL above a cusp does not give the correct structure. After "blowing up" suitable points of PL above the cusps one obtains a new bundle $\tilde{PL} \to Z$. The fibre above any cusp is isomorphic with the reduction B with all its structure, in other words it is a generalized Drinfeld module (rank two, level f) over the field K_∞. The ring A acts on a big open part $\tilde{PL}^*$ of $\tilde{PL}$ (i.e. the complement of the infinite section and for every cusp the irreducible components which are no "end lines").
The isomorphism of sheaves of A-modules $(f^{-1}A/A)^2 \to ker(\phi_f, L)$ on the

open subset $M^2(f) \otimes K_\infty$ of Z is extended to an isomorphism $(f^{-1}A/A)^2 \to ker(f\cdot, \tilde{PL}^*)$. The *universal Drinfeld module of rank two and with level f* is the object $\tilde{PL} \to Z$ with the additional data:
(1) A structure of algebraic group on $\tilde{PL}^*$
(2) The group law $\tilde{PL}^* \times \tilde{PL}^* \to \tilde{PL}^*$ extends to an algebraic action $\tilde{PL}^* \times \tilde{PL} \to \tilde{PL}$.
(3) An action of A on $\tilde{PL}^*$. Above an ordinary point of Z this comes from a Drinfeld module of rank two. Above a cusp the fibre is isomorphic to $\mathbf{G}_{a,K_\infty} \times f^{-1}A/A$, where the first factor has a Drinfeld module structure of rank one.
(4) An isomorphism of the constant sheaf $(f^{-1}A/A)^2$ with $ker(f\cdot, \tilde{PL}^*)$.

For a suitable definition of a generalized Drinfeld module (rank two and level f) above any K_∞-scheme, one can show that the universal Drinfeld module described above is indeed universal.

4.4.5 Remarks

One can describe the line bundle L on $X(f)$ (which is a component of $\hat{M}^2(f) \otimes K_\infty$) in terms of the cusps and the "elliptic points" of $X(f)$. The set of elliptic points, in our terminology this is the preimage of the point $\xi \in X(1)$ under the canonical map $X(f) \to X(1)$, will be denoted by **elliptic**. The term $\phi_T \in End(L)$ has the form $T + c_1\tau + c_2\tau^2$ with c_1 a section of L^{1-q} and c_2 a section of L^{1-q^2}. The section c_1 has zeroes of order one at the set **elliptic**. Hence $L^{1-q} \cong O(\textbf{elliptic})$.
The section c_2 has zeroes of order $q^n(q-1)$ at the set **cusps**. Hence $L^{1-q^2} \cong O(\textbf{cusps})^{q^n(q-1)}$. We may conclude from this that $L^{1+q} \cong O(\textbf{cusps})^{-q^n}$. (There is indeed the possibility of replacing L by $L \otimes M$ where M is a line bundle of order $q-1$. This seems to correspond with an "automorphy factor" of the form det^m). Hence $L^2 \cong O(\textbf{elliptic})O(\textbf{cusps})^{-q^n}$.

5 Modular forms

A *modular function of weight k and type m* for a group Γ of finite index in $Gl(Y)$ is a meromorphic function f on Ω satisfying

$$f(\frac{az+b}{cz+d}) = (cz+d)^k(ad-bc)^{-m}f(z),$$

for every $\begin{pmatrix} a & b \\ c & d \end{pmatrix} \in \Gamma$. For every point $c \in \mathbf{P}^1(K)$ one considers the subgroup Γ_c of Γ consisting of the matrices with a double eigenvalue 1 and having c as

fixed point.
The subgroup Γ_∞ can be represented as $\{z \mapsto z + h | \, h \in H\}$, where $H \subset K$ is a discrete subgroup such that $H \backslash K_\infty$ is compact. Let, as before, $e_H : \mathbf{C} \to \mathbf{C}$ be given by the formula

$$e_H(z) = z \prod_{h \in H, \, h \neq 0} (1 - \frac{z}{h}).$$

This function identifies the analytic space $H \backslash \mathbf{C}$ with $\mathbf{C}$. Moreover for suitable R one has that $e_H^{-1}(R < |z| < \infty) \subset \Omega$. The restriction of the modular function f to $e_H^{-1}(R < |z| < \infty)$ is H-invariant and is therefore equal to a Laurent series in the function $t = e_H^{-1}$, i.e. equal to a convergent infinte sum $\sum_{n=-\infty}^{+\infty} a_n t^n$. The modular function is called holomorphic (meromorphic) at the cusp ∞ if $a_n = 0$ for $n < 0$ ($a_n = 0$ for $n << 0$ respectively).

For any other cusp of Γ represented by a $c \in K$ one chooses a $g \in Gl(2, K)$ with $g(\infty) = c$. The modular function f is called holomorphic (meromorphic) at the cusp c if the function $z \mapsto f(g(z))$ is holomorphic (meromorphic) at the cusp ∞ for the group $g \Gamma g^{-1}$.
The modular function f is called a *modular form* if moreover, f is holomorphic on Ω and is holomorphic at the cusps of Γ.

Examples
A meromorphic differential form on Ω can be written in the form $f(z)dz$. For

$$g = \begin{pmatrix} a & b \\ c & d \end{pmatrix} \in Gl(2, K_\infty) \text{ one has } g(f(z)dz) = f(g(z)) \frac{ad - bc}{(cz + d)^2} dz.$$

Let Γ be a subgroup of finite index in $Gl(Y)$. The differential $f(z)dz$ is Γ-invariant if and only if f is a modular function for Γ of weight 2 and type 1. In this case $f(z)dz$ represents a meromorphic differential on $\Gamma \backslash \Omega$. If f is moreover meromorphic at the cusps then $f(z)dz$ extends to an analytic meromorphic differential on the projective curve $\overline{\Gamma \backslash \Omega}$. By GAGA such a differential is also an algebraic meromorphic differential.
Suppose now that $f(z)$ is a modular form of weight 2 and type 1. Then $f(z)dz$ is a differential form on $\overline{\Gamma \backslash \Omega}$ which has at most poles at the cusps of a prescribed order. The prescription of this order depends on the choice of $\Gamma \subset Gl(Y)$.

A differential form of degree k on Ω is an expression $f(z)(dz)^k$. This expression is Γ-invariant if and only if f is a modular function of weight $2k$ and type k. An invariant $f(z)(dz)^k$, with f meromorphic on Ω and at the cusps, is a meromorphic differential of degree k on $\overline{\Gamma \backslash \Omega}$.
The sheaf of the holomorphic differential forms on $\overline{\Gamma \backslash \Omega}$ is a line bundle $\mathcal{L}$. The

sheaf of the holomorphic forms of degree k is equal to the line bundle $\mathcal{L}^k$. In a similar way one can construct, for any k, m, a line bundle, which we denote for the moment by $\mathcal{L}_{k,m}$, on $\overline{\Gamma\backslash\Omega}$ such that the space of the modular forms of weight k and type m are sections of this line bundle with prescribed poles at the cusps.

We cite from [G] more precise information:

Let $\Gamma = Gl(Y)$. The line bundle of the differentials on $\overline{\Gamma\backslash\Omega}$ is denoted by $\mathcal{L}$. The space of the modular forms of weight $2k$ and type k is equal to the sections of $\mathcal{L}^k$ with poles at the cusps of order $\leq \frac{kq}{q-1}$ and at the elliptic points of order $\leq \frac{kq}{q+1}$.

Let $\Gamma = \Gamma_Y(\underline{n})$ with $\underline{n}$ a proper ideal of A. The line bundle of the differentials on $\overline{\Gamma\backslash\Omega}$ is again denoted by $\mathcal{L}$. The space of the modular forms of weight $2k$ and type k is equal to the sections of $\mathcal{L}^k$ with poles at the cusps of order $\leq 2k$.

References

[D-R] P. Deligne et M. Rapoport - *Les schémas de modules de courbes elliptiques* - In Modular Functions of One Variable 2, Lect. Notes in Math. 349, 1973, Springer Verlag.

[G] E.U. Gekeler - *Drinfeld modular curves* - Lect. Notes in Math. 1231, 1986, Springer Verlag.

[G-H-vdP] L. Gerritzen, F. Herrlich and M. van der Put - *Stable n-pointed trees of projective lines* - Proceedings KNAW series A 91 (2), p. 131-163, 1988.

[G-vdP] L. Gerritzen and M. van der Put - *Schottky groups and Mumford curves* - Lect. Notes in Math. 817, 1980, Springer Verlag.

[S] J-P. Serre - *Arbres, Amalgames, Sl_2* - Astérisque 46, 1977.

Lecture 9

Algebraic compactification and modular interpretation

Marius van der Put and Jaap Top
Department of Mathematics, University of Groningen
P.O.Box 800, 9700 AV Groningen, The Netherlands
mvdput@math.rug.nl and top@math.rug.nl

1 Introduction

In section 9 of Drinfeld's seminal 1974 paper [D], results concerning the algebraic compactification of the moduli space of rank 2 Drinfeld modules may be found. We will state part of this here and give some comments. Next, some proofs and supplements to his results will be given. In particular we will give a precise description how the universal Drinfeld module with a level structure degenerates over a cusp. This appears to be new; it is quite analogous to the generalized elliptic curves (n-gons) which Deligne and Rapoport use in [DR] although it seems technically more tricky. The results are strongly motivated by the descriptions found in the (rigid) analytic theory (Lecture 8). In our present setup, compared with Drinfeld's work matters will be simplified in two ways. Firstly, only Drinfeld $\mathbf{F}_q[T]$-modules are considered. Secondly, when dealing with level (f)-structures we restrict to $\mathbf{F}_q[T][\frac{1}{f}]$-schemes.

The first simplification allows one to avoid considerations concerning class groups (in our setup projective modules are free). An additional benefit is that Drinfeld $\mathbf{F}_q[T]$-modules are in a sense more explicit. Although the necessary modifications needed to cover the case of more general rings could be rather involved or messy, this first simplification does not seem to be a very essential one.

With the second restriction one avoids an explanation of "level structures in a characteristic dividing the level". Of course Drinfeld's idea of introducing what is now called a Drinfeld level structure has been very important; it had a great impact on reduction theory of classical modular curves (compare [KM], etc.). Nevertheless, even without this technical aspect it seems worth while to try to describe a compactification.

1.1 Drinfeld's results

Unlike in the other sections, here A will be general; so $C/\mathbf{F}_q$ is a smooth, complete absolutely irreducible curve, ∞ is a place of the function field $\mathbf{F}_q(C)$ and $A \subset \mathbf{F}_q(C)$ is the ring of functions which are regular outside ∞.

Suppose $S \to \mathrm{Spec}(A)$ is an A-scheme, $\mathcal{L}$ is a line bundle on S, and $\phi : A \to \mathrm{End}_S(\mathcal{L})$ is a rank 2 Drinfeld A-module over S.

For $I \subset A$ a nonzero ideal, Drinfeld defines a level I-structure on ϕ to be a homomorphism of A-modules

$$\varphi : (A/I)^2 \to \mathcal{L}(S),$$

such that $\sum_{f \in (A/I)^2} \varphi(f)$, seen as effective Cartier divisor over S, is the same as $\mathrm{Ker}(\phi(I))$.

We will not explain this; for more information about such a "Drinfeld level structure" one may consult [KM]. If the points in $\mathrm{Spec}(A)$ on which I vanishes are *not* in the image of the structure morphism $S \to \mathrm{Spec}(A)$, then a level I-structure is simply an isomorphism from the constant group scheme $\underline{(A/I)^2}_S$ over S to $\mathrm{Ker}(\phi(I))$.

Drinfeld's result is as follows.

Theorem 1.1 (Drinfeld) *Suppose that at least two prime ideals of A contain I.*

Then the functor assigning to an A-scheme S the set of isomorphism classes of rank 2 Drinfeld A-modules over S with level I-structure is representable by an A-scheme M_I^2.

This M_I^2 is of finite type over A, with fibres of dimension 1, *and $M_I^2 \to$ Spec(A) is smooth outside the primes containing I.*

Moreover, there exists a (unique upto isomorphism) scheme $\overline{M}_I^2$ of finite type over A, containing M_I^2 as an open and everywhere dense subscheme, such that (again) $\overline{M}_I^2 \to$ Spec(A) is smooth outside primes containing I, and $\overline{M}_I^2 \to$ Spec(A) is proper.

Remarks.

1. The assumption that I is contained in at least two different prime ideals is used by Drinfeld because he wants to construct $M_I^2, \overline{M}_I^2$ over *all* of $\mathrm{Spec}(A)$. The idea is that if $I \subset \underline{m} \cap \underline{n}$, then the level I-structure yields a level $\underline{m}$-structure as well, and this structure survives over $\underline{n} \in \mathrm{Spec}(A)$.

   A similar idea is well known in the theory of elliptic curves. Here one uses the 'Legendre family' given by $y^2 = x(x-1)(x-\lambda)$ to describe curves over rings in which 2 is invertible and the 'Hessian family' given by $x^3 + y^3 + z^3 - 3\mu xyz = 0$ in cases where 3 is invertible. After gluing over rings in which both are invertible (and if necessary taking quotients in order to get rid of level 2– and level 3–structures in the families above), one finally obtains moduli spaces over all of $\mathbb{Z}$; compare [KM].

2. If one inverts the primes containing I, a ring B is obtained. The same proof sketched by Drinfeld for the theorem above shows, that whenever I is such that Drinfeld modules with level I-structure have no automorphisms, the restriction of our functor to B-schemes is representable. The corresponding moduli space and compactification are again denoted $M_I^2, \overline{M}_I^2$, respectively. These are smooth schemes over $\mathrm{Spec}(B)$. In fact, they are simply the restrictions (base changes) to $\mathrm{Spec}(B) \subset \mathrm{Spec}(A)$ of the A-schemes which exist by Drinfeld's theorem.

1.2 Some examples

From now on we restrict ourselves to the case $A = \mathbf{F}_q[T]$. Motivated by the analytic compactification (Lecture 8), we try to classify rank 2 Drinfeld modules with a level (f)-structure and their degenerations, for $f = T$ and $f = T^2$.

1.2.1 Example: level T

The general way to describe a rank 2 Drinfeld A-module is by giving the endomorphism $\phi_T = T + A_1\tau + A_2\tau^2$. This makes sense over the ring $A[\frac{1}{T}, A_1, A_2, \frac{1}{A_2}]$ and as a line bundle on its spectrum one takes the trivial bundle (which corresponds to the structure sheaf). A level (T)-structure means that two solutions X_1, X_2 of $\phi_T X = 0$ are given, which are independent over $\mathbf{F}_q$. This independence says that no $a, b \in \mathbf{F}_q$ exist, not both 0, such that $aX_1 + bX_2 = 0$. In other words, it precisely states that the projective point $(X_1 : X_2) \in \mathbf{P}^1$ is not $\mathbf{F}_q$-rational, hence that $X_1^q X_2 - X_1 X_2^q \neq 0$. We can therefore construct rank 2 modules with level (T)-structure precisely over rings which are algebras over

$$\mathbf{F}_q[T, T^{-1}][A_1, A_2, X_1, X_2]_{X_1X_2^q - X_1^qX_2}/I,$$

where I is the ideal generated by the two elements $\phi_T(X_i) = TX_i + A_1X_i^q + A_2X_i^{q^2}$. This ring is in fact equal to

$$\mathbf{F}_q[T, T^{-1}][X_1, X_2]_{X_1X_2^q - X_1^qX_2},$$

and $A_1 = T\frac{X_1^{q^2}X_2 - X_1X_2^{q^2}}{(X_1X_2^q - X_1^qX_2)^q}$ and $A_2 = T\frac{X_1X_2^q - X_1^qX_2}{(X_1X_2^q - X_1^qX_2)^q}$. This ring is not quite what we want, since it classifies A-Drinfeld modules of rank 2 and level T rather than isomorphism classes of them. The multiplicative group $\mathbf{G}_m =$

$\mathrm{Spec}(\mathbf{F}_q[u,u^{-1}])$ acts on the ring above by $X_i \mapsto u \otimes X_i$. This precisely describes all possible isomorphisms. On the sections A_1, A_2 the action is given by $A_1 \mapsto u^{1-q} \otimes A_1$ and $A_2 \mapsto u^{1-q^2} \otimes A_2$. Dividing the algebra above by this action gives the affine ring of $\mathrm{Proj}(\mathbf{F}_q[T,T^{-1}][X_1,X_2]) \setminus V(X_1^q X_2 - X_1 X_2^q)$. In other words, $M^2_{(T)}$ is the projective line over $\mathbf{F}_q[T,T^{-1}]$ minus the $q+1$ points $\mathbf{P}^1(\mathbf{F}_q)$. The scheme $\overline{M}^2_{(T)}$ must be the projective line over $\mathbf{F}_q[T,T^{-1}]$. The cusps are the $q+1$ points above.

The formulas for ϕ_T and A_1, A_2 do not make sense at the cusps. We want to change the formulas in such a way that locally at a cusp ϕ looks like a "Tate-Drinfeld module" as described in Lecture 8. The obvious way to do this is to replace ϕ_T by $(X_1X_2^q - X_1^qX_2)^{-1}\phi_T(X_1X_2^q - X_1^qX_2)$. This yields

$$\phi_T = T + T\frac{X_1^{q^2}X_2 - X_1X_2^{q^2}}{(X_1X_2^q - X_1^qX_2)}\tau + T(X_1X_2^q - X_1^qX_2)^{q^2-q}\tau^2.$$

Outside the cusps the level structure is given by $\frac{X_i}{X_1X_2^q - X_1^qX_2}$ with $i = 1,2$. The line bundle L on the projective line over $\mathbf{F}_q[T,T^{-1}]$ must be such that A_1 is a section of L^{1-q} and A_2 is a section of L^{1-q^2} and the $\frac{X_i}{X_1X_2^q - X_1^qX_2}$ are meromorphic sections of L. The degree of L is $-q$. We take $L = O(-q)$. Clearly A_1 is a section of $O(-q)^{1-q} = O(q^2-q)$ and its set of zeroes is $\mathbf{F}_{q^2}\setminus\mathbf{F}_q$. Further A_2 is a section of $O(-q)^{1-q^2} = O(q^3-q)$ with zeroes of order q^2-q at the points $\mathbf{P}^1(\mathbf{F}_q)$.

The $\frac{X_i}{X_1X_2^q - X_1^qX_2}$ with $i = 1,2$ are meromorphic sections of $O(-q)$ with poles of order at most one at the points of $\mathbf{P}^1(\mathbf{F}_q)$. On the open part $X_1X_2^q - X_1^qX_2 \neq 0$ we have the level structure $\lambda : (A/TA)^2 \to O(-q)(\mathbf{P}^1_{\mathbf{F}_q[T,T^{-1}]})$ which maps (a,b) to the section $\frac{aX_1 + bX_2}{X_1X_2^q - X_1^qX_2}$. For every cusp there is a unique line in $(A/TA)^2$ for which λ extends at the cusp. For the cusp 0, i.e. the point $(1:0)$ this line is $0 \times A/TA$. Thus we have found a structure like a rank 2 Drinfeld module on the line bundle $O(-q)$ on $\mathbf{P}^1_{\mathbf{F}_q[T,T^{-1}]}$, which locally at the cusps $\mathbf{P}^1(\mathbf{F}_q)$ behaves exactly like what was found in the analytic theory (Lecture 8).

We now suppose that we have chosen coordinates in such a way, that $X = 0$ defines a cusp. Moreover, we will assume that the sections in $\mathrm{Ker}(\phi_T)$ which extend over this cusp, in fact reduce to $\mathbf{F}_q$. Since the reduced Drinfeld module is given by $\bar{\phi}_T = T + c\tau$ for some c, this implies $0 = \bar{\phi}_T(1) = T + c$, hence $c = -T$. Finally, we demand that the "original" module has the form $\phi_T = T - T\tau + XA_2\tau^2$. By Hensel's lemma, the section 1 lifts uniquely to a T-torsion section of ϕ over the completion of the local ring at the cusp. In the analytic theory and also above, we found an other generator of the torsion with a pole of order 1 at the cusp. If we want this other generator to be $1/X$, we obtain the Drinfeld module with

$$\phi_T = T - T\tau - TX^{q^2-q}(X^{q-1} - 1)\tau^2.$$

This is the "Tate-Drinfeld module of rank 2 and level T", as will be defined in Definition 2.4 below.

1.2.2 Example: level T^2

We now consider rank 2 Drinfeld A-modules with level T^2 structure. The analytic theory shows that the moduli spaces $M^2_{(T^2)} \otimes K_\infty$ is not connected (and has in fact q connected components). We will see this via an algebraic approach as well for $M^2_{(T^2)} \otimes K^{sep}$. We note that $M^2_{(T^2)} \otimes K$ *is* connected. This follows from a more general result proven in section 5.

We have $A/T^2A \cong \mathbf{F}_q[\epsilon]$, with $\epsilon^2 = 0$. The diagram

$$\begin{array}{ccc} \mathbf{F}_q[\epsilon] \times \mathbf{F}_q[\epsilon] & \xrightarrow{\sim} & \operatorname{Ker} \phi_{T^2} \\ \cup \ \downarrow \cdot\epsilon & & \cup \ \downarrow \phi_T \\ \epsilon \cdot (\mathbf{F}_q \times \mathbf{F}_q) & \xrightarrow{\sim} & \operatorname{Ker} \phi_T \\ \cong & & \\ \mathbf{F}_q \times \mathbf{F}_q & & \end{array}$$

shows that rank 2 Drinfeld modules with T^2-structure also carry a level T-structure. Hence it is natural to describe $M^2_{(T^2)}$ as a covering of $M^2_{(T)}$.

To this end, start with the ring $R = \mathbf{F}_q[T, T^{-1}][X_1, X_2]_{X_1X_2^q - X_1^qX_2}$ from the previous example. Recall that over R the rank 2 Drinfeld module ϕ is given by $\phi_T = T + A_1\tau + A_2\tau^2$, and A_1, A_2 are as follows. Write

$$\delta = X_1X_2^q - X_1^qX_2 = -X_2 \prod_{\alpha \in \mathbf{F}_q} (X_1 - \alpha X_2)$$

and

$$\delta_2 = X_1^{q^2}X_2 - X_1X_2^{q^2} = X_2 \prod_{\alpha \in \mathbf{F}_{q^2}} (X_1 - \alpha X_2).$$

Then $A_1 = T\frac{\delta_2}{\delta^q} = -T\prod_{\alpha \in \mathbf{F}_{q^2} \setminus \mathbf{F}_q}(X_1 - \alpha X_2) \cdot \frac{1}{\delta^{q-1}}$ and $A_2 = T \cdot \frac{1}{\delta^{q-1}}$.

To obtain a level T^2 structure one precisely needs Y_1, Y_2, possibly in some extension ring of R, which satisfy $\phi_T(Y_i) = X_i$. The fact that the X_i are independent over $\mathbf{F}_q$ then implies that the Y_i generate a free module over $\mathbf{F}_q[\epsilon]$ of rank 2. Clearing denominators one finds that Y_i is a zero of

$$\delta^{q-1}X_i - \delta^{q-1}TY_i + T \prod_{\alpha \in \mathbf{F}_{q^2} \setminus \mathbf{F}_q} (X_1 - \alpha X_2)Y_i^q - TY_i^{q^2}$$

for $i = 1, 2$. These two polynomials define a homogeneous ideal J in $R[Y_1, Y_2]$. So as before, rings over which a rank 2 A-Drinfeld module with level T^2-structure are defined, are precisely the $\tilde{R} = R[Y_1, Y_2]/J$-algebras.

An isomorphism of such a Drinfeld module is given by the usual $\mathbf{G}_m$-action on the X_i and Y_i, so we find a natural model for $\overline{M}^2_{(T^2)}$ in $\mathbf{P}^3_{A[\frac{1}{T}]}$, defined by the ideal J. The projection on the X_i coordinates is a well defined morphism, corresponding to the natural map from a level T^2 structure to the underlying level T-structure. From the analytic theory one knows that on each absolutely irreducible component, this projection has degree q^3, and there are q such components. This nicely agrees with the fact that visibly our projection map has total degree q^4.

The group $\mathbf{F}_q \times \mathbf{F}_q \times \mathbf{F}_q \times \mathbf{F}_q = \operatorname{Ker}\phi_T \times \operatorname{Ker}\phi_T$ acts on our model. This is obvious from the equations $\phi_T(Y_i) = X_i$ and $\phi_T(X_i) = 0$; the action is given by $Y_1 \mapsto Y_1 - aX_1 - bX_2$ and $Y_2 \mapsto Y_2 - cX_1 - dX_2$. The projection to the X-coordinates is precisely the quotient map for this action. This means that the map from $\overline{M}^2_{(T^2)}$ to $\overline{M}^2_{(T)}$ is unramified except possibly over the cusps.

Note that over the cusps of $\overline{M}^2_{(T)}$, the closure in $\mathbf{P}^3$ of our model is singular. Hence one cannot immediately read off the total number of cusps for level T^2 from our model. From the analytic theory we know this number to be $q^3(q+1)$, and the model given here contains only $q^2(q+1)$ (singular!) points over the cusps for level T.

Write $M = \mathbf{F}_q[\epsilon]e_1 + \mathbf{F}_q[\epsilon]e_2$ for a fixed free rank 2 $\mathbf{F}_q[\epsilon]$-module. The level structure on our Drinfeld module ϕ is given by $\varphi : e_i \mapsto Y_i$. The action of $\mathbf{F}_q \times \mathbf{F}_q \times \mathbf{F}_q \times \mathbf{F}_q$ changes this to a level structure $\tilde{\varphi}$, with $e_1 \mapsto Y_1 - aX_1 - bX_2$ and $Y_2 \mapsto Y_2 - cX_1 - dX_2$. Hence the action yields an automorphism $\tilde{\varphi}^{-1}\varphi$ of M, given by a matrix $\begin{pmatrix} 1+a\epsilon & c\epsilon \\ b\epsilon & 1+d\epsilon \end{pmatrix}$. From the analytic theory we know that the two level structures define points on the same component of the moduli space precisely when this matrix is in the $Gl(2, A)$-orbit of the identity matrix. This is the case precisely when $a + d = 0$.

The above observation allows one to find the components of our model, at least for $q = 2$. Consider the affine open part $X_2 \neq 0$, and write $x = X_1/X_2, y = Y_1/X_2, \xi = Y_2/X_2$. We will work over the field $L = \overline{\mathbf{F}_2(T)}(x)$. The equation for y reads

$$x^2(x+1) - Tx(x+1)y + T(x^2+x+1)y^2 - Ty^4 = 0.$$

This defines a degree 4 field extension $L(y)/L$ which is Galois with group $\mathbf{F}_2 \times \mathbf{F}_2$ (the equation for y is equivalent to $\phi_T(y) = x$, from which the assertion is obvious).

Proving that the moduli space in this case consists of $q = 2$ components now boils down to showing that the remaining polynomial

$$x(x+1) - Tx(x+1)\xi + T(x^2+x+1)\xi^2 - T\xi^4$$

factors in $L(y)[\xi]$. Suppose η is a zero of an irreducible factor. Then (y,η) corresponds to a level structure. Any zero of the polynomial is of the form $\eta + cx + d$ with $c, d \in \mathbf{F}_2$. Hence $\eta + cx + d$ is a zero of the same factor precisely when the corresponding level structure $(y, \eta + cx + d)$ gives a point on the same component as (y,η). We have seen above that this is the case when $d = 0$. This means that the zeroes of a factor are $\eta, \eta + x$ and hence a factorization should have the form

$$(\xi^2 + x\xi + a)(\xi^2 + x\xi + b)$$

for certain $a, b \in L(y)$. Comparing coefficients shows that a, b are in fact in a quadratic extension of L. These subfields are easily found, and one obtains $a = (x+1)g + (y^2+y)/x$ and $b = x + 1 + a$, in which g satisfies $g^2 + g = T^{-1}$.

This g is not an element of $K = \mathbf{F}_2(T)$, but it is contained in $K_\infty = \mathbf{F}_2((1/T))$. Namely, here $g = \sum_{n>0} T^{-2^n}$. This confirms the fact that in the analytic theory all components of the moduli space are defined over K_∞.

We now try to compute the Tate-Drinfeld module of rank 2 and level T^2, as we did for level T. For this, we start with the rank 1 module given by $\phi_T = T + c\tau$. As before, we want that 1 generates the T^2-torsion (as a module over $A/(T^2)$). This means that $(T + c\tau)(T + c\tau)(1) = 0$, but $(T + c\tau)(1) \neq 0$. In other words $c \neq -T$ but $c^{q+1} + c(T^q + T) + T^2 = 0$. Factoring $c^{q+1} + c(T^q + T) + T^2 = (c+T)(c(c+T)^{q-1} + T)$, this implies that c is any root of $c(c+T)^{q-1} + T = 0$. Note that for $q = 2$, this defines the same extension of $\mathbf{F}_2(T)$ which we need to separate the components of $M^2_{T^2}$. This is not a coincidence.

Next, we want $\phi_T = T + c\tau + eX^{q^3-q^2}\tau^2$ and $\phi_{T^2}(1/X) = 0$. This becomes a polynomial in X and e after multiplication by X^{q^2}. In fact, $X^{q^2}\phi_{T^2}(1/X) \equiv ec^{q^2} + c^{q+1} \bmod X$. Using Hensel's lemma again, this implies that a unique $e \in A[c][[x]]$ exists such that the T^2-torsion of the corresponding Drinfeld module is indeed generated by $1/X$ and a unit $\equiv 1 \bmod X$.

2 Tate-Drinfeld modules

2.1 Drinfeld modules of rank one and class field theory

In lecture 4 a general survey of class field theory over function fields is given. In the sequel we will provide more details about the case $A = \mathbf{F}_q[T]$ and abelian extensions of the field $K = \mathbf{F}_q(T)$ which are totally split at ∞. Those results are needed for the description of the Tate-Drinfeld modules of rank two with level f.

Let R be an A-algebra with structure homomorphism $\gamma : A \to R$. A rank one Drinfeld module $\psi : A \to R\{\tau\}$ is determined by $\psi_T = \gamma(T) + c\tau$ with $c \in R^*$. Let $f \in A$ be a monic polynomial of degree $n \geq 1$. A (strong) level f-structure is an isomorphism λ of (the constant sheaf) $f^{-1}A/A$ to $ker(\psi_f, R)$ (as a sheaf on $Spec(R)$). Since the polynomial $\psi_f(z)$ has the form

$$\gamma(f)z + *z^q + \ldots + c^{1+q+\ldots+q^{n-1}} z^{q^n}$$

the existence of such a level structure implies that $\gamma(f)$ is invertible in R. In the sequel we suppress γ from the notation and we will consider $A[1/f]$-algebras R. The triple (R, ψ', λ') is called equivalent to (R, ϕ, λ) if there is an $r \in R^*$ such that $\psi' = r^{-1}\psi r$ and $\lambda' = r^{-1}\lambda$. This defines a functor $\mathcal{M}^1(f)$ from the category of the $A[1/f]$-algebras to the category of sets, by $\mathcal{M}^1(f)(R)$ is the set of equivalence classes of Drinfeld modules of rank one and with level f above R. (The functor extends to the category of $A[1/f]$-schemes). As we know, this functor is representable and we will make this explicit. For this purpose we study the Carlitz module given by $\psi : A \to A\{\tau\}$ with $\psi_T = T + \tau \in A\{\tau\}$.

From an analytic point of view the Carlitz module is derived from a lattice $\xi A \subset \mathbf{C}$ for a certain $\xi \in \mathbf{C}$. The Carlitz exponential function

$$e = e_{\xi A} = z \prod_{0 \neq a \in A} (1 - \frac{z}{\xi a}) = z + \sum_{0 \neq a \in A} (\frac{1}{\xi a})^{q-1} z^q + *z^{q^2} + \ldots$$

is supposed to have the property $e(Tz) = Te(z) + e(z)^q$. The sum $\alpha := \sum_{0 \neq a \in A} \frac{1}{a^{q-1}} \in K_\infty$ satisfies $|\alpha + 1| < 1$. A calculation shows then

$$\xi = T \sqrt[q-1]{1 - T^{1-q}} \sqrt[q-1]{-\alpha} \sqrt[q-1]{-T}.$$

In particular $\xi \in K_\infty(\sqrt[q-1]{-T})$. Thus e induces an isomorphism between $K_\infty(\sqrt[q-1]{-T})/\xi A$ and $K_\infty(\sqrt[q-1]{-T})$.

The solutions of the equation $\psi_f(z) = 0$ in the field $K_\infty(\sqrt[q-1]{-T})$ are the $e(\frac{g\xi}{f})$ with $g \in A$ of degree $< n$. Hence the equation splits in $K_\infty(\sqrt[q-1]{-T})$. One defines the polynomial

$$\chi(f) := \prod_{a \in (A/fA)^*} (z - e(\frac{a\xi}{f})),$$

note that $\chi(1) = z$. It is clear that the degree of $\chi(f)$ equals $\#(A/fA)^*$. Also, when f is not constant,

$$\chi(f) = \prod_{\substack{a \in (A/fA)^* \\ a \equiv 1 \bmod f}} \prod_{c \in \mathbf{F}_q^*} (z - ce(\frac{a\xi}{f})) = \prod_{\substack{a \in (A/fA)^* \\ a \equiv 1 \bmod f}} (z^{q-1} - e(\frac{a\xi}{f})^{q-1})$$

hence $\chi(f)$ is a polynomial in z^{q-1}. We now show that $\chi(f) \in A[z^{q-1}]$. Since the zeroes of $\chi(f)$ are also zeroes of the monic polynomial $\psi_f(z)$ which has coefficients in A, they are integral over A. Hence we only need to show that $\chi(f) \in K(z)$. Note that $\psi_f(z) = \prod_{d|f,\, d \text{ monic}} \chi(d)$. Let $\mu : A \setminus \{0\} \to \{0, 1, -1\}$ denote the Möbius function, given by $\mu(c) = 1$ for $c \in \mathbf{F}_q^*$ and $\mu(f_1^{e_1} \dots f_s^{e_s})$ (with irreducible inequivalent $f_1, \dots, f_s$) is equal to 0 if some $e_i > 1$ and is equal to $(-1)^s$ if all $e_i = 1$. Then $\chi(f) = \prod_{d|f,\, d \text{ monic}} \psi_{f/d}^{\mu(d)}$. This implies $\chi(f) \in K(z)$, and hence $\chi(f) \in A[z^{q-1}]$ for $f \in A$ not constant, as we wanted to prove.

We note that $\chi(f)$ is the equivalent for the Carlitz module (A, ψ) of the classical cyclotomic polynomial Φ_n. In fact,

Lemma 2.1 *1. $\chi(f)$ is the minimal monic polynomial of $e(\frac{\xi}{f})$ over K.*

2. The splitting field L of $\psi_f(z)$ over K is equal to $L = K(e(\frac{\xi}{f})) \cong K[z]/(\chi(f))$.

3. The Galois group of L/K acts A-linearly on $ker(\psi_f, L)$. The resulting homomorphism $Gal(L/K) \to (A/fA)^*$ is an isomorphism. We will identify the two groups.

4. The integral closure of A in L is $A[z]/(\chi(f))$.

5. The finite ring extension $A[1/f] \subset A[1/f, z]/(\chi(f))$ is unramified.

6. $e(\frac{\xi}{f})$ is invertible in $A[1/f, z]/(\chi(f))$ for non-constant f.

7. The extension L/K is the abelian extension of K which by class field theory corresponds to the subgroup

$$\left((1 + \frac{1}{T}\mathbf{F}_q[[\frac{1}{T}]]) \times \prod_{\pi | f} (1 + fA_\pi) \prod_{\text{other } \pi} A_\pi^* \right) / \mathbf{F}_q^*$$

of the idele class group $\mathcal{I}_K/K^* \cong \left(\mathbf{F}_q[[\frac{1}{T}]]^* \times \prod_\pi A_\pi^*\right)/\mathbf{F}_q^*$. Here the π are primes of A and A_π denotes the corresponding completion.

Proof. (1) It is obvious that $\chi(f)$ is a monic polynomial with $e(\frac{\xi}{f})$ as a root. For constant f it is z, so we may assume $n = \deg(f) > 0$. Consider any prime π dividing f. Note that modulo π, the Carlitz module ψ reduces to a rank one module $\overline{\psi}$ over $A/\pi A$. The $\overline{\psi}_f$-torsion is a free module over $A/\pi A$ of rank at most 1, hence since $\overline{\psi}_f(z)$ is not separable, it is 0. This implies that $\psi_f(z) \equiv z^{q^n} \bmod \pi$. Hence the divisor $\chi(f)$ of $\psi_f(z) \equiv z^{q^n} \bmod \pi$ also reduces modulo π to a power of z, which shows that all its coefficients except the leading one are divisible by π. Moreover, the constant term of $\psi_f(z)/\chi(1)$ is f, hence the divisor $\chi(f)$ of this has a constant term which is a divisor of f. In case π can be chosen such that it divides f exactly once, this implies that $\chi(f)$ is an Eisenstein polynomial at the prime π.

If $f = \pi^{n+1}$ for $n > 0$ and π irreducible, then observe $\chi(f) = \chi(\pi)(\psi_{\pi^n}(z))$. This is true because both sides have as zeroes precisely the elements of $\mathrm{Ker}(\psi_f)$ which are not in $\mathrm{Ker}(\psi_{\pi^n})$. It follows that $\chi(f)(0) = \chi(\pi)(0)$, hence one concludes that also in the present case $\chi(f)$ is Eisenstein at π.

The general case now easily follows: write $f = \pi_1^{n_1} \cdot \ldots \cdot \pi_t^{n_t}$. The splitting field L of $\chi(f)$ contains the splitting fields L_i of $\chi(\pi_i^{n_i})$. These fields are linearly disjoint over K, because L_i is totally ramified at π_i while L_j is unramified at π_i for $j \neq i$. Hence

$$\#(A/fA)^* \geq [L:K] \geq \prod [L_i : K] = \prod \#(A/\pi_i^{n_i}A)^* = \#(A/fA)^*$$

which shows $[L:K] = \#(A/fA)^*$ and hence $\chi(f)$ is irreducible.
(2 and 3) are immediate using the argument above.
(4) We may assume that f is not constant. It suffices to show that $\chi(f) = 0$ defines a smooth affine variety in $\mathbf{A}^2_{\mathbf{F}_q}$. In other words, we want to verify that $\chi(f)$ and its derivatives to T and z have no common zero. Suppose a common zero would exist. Then we have a maximal ideal $M \subset A$ containing $\chi(f)$ and its derivatives. Since $\psi_f(z) = g \cdot \chi(f)$ for some g, one sees by taking the derivative to z that $f \in M$. Hence $\psi_f(z) \equiv z^{q^n} \bmod M$, and therefore $z \in M$. A similar computation, now using that the derivative to T of $\psi_f(z)/z$ is in M, shows that the derivative f' of f to T is in M. This already leads to a contradiction whenever all roots of f are simple.

In case $f = \pi^m$ and π irreducible, we computed in (1) that $\chi(f) = u\pi + z*$ for a unit $u \in \mathbf{F}_q^*$, hence again taking the derivative to T shows that $\pi' \in M$. Since $\pi \in M$ as well, this again gives a contradiction.

Finally, the remaining case is that $f = \pi^m g$ with π irreducible not dividing g, and $\pi \in M$. Since the zeroes of $\chi(f)$ are also zeroes of $h := \chi(\pi^m)(\psi_g(z))$, the latter polynomial is a multiple of the former hence an element of M. Modulo π, so modulo M as well, we have seen that h is a power of $\psi_g(z)$, hence also $\psi_g(z) \in M$. Again taking the derivative to T it follows that $\pi' \in M$, a contradiction.

(5 and 6) are again immediate: If f is not constant, $\psi_f(z)/z$ is a monic polynomial with constant term f. It is separable over any field in which $f \neq 0$.
(7) The computation in (1) above shows that the conductor of our extension L/K is as asserted, at each prime dividing f. Moreover, by (5) L/K is unramified at the primes of A not in f. At the infinite prime one may argue as follows. Consider the rank one module $\varphi : A \to K\{\tau\}$ given by $\varphi_T = T - T\tau$. A calculation shows that this module corresponds to a lattice λA with $\lambda \in K_\infty$. Hence all torsion of φ is inside K_∞. Take u in an extension of K satisfying $u^{q-1} = -1/T$. Then ψ and φ are isomorphic over $K(u)$. More explicitly, $\psi_f = u^{-1}\varphi_f u$ for any $f \in A$. This means that α is an f-torsion point for ψ precisely when $u\alpha$ is f-torsion for φ. Hence for any non-constant f the f-torsion generates the extension $K_\infty(u)$ of K_∞. This extension is totally ramified of degree $q-1$, which is what we wanted to prove.

Finally, consider a monic irreducible $\pi \in A$ which does not divide f. Then $\pi \bmod f \in (A/fA)^*$ corresponds in (3) to the automorphism σ_π of L defined by $\sigma_\pi(e(\frac{\xi}{f})) = e(\frac{\pi\xi}{f}) = \psi_g(e(\frac{\xi}{f}))$. Write $m = \deg(\pi)$ and let M be a prime of L over π. From (1) we know that $\psi_g(z) \equiv z^{q^m} \bmod M$. This shows that σ_π induces the Frobenius on the residue field modulo M, completing the proof of (7). □

Remark. Consider the rank one module φ given by $\varphi_T = T - T\tau$. The proof of Lemma 2.1 (7) implies, that the torsion of this module generates the maximal abelian extension of K which is totally split at infinity. The Galois group of this extension is isomorphic to $\hat{A}^*/\mathbf{F}_q^*$. The f-torsion of φ generates the abelian extension of K corresponding for $f(0) \neq 0$ to the quotient $(A/fA)^*$. In case $f(0) = 0$, the quotient $(A/fA)^*/\mathbf{F}_q^*$ is obtained. The drawback of using the module φ is that it is in fact a module over $A[1/T]$. Hence, for instance, one cannot expect that torsion points are integral at the place $T = 0$. One way to cope with this, is to use a second module with also period in K_∞ but now defined over $A[1/(T-1)]$.

The ring $A[1/f, z]/(\chi(f))$ with the Drinfeld structure $\psi_T = T + \tau$ is somewhat to big for our purposes. We normalize the Drinfeld structure to $\phi := d^{-1}\psi d$ where $d := e(\frac{\xi}{f})$. Put $c = d^{q-1}$. The minimal monic equation of c over K is $\chi(f)^*$, defined by $\chi(f)^*(z^{q-1}) = \chi(f)(z)$.
Then $\phi_T = T + c\tau$ and the generator of the level f-structure is now 1. The relation $\phi_f(1) = 0$ is a monic polynomial equation for c of degree $\frac{q^n-1}{q-1}$.
Let us introduce a variable C and define formally $\tilde{\phi} : A \to A[C]\{\tau\}$ by $\tilde{\phi}_T = T + C\tau$. The expression $\tilde{\phi}_f(1)$ is a monic polynomial in C of degree $\frac{q^n-1}{q-1}$. For a monic divisor d of f one finds that $\tilde{\phi}_d(1)$ is a divisor of $\tilde{\phi}_f(1)$. Further for any irreducible monic $f \in A$ one has $\tilde{\phi}_f(1) = \chi(f)^*$. It follows that the polynomial

$\tilde{\phi}_f(1)$ in C is the product of the $\chi(d)^*$ over all monic divisors d of f.

Thus we obtain the Drinfeld module $(A[1/f, c], \phi, can)$ which has a canonical level structure given by $can : 1/f \in f^{-1}A/A \mapsto 1 \in A[1/f, c]$. From Lemma 2.1 above one easily derives the following properties:

1. The subgroup $\mathbf{F}_q^*$ of the Galois group of L/K acts on $e(\frac{\xi}{f})$ as multiplication and thus $A[c] \subset K_\infty$.
2. The finite extension $A^f := A[1/f, c] \supset A[1/f]$ is unramified.
3. c is invertible in A^f.
4. The field of quotients K^f of A^f is an extension of K with Galois group $(A/fA)^*/\mathbf{F}_q^*$ and is totally split at ∞.
5. The field K^f is the field associated by class field theory to the group $(A/fA)^*/\mathbf{F}_q^*$.

Proposition 2.2 *The functor $\mathcal{M}^1(f)$ is represented by $A^f = A[1/f, c]$ (or its spectrum). The universal rank one Drinfeld module with level f is (A^f, ϕ, can).*

Proof. Let an element $B \in \mathcal{M}^1(f)(R)$ be given. The equivalence class B contains a unique Drinfeld module (R, ψ, λ) such that λ maps $1/f \in f^{-1}A/A$ to $1 \in R$. Write $\psi_T = T + \tilde{c}\tau$ with $\tilde{c} \in R^*$. Then $\psi_f(1) = 0$ and for every proper divisor d of f one has that $\psi_d(1) \neq 0$. As shown above, this implies that $\chi(f)^*(\tilde{c}) = 0$. Hence there is a A-homomorphism $h : A^f \to R$, given by $h(c) = \tilde{c}$, which transports the Drinfeld structure above A^f to (R, ψ, λ). Clearly h is unique. □

Examples:
We give the polynomial $\chi(f)^*$ which $c \in A[1/f, c]$ satisfies, for some f's.
(1) Let $f = T^2 + 1 \in \mathbf{F}_3[T]$. Then c must satisfy the equation $c^4 + c(T^3 + T) + T^2 + 1 = 0$. This defines indeed an étale extension of $A[1/f]$ of degree 4 in which c is invertible.
(2) For $f = T \in \mathbf{F}_q[T]$ one has $c = -T$.
(3) Take $f = T^2 \in \mathbf{F}_q[T]$. Let $z \neq 0$ be a solution of $\phi_T(z) = 0$ and let $\phi_T(1) = z$. Then $z^{q-1} = \frac{-T}{c}$ and $T + c = z$. Hence $c(c+T)^{q-1} + T = 0$, as was already shown in section 1. This equation is irreducible and separable. In fact $A[1/T] \subset A[1/T][c]$ is étale of degree q.
(4) Put $f = T(T+1) \in \mathbf{F}_q[T]$. Let $z \neq 0$ satisfy $\phi_T(z) = 0$ and $\phi_{T+1}(1) = z$. This yields the equation $c(c + T + 1)^{q-1} + T = 0$. A trivial solution is $c = -T$. The irreducible monic polynomial $\chi(T(T+1))^*$ is $\frac{c(c+T+1)^{q-1}+T}{c+T}$.

2.2 The Tate-Drinfeld module without level

Let an A-algebra R be given with structure homomorphism $\gamma : A \to R$. A Drinfeld module of rank two over R is given by a homomorphism $\phi : A \to R\{\tau\}$ with $\phi_T = \gamma(T) + c_1\tau + c_2\tau^2$ such that $c_1 \in R, c_2 \in R^*$. We will attach to this a *geometric object*, namely the projective line $\mathbf{P}^1_R$ over R with its infinite section ∞, the line bundle $\mathbf{A}^1_R = \mathbf{P}^1_R \setminus \{\infty\}$ and the A-action on this line bundle given by ϕ. Let $(t_0 : t_1)$ denote the homogenous coordinates of the projective line over R. For $f \in A$ of degree $n \geq 0$ one can extend this action by the formula

$$f * (t_0 : t_1) = (t_0^{q^{2n}} : t_0^{q^{2n}} \phi_f(\frac{t_1}{t_0})).$$

This is seen as an analogue of an elliptic curve over R with its natural action of $\mathbb{Z}$.

Tate's elliptic curve E over $\mathbb{Z}[[q]]$ lies in $\mathbf{P}^2_{\mathbb{Z}[[q]]}$ and is given by the affine equation $y^2 + xy = x^3 + Ax + B$ with certain $A, B \in q\mathbb{Z}[[q]]$. Above the open set $q \neq 0$ of $Spec(\mathbb{Z}[[q]])$ this defines a "true" elliptic curve. Above $q = 0$ the affine equation reduces to $y^2 + xy = x^3$, which defines a rational curve with an ordinary double point. After removal of this double point the group structure above $q = 0$ is the multiplicative group over $\mathbb{Z}$.

The obvious analogue for Drinfeld modules is the *Tate-Drinfeld module of rank two* given by $\phi : A \to A[[x]]\{\tau\}$ with $\phi_T = T + \tau + x\tau^2$. The geometric object corresponding to this is the projective line $\mathbf{P}^1_{A[[x]]}$ with its section at infinity ∞ and the line bundle $\mathbf{A}^1_{A[[x]]} = \mathbf{P}^1_{A[[x]]} \setminus \{\infty\}$ with its A-action given by ϕ. The action of $f \in A$ of degree $n \geq 0$ extends to the open set $\mathbf{P}^1_{A[[x]]} \setminus V(x, t_0)$ by the formula

$$f * (t_0 : t_1) = (t_0^{q^{2n}} : t_0^{q^{2n}} \phi_f(\frac{t_1}{t_0})).$$

Above the open subset $x \neq 0$ of $Spec(A[[x]])$ the structure above is an ordinary Drinfeld module of rank two. Above $x = 0$ the structure reduces to a rank one Drinfeld module, namely the *Carlitz module* given by $\psi_T = T + \tau$. This seems to be the canonical choice for the Tate-Drinfeld module of rank two without level structure.

2.3 The Tate-Drinfeld module with level (f)

Tate's elliptic curve E, described above, has no level structure. Fix an integer $n \geq 3$. A level n-structure would be an isomorphism of (the constant sheaf) $(n^{-1}\mathbb{Z}/\mathbb{Z})^2$ with the group of the n-torsion points. We have to extend the base

$\mathbb{Z}[[q]]$ in order to obtain enough torsion points. Above the open subset $q \neq 0$ one needs the extension $\mathbb{Z}[1/n, \zeta_n][[q^{1/n}]]$ of $\mathbb{Z}[[q]]$ for this. Above $q^{1/n} = 0$, the group of the n-torsion points is equal to the group of the n-torsion points of the multiplicate group (over $\mathbb{Z}[1/n, \zeta_n]$). This group is isomorphic to $n^{-1}\mathbb{Z}/\mathbb{Z}$. In order to obtain also above $q^{1/n} = 0$ enough n-torsion points one has to blow up the special fibre of $E \otimes \mathbb{Z}[[1/n, \zeta_n][[q^{1/n}]]$. This defines the Néron model of *the Tate curve with level n*.
The universal family $\mathcal{E}$ of "generalized elliptic curves of level n" lives above the modular curve $X(n)$ corresponding to level n. For any cusp c of $X(n)$ the completion of the fibre of $\mathcal{E}$ at c is the Néron model of the Tate curve of level n, described above.

Fix a monic $f \in A$ of degree $n \geq 1$. The Tate-Drinfeld module with level f-structure should have approximately the form $\phi : A \to R[[x]]\{\tau\}$, with $\phi_T = T + c_1\tau + c_2\tau^2$ with $c_1 \in R[[x]]^*$ and $0 \neq c_2 \in xR[[x]]$. For $x = 0$ the structure should reduce to the universal rank one Drinfeld module with level f. Thus we have $R = A^f = A[1/f, c]$ and we may normalize the situation such that $c_1 = c$. Above the open set $x \neq 0$ the set $ker(\phi_f, A^f[[x]][x^{-1}])$ should be isomorphic to the A-module $(f^{-1}A/A)^2$. The element $1 \in R[[x]]/(x)$ is a generator of the f-torsion of the rank one Drinfeld module. This generator lifts in a unique way to a generator of $ker(\phi_f, A^f[[x]]) \cong f^{-1}A/A$. We try to normalize the situation by imposing that x^{-1} is a generator of the factor module $ker(\phi_f, A^f[[x]][x^{-1}])/ker(\phi_f, A^f[[x]])$. According to the next lemma this is possible and defines a unique c_2. We first note that c_2 must have the form $ex^{q^n(q-1)}$ with $e \in A^f[[x]]^*$. Indeed, replace A^f by any quotient field $k = A^f/\underline{m}$. A calculation over the field k shows that c_2 mod $\underline{m}$ has order $q^n(q-1)$ with respect to the variable x.

Lemma 2.3 *The equation $x^{q^n}\phi_f(x^{-1}) = 0$ can be seen as a polynomial equation for e over the ring $A[c][x]$. It has a unique solution in $A^f[[x]]$. This solution has the form $e \in -c^{1+q-q^n} + x^{q^n-q^{n-1}}A^f[[x]]$.*

Proof. Put $f = \sum_{i=0}^n f_i T^i$ (with $f_n = 1$) and put

$$P(e) = c^{-q^2-q^3-\dots-q^n} \sum_{i=0}^{n} f_i x^{q^n}(T + c\tau + ex^{q^n(q-1)}\tau^2)^i x^{-1}.$$

One computes that $(T+c\tau+ex^{q^n(q-1)}\tau^2)^i x^{-1}$ has the form $c^{1+q+\dots+q^{i-1}}x^{-q^i}+\dots$ for $i = 1, \dots, n-1$ and $c^{1+q+\dots+q^{n-1}}x^{-q^n} + ec^{q^2+\dots+q^n}x^{-q^n} + *x^{-q^{n-1}} + \dots$ for $i = n$. Hence $P(e)$ is a polynomial in e with coefficients in the ring $A[c][x]$. From the above it follows that $P(e) \in (e + c^{1+q-q^n}) + x^{q^n-q^{n-1}}A^f[[x]][e]$ and that the derivative satisfies $P'(e) \in 1+x^{q^n-q^{n-1}}A^f[[x]][e]$. If e exists then $e \in -c^{1+q-q^n}+$

$x^{q^n-q^{n-1}}A^f[[x]]$. Now we observe that $P(-c^{1+q-q^n}) \in x^{q^n-q^{n-1}}A^f[[x]]$ and that $P'(-c^{1+q-q^n}) \in 1 + x^{q^n-q^{n-1}}A^f[[x]]$. Applying Newton's approximation one finds the unique solution $e \in A^f[[x]]$ of the equation. □

Definition 2.4 *The Tate-Drinfeld module of rank two and level f is defined by $\phi : A \to A^f[[x]]\{\tau\}$ with $\phi_T = T + c\tau + ex^{q^n(q-1)}\tau^2$, where A^f, c and e are defined in* (2.1) *and* (2.2).

By Newton's approximation, the canonical map

$$ker(\phi_f, A^f[[x]]) \to ker(\phi_f, A^f[[x]]/(x)) \cong f^{-1}A/A$$

is an isomorphism. Further the elements $x^{-1}, \phi_T x^{-1}, \ldots, (\phi_T)^{n-1}x^{-1}$ in $ker(\phi_f, A^f[[x]][x^{-1}])$ have the form

$$x^{-1}, cx^{-q} + \ldots, c^{1+q}x^{-q^2} + \ldots, \ldots, c^{\frac{q^{n-1}-1}{q-1}}x^{-q^{n-1}} + \ldots$$

They are linearly independent over $\mathbf{F}_q$ in the factor module $ker(\phi_f, A^f[[x]][x^{-1}])/ker(\phi_f, A^f[[x]])$. Thus the A-module $ker(\phi_f, A^f[[x]][x^{-1}])$ is isomorphic to $(f^{-1}A/A)^2$ as required.

The Tate-Drinfeld module that we have constructed above has a certain universal property. We formulate a weak form of this.

Proposition 2.5 *Let R be an $A[1/f]$-algebra without zero divisors and let a Drinfeld structure $\psi : A \to R[[y]]\{\tau\}$ be given by $\psi_T = T + c_1\tau + c_2\tau^2$. Suppose further that*
(1) $c_1 \in R[[y]]^*$ and $c_2 \in yR[[y]]$.
(2) $ker(\psi_f, R[[y]])$ is isomorphic to $f^{-1}A/A$.
(3) $ker(\psi_f, R[[y]][y^{-1}])$ is isomorphic to $(f^{-1}A/A)^2$.
Then there is a homomorphism of A-algebras $h : A^f[[x]] \to R[[y]]$ such that $h(\phi)$ is equivalent over $R[[y]]$ with ψ.

Proof. The canonical map $ker(\psi_f, R[[y]]) \to ker(\psi_f, R[[y]]/(y))$ is an isomorphism. After changing ψ we may suppose that the last A-module is generated by $1 \in R$. The universality of the rank one module (A^f, ϕ mod (x)) induces a unique A-linear homomorphism $h : A^f \to R$. Let $v \in R[[y]][y^{-1}] \setminus R[[y]]$ have "minimal negative" order among the elements of $ker(\psi_f, R[[y]][y^{-1}])$. Then we extend the homomorphism h to $h : A^f[[x]] \to R[[y]]$ be sending x to v^{-1} (which belongs indeed to $yR[[y]]$). Then $h(\phi)$ and ψ have the same f-torsion groups on $R[[y]][y^{-1}]$. From this it follows that $h(\phi) = \psi$. □

Remarks

The condition in the proposition that R has no zero divisors will be removed in

the sequel by means of the notion of generalized Drinfeld module (of rank two and with level (f)-structure).
The homomorphism h of the proposition induces of course an isomorphism $\tilde{h}$ between the A-modules $ker(\phi_f, A^f[[x]][x^{-1}])$ and $ker(\psi_f, R[[y]][y^{-1}])$. And in fact h is determined by $\tilde{h}$.

3 Models for the projective line over a discrete valuation ring

Let R be a discrete valuation ring with field of fractions L and residue field l. The projective line over L, i.e. $\mathbf{P}^1_L$, has a standard model $\mathbf{P}^1_R$ over R. Attached to this model there is a reduction map for points, namely the map $\mathbf{P}^1_L(L) \tilde{\to} \mathbf{P}^1_R(R) \to \mathbf{P}^1_l(l)$. The models M of $\mathbf{P}^1_L$ over R that we consider here are supposed to have the properties:
(a) M is a proper, flat scheme over R.
(b) The generic fibre $M \otimes_R L$ is isomorphic to $\mathbf{P}^1_L$.
(c) The special fibre $M_s := M \otimes_R l$ has at most ordinary double points as singularities.

Since the genus of the special fibre M_s is 0, this fibre must be a finite tree of projective lines over the residue field l. For any model M one defines a reduction map r_M for points

$$r_M : \mathbf{P}^1_L(L) \tilde{\to} M(R) \to M_s(l)$$

Lemma 3.1 *Let F be a finite subset of $\mathbf{P}^1_L(L)$ with cardinality ≥ 3. Then there is a unique model M which has the following properties:*
(1) The reduction map r_M is injective on F and $r_M(F)$ does not contain any double point of M_s.
(2) The special fibre M_s together with the points $r_M(F)$ form a "stable tree". This means that every irreducible component of M_s contains at least three special points. The special points of M_s are defined to be the union of the double points of M_s with $r_M(F)$.

Proof. The statement is well known, see [GHP]. We will sketch another proof of the lemma.

Suppose for convenience that $\{0, 1, \infty\} \subset F$. The standard model gives as reduction a projective line over l and F is mapped to some points on this line. If the reduction map (restricted to F) is injective then we are finished. If the reduction map (restricted to F) is not injective then there is a point a on the projective line which is the image of more than one point of F. One blows up the point a, seen as a point of the surface M. The new special fibre is a tree with two lines; on each line there are at least two special points. If F is mapped injectively to the regular points of this tree then we stop. If not then one blows up another point et cetera. One stops when condition (1) is satisfies. Condition (2) does not hold necessarily, since one may have introduced irreducible components which carry only two special points. Those irreducible components are blown down until (2) is also satisfied. It is an exercise to show the uniqueness of the model M for F. □

The lemma is applied to the field $\mathbf{F}_q(x)$ with as valuation ring the local ring at $x = 0$, i.e. the local ring $\mathbf{F}_q[x]_{(x)}$. For the set F we choose

$$\{\mathbf{F}_q + \mathbf{F}_q x^{-1} + \mathbf{F}_q x^{-q} + \ldots + \mathbf{F}_q x^{-q^{n-1}}\} \cup \{\infty\}.$$

The special fibre RT of the model, given by the lemma and the F above, has been described in Lecture 8 in detail.

In the sequel we will also use the concept of *stable n-pointed tree*. Let S be a scheme and let $n \geq 3$. A stable n-pointed tree over S is a morphism $\alpha : C \to S$ which has the following properties:

1. $\alpha : C \to S$ is a flat and projective morphism of relative dimension one.
2. There are given n sections $f_1, \ldots, f_n : S \to C$.
3. For every geometric point $\xi = Spec(k)$ of S, the geometric fibre C_ξ is a stable tree of projective lines over the field k, i.e.
   (a) C_ξ is reduced
   (b) Every irreducible component of C_ξ is a projective line over k.
   (c) The only singular points of C_ξ are ordinary double points.
   (d) The images of the f_i in C_ξ are distinct and do not contain a double point.
   (e) On every irreducible component of C_ξ there are at least three special points. By a special point we mean either a double point or an image of some section f_i.

4 A Néron model of the Tate-Drinfeld module

The Tate-Drinfeld module, rank two and level f, was given by $\phi_T = T + c\tau + ex^{q^n(q-1)}\tau^2$ with $c \in A^f \subset K^f$ and $e \in A^f[[x]]^*$. Put $F = ker(\phi_f, A^f[[x]][x^{-1}]) \cup \{\infty\}$. We can consider this over the discrete valuation ring $K^f[[x]]$. The previous section produces a model M_0 of $\mathbf{P}^1_{K^f[[x]]}$ over $K^f[[x]]$ with respect to F.

Let $\underline{m}$ denote a maximal ideal of A^f with residue field, say k. Write $\bar{F}$ for the image of F in $\mathbf{P}^1_{k[[x]]}$. Section 3 produces again a model M_k of $\mathbf{P}^1_{k[[x]]}$ over $k[[x]]$ with respect to $\bar{S}$. One can combine all those models in a model M over $A^f[[x]]$ of $\mathbf{P}^1_{A^f[[x]]}$ with respect to F which has the properties:
(1) M is projective and flat over $Spec(A^f[[x]])$.
(2) $M \otimes A^f[[x]][x^{-1}]$ is isomorphic to $\mathbf{P}^1_{A^f[[x]][x^{-1}]}$.
(3) $M \otimes K^f[[x]]$ is isomorphic to M_0.
(4) $M \otimes k[[x]]$ is isomorphic to M_k for every maximal ideal of A^f.
In particular, for every geometric point ξ of $Spec(A^f[[x]])$ the fibre M_ξ is either an ordinary Drinfeld module of rank two with level f-structure and completed to a projective line or a generalized Drinfeld module over a field as defined in Lecture 8.

The existence of M can be proven as follows. Consider two sections f_1, f_2 of $\mathbf{P}^1_{A^f[[x]]}$. They are two-dimensional smooth subvarieties of the smooth three-dimensional space. If their intersection is not empty then they intersect transversely in a space isomorphic to $Spec(A^f)$. This is easily seen from the special form of the sections in F, namely for $f_* \in F, f_* \neq 0, \infty$ the leading coefficient is invertible in A^f. One blows up all the non-empty intersections $f_1 \cap f_2$. The result is the model M that has the required properties. We will call M the *Néron model of the Tate-Drinfeld module.*
It has a lot of structure induced by its fibre $\mathbf{P}^1_{A^f[[x]][x^{-1}]}$.

1. The open set M^* of M is defined by deleting the section ∞ and the closure of the irreducible components of $M \otimes K^f[[x]]/(x)$ on which no sections $\neq \infty$ of F lie. Then M^* has an induced structure of an algebraic group scheme.

2. The group law $M^* \times M^* \to M^*$ extends to an action of group schemes $M^* \times M \to M$. There is an A-action on the group scheme M^*.

3. Above $x = 0$ the group scheme M^* is isomorphic to $\mathbf{G}_{a,A^f} \times f^{-1}A/A$. The first term is the additive group over $A^f = A^f[[x]]/(x)$; it has the structure of the universal rank one Drinfeld module with level f. The second term is the constant group above A^f with as fibre the A-module $f^{-1}A/A$. The A-action on M^* above $x = 0$ coincides with the one given by this rank one Drinfeld module and the obvious action on $f^{-1}A/A$.

4. The kernel of $f\cdot$ on M^* is equal to the union of the section $f_* \in F, f_* \neq \infty$.

5. For every geometric point ξ of $Spec(A^f[[x]])$ the fibre M_ξ is either an ordinary Drinfeld module of rank two and level f completed with a section ∞, or a generalized Drinfeld module (rank two and level f) over the field corresponding to ξ.

5 The compactification of $M^2(f) \otimes A[1/f]$

In [D] ,§9 the compactification of M^2, this is the projective limit of all the Drinfeld modular curves M_I^2, is produced in an adelic way from M^1, which is the projective limit of the schemes M_I^1. A direct consequence is the existence of a morphism $M^2 \to M^1$. In particular one obtains, for any nontrivial ideal I of A, an embedding of the coordinate ring $O(M_I^1)$ into $O(M_I^2)$, the one of M_I^2. Unfortunately we are unable to understand this construction. To the best of our knowledge, this construction of Drinfeld has not been developed in more detail.
One of our goals in this section is to give an explicit construction of the compactification of M_I^2 and the morphism $M_I^2 \to M_I^1$ in the special case $A = \mathbb{F}_q[T]$ and over the ring $A[1/f]$, where $I = (f)$.
The other aim of this section is to make the Néron structure above the cusps of the compactification $\bar{M}^2_{(f)}$ explicit and to formulate a moduli functor which is represented by $\bar{M}^2_{(f)}$.

Put $\mathcal{Z} := M^2(f) \otimes A[1/f]$. Our aim is to construct its compactification $\hat{\mathcal{Z}}$ together with all its Drinfeld data. Let us collect the information that we have about $\mathcal{Z}$.

According to lecture 2, the following structure lives above $\mathcal{Z}$: a line bundle L, a rank two Drinfeld module $\phi : A \to End(L)$ with $\phi_T = T + c_1\tau + c_2\tau^2$, $c_1 \in L^{1-q}$, $c_2 \in L^{1-q^2}$ and c_2 nowhere vanishing, plus a level structure $\lambda : (f^{-1}A/A)^2 \to ker(\phi_f, L)$. We also know that $\mathcal{Z}$ is affine and that $\mathcal{Z} \to Spec(A[1/f])$ is smooth of relative dimension 1. For every $\sigma \in Gl(2, A/fA)$ one can consider the Drinfeld module $(\mathcal{Z}, L, \phi, \lambda\sigma)$. From the universality of $(\mathcal{Z}, L, \phi, \lambda)$ one obtains an automorphism $\tilde{\sigma}$ of $\mathcal{Z}$ such that $(\mathcal{Z}, \tilde{\sigma}L, \phi, \tilde{\sigma}\lambda)$ is equivalent with $(\mathcal{Z}, L, \phi, \lambda\sigma)$. The kernel of the homomorphism $\sigma \mapsto \tilde{\sigma}$ is equal to $\mathbf{F}_q^* id$. In this way $Gl(2, A/fA)/\mathbf{F}_q^*$ acts as a group of automorphisms on $\mathcal{Z}$. The quotient of $\mathcal{Z}$ by this action is $\mathbf{A}^1_{A[1/f]}$. The morphism $\mathcal{Z} \to \mathbf{A}^1_{A[1/f]}$ is explicitly given by the map $j : \mathcal{Z} \to \mathbf{A}^1_{A[1/f]}$ with the formula $j = \frac{c_1^{q+1}}{c_2}$.
According to lecture 8, the curve $\mathcal{Z} \otimes K_\infty$ coincides with the analytic object $Gl(2, A)\backslash(\Omega \times Gl(2, A/fA))$. The analytic compactification Z and the Drinfeld data on it have been described in (4.4.4) of lecture 8. We recall that Z has $\#(A/fA)^*/\mathbf{F}_q^*$ components.

An explicit description of $\mathcal{Z}$.
The section $\lambda(1/f, 0)$ of L vanishes nowhere and we can therefore identify L with $O_{\mathcal{Z}}$ and this section with $1 \in O(\mathcal{Z}) = O_{\mathcal{Z}}(\mathcal{Z})$. The section $\lambda(0, 1/f)$ is then an element $y \in O(\mathcal{Z})$ and also c_1, c_2 are elements of $O(\mathcal{Z})$. Further, for

any non-zero element $\xi \in (f^{-1}A/A)^2$ the element $\lambda(\xi) \in O(\mathcal{Z})$ is invertible. This easily leads to the following description of the ring $O(\mathcal{Z})$:
$O(\mathcal{Z}) = A[1/f][c_1, c_2, c_2^{-1}, y]_{loc}$ with ϕ given by $\phi_T = t + c_1\tau + c_2\tau^2$. The only relations (between the c_1, c_2, y) in this ring are given by $\phi_f(1) = 0$ and $\phi_f(y) = 0$. The ring is localized at the finite set consisting of all elements $\phi_{g_1}(1) + \phi_{g_2}(y)$ where $g_1, g_2 \in A$ are polynomials of degree $< n$, not both zero.

Definition of the scheme $\mathcal{Y}$

The next thing that we want to construct is an $A[1/f]$-algebra R such that its associated affine scheme $\mathcal{Y}$ over $A[1/f]$ will later be "an affine neighbourhood of a cusp of $\hat{\mathcal{Z}}$ ".
Consider the ring $R = A[1/f][c, c^{-1}, e, x]_{loc}$ and ϕ given by $\phi_T = T + c\tau + ex^{q^n(q-1)}\tau^2$, having the only relations $\phi_f(1) = 0$ and $x^{q^n}\phi_f(x^{-1}) = 0$. Note that this makes sense since by Lemma 2.3 $x^{q^n}\phi_f(x^{-1})$ is indeed an element of $A[1/f][c, c^{-1}, e, x]$. The localization is taken with respect to the finite set $x^{q^{m_2}}(\phi_{g_1}(1) + \phi_{g_2}(x^{-1}))$, where $g_1, g_2 \in A$ are polynomials (not both zero) of degrees $m_1, m_2 < n$. In case $g_2 = 0$ the expression $x^{q^{m_2}}$ is by definition 1. Again, all the expressions are well defined.

Lemma 5.1

(1) *R/xR with its Drinfeld data is isomorphic to $A^f[[x]]/(x)$ with the Drinfeld data induced by the Tate-Drinfeld module $A^f[[x]]$ defined in* (2.3).
(2) *The completion of R with respect to the ideal xR with its Drinfeld data is isomorphic to the Tate-Drinfeld module $A^f[[x]]$.*
(3) *e is invertible in R.*
(4) *$R[x^{-1}]$ with its Drinfeld data is isomorphic to $O(\mathcal{Z})[c_1^{-1}]$.*
(5) *R is smooth over $A[1/f]$ of relative dimension one.*

Proof. (1) According to the explicit expression for $x^{q^n}\phi_f(x^{-1})$ in (2.2), R/xR is isomorphic to $A[1/f][c, c^{-1}, e]_{loc}/(\phi_f(1), e + c^{1+q-q^n})$. Thus R/xR is isomorphic to the ring $A^f[[x]]/(x) = A^f$ (including its Drinfeld data).
(2) The same thing holds for R/x^kR for any $k \geq 1$, i.e. R/x^kR is isomorphic to $A^f[[x]]/(x^k)$ (including the Drinfeld data). We conclude that the completion of R with respect to the ideal xR (i.e. the projective limit of the R/x^kR) is isomorphic to $A^f[[x]]$ (including the Drinfeld data). We note further that R/xR is smooth over $A[1/f]$ and that e is invertible in R/xR.
(3) Suppose that e is not invertible in R and let $M \subset R$ be a maximal ideal which contains e. Put $k = R/M$. From (2) above it follows that x is invertible in k. Since $e = 0$ in k, the Drinfeld structure on k is that of a rank one Drinfeld module. The f-torsion of this module contains the images in k of the $\phi_{g_1}(1) + \phi_{g_2}(x^{-1}) \in R[x^{-1}]$. This module is still isomorphic to $(f^{-1}A/A)^2$,

which contradicts the observation that the Drinfeld module over k has rank 1. Thus e is invertible in R.
(4) Consider the $A[1/f]$-homomorphism $O(\mathcal{Z})[c_1^{-1}] \to R[x^{-1}]$, given by $c_1 \mapsto c$, $c_2 \mapsto ex^{q^n(q-1)}$, $y \mapsto x^{-1}$. This has an inverse given by $c \mapsto c_1$, $x \mapsto y^{-1}$, $e \mapsto c_2 y^{q^n(q-1)}$. It is clear that this isomorphism of $A[1/f]$-algebras preserves the Drinfeld data.
(5) From (4) it follows that $R[x^{-1}]$ is smooth over $A[1/f]$. Combining this with the isomorphism $R/xR \cong A^f$ found in (1) and the fact (proved in section 2.1) that A^f is unramified over $A[1/f]$, we conclude that R is smooth over $A[1/f]$. □

Lemma 5.2
(1) *$\mathcal{Y} \otimes K_\infty$ has $\#(A/fA)^*/\mathbf{F}_q^*$ components, as is shown in lecture 8.*
(2) *$\mathcal{Y} \otimes K_\infty$ has an embedding as an open subset of Z. The closed subset, given by $x = 0$, is then identified with a set C consisting of $\#(A/fA)^*/\mathbf{F}_q^*$ cusps of Z. On each component of $\mathcal{Y} \otimes K_\infty$ there lies one point of C.*
(3) *The stabilizer $H \subset Gl(2, A/fA)/\mathbf{F}_q^*$ of $\mathcal{Y} \otimes K_\infty$ as subset of Z is (up to conjugation) the subgroup of the matrices $(h_{i,j})$ with*

$$h_{1,1} \in (A/fA)^*,\ h_{1,2} = 0,\ h_{2,1} \in A/fA,\ h_{2,2} \in \mathbf{F}_q^*.$$

(4) *$\mathcal{Y}$ is connected.*
(5) *$\mathcal{Z}$ is connected.*

Proof. (1) The open subset of $\mathcal{Y} \otimes K_\infty$, given by $x \neq 0$, is identified with the open subset of $\mathcal{Z} \otimes K_\infty$ defined by $c_1 \neq 0$. The latter is known to have this number of components.
(2) The open subset $x \neq 0$ of $\mathcal{Y} \otimes K_\infty$ is identified with the open subset $c_1 \neq 0$ of $\mathcal{Z} \otimes K_\infty$ by lemma (5.1). Moreover one chooses an isomorphism of $\mathcal{Z} \otimes K_\infty$ with the affine part of Z. In this way $\mathcal{Y} \otimes K_\infty$ is embedded in Z. The closed subset $x = 0$ of $\mathcal{Y} \otimes K_\infty$ has affine ring $A^f \otimes K_\infty$ and thus consists of $\#(A/fA)^*/\mathbf{F}_q^*$ points which are rational over K_∞. Clearly this is identified with a set C of cusps of Z. We note that the group $G := Gl(2, A/fA)/\mathbf{F}_q^*$ acts on Z and $\mathcal{Z}$ in a compatible way. The subgroup which preserves the components of Z is equal to the image of $\{M \in Gl(2, A/fA) |\ det(M) \in \mathbf{F}_q^*\}$ in G.

In order to determine the nature of this set C we study the localization with respect to x of the formal completion at $x = 0$ of $\mathcal{Y} \otimes K_\infty$. This is the Drinfeld module $\phi : A \to (A^f \otimes K_\infty)[[x]][x^{-1}]\{\tau\}$ with $\phi_T = T + c\tau + ex^{q^n(q-1)}\tau^2$ and with level structure given by $(1/f, 0)$ maps to the unique element in $ker(\phi_f, (A^f \otimes K_\infty)[[x]])$ with image 1 in $(A^f \otimes K_\infty)[[x]]/(x)$ and $(0, 1/f)$ maps to x^{-1}. On this structure the Galois group of $A^f/A[1/f]$ acts. This action is trivial on the

level structure and is nontrivial on ϕ_T. We will change ϕ into an equivalent ψ such that the action of the Galois group is trivial on ψ and acts nontrivially on the level structure. There is a $u \in (A^f \otimes K_\infty)^*$ such that $u^{1-q}c = -T$. Then $\psi := u\phi u^{-1}$ has the form $\psi_T = T - T\tau + ax^{q^n(q-1)}\tau^2$ with $a \in (A^f \otimes K_\infty)[[x]]^*$. This ψ is invariant under the Galois group of $A^f/A[1/f]$ because the coefficient $-T$ of τ is invariant. We note that this Galois group is isomorphic to the group $(A/fA)^*/\mathbf{F}_q^*$. The action of this Galois group on the level permutes the points of C.
For a fixed point c_0 of C its stabilizer in G can be identified with the subgroup H_0 of G consisting of the matrices $(h_{i,j})$ with $h_{1,1}, h_{2,2} \in \mathbf{F}_q^*$, $h_{1,2} = 0$, $h_{2,1} \in A/fA$.
It can be calculated that every element in the Galois group of $A^f/A[1/f]$ can be lifted to an element of H, defined in the lemma. This yields a homomorphism of this Galois group to H/H_0, which can be shown to be an isomorphism. As a consequence one finds that there lies a point of C on every component of Z.
(3) The group H stabilizes the set C. A counting of the cusps shows that H is in fact the stabilizer of C. The open subset $\mathcal{Y} \otimes K_\infty \setminus C$ is invariant under the whole group G. Thus H is also the stabilizer of $\mathcal{Y} \otimes K_\infty$.
(4) and (5). The element $x \in O(\mathcal{Y}) = R$ cannot be identical zero on any component of $\mathcal{Y}$ since the closed subset $x = 0$ of $\mathcal{Y}$ has relative dimension 0. Similarly, $c_1 \in O(\mathcal{Z})$ cannot be identical zero on a component of $\mathcal{Z}$. Moreover the open subsets $x \neq 0$ of $\mathcal{Y}$ and $c_1 \neq 0$ of $\mathcal{Z}$ are isomorphic. Therefore (4) and (5) are equivalent.
In order to prove (4) we remark that it suffices to show that R has no idempotents $\neq 0, 1$. Let $h \in R$ be an idempotent. Then the image of h in $R/xR = A^f$ is 0 or 1. We may suppose that the image is 0 and so $h \in xR$.
The morphism $R \to R \otimes_{A[1/f]} K_\infty$ is injective, since R is a flat $A[1/f]$-module. Therefore h can be considered as an idempotent function on $\mathcal{Y} \otimes K_\infty$. Since $h \in xR$, the function h has a zero on each component of $\mathcal{Y} \otimes K_\infty$. Thus h is identically 0. □

Construction of $\hat{\mathcal{Z}}$.
This space is obtained by gluing copies of $\mathcal{Y}$ to $\mathcal{Z}$. Let $\mathcal{U}$ denote the open subset of $\mathcal{Z}$, defined by $c_1 \neq 0$. From the proof of lemma (5.1) we obtain an explicit isomorphism h of the open subset $x \neq 0$ of $\mathcal{Y}$ with $\mathcal{U}$. For the moment we denote the result of gluing $\mathcal{Y}$ and $\mathcal{Z}$ over $\mathcal{U}$ (by means of h) by $\tilde{\mathcal{Z}}$. We note that $\tilde{\mathcal{Z}}$ is separated. For this we have to verify that the map $O(\mathcal{Y}) \otimes_{A[1/f]} O(\mathcal{Z}) \to O(\mathcal{U})$ is surjective. This is a consequence of $O(\mathcal{Y})[x^{-1}] = O(\mathcal{U})$ and $x^{-1} \in O(\mathcal{Z})$. It is further clear that $\tilde{\mathcal{Z}}$ is smooth over $A[1/f]$ of relative dimension one.
The (chosen) isomorphism of $\mathcal{Z} \otimes K_\infty$ with the affine part of Z extends in a unique way to an embedding of $\tilde{\mathcal{Z}} \otimes K_\infty$ in Z.

From (5.3) one concludes that the subgroup H of G which stabilizes $\tilde{\mathcal{Z}} \otimes K_\infty$ consists of the triangular matrices $(g_{i,j}) \in G$ with diagonal entries $g_{1,1} \in (A/fA)^*$, $g_{2,2} \in \mathbf{F}_q^*$ and further $g_{1,2} = 0$, $g_{2,1} \in A/fA$. Let S denote a set of representatives of G/H. Then $\cup_{\sigma \in S} \sigma(C)$ is precisely the set of all cusp of Z. As a consequence Z is the union of the $\sigma(\tilde{\mathcal{Z}} \otimes K_\infty)$ taken over all $\sigma \in S$. Further, any two of those open subsets (with distinct σ's) intersect in $\mathcal{Z} \otimes K_\infty$. This is used in the construction of $\hat{\mathcal{Z}}$.

For each $\sigma \in S$ we consider a copy $\mathcal{Y}_\sigma$ of $\mathcal{Y}$. Every $\mathcal{Y}_\sigma$ is glued to $\mathcal{Z}$ by using the isomorphism σh between the open subset $x \neq 0$ and $\mathcal{U}$. The resulting space is called $\hat{\mathcal{Z}}$. Again, by gluing one obtains a line bundle, again denoted by L, on $\hat{\mathcal{Z}}$ and a homomorphism $\phi : A \to End(L)$. The finite morphism $j : \mathcal{Z} \to \mathbf{A}^1_{A[1/f]}$ extends to a morphism $j : \hat{\mathcal{Z}} \to \mathbf{P}^1_{A[1/f]}$.

Theorem 5.3
(1) *$\hat{\mathcal{Z}}$ is separated, connected and smooth of relative dimension one over $A[1/f]$.*
(2) *$j : \hat{\mathcal{Z}} \to \mathbf{P}^1_{A[1/f]}$ is a finite morphism. In particular $\hat{\mathcal{Z}}$ is the integral closure of $\mathbf{P}^1_{A[1/f]}$ in the field of fractions of $O(\mathcal{Z})$. The group $Gl(2, A/fA)/\mathbf{F}_q^*$ acts on $\hat{\mathcal{Z}}$ and the quotient of $\hat{\mathcal{Z}}$ by this action is $\mathbf{P}^1_{A[1/f]}$.*
(3) *$\hat{\mathcal{Z}}$ is proper over $A[1/f]$.*
(4) *$\hat{\mathcal{Z}} \otimes K_\infty$ is isomorphic to Z.*
(5) *The ring of global functions $O(\hat{\mathcal{Z}})$ on $\hat{\mathcal{Z}}$ is isomorphic to A^f. In particular, there is a morphism $\hat{\mathcal{Z}} \to M^1_{(f)}$.*
(6) *The connected components of $\hat{\mathcal{Z}} \otimes_{A[1/f]} A^f$ are absolutely irreducible and remain irreducible after tensoring with K_∞ over the subring A^f of K_∞.*

Proof. (1) The definition of the space $\hat{\mathcal{Z}}$ yields an obvious affine covering $\{U_i\}$, i.e. $\mathcal{Z}$ and the $\mathcal{Y}_\sigma$ with $\sigma \in S$. All the intersections $U_i \cap U_j$ (with $i \neq j$) are equal to the affine space $\mathcal{U}$. For $i \neq j$ one has to verify that the map $m_{i,j} : O(U_i) \otimes_{A[1/f]} O(U_j) \to O(\mathcal{U})$ is surjective. In the case of $\mathcal{Y}_\sigma$ and $\mathcal{Z}$ this follows from $O(\mathcal{U}) = O(\mathcal{Y}_\sigma)[x^{-1}]$ and $x^{-1} \in O(\mathcal{U})$. In the case of $\mathcal{Y}_{\sigma_1}$ and $\mathcal{Y}_{\sigma_2}$ with $\sigma_1 \neq \sigma_2$ one has again $O(\mathcal{U}) = O(\mathcal{Y}_{\sigma_1})[x^{-1}]$. From $\sigma_1 H \neq \sigma_2 H$ one can deduce that the images of $ker(\phi_f, O(\mathcal{Y}_{\sigma_1})$ and $ker(\phi_f, O(\mathcal{Y}_{\sigma_2})$ in $O(\mathcal{U})$ generate $ker(\phi_f, O(\mathcal{U}))$. It follows that x^{-1} lies in the image of $m_{i,j}$ and thus $m_{i,j}$ is again surjective. This proves that $\hat{\mathcal{Z}}$ is separated. Since each U_i is smooth and of relative dimension one over $A[1/f]$, also $\hat{\mathcal{Z}}$ is smooth over $A[1/f]$ of relative dimension one.
(2) The preimage $\mathcal{F}$ of the infinite section of $\mathbf{P}^1_{A[1/f]}$ is equal to the disjoint union of the closed subspaces $x = 0$ of the $\mathcal{Y}_\sigma$ and thus isomorphic to $\#S$ copies of $Spec(A^f)$. The ramification is everywhere $q^n(q-1)$. For every geometric

point ξ of $Spec(A^f)$ the fibre $\mathcal{F}_\xi$ has the correct number of points. From this one concludes that j is finite. The remaining part of (2) is obvious.
(3) follows from (2).
(4) is obvious from the construction.
(5) and (6) From $\hat{\mathcal{Z}}$ connected, smooth and proper over $A[1/f]$, one draws the conclusion that the domain $O(\hat{\mathcal{Z}})$ is finite and unramified over $A[1/f]$. The group $G := Gl(2, A/fA)/\mathbf{F}_q^*$ acts on $\hat{\mathcal{Z}}$ and on $O(\hat{\mathcal{Z}})$. Define the normal subgroup H of G by $H := \{B \in Gl(2, A/fA) |\ det(B) \in \mathbb{F}_q^*\}/\mathbb{F}_q^*$. We claim that H acts trivially on $O(\hat{\mathcal{Z}})$. Indeed, the analytic theory tells us that $O(\hat{\mathcal{Z}}) \otimes K_\infty$ is the ring of the global sections of $\hat{\mathcal{Z}} \otimes K_\infty$. The group G also acts on the analytic curve and one easily sees that the group H stabilizes every component of $\hat{\mathcal{Z}} \otimes K_\infty$. Thus H acts trivially on $O(\hat{\mathcal{Z}}) \otimes K_\infty$ and also on $O(\hat{\mathcal{Z}})$. The quotient group G/H is isomorphic to $A^*/\mathbb{F}_q^*$ and the invariants under this group acting upon $O(\hat{\mathcal{Z}})$ equals $A[1/f]$ by (2). Finally, the analytic theory also tells us that $O(\hat{\mathcal{Z}}) \otimes K_\infty$ is a product of copies of the field K_∞. Thus the field $O(\hat{\mathcal{Z}}) \otimes K$ is an abelian extension of K, totally split at the place ∞, and the ring $O(\hat{\mathcal{Z}})$ is a finite unramified extension of $A[1/f]$. The class field theory, explained in section 2.1, implies that $O(\hat{\mathcal{Z}})$ is isomorphic to A^f. The remaining statements of (5) and (6) are easy consequences of the proof given above. □

The line bundle L above $\hat{\mathcal{Z}}$ is completed in the obvious way to a bundle of projective lines which we will denote by PL. Let F denote the set $(f^{-1}A/A)^2 \cup \{\infty\}$. Outside the cusps there is for every $\xi \in F$ a section of PL. Each section has an obvious extension to $\hat{\mathcal{Z}}$. Those sections still intersect above the cusps. As in section 4, one can blow up the (non-empty) intersections of sections. Let us call $pr : M \to \hat{\mathcal{Z}}$ the result. M has a lot of structure induced by the Drinfeld data above the open part $\mathcal{Z}$.

1. $pr : M \to \hat{\mathcal{Z}}$ is a stable F-pointed tree, where F is the set $(f^{-1}A/A)^2 \cup \{\infty\}$ and where $\lambda : F \to \{$ sections of $pr\}$ is fixed.

2. Let M^* denote the open subset of M obtained by deleting the infinite section and above every cusp of $\hat{\mathcal{Z}}$ the irreducible components which do not carry points of $\lambda((f^{-1}A/A)^2)$. Then $pr^* : M^* \to \hat{\mathcal{Z}}$ has the structure of a group scheme with an A-action on it.

3. The group law $M^* \times M^* \to M^*$ extends to an action of a group scheme $M^* \times M \to M$.

4. The restriction of λ to the map $\lambda^* : (f^{-1}A/A)^2 \to \{$ the sections of $pr^*\}$ is an isomorphism between the constant sheaf of A-modules on $\hat{\mathcal{Z}}$ and the kernel of the multiplication by f on M^*.

5. Above $\mathcal{Z}$, one has that M^* with its group structure and A-action is identical with the Drinfeld data above $\mathcal{Z}$.

6. The formal fibre of M above any cusp is isomorphic to the Néron model of the Tate-Drinfeld module studied in section 4.

6 Generalized Drinfeld modules over a scheme

It is possible to produce a more or less natural functor $\mathcal{F}$ from the collection of $A[1/f]$-schemes to the category of sets which is represented by the scheme $\hat{\mathcal{Z}}$ (with its Drinfeld data). For a scheme S over $A[1/f]$, $\mathcal{F}(S)$ is the set of isomorphy classes of generalized Drinfeld modules of rank two and level f above S. We propose the following definition of a *generalized Drinfeld module of rank two and with level f over S*:

1. A stable F-pointed tree $pr : C \to S$, where F is the set $(f^{-1}A/A)^2 \cup \{\infty\}$ and where $\lambda : F \to \{$ sections of $pr\}$ is fixed.

2. Let C^* denote the open subset of C such that every geometric fibre of $pr^* : C^* \to S$ is equal to the geometric fibre of pr where $\lambda(\infty)$ and the irreducible components which do not carry points of $\lambda((f^{-1}A/A)^2)$ are deleted. Then $pr^* : C^* \to S$ is supposed to have a structure of group scheme with an A-action.

3. The group law $C^* \times C^* \to C^*$ extends to an action of a group scheme $C^* \times C \to C$.

4. The restriction of λ to the map $\lambda^* : (f^{-1}A/A)^2 \to \{$ the sections of $pr^*\}$ is an isomorphism between the constant sheaf of A-modules on S and the kernel of the multiplication by f on C^*.

5. For every geometric point ξ of S the fibre C_ξ is either an ordinary Drinfeld module of rank two and level f completed with a section ∞, or a generalized Drinfeld module (rank two and level f) over the field corresponding to ξ.

6. Let the closed point $s \in S$ have a fibre C_s which is a not a projective line. Then there exists a homomorphism $\alpha : A^f[[x]] \to \hat{O}_{S,s}$ such that $M \otimes_{A^f[[x]]} \hat{O}_{S,s}$ is isomorphic as scheme to $C \times_S Spec(\hat{O}_{S,s})$.

The last condition is not very natural. It seems possible to replace this by a more natural one, at the cost of rather technical statements and proofs.
It is obvious what is meant by an isomorphism of generalized Drinfeld modules over S. Thus the functor $\mathcal{F}$ is well defined.

The definition above is composed such that $M \to \hat{\mathcal{Z}}$ is the universal generalized Drinfeld module of rank two and with level f. We will not give a proof of this.

References

[DR] P. Deligne and M. Rapoport, *Les schémas de modules de courbes elliptiques,* in Modular Functions of One Variable II (P. Deligne and W. Kuijk, eds), Springer-Verlag, Berlin etc., LNM 349, 1973 (2nd corrected printing 1986), pp. 143–316.

[D] V. G. Drinfeld, *Elliptic Modules,* Math. USSR Sbornik **23** (1974), pp. 561–592.

[GHP] L. Gerritzen, F. Herrlich and M. van der Put, *Stable n-pointed trees of projective lines,* Proc. KNAW Series A **91** (1988), pp. 131-163.

[KM] N. Katz and B. Mazur, Arithmetic Moduli of Elliptic Curves. Princeton Univ. Press, Princeton, New Jersey, 1985.

Lecture 10

A survey of Drinfeld modular form

Gunther Cornelissen [1]
University of Gent
Department of Pure Mathematics and Computer Algebra
Galglaan 2, B-9000 Gent, Belgium
e-mail gc@cage.rug.ac.be

The possibility of a theory of modular forms for function fields was perhaps implicit in Drinfeld's original work, but the notion was only really introduced by Goss and Gekeler in the beginning of the eighties. A lot of it developed along "classical" lines (by the word "classical" we will always refer to modular forms for subgroups of $SL(2,\mathbf{Z})$), but there are also a few very intriguing novel facts to the function field side, and on these we want to concentrate. People who want to compare with the theory of classical modular forms are warmly refered to the well known classical literature.

One can see this paper as a kind of (very personal) filter applied to the original works of Gekeler, Goss and Teitelbaum. It should serve as an introduction to (and motivation for) reading their original papers in the reference list, which inevitably contain much more information then can be presented here.

In this paper, most interest will go to the "analytic" theory of modular forms and curves, and thus I have neglected parts of the intriguing theory of "algebraic" modular forms, developed in particular by Goss in 1980. For compensation, one should at least glimpse at his papers.

Most novelties over function fields occur in connection with (a) the consideration of an arbitrary "ground ring" A, which already makes the moduli space without level structure look complicated; (b) the characteristic-p situation, which causes loss of information in taking derivatives of objects; (c) the analytic reduction of the upper half space to a combinatorial object, namely the Bruhat-Tits tree, and the interpretation of objects "down there"; (d) the fact that "modular forms" are only half of the story, the other side being complex valued Jacquet-Langlands like automorphic forms – and the interaction between these two notions.

The organization of the paper will be as follows: in the first paragraph I will show how modular forms arise naturally in the search for a genus formula for the moduli space without level for $GL(2,A)$, and a few modular forms are introduced. The second paragraph contains a sketch of the explicit calculation of these genera. In the third part, function fields of modular curves are

[1] supported as "aspirant" by the Belgian National Fund for Scientific Research (NFWO)

described by means of modular forms, and the fourth paragraph calculates the Hilbert function and presentation of a few rings of modular forms. In paragraph five, we present a few highlights in the theory of modular forms modulo primes and its connection with supersingular invariants. Paragraph six sketches the interpretation of modular forms on the tree via Teitelbaum's residue-map, and its application to modular forms of low weight. In paragraph seven, the absolute value of some modular forms is studied as a complex-valued object on the tree.

Needless to say, I do not claim originality in this survey, except for minor generalizations here and there, in particular in (1.5.3), (4.3) and (4.6.2).

List of notations:

$\mathbf{F}_q$ = finite field of characteristic $\mathrm{char}(\mathbf{F}_q) = p$
K = function field of a complete connected curve $\mathcal{C}$ over $\mathbf{F}_q$
$Z_K(S) = \zeta_K(s)$ = the zeta-function of K, $(S = q^{-s})$ and
$P(S)$: the numerator of $Z_K(S)$; in particular,
$P(1)$ = the class number of K.
∞ = a fixed place of K, of degree δ, with associated absolute value $|\cdot|$
K_∞ = completion of K w.r.t. $|\cdot|$, and
π = a uniformizing element for $|\cdot|$
$\mathcal{O}_\infty$ = the ring of integers in K_∞
C = completion of an algebraic closure of K_∞
$\Omega = C - K_\infty$ = Drinfeld's "half" plane
A = the ring of elements of A that are regular outside ∞
$\delta \cdot P(1)$ is the class number of this ring
$GL(2,A)$ = the invertible 2×2 matrices over A
$\mathcal{Z}$ = the center of $GL(2,A)$
$\mathfrak{n}, \mathfrak{p}, \ldots$ (Fraktur letters) = places of A, of degree $\deg(\mathfrak{n})$,
where $\mathfrak{p}$ = prime place

$$\Gamma(\mathfrak{n}) = \{\gamma \in GL(2,A) : \gamma \equiv \begin{pmatrix} 1 & 0 \\ 0 & 1 \end{pmatrix} \mod \mathfrak{n}\}$$

$$\Gamma_0(\mathfrak{n}) = \{\gamma \in GL(2,A) : \gamma \equiv \begin{pmatrix} * & * \\ 0 & * \end{pmatrix} \mod \mathfrak{n}\}$$

$\Gamma \subseteq GL(2,A)$ = an arithmetic group (*i.e.* it contains $\Gamma(\mathfrak{n})$ for some $\mathfrak{n}$)
$\gamma \cdot z = \frac{az+b}{cz+d}$ = the action of an element $\gamma \in GL(2,A)$ on $z \in \Omega$ via fractional linear transformation
$M_\Gamma = \Gamma\backslash\Omega$ the quotient space for this action
$\bar{M}_\Gamma = \Gamma\backslash\Omega \cup \mathbf{P}^1(K)$ = the compactification of M_Γ
$\bar{M}_\Gamma - M_\Gamma = \Gamma\backslash\mathbf{P}^1(K)$ the cusps of Γ
$\infty = (1:0) \in \mathbf{P}^1(K)$ the cusp at infinity
$\mathrm{Stab}_G(x)$ = the stabiliser subgroup of x in G

ϕ = an A-Drinfeld module of rank two over C, given by a map $\phi : A \to C\{\tau\}$, where $C\{\tau\}$ is the "twisted" polynomial ring with multiplication rule $\tau x = x^q \tau$.
${}_{\mathfrak{n}}\phi$ = the $\mathfrak{n}$-torsion of ϕ
ρ = the Carlitz module = the rank one Drinfeld module over $\mathbf{F}_q[T]$ given by $\rho_T = T + \tau$, and $\bar{\pi}$ = it's period, i.e. ρ corresponds to the lattice $\bar{\pi}\mathbf{F}_q[T]$.
$g(X)$ = the geometric genus of a projective curve X
$\mathrm{Div}(X)$ = the group of divisors on a projective curve X, $\deg D$ = the degree of a divisor D
$\mathcal{D}$ = the canonical bundle on a fixed curve
$\mathcal{O}(D)$ = the line bundle associated with a divisor D on a curve
dim = the dimension of a C−vector space
$[\cdot]$ = the integer part, and $\langle\cdot\rangle$ = the fractional part of a real number

1. From modular curves to modular forms

(1.1) Since Riemann, understanding algebraic curves has become almost synonymous with the study of meromorphic functions and regular differentials (say, tangents) on them; for instance the number of independent differentials on a curve equals its fundamental invariant, the genus.

The same general philosophy should apply to the Drinfeld modular curves $\bar{M}_\Gamma$ associated with arithmetic groups $\Gamma \subseteq GL(2, A)$. The modular curves are (compactified) quotients of the Drinfeld "half" space Ω, so it seems only natural to look for functions and differentials "up" on Ω, let them descend on $\Gamma\backslash\Omega$ and extend them to the compactification.

(1.2) To make this descending procedure possible, we will first indicate how the analytic structures of Ω and $\bar{M}_\Gamma$ are related. There are two kinds of special points on $\bar{M}_\Gamma$ where the analytic structure ramifies — the local parameter at all other points is just $z \in \Omega$ itself.

(1.2.1) First of all, the *cusps* $\bar{M}_\Gamma - M_\Gamma$. Let $s \in \Gamma\backslash\mathbf{P}^1(K)$ be a cusp, and $\nu \in GL(2, K)$ such that $s = \nu \cdot \infty$. The stabilizer $\mathrm{Stab}_{\nu^{-1}\Gamma\nu}(\infty)$ contains a maximal subgroup of matrices $\begin{pmatrix} 1 & b \\ 0 & 1 \end{pmatrix}$, where b runs through some fractional ideal $\mathfrak{b} \subseteq A$. Then we set $t_s(z) = \bar{\pi}^{-1} e_{\mathfrak{b}}^{-1}(z)$. (Note : The normalising factor $\bar{\pi}$ is introduced for rationality questions of expansions, and this is not done everywhere in the literature). The *genuine* analytic parameter on $\bar{M}_\Gamma$ can be a strict power of t_s: we set w_s = the order of the group of homotheties $z \mapsto az$ contained in $\mathrm{Stab}_{\nu^{-1}\Gamma\nu}(\infty)$, then $t_s^{w_s}$ is the local analytic parameter on $\bar{M}_\Gamma$. For instance, if $\Gamma = GL(2, A)$, then $w_s = q - 1$ for all cusps s. For $\Gamma = GL(2, A)$, the number of such cusps is $\delta P(1)$, the class number of A.

(1.2.2) There is one more "special" kind of points, those $e \in \Omega$ for which the stabilizer $Stab_\Gamma(e)$ of $e \in \Omega$ in Γ (for the action of fractional transform) is strictly bigger then the trivial scalar matrices $\mathcal{Z}(\Gamma)$. Such points are called *elliptic*; the set of Γ-equivalence classes of elliptic points is denoted by $\mathcal{E}$.

The local parameter at an elliptic point $e \in \mathcal{E}$ on $\bar{M}_\Gamma$ is $t_e = (z-e)^{q+1}$. If $\gamma_e = \begin{pmatrix} a & b \\ c & d \end{pmatrix}$ is a non-scalar matrix such that $\gamma_e.e = e$ for some elliptic point, then $ce^2 + (d-a)e - b = 0$, and this equation has the same discriminant as the characteristic equation of γ_e, which has two distinct roots of unity as eigenvalues (γ_e is of finite order, since it fixes the image of e under the "building map" $\Omega \to \mathcal{T}$. This order is prime to p, since p-order elements in $GL(2, A)$ are conjugate to elements of the Borel group $\mathcal{B}$, which don't fix any element). Hence e lies in the constant extension $K(e) = \mathbf{F}_{q^2} \otimes K$.

Also note that if δ is even, $K(e) \subset \mathbf{F}_{q^\delta}[[X]] \cong K_\infty$, but $e \in \Omega$ cannot be in K_∞, so there are no elliptic points in this case. On the other hand, for odd δ, the number of elliptic points can be calculated using some arithmetic in $K(e)$ ([5], V.4.5), and one finds that $|\mathcal{E}| = P(-1)$ for δ odd.

(1.3) We are now ready to describe the descending procedure for functions and differentials:

A meromorphic function $f : \Omega \to C$ descends nicely if $f(\gamma \cdot z) = f(z)$ for all $z \in \Omega$ and $\gamma = \begin{pmatrix} a & b \\ c & d \end{pmatrix} \in \Gamma$.

As for differentials, $f(\gamma \cdot z) d(\gamma \cdot z) = f(\gamma \cdot z)(cz+d)^{-2} \det \gamma dz$, so $f(z)dz$ is invariant if

$$(*) \quad f(\gamma \cdot z) = \det \gamma^{-1} (cz+d)^2 f(z) \text{ for all } z \in \Omega, \gamma \in \Gamma.$$

In order to extend these functions and differentials nicely to the cusps $\bar{M}_\Gamma - M_\Gamma$, we will need an additional mero- or holomorphy condition at those points.

As will become clear in a moment, it is sensible to extend $(*)$ to include "multidifferentials" $f(z)dz^{\otimes k}$, since it will turn out that some of these are easier to construct and/or control. Hence we arrive at the following definition:

(1.4) Definition A meromorphic (in the sense of rigid analysis) function $f : \Omega \to C$ which is invariant under Γ, i.e. $f(\gamma \cdot z) = f(z)$, $\forall z \in \Omega, \gamma \in \Gamma$; and such that f is developable in a Laurent series in the parameter t_s at the cusps of Γ, is called a *modular function for* Γ. A (rigid) holomorphic function $f : \Omega \to C$ which satisfies

$$(1.4.1) \quad f(\gamma \cdot z) = \det \gamma^{-l} (cz+d)^k f(z), \ \forall z \in \Omega, \gamma \in \Gamma$$

and which is developable in a Taylor series in the parameter t_s at the cusps of Γ, is called a *modular form of weight k and type l for* Γ.

(1.4.2) Note that, although a modular form f is not invariant under Γ, it's order $\mathrm{ord}_{x} f$ at a point $x \in \Omega$ is (which is the lowest non-trivial power of $z - x$ occuring in the Taylor series development of f around x — also, we define the order of f at a cusp s to be the lowest non-trivial power of the parameter t_s occuring in the Taylor series of f in t_s. Note: this definition is in accordance with [6], but not with [5], where the parameter $t_s^{w_s}$ is used. But if one considers modular forms f of type $l \neq 0$, only $t_s^{-l} f$ admits a series development in $t_s^{w_s}$.) We define the "divisor" of a modular form f for Γ to be the "rational" divisor $\mathrm{div} f \in \mathrm{Div}(\bar{M}_\Gamma) \otimes \mathbf{Q}$ on the modular curve $\bar{M}_\Gamma$ given by

$$\mathrm{div} f = \sum_{x \in \Gamma \backslash \Omega - \mathcal{E}} \mathrm{ord}_{x} f + \sum_{e \in \mathcal{E}} \frac{\mathrm{ord}_{e} f}{q+1} + \sum_{s \in \bar{M}_\Gamma - M_\Gamma} \frac{\mathrm{ord}_{s} f}{w_s}.$$

(1.4.3) The C-vector space of modular forms of weight k and type l for Γ is denoted by $M_{k,l}(\Gamma)$. If $l = 0$, we write $M_k(\Gamma)$ instead. We will also denote the graded rings of modular forms as follows :

$$M(\Gamma) := \bigoplus_{\substack{k \in \mathbf{Z}_{>0} \\ l \in \mathbf{Z}/(q-1)}} M_{k,l}(\Gamma), \quad M_0(\Gamma) := \bigoplus_{k \in \mathbf{Z}_{>0}} M_k(\Gamma).$$

A superscript $M^n_{*,*}$ will denote the space of n-fold cusp forms, i.e. forms that vanish of order $\geq n$ at all cusps.

(1.4.4) *Remark* If $\gamma = \mathrm{diag}(\epsilon, \epsilon) \in \Gamma$ for some $\epsilon \in \mathbf{F}_q^*$ of multiplicative order m, and if $k - 2l \neq 0 \bmod m$, then $M_{k,l} \equiv 0$. In particular, $M_k(GL(2, A)) = 0$ if $k \neq 0 \bmod q - 1$. This can be seen by substituting γ in (1.4.1).

(1.5) *Examples* We will give some examples of non-trivial modular forms.

(1.5.1) **Eisenstein series** (Goss, [10]) The function

$$E_k(z) := \sum_{(0,0) \neq (a,b) \in A^2} \left(\frac{1}{az + b}\right)^k,$$

is called the Eisenstein series of weight k for $GL(2, A)$. The transformation equation (1.4.1) is easy to check, for holomorphy at the cusps, see *loc. cit.*

(1.5.2) **Coefficient forms** If ϕ^{Λ_z} is the rank two Drinfeld module associated with the lattice $\Lambda_z = zA \oplus A$ for some $z \in \Omega$, then for $a \in A$,

$$\phi_a^{\Lambda_z} = a + \sum_{i=1}^{2\deg(a)} l_i(z, a) \tau^i,$$

the "coefficient forms" $l_i(z,a)$ are modular forms of weight $q^i - 1$ for Γ. To check the transformation formula, use the fact that $\phi^{c.\Lambda} = c\phi^\Lambda c^{-1}$ for any rank two lattice Λ and scaling constant $c \in C^*$. For holomorphy, once can use the following "mysterious" connection between this and the first example:

$$aE_{q^k-1}(z) = \sum_{i+j=k} E_{q^i-1}(z) l_j^{q^i}(z,a),$$

where we set $E_0 \equiv -1$. (see [5], II.2.11). In particular, the leading term

$$\Delta_a := l_{2\deg(a)}(z,a)$$

is a modular forms of weight $q^{2\deg(a)} - 1$ for $GL(2,A)$. Furthermore, by the definition of a rank two Drinfeld module, $\Delta_a(z) \neq 0$ if $z \in \Omega$. This means Δ_a has its divisor supported at the cusps.

As far as notation is concerned, if $A = \mathbf{F}_q[T]$, we let $g(z) = \bar{\pi}^{1-q} l_1(z,T)$ and $\Delta(z) = \bar{\pi}^{1-q^2} l_2(z,T)$. Let me also remark that while reading the paper [6] by Gekeler, one should beware of the change of notation for g, Δ on page 683. Here, we have normalised as in the second part of that paper.

(1.5.3) **Derivatives** The derivative f' of a modular form f with respect to z isn't always a modular form. If $f \in M_{k,l}(\Gamma)$ and $\gamma \in \Gamma$, one has

$$f'(\gamma.z) = \det\gamma^{-(l+1)}(cz+d)^{k+2} f'(z) + kc\det\gamma^{-(l+1)}(cz+d)^{k+1} f(z).$$

But on the other hand, if we choose a non-constant $a \in A$, and let $E_a(z) := \bar{\pi}^{-1}\Delta_a'/\Delta_a$, then

$$E_a(\gamma.z) = \det\gamma^{-1}(cz+d)^2 E_a(z) - c\bar{\pi}^{-1}\det\gamma^{-1}(cz+d).$$

If we define an operator (the "Serre-derivative") on $M_{k,l}(\Gamma)$ by

(1.5.4) $\quad \partial_a f(z) := \bar{\pi}^{-1} f'(z) - kE_a(z) f(z),$

then it follows from the above calculations that $\partial_a f$ transforms like a form of weight $k+2$ and type $l+1$. Also, it is holomorphic where f is. (Since $\Delta_a(z) \neq 0$ if $z \in \Omega$, the only possible problem would occur at the cusps. But one sees easily from the product development of Δ_a given in [5], VI (4.12), that its "logarithmic derivative" $E_a(z)$ is holomorphic. — Note that for the uniformisers t_s one has $t_s'/t_s = -t_s$ since the derivative of an "exponential" function $e_\Lambda(z)$ is trivial.)

The derivation operator ∂_a gives a way of constructing forms of non-trivial type from examples of type 0. In particular, for $A = \mathbf{F}_q[T]$, we set $\partial := \partial_T$, and $h := \partial g \in M_{q+1,1}(GL(2,\mathbf{F}_q[T]))$. One then has (cf. Gekeler, [6]) : $h^{q-1} = -\Delta$, and

$$h(z) := \sum_{\substack{c,d \in A \\ (c,d)=1}} \frac{t_\infty(\gamma_{c,d}.z)}{(cz+d)^{q+1}},$$

is a kind of Poincaré-series, where $\gamma_{c,d}$ is any chosen matrix of determinant one and bottom row (c,d).

(1.5.5) **Division values** Let $\mathfrak{n} \subset A$ be an ideal, and $u = (u_1, u_2) \in \mathfrak{n}^{-1}A^2 - A^2$. Define $e_u(z) := e_{\Lambda_z}(u_1 z + u_2)$. Then e_u^{-1} is a modular form of weight one for the principal congruence subgroup $\Gamma(\mathfrak{n}) = \{\gamma \in GL(2,A) : \gamma \equiv \mathbf{1} \bmod \mathfrak{n}\}$. As a matter of fact,

$$e_u^{-1}(z) = \sum_{\substack{0 \neq a \in K^2 \\ a \equiv u \bmod A}} \frac{1}{a_1 z + a_2}$$

is a kind of Eisenstein series of weight one for $\Gamma(\mathfrak{n})$, where the non-archimedean convergence of the right hand side is easy. Furthermore, the values of $e_u(z)$ for different u give the $\mathfrak{n}$-torsion of the Drinfeld module ϕ^{Λ_z} corresponding to the lattice $\Lambda_z = zA \oplus A$. (Note that classically, x-coordinates of division points are modular forms of weight *two*, connected with Heckes "non-modular" Eisenstein series of weight two).

This example connects up with the previous ones via the fact that for a principal ideal $\mathfrak{n} = (n), \Delta_n$ and $\prod_u e_u^{-1}(z)$ differ only up to a constant, where u runs through a system of representatives of $n^{-1}A^2/A^2$. Similarly, E_{q-1} is a scalar multiple of $\sum e_u^{1-q}$. Also, for $A = \mathbf{F}_q[T]$, h is homothetic to

$$\sum_{u,v \in \mathbf{F}_q^2/\mathbf{F}_q^*} (u_1 v_2 - u_2 v_1) e_{T^{-1}u}^{-q} e_{T^{-1}v}^{-1},$$

which follows from [6], (9.3).

The expansion of e_u at the cusps of $\Gamma((n))$ can be calculated (*cfr.* Gekeler [5], VI); and from this a product expansion of the "discriminant function" Δ_n follows. For $A = \mathbf{F}_q[T]$, it takes on the following form:

$$\bar{\pi}^{1-q^2}\Delta = -t_\infty^{q-1} \prod_{a \in A \text{ monic}} (\rho_a(t_\infty^{-1})t^{q^{\deg(a)}})^{(q-1)(q^2-1)}.$$

This is indeed analogous to Jacobi's formula for the classical discriminant function, if one regards $\rho_a(X)$ as similar to the cyclotomic $X^n - 1$. For general A, a similar formula exists, albeit somewhat involved ([5],VI). But one can calculate exactly the divisor of Δ_n, and hence its degree. Let $d = \deg(n)$. If we suppose that $f \in M_k(\Gamma)$ is a modular form of weight k, then $f^{1-q^{2d}}\Delta_n^k$ is a modular function, hence of degree zero. And since degree is additive, we find that

$$(1.5.6) \quad f \in M_k(GL(2,A)) \;:\; \deg(\operatorname{div} f) = \frac{k(q^\delta - 1)}{(q^2-1)(q-1)} P(q),$$

using the formula for deg (div Δ_a).

Suppose $A = \mathbf{F}_q[T]$. The divisors of the modular functions $e_u e_v^{-1}$ are supported at the cusps of $\Gamma(\mathfrak{n})$, and one can show that they generate a group of finite index in the group of cuspidal divisors of degree zero. Hence the group generated by the cusps in the Jacobian of $\Gamma(\mathfrak{n})$ is finite (*loc. cit.*); this is the theorem of Manin-Drinfeld for function fields ([12], IV.2).

(1.6) There are Hecke-operators acting on modular forms, just like classically. A fact which is "badly understood" is that for $A = \mathbf{F}_q[T]$, the eigenvalues of the Hecke operator $T_\mathfrak{p}$ for a prime $\mathfrak{p}$ on g and Δ are *the same*. Also, something like $T_{\mathfrak{p}^2} = T_\mathfrak{p}^2$ is different from the classical situation.

2. The genus of modular curves

(2.1) The modular curve for $GL(2, \mathbf{F}_q[T])$ is easy to describe : if $\phi_T^{\Lambda_z} = T + g(z)\tau + \Delta(z)\tau^2$ is the rank two Drinfeld module of lattice Λ_z, then ϕ^{Λ_z} is isomorphic to $\phi^{\Lambda_{z'}}$ if and only if there exists a constant $c \in C^*$ such that $g(z) = c^{1-q}g(z')$ and $\Delta(z) = c^{1-q^2}\Delta(z')$, and hence

$$j : M_{GL(2,\mathbf{F}_q[T])} \to C : z \mapsto j(z) := \frac{g^{q+1}(z)}{\Delta(z)}$$

is an isomorphism. In particular, the modular curve $\bar{M}_{GL(2,\mathbf{F}_q[T])}$ is of genus zero. But already for a more general A, such an easy description of moduli of rank two Drinfeld modules is not available. As a matter of fact, already the construction of Drinfeld modules is non-trivial, *cfr.* Dummit and Hayes [3].

(2.2) To extract information about the modular curve for a general $\Gamma = GL(2, A)$, we will use the dictionary between modular curves and modular forms.

(2.2.1) As we saw, the local parameters at the cusps are t_s^{q-1}, and for differential forms this gives the formula

$$f(z)dz^{\otimes k} = f(z(t_s^{q-1}))(t_s^{q-1})^{\frac{-kq}{q-1}} d(t_s^{q-1})^{\otimes k}.$$

(2.2.2) At elliptic points $e \in \mathcal{E}$ we have a uniformizer $t_e = (z-e)^{q+1}$. Hence for differential forms we find locally at elliptic points :

$$f(z)dz^{\otimes k} = f(z(t_e))t_e^{\frac{-kq}{q+1}} dt_e^{\otimes k}.$$

(2.3) If f is a modular form of weight $2k$ and type k for $GL(2, A)$, we find that $f(z)dz$ is a k-fold differential form on $\bar{M}_\Gamma$ which (by the relation between the uniformizers on Ω and $\bar{M}_\Gamma$ which was just established)

- is holomorphic at $z \in \Gamma\backslash\Omega - \mathcal{E}$
- has a pole of order $\leq \frac{kq}{q+1}$ at $z \in \mathcal{E}$
- has a pole of order $\leq \frac{kq}{q-1}$ at cusps z.

If we view $\Delta_a(z)dz$ as a section of the divisor of meromorphic $\frac{q^{2d}-1}{2}$-fold differential forms satisfying the above three criteria, we find that

$$\deg(\mathrm{div}\Delta_a) = \frac{q^{2d}-1}{2}((2g-2) + |\mathcal{E}|\frac{q}{q+1} + |\bar{M}_\Gamma - M_\Gamma|\frac{q}{q-1}),$$

where g is the genus of $\bar{M}_\Gamma$. Hence using formula (1.5.6) for the left hand side, and the expression for the number of cusps and elliptic points (1.2) on the right hand side, we find that

(2.3) Theorem (Gekeler, [5]) *The genus of the Drinfeld modular curve for $GL(2,A)$ equals*

$$g(\bar{M}_{GL(2,A)}) = 1 + \frac{1}{q^2-1}(\frac{q^\delta - 1}{q-1}P(q) - \frac{q(q+1)}{2}\delta P(1) + \eta),$$

where $\eta = 0$ if δ is even, and $\eta = -\frac{q(q-1)}{2}P(-1)$ if δ is odd, where as before, P is the numerator polynomial of the zeta-function of K.

(2.4) *Remark* The form Δ_a was preferable, because its divisor is so well understood. On the other hand, the distribution of the zeroes of e.g. the Eisenstein series is still not very well known, but for weight $q^k - 1$ and $A = \mathbf{F}_q[T]$, see [1]. For instance, one has

(2.4.1) Proposition ([5], VII.3.3) *The order of E_{q-1} at all elliptic points is one.*

(2.5) *Remark* The above theorem (2.3) also calculates the first Betti number of the groups $GL(2,A)$, generalising Serre [13] ([5], V.A.11).

(2.6) For arithmetic subgroups $\Gamma \subseteq GL(2,A)$, one can now calculate the genus of the corresponding modular curve as is done classically, via the Hurwitz-formula.

(2.6.1) ([5],VII.5) For a principal congruence subgroup $\Gamma(\mathfrak{n})$ of level $\mathfrak{n}$, the group of the covering $\bar{M}_\Gamma \to \bar{M}_{GL(2,A)}$ is $G(\mathfrak{n}) := GL(2,A)/\Gamma(\mathfrak{n}) \cdot \mathcal{Z}$ (where $\mathcal{Z}$ is the center of $GL(2,A)$) of order

$$|G(\mathfrak{n})| = q^{3\deg(\mathfrak{n})} \prod_{\substack{\mathfrak{p}|\mathfrak{n} \\ \mathfrak{p}\ \text{prime}}} (1 - \frac{1}{q^{2\deg(\mathfrak{p})}}),$$

and the ramification is tame of order $q+1$ at the elliptic points. There is wild ramification at the cusps, with only a non-trivial first ramification group of order $q^{\deg(\mathfrak{n})}$. Applying Hurwitz, we find

$$g(\bar{M}_{\Gamma(\mathfrak{n})}) = 1 + |G(\mathfrak{n})|\left(\frac{q^{\delta}-1}{(q-1)(q^2-1)}P(q) - \frac{\delta P(1)}{q^{\deg(\mathfrak{n})}(q-1)}\right).$$

Note that apparently, there is a change of notation in [5], VII.5.8: from there on, h is the class number of A (whereas it used to be the class number of K).

(2.6.2) An even more tedious calculation using the covering $\bar{M}_{\Gamma(\mathfrak{p})} \to \bar{M}_{\Gamma_0(\mathfrak{p})}$ show that the genus of $\bar{M}_{\Gamma_0(\mathfrak{p})}$ for the Hecke group $\Gamma_0(\mathfrak{p}) := \{\gamma \in GL(2,A) : \gamma \equiv \begin{pmatrix} * & * \\ 0 & * \end{pmatrix} \bmod \mathfrak{p}\}$ of *prime* level $\mathfrak{p}$ equals

$$g(\bar{M}_{\Gamma_0(\mathfrak{p})}) = 1 + \frac{1}{q^2-1}\left(\frac{q^{(\deg(\mathfrak{p})}+1)(q^{\delta}-1)}{q-1}P(q) - \delta q(q+1)P(1) + 2\eta\right),$$

where η is as in theorem (2.3).

$g(K)$	δ	$g(\bar{M}_{GL(2,A)})$	$\deg(\mathfrak{p})$, $\mathfrak{p}$ prime	$g(\bar{M}_{\Gamma(\mathfrak{p})})$	$g(\bar{M}_{\Gamma_0(\mathfrak{p})})$
0	1	0	1	0	0
			2	q^4-q^3-q	0
			3	—	q
	2	0	1	q^2-q-1	0
			2	—	q
	3	0	1	q^3+q^2-2q-2	q
	4	q	—	—	—
1	1	0	1	$(q^4-q^2-q)+$ $+(q^2-q-1)(P(1)-q-1)$	q^2
	2	q^2	—	—	—
	3	$q(q^2+P(1)-1)$	—	—	—

A table of low genera

3. Function fields of some modular curves

(3.1) Modular forms also give a natural construction of modular functions: the quotient of two modular forms of the same weight is a meromorphic Γ-invariant function. We have already encountered "the" modular function j in (2.1) as an example of this. More generally, a Drinfeld module is determined by its coefficient forms (1.5.2), so it seems only natural to consider suitable quotients of them as functions on the modular curve $\bar{M}_{GL(2,A)}$. As a matter of fact,

(3.1.1) Proposition ([5], VII.1.3) *The function field of $\bar{M}_{GL(2,A)}$ is the field F consisting of quotients of homogeneous polynomials of the same degree in the coefficient forms (1.5.2) $l_i(a,z)$, (or E_{q^i-1}, which is the same) for $1 \le i \le 2\deg(a)$ and an arbitrary non-constant $a \in A$.*

Note that the term "same degree" in the above proposition is taken with respect to the natural grading by weight. (weight$(l_i) = q^i - 1$).

The proof consists essentially in checking that if all functions in F take on the same value at two different points of Ω, then these points are $GL(2,A)$-equivalent.

(3.1.2) *Remark* Application of the Riemann-Roch theorem to divisors of the form $k.\infty$ on K shows immediately that there exists a non-constant $a \in A$ of degree $= [\max\{\frac{g(K)}{\delta}, 2\frac{g(K)-1}{\delta}\}]$, where $[\cdot]$ indicates the integer part.

(3.2) Proposition (loc. cit.) *The function field of of $\bar{M}_{\Gamma(\mathfrak{n})}$ is generated over the function field of $\bar{M}_{GL(2,A)}$ by $E_{q-1}(z)e_u^{q-1}(z)$, where u runs through a set of representatives of $\mathfrak{n}^{-1}A^2/A^2$.*

Again, the proof consists in verifying that two points of Ω at which all the above functions agree are $\Gamma(\mathfrak{n})$-equivalent. Also note that, for $A = \mathbf{F}_q[T]$, the above two proposition are the analogues of what is very well known classically ([14], 6.1), which is why we leave out the details.

4. From modular curves to modular forms

(4.1) We have seen that modular forms seem to be very effective in studying the C-geometry of modular curves. Let's consider them for a moment as objects of study on their own; a natural question would be to ask about the structure of the rings $M(\Gamma)$ of modular forms for different Γ. (As a matter of fact, this remains linked with more subtle geometric aspects of the modular curves, namely some of their smooth embeddings in projective spaces).

A few answers along classical lines are known: first of all, the graded parts $M_{k,l}(\Gamma)$ are finite-dimensional C- vector spaces, and an application of the Riemann-Roch theorem on the modular curve will allow the calculation of their dimension in most cases (a few very interesting ones left aside for the moment). I will start by sketching the results for $\Gamma = GL(2,A)$, where A is general and the type $l = 0$; and then for $A = \mathbf{F}_q[T]$ and arbitrary type (although a detailed analysis of the behaviour of ∂E_{q-1} would certainly lead to a satisfactory formula in the case of arbitrary weight and type).

Because of the funny way in which parameters on Ω and $\bar{M}_\Gamma$ are linked, we found the divisor of a modular form on $\bar{M}_\Gamma$ having rational coefficients. But if we want to apply the Riemann-Roch theorem we will need "integral" divisors. Fortunately, if we define the integral part of a rational divisor $D =$

$\sum_{x\in \bar{M}_\Gamma} a_x x \in \mathrm{Div}(\bar{M}_\Gamma)\otimes \mathbf{Q}$ to be $[D]=\sum_{x\in\bar{M}_\Gamma}[a_x]x$, where $[a_x]$ is the integral part of a_x, then it is easy to see that $H^0(\bar{M}_\Gamma, D) = H^0(\bar{M}_\Gamma,[D])$ (Shimura, [14], 2.21).

If f is a non-zero modular form for $GL(2,A)$, its weight k is divisible by $q-1$, as we noted in (1.4.4). Here is the basic trick: we send a modular form f of weight k and type zero to the modular function $f.E_{q-1}^{-\frac{k}{q-1}}$, which establishes an isomorphism of $M_k(\Gamma)$ with $H^0(\bar{M}_\Gamma,[k\mathrm{div}E_{q-1}])$. To apply Riemann-Roch, we want to know the degree of this divisor; from (1.5.6) we know the degree of the rational divisor $\deg(\mathrm{div}E_{q-1})$, but there is no definite general way in which $\deg D$ and $\deg[D]$ are linked for arbitrary D. Fortunately we also know that the order of E_{q-1} at cusps s is always divisible by $q-1$ (E_{q-1} has a Taylor series in t_s^{q-1} at all cusps, because it is of type zero), so that the "cuspidal" part of $[\mathrm{div}E_{q-1}]$ is integral. But even more is known, all zeroes at elliptic points are simple by (2.4.1). Hence

$$\deg[k\mathrm{div}E_{q-1}] = k\deg(\mathrm{div}E_{q-1}) - \langle\frac{k}{q+1}\rangle|\mathcal{E}|,$$

where $\langle x\rangle = x-[x]$ is the fractional part of x. We see that unless $k=1, q=2$, this degree is bigger than $2g(\bar{M}_\Gamma)-2$, so that the divisor is non-special, and the Riemann-Roch theorem gives:

(4.2) Theorem (Gekeler, [5], VII.4.6) *If $(q,k)\neq(2,1)$, the dimension of the vectorspace of modular forms of weight k for $GL(2,A)$ is zero if $q-1$ doesn't divide k, and otherwise*

$$\dim M_k(GL(2,A)) = 1 - g(\bar{M}_{GL(2,A)}) + \frac{k(q^\delta-1)}{(q-1)(q^2-1)}P(q) - \mu,$$

where $\mu=0$ when δ is even, and $\mu=\langle\frac{k}{q^2-1}\rangle P(-1)$ if δ is odd.

If we now let $A=\mathbf{F}_q[T]$, and f is a modular form of weight k and type l for $GL(2,\mathbf{F}_q[T])$, then $k-2l\equiv k-l(q+1)\equiv 0 \bmod q-1$, as we saw in (1.4.4). Therefore

$$M_{k,l}(GL(2,\mathbf{F}_q[T])\to H^0(\bar{M}_{GL(2,\mathbf{F}_q[T])},[\frac{k-l(q+1)}{q-1}\mathrm{div}g + l\mathrm{div}h]),$$

given by $f\mapsto fh^{-l}g^{\frac{l(q+1)-k}{q-1}}$ is an isomorphism. We note that (1.5.6) again allows the calculation of the degree of div h, and we see that $\mathrm{ord}_\infty h = 1$ from the series development. Furthermore, as $h=\partial g$, and the zero of g at the elliptic point is simple ($g'(e)\neq 0$), it follows that $h(e)\neq 0$. Similar to (4.2), one arrives at:

(4.3) Proposition *The dimension of the vectorspace of modular forms of weight k and type l for $GL(2, \mathbf{F}_q[T])$ is zero unless $k \equiv 2l \bmod q-1$, when*

$$\dim M_{k,l}(GL(2, \mathbf{F}_q[T]) = 1 + \frac{k}{q^2-1} - \langle \frac{k - l(q+1)}{q^2-1} \rangle - \langle \frac{l}{q-1} \rangle.$$

(4.4) The situation for a principal congruence subgroup is even more advantageous (compare [10], [5], VII.6): there are no elliptic points for $\Gamma(\mathfrak{n}), \mathfrak{n} \subset A$, since all elements of finite order in it are of order divisible by p (see the discussion of elliptic points). Also, since all elements in $\Gamma(\mathfrak{n})$ have determinant one, the type doesn't play a role.

The parameter at cusps s is just t_s, and we see that if $f \in M_2(\Gamma(\mathfrak{n}))$, then $f(z)dz = f(z(t_s))t_s^{-2}dt_s$ is a differential form on the modular curve which is everywhere holomorphic; except at the cusps, where it can have a pole of order 2.

If f is a modular form of weight k for $\Gamma(\mathfrak{n})$, then $fe_u^k \in H^0(\bar{M}_{\Gamma(\mathfrak{n})}, \mathcal{M}^{\otimes k})$, where $\mathcal{M} := \operatorname{div}(e_u^{-1})$ is the linebundle of modular forms of weight one. By the above identification of modular forms of weight 2 and differential forms, we see that if we let $\mathcal{D}$ denote the canonical bundle on $\bar{M}_{\Gamma(\mathfrak{n})}$, and Σ the "divisor of cusps" $\sum_{x \text{ cusp}} x$, we have

$$\mathcal{M}^{\otimes 2} \cong \mathcal{D} \otimes \mathcal{O}(2\Sigma).$$

One can immediately calculate the degree of $\mathcal{M}$ from this to find

$$\deg \mathcal{M} = \frac{q^\delta - 1}{(q-1)(q^2-1)} P(q)|G(\mathfrak{n})|,$$

and it is easy to see that this degree is $> 2g(\bar{M}_{\Gamma(\mathfrak{n})}) - 2$ if $k > 1$, hence an application of Riemann-Roch and duality can be made.

If $k = 1$, there is no obvious (degree) reason why the "speciality index" $h^1(\bar{M}_{\Gamma(\mathfrak{n})}, \mathcal{M})$ should vanish. Note however that we can interpret it in terms of modular forms: via duality, $H^1(\bar{M}_{\Gamma(\mathfrak{n})}, \mathcal{M}) = H^0(\bar{M}_{\Gamma(\mathfrak{n})}, \mathcal{M}^{\otimes -1} \otimes \mathcal{D}) = H^0(\bar{M}_{\Gamma(\mathfrak{n})}, \mathcal{M} \otimes \mathcal{O}(-2\Sigma)) = M_{1,0}^2(\Gamma(\mathfrak{n}))$, which is the space of so-called *double cusp forms of weight one.*

(4.4.1) Proposition ([5], VII.6) *The dimension of the space of modular forms of weight $k > 1$ for a principal congruence subgroup $\Gamma(\mathfrak{n})$ is*

$$\dim M_k(\Gamma(\mathfrak{n})) = 1 - g(\bar{M}_{\Gamma(\mathfrak{n})}) + \frac{q^\delta - 1}{(q-1)(q^2-1)} P(q)|G(\mathfrak{n})|,$$

and the dimension of $M_1(\Gamma(\mathfrak{n}))$ is the number of cusps of $\Gamma(\mathfrak{n})$ plus the number of double cusp forms of weight one.

(4.5) A very similar argument works for the case of a Hecke group $\Gamma_0(\mathfrak{p})$, say of prime level $\mathfrak{p}$.

(4.5.1) Proposition ([5], VII.6) *If $(q,k) \neq (2,1)$, the dimension of the space of modular forms of weight k for a Hecke subgroup $\Gamma_0(\mathfrak{p})$ of prime level $\mathfrak{p}$ is zero unless $q-1$ divides k, when it is*

$$\dim M_k(\Gamma_0(\mathfrak{p})) = 1 - g(\bar{M}_{\Gamma_0(\mathfrak{p})}) + k\frac{(q^\delta - 1)(q^{\deg(\mathfrak{p})} - 1)}{(q-1)(q^2-1)}P(q) - 2\mu,$$

where μ is as in (4.2).

$g(K)$	δ	$\dim M_{k(q-1)}(GL(2,A))$
0	1	$1 + [\frac{k}{q+1}]$
	2	$1+k$
	3	$1 + k\frac{q^2+q+1}{q+1} - \langle\frac{k}{q+1}\rangle$
	4	$1 - q + k(q^2+1)$
1	1	$1 + k\frac{P(q)}{q+1} - \langle\frac{k}{q+1}\rangle P(-1)$
	2	$1 - q^2 + kP(q)$

$g(K)$	δ	$\deg(\mathfrak{p})$, $\mathfrak{p}$ prime	$\dim M_k(\Gamma(\mathfrak{p}))$ $k>1$	$\dim M_{k(q-1)}(\Gamma_0(\mathfrak{p}))$
0	1	1	$kq+1$	$k+1$
		2	$k(q^4+q^2+1)+$ $-q^4+q^3+q$	$1 + k\frac{q^2+1}{q+1} - 2\langle\frac{k}{q+1}\rangle$
		3	—	$1 - q + k(q^2-q+1)$
	2	1	$k(q^2+1) - q^2 + q$	$1 + k(q+1)$
		2	—	$1 - q + k(q^2+1)$
	3	1	$k(q^3+q^2-q-1)+$ $-q^3-q^2+2q+2$	$1 - q + k(q^2+q+1)$
1	1	1	$k(q^4-q^3+(P(1)-1)q^2+$ $+q) - q^4 + q^3 + q^2+$ $-q - P(1)(q^2-q-1)$	$1 - q^2 + kP(q)$

Tables of dimensions of spaces of modular forms

(4.6) The full structure of the algebra of modular forms is only known in the most "trivial" cases. One has for instance, along classical lines:

(4.6.1) Proposition (Goss [9] and Gekeler [6] (5.12), (5.13)) *The algebra of modular forms for $GL(2,\mathbf{F}_q[T])$ is isomorphic to polynomial ring $C[X,Y]$, doubly graded by weight $\in \mathbf{Z}_{>0}$ and type $\in \mathbf{Z}/(q-1)$, where weight$(X) = q-1$, weight$(Y) = q+1$, type$(X) = 0$ and type$(Y) = 1$; the isomorphism is given by sending X to g and Y to h.*

The algebra of modular forms of type zero for $GL(2,\mathbf{F}_q[T])$ is isomorphic to the graded polynomial ring $C[X,Y]$, where weight$(X) = q-1$ and weight$(Y) = q^2-1$; the isomorphism is given by sending X to g and Y to Δ.

The proof of these two statements is analogous and classical, so we will stick to a sketch of the first one. It suffices to show that g and h are algebraically independent. Because then, one can calculate the dimension of the (k,l)-graded part $\dim C[g,h]_{k,l}$ of the subring $C[g,h]$ of $M(GL(2,\mathbf{F}_q[T]))$, and it is easy to see that it coincides with $\dim M_{k,l}(GL(2,\mathbf{F}_q[T]))$, calculated in (4.3).

Suppose there exists an algebraic relation between g,h. We might as well suppose it is homogeneous with respect to the natural grading by weight, and of minimal weight d. If $g^{\frac{d}{q-1}}$ occurs, then we evaluate the expression at ∞, and since $h(\infty) = 0$, this would imply that $g(\infty) = 0$, a contradiction. On the other hand, if $h^{\frac{d}{q+1}}$ occurs, we evaluate at an elliptic point instead, and using $g(e) = 0$, we would find $h(e) = 0$, but $h^{q-1}(e) = -\Delta(e) \neq 0$. Hence the algebraic relation is divisible by gh, and performing this division, we would find a relation of smaller degree, contradicting the minimality of d. This proves the theorem. □

(4.6.2) Proposition *The ring of modular forms of type zero for the Hecke group $\Gamma_0((T))$ of level T over $\mathbf{F}_q[T]$ is isomorphic to the graded ring $C[X,Y]$, where X,Y are variables of weights $q-1$; the isomorphism is given by sending X to $g(z)$ and Y to $g(Tz)$.*

The fact that $g(Tz)$ is a modular form for $\Gamma_0((T))$ follows as in [11],III prop. 17 *et seq*. The same technique of the previous proof applies, where we now first evaluate a possible algebraic relation between $g(z)$ and $g(Tz)$ at an elliptic point e, and then at $T^{-1}e$. All we have to note is that g has its unique zero at e, and that $T^{-1}e$ cannot be an elliptic point: since e satisfies a quadratic equation of discriminant $D(e) \in \mathbf{F}_q$ (see (1.2.2)), one has $D(T^{-1}e) = T^2D(e) \notin \mathbf{F}_q$, hence $T^{-1}e$ is not elliptic. □

5. Congruences for modular forms and supersingular invariants

(5.1) Untill now, we have not shown very much interest in the coefficients of modular forms, by which we mean the coefficients of the Taylor series at different cusps. Nevertheless, it is known classically that after suitable normalisation, these have a significant arithmetical relevance. Although some of the interesting coefficients in the function field case will occur for automorphic, complex valued forms (think of number of representations of polynomials by quadratic forms), it is still possible to normalise say Eisenstein series, and extract information about supersingular invariants, along the same lines as the classical study of Swinnerton-Dyer in the Antwerp volumes [15]. Throughout this paragraph, we will assume $A = \mathbf{F}_q[T]$.

(5.2) If $\mathfrak{p}$ is a prime (irreducible polynomial) of $\mathbf{F}_q[T]$, then we let $M_\mathfrak{p}$ be the ring of modular forms f for $GL(2, \mathbf{F}_q[T])$ for which the Taylor series of f at the cusp ∞ lies in the localisation $\mathbf{F}_q(T)_\mathfrak{p}[[t_\infty]]$, i.e. such that all coefficients lie in K, and have denominator prime to $\mathfrak{p}$. It then makes sense to consider the ring

$$\bar{M}_\mathfrak{p} = \{f \in \mathbf{F}_q[T]/(\mathfrak{p})[[t_\infty]] : f \equiv f' \bmod \mathfrak{p} \text{ for some } f' \in M_\mathfrak{p}\}$$

of modular forms modulo $\mathfrak{p}$ (congruence is meant termwise in t_∞).

(5.3) Here is the prototype: we will let

$$g_k := (-1)^{k+1}\bar{\pi}^{1-q^k}(T^{q^k} - T)(T^{q^{k-1}} - T)...(T^q - T)E_{q^k-1},$$

where E_{q^k-1} is the Eisenstein series for $A = \mathbf{F}_q[T]$ of weight $q^k - 1$. It turns out that the series development of g_k at the cusp ∞ has A-integral coefficients, i.e. belongs to $A[[t_\infty]]$. This is true in particular for g and hence for h.

(5.3.1) Theorem (Gekeler, [6]) *Let $\mathfrak{p} \in \mathbf{F}_q[T]$ be a prime of degree k. Then $g_k \equiv 1 \bmod \mathfrak{p}$, and if we let $B_k \in A[X, Y]$ denote the polynomial such that $g_k = B_k(g, h)$ (which exists by (4.6.1)), then*

$$\bar{M}_\mathfrak{p} \cong \mathbf{F}_q[T]/(\mathfrak{p})[X, Y]/(b_k(X, Y) - 1),$$

where b_k is obtained from B_k by reducing coefficients modulo $\mathfrak{p}$. Also, $\bar{M}_\mathfrak{p}$ is a normal ring.

Although the result is entirely similar to the one by Swinnerton-Dyer, the proof is slightly different; whereas the classical proof uses a double application of the Serre-derivative to the Eisenstein series, it is known that for function fields (*loc. cit.*) $\partial^2 g = \partial h = 0$, so $\partial^2 \equiv 0$ on the whole of $M(GL(2, A))$.

(5.4) We will consider the formal "Tate"-Drinfeld module $\phi_T^{Ta} = TX + g(t_\infty)X^q + \Delta(t_\infty)X^{q^2}$ over $\mathbf{F}_q[T]((t_\infty))$, where we subsitute $g(t_\infty), \Delta(t_\infty)$ for their expansions in t_∞. One can reduce this formal module modulo $\mathfrak{p}$ to find a Drinfeld module over $\mathbf{F}_q[T]/(\mathfrak{p})((t_\infty))$ (Note that the series expansions start of as $g = 1 + ..., \Delta = t_\infty^{q-1} + ...$). Then $\phi_\mathfrak{p}^{Ta} \bmod \mathfrak{p}$ will have lowest degree term $H_\mathfrak{p}(t_\infty)X^{q^{\deg \mathfrak{p}}}$ for some power series $H_\mathfrak{p} \in \mathbf{F}_q[T]/(\mathfrak{p})((t_\infty))$, and it turns out that $H_\mathfrak{p} \in \bar{M}$, namely $H_\mathfrak{p} \equiv g_k \bmod \mathfrak{p}$ (*loc. cit.*). $H_\mathfrak{p}$ is called the ***Hasse-invariant***. The name arises as follows: if one gives t_∞ a fixed value such that the series for g, Δ converge, then the reduction ϕ_T^{Ta} mod $\mathfrak{p}$ is a Drinfeld module ϕ' of finite characteristic $\mathfrak{p}$, which will be *supersingular* $\iff H_{\mathfrak{p}}(t_\infty) = 0$ $\iff$ the $\mathfrak{p}$-torsion ${}_{\mathfrak{p}}\phi' = 0$ is trivial $\iff$ the endomorphism ring of ϕ' is non-commutative. This observation opens up the possibility of studying supersingular Drinfeld modules via reductions of Eisenstein series.

6. From modular forms to C-valued harmonic cochains

(6.1) The upper half space Ω has a cannonical reduction, which is the (intersection dual of) the Bruhat-Tits tree $\mathcal{T}$. One might ask what remains of modular forms on Ω when they are reduced to $\mathcal{T}$. It turns out that the result is analogous to the classical theory of periods of modular forms ([12], VI).

(6.2) Suppose f is a weight two modular form for some group Γ. Then as we saw, $f(z)dz$ is a holomorphic differential form on Ω. It was shown by Drinfeld how to "descend" with such a differential form to the tree: Let e be an edge of $\mathcal{T}$ corresponding to the interior of an analytic annulus $D^0_{x,n} = \{z \in \Omega : |\bar{\pi}_\infty|^{n+1} < |z - x| < |\bar{\pi}_\infty|^n\}$, and assume that the inner circle $|z - x| = |\bar{\pi}_\infty|^{n+1}$ reduces to the *origin* of e. The restriction of $f(z)dz$ to D^0_i admits a series development $f(z)dz = \sum a_k t^k dt$, where $t = \bar{\pi}^{-n}(z - x)$ is the boundary parameter. We define the residue of $f(z)dz$ at e to be $\mathrm{res}(f(z)dz, e) := a_{-1}$. The residue theorem applied to the arc $D^0_{x,n}$ implies that $\sum \mathrm{res}(f(z)dz, e) = 0$, where the sum is over all edges starting at an arbitrary vertex, i.e. the function res is harmonic; and it is Γ-invariant for the natural action of Γ on $\mathcal{T}$. To be able to generalize this construction, we recall

(6.3) Definition For an abelian group B (written additively), we let $\underline{H}(\mathcal{T}, B)$ denote the set of *harmonic cochains* on $\mathcal{T}$, i.e. functions from the set of edges of $\mathcal{T}$ to B, such that if the orientation of an edge is reversed, so is the value of the cochain; and the sum of the values of the cochain over all edges leaving a fixed vertex is zero. If Γ is an arithmetic group, we let $\underline{H}(\mathcal{T}, B)^\Gamma$ denote Γ-invariant cochains, and we will let $\underline{H}_!(\mathcal{T}, B)^\Gamma$ denote the set of *cuspidal harmonic cochains for* Γ, i.e. cochains with compact support modulo Γ.

(6.4) Let V be a two-dimensional vectorspace over C with the standard action of Γ. Set

$$W_{k,l} = [(\det)^{l-1} \otimes \mathrm{Sym}^{k-2}(V^*)]^*,$$

for k, l are positive integers (one should understand "det" as C with Γ-action by multiplication with the determinant). This definition is very formal, but can be understood practically as follows : suppose $\varphi \in H(\mathcal{T}, W_{k,l})$, and choose a basis X, Y of V^*. For any edge e, $\varphi(e) \in W_{k,l}$ can be seen as a function on the space of $\mathrm{Sym}^{k-2}(V^*)$ spanned by monomials of degree $k-2$ in the variables

X, Y. So a harmonic cochain like this is really just a map, associating to any edge of $\mathcal{T}$ a "vector" of $k-1$ values in C, namely the values $\varphi(e)(X^jY^{k-2-j})$ for $j = 0, \ldots, k-2$ (but with the right Γ-action).

(6.5) Suppose $f \in M_{k,l}(\Gamma)$ is a modular form of weight $k > 1$ and type l. Define a map

$$\text{Res} : M_{k,l}(\Gamma) \to \underline{H}_!(\mathcal{T}, W_{k,l})^\Gamma$$

by the following :

$$\text{Res}(f, e)(X^jY^{k-2-j}) = \text{res}(z^j f(z)dz, e).$$

(The expression $z^j f(z)dz$ should be seen as a differential form on Ω, it makes no sense on the modular curve.) This is again a harmonic cocycle by the rigid residue theorem, and one computes immediately that $\text{Res}(f, \gamma \cdot e) = \gamma \cdot \text{Res}(f, e)$, which implies Γ-equivariance. Also, the resulting cochain is cuspidal, since Teitelbaum ([16],3) has shown the "strange" positive characteristic phenomenon that every Γ-equivariant cochain taking values in a group killed by $p = \text{char}(\mathbf{F}_q)$ is cuspidal.

(6.6) The prototype of non-cusp forms, the Eisenstein series have residue $= 0$ since the associated differential forms are holomorphic on Ω. On the other hand, Teitelbaum has constructed an explicit inverse to the map Res for cusp forms, using a "non-archimedean" integral, and shown this inverse to be injective. He can then use the dimension formulae for $M_{k,l}(\Gamma)$ and an explicit construction of sufficiently many harmonic cochains to make Res bijective:

(6.7) Theorem (Teitelbaum, [16]) *The map*

$$Res : M^1_{k,l}(\Gamma) \to \underline{H}_!(\mathcal{T}, W_{k,l})^\Gamma$$

is an isomorphism for $k > 1$.

(6.8) *Remark* In *loc. cit.*, Teitelbaum uses the interpretation of cocycles through modular forms to obtain results about the relative homology of Γ.

(6.9) The most striking application of theorem 6.7 however, is already a few years old, but remains unpublished untill now. Here it is:

(6.9.1) Theorem (Gekeler — unpublished) *Let $\Gamma \subseteq GL(2, A)$ be an arithmetic subgroup. Then there are no cusp forms of weight one for Γ whatsoever, i.e. $M^1_{1,l}(\Gamma) = 0, \forall l$.*

Proof. Suppose $f \in M^1_{1,l}(\Gamma)$, then it's q-th power $f^q \in M^1_{q,l}(\Gamma)$. Choose an edge $e \in \mathcal{T}$, corresponding to a annulus $D^0_{x,n}$ with parameter t on Ω as in (6.2). If $f(z)dz = \sum_{k \in \mathbf{Z}} a_k t^k dt$ on $D^0_{x,n}$, then $z^i f^q(z)dz = \sum_{k \in \mathbf{Z}} a_k^q t^{kq+i} dt$.

In particular, $\mathrm{res}(z^i f^q(z)dz, e) = 0$ for $0 \leq i \leq q-2$. Hence $\mathrm{Res}(f^q) \equiv 0$, and by Teitelbaum's theorem, $f^q \equiv 0$, so $f \equiv 0$. □

(6.9.2) Corollary *One can complement the results in paragraph 4 as follows:*
$\dim M_1(\Gamma(\mathfrak{n}))$ is exactly equal to the number of cusps of $\Gamma(\mathfrak{n})$. Furthermore, (4.2) and (4.5.1) remain valid if $(q,k) = (2,1)$.

(6.9.3) *Remark* The theorem might seem disturbing at first; classical modular forms in $SL_2(\mathbf{Z})$ of weight one are intimately connected with Artin's conjecture for representations of $\mathrm{Gal}(\bar{\mathbf{Q}}/\mathbf{Q})$ in $GL(2,\mathbf{C})$, and counting, constructing and interpreting $M_1(\Gamma)$ over the integers is very deep. Notice however that the classical "equivalence" between modular forms and automorphic forms/Galois representations doesn't remain valid in a direct way for function fields: whereas modular forms and modular curves are the natural characteristic-p defined moduli objects, defined starting from Drinfeld's "fine" space Ω, the automorphic side is about *complex* valued harmonic forms on the tree $\mathcal{T}$.

7. From modular forms to complex valued harmonic cochains

(7.1) The "automorphic" side of function fields arithmetic involves complex harmonic analysis on the tree $\mathcal{T}$. And much of the succes of the "modular" theory over function fields lies in the intimate relation between these distinct concepts of modular and automorphic form, *see* Gekeler & Reversat [8]. We will only touch upon a few recent applications of it to questions about modular forms and modular curves.

(7.2) For this, we need the r-map, introduced by van der Put in his study of invertible maps on rigid analytic spaces. It should be considered as a "good" substitute for the ordinary logarithmic derivative in the case of positive characteristic (recall e.g. that ∂^2 is bad, since it kills all of $M(GL(2,\mathbf{F}_q[T]))$!). The map r associates to every *invertible* holomorphic function on Ω a $\mathbf{Z}$-valued harmonic cochain. If e is an edge of $\mathcal{T}$, then $q^{r(f)(e)}$ is (by definition) the quotient of the maximal absolute value of f attained on the circle corresponding to the endpoint of e, by the maximal absolute value of f attained on the circle corresponding to the origin of e. It is connected with the "characteristic-p" cochains of the previous paragraph via

$$\mathrm{res}(f'(z)/f(z)dz, e) = r(f)(e) \bmod p,$$

which explains its relation with the logarithmic derivative. Also, if two points $z_1, z_2 \in \Omega$ reduce to vertices v_1, v_2 on $\mathcal{T}$, then the value

(*) $\log_q |f(z_2)/f(z_1)|$ = the sum of the values of $r(f)$ along the path from v_1 to v_2 in $\mathcal{T}$.

For proofs of all of this, see [4],I.8.9, [8], (1.7).

(7.3) A modular form f takes on values in C. But if we take its absolute value, then $|f| : \Omega \to \mathbf{R}$ takes on real values. And via (*), we can study this absolute value as an $r(f)$-weighted distance function on the tree, at least if f is invertible. e.g. the discriminant function:

(7.4) Theorem (Gekeler [7], (2.8)) *Let $z \in \Omega$ reduce to a vertex $[\pi^{-k}\mathcal{O}_\infty \oplus \mathcal{O}_\infty]$ of $\mathcal{T}$. Then*

$$\log_q |\Delta(z)| = \begin{cases} q^2 + q - q^{1-k} & \text{if } k \leq 1 \\ q^2 + q + k(q^2-1) - q^{k+1} & \text{if } k \geq 1 \end{cases}$$

I have only indicated the formula for points reducing to special vertices, but actually, this allows the calculation of the value everywhere. It would also be interesting to know $|e_u|$ for division values e_u.

(7.5) There are immediate applications of the above result to the question of extracting "roots" out of functions like Δ and $\Delta(z)/\Delta(nz)$ in the ring of invertible functions on Ω, and in the function field of $\bar{M}_{\Gamma_0((n))}$, and from this, one can proceed to study the group of divisor classes supported at the cusps of $\bar{M}_{\Gamma_0((n))}$, e.g.

(7.6) Theorem (Gekeler [7], 3.23) *The cuspidal divisor class group of $\bar{M}_{\Gamma_0(\mathfrak{p})}$ for a prime ideal $\mathfrak{p}$ is cyclic of order $(q^{\deg \mathfrak{p}} - 1)/(q^2-1)$ if $\deg \mathfrak{p}$ is even and $(q^{\deg \mathfrak{p}} - 1)/(q-1)$ if $\deg \mathfrak{p}$ is odd.*

References

[1] G. Cornelissen, *Sur les zéros des séries d'Eisenstein de poids $q^k - 1$ pour $GL_2(F_q[T])$*, C. R. Acad. Sci. Paris Sér. I Math. **321** (1995), 817–820.

[2] ———, *Drinfeld modular forms of weight one*, this volume, 1996.

[3] D. S. Dummit and D. Hayes, *Rank-one Drinfeld modules on elliptic curves*, Math. Comp. **62** (1994), no. 206, 875–883.

[4] J. Fresnel and M. van der Put, *Géometrie analytique rigide et applications*, Progress in Mathematics, vol. 18, Birkhäuser, 1981.

[5] E.-U. Gekeler, *Drinfeld modular curves*, Lecture Notes in Math., vol. 1231, Springer, Berlin - Heidelberg - New York, 1986.

[6] ______, *On the coefficients of Drinfeld modular forms*, Invent. Math. **93** (1988), 667–700.

[7] ______, *On the Drinfeld discriminant function*, to appear in *Compos. Math.*, 1995.

[8] E.-U. Gekeler and M. Reversat, *Jacobians of Drinfeld modular curves*, J. Reine Angew. Math. **476** (1996), 27–93.

[9] D. Goss, *Modular forms for $F_r[T]$*, J. Reine Angew. Math. **31** (1980), 16–39.

[10] ______, *$\bar{\pi}$-adic Eisenstein series for function fields*, Compositio Math. **41** (1980), 3–38.

[11] N. Koblitz, *Introduction to elliptic curves and modular forms*, 2 ed., Graduate Texts in Math., vol. 97, Springer, Berlin - Heidelberg - New York, 1993.

[12] S. Lang, *Introduction to modular forms*, Grundlehren der math. Wiss., vol. 222, Springer, Berlin - Heidelberg - New York, 1976.

[13] J.-P. Serre, *Trees*, Springer, Berlin - Heidelberg - New York, 1980.

[14] G. Shimura, *Introduction to the arithmetic theory of automorphic functions*, Publ. of the Math. Soc. of Japan, vol. 11, Iwanami Shoten and Princeton University Press, 1971.

[15] H.P.F. Swinnerton-Dyer, *On l-adic representations and congruences for coefficients of modular forms*, in: Modular functions of one variable III, Lecture Notes in Math., vol. 350, Springer, Berlin - Heidelberg - New York, 1973, 1—55.

[16] J. T. Teitelbaum, *The Poisson kernel for Drinfeld modular curves*, J. Amer. Math. Soc. **4** (1991), no. 3, 491–511.

Lecture 11

Automorphic forms and Drinfeld's reciprocity law

Marius van der Put and Marc Reversat
Department of Mathematics, P.O.Box 900, 9700 AV Groningen, The Netherlands
email mvdput@math.rug.nl
and Laboratoire de Mathématiques Emile Picard, U.M.R. 5580 Université Paul Sabatier-Toulouse 3, 118 route de Narbonne F-31062 Toulouse cedex 4, France,
email reversat@picard.ups-tlse.fr

Introduction

The purpose of this lecture is to explain §10 and §11 of Drinfeld's paper [Dr]. At the time [Dr] was written there was no étale cohomology available for rigid analytic spaces. This (and other matters) made the paper [Dr] and the proof of the reciprocity law difficult to understand. Parts of [Dr] are already clearly exposed in [De-Hu]. Nowadays étale cohomology and other ingredients of the proof are better understood and worked out in some recent papers (e.g. [Be], [dJ-vdP], [Hu] and [Ge-Re]). This enables us to give a fuller account of Drinfeld's reciprocity law.

We will sketch the contents of this lecture. The main object is a representation $\underline{H}$ of the group $Gl(2,\mathcal{A}_f) \times Gal(K^{sep}/K)$ which is obtained from the first l-adic cohomology groups on the projective system of the Drinfeld modular curves $\bar{M}_I^2$. A more explicit description of this representation is obtained by comparing $\underline{H}$ with the analytic étale cohomology groups. The analytic étale cohomology groups can be made visible by means of harmonic cochains on the Bruhat-Tits tree. Those groups of harmonic cochains have a translation into cuspidal automorphic forms. This leads to a bijecion between a class of automorphic representations and families of two-dimensional Galois representations. This bijection is called "Drinfeld's reciprocity law".
A special feature of this reciprocity is that it associates to an elliptic curve E over K with split multiplicative reduction over K_∞, a certain irreducible admissible automorphic representation. In lecture 12, it will be shown that this automorphic representation is obtained from a certain theta function. This has as consequence that E is a quotient of the Jacobian variety of a Drinfeld modular curve over K assosiated to some level. The last statement is the Shimura-Taniyama-Weil conjecture for function fields.
The aim of this lecture is to give "complete" proofs. For most of the results presented here, we give either a complete proof or precise and accessible references in the literature. The part of this lecture which escapes this procedure is

the one concerned with automorphic representations. Here we try to make the definitions and the statements clear and refer in a more general way to [Gel, Ha, J-L, De 1, De 2] for complete information and "complete proofs".

1 Cohomology on Drinfeld's modular curves

1.1 *The cohomology group* $\underline{H}$

As in previous lectures K is a function field, containing the finite field $\mathbb{F}_q$. A place ∞ of K is chosen. $A \subset K$ is the ring of the elements which are regular outside ∞. Further K_∞ is the completion of K at the place ∞ and $\mathbf{C}$ is the completion of the algebraic closure of K_∞.
For a nontrivial ideal I of A, one can consider the functor, which associates to any K-scheme S the set of the isomorphy classes of the rank two Drinfeld modules above S with a level I structure. This functor is representable by an affine smooth curve over K, which will be denoted by M_I^2. (See lecture 2). The compactification of M_I^2 will be denoted by $\bar{M}_I^2$. The family $\{\bar{M}_I^2\}_I$ forms a projective system of smooth projective curves over K. The projective limit of this family will be denoted by $\bar{M}^2$ and the projective limit of the family of affine curves $\{M_I^2\}_I$ will be denoted by M^2. The two projective limits are well defined K-schemes. The main object, studied in this lecture, is

$$\underline{H} := H^1_{et}(\bar{M}^2 \otimes K^{sep}, \overline{\mathbb{Q}}_l).$$

This expression needs a bit of explaining.

1. The term H^1_{et} stands for the first *algebraic* étale cohomology group. The letter l denotes a prime different from p. Further $H^1_{et}(-, \overline{\mathbb{Q}}_l)$ is defined as $H^1_{et}(-, \hat{\mathbb{Z}}_l) \otimes_{\hat{\mathbb{Z}}_l} \overline{\mathbb{Q}}_l$ and finally $H^1_{et}(-, \hat{\mathbb{Z}}_l)$ is defined as the projective limit of the groups $H^1_{et}(-, \mathbb{Z}/l^r\mathbb{Z})$.

2. It can be shown that $H^1_{et}(\bar{M}^2, \mathbb{Z}/l^r\mathbb{Z})$ is the direct limit of the groups $H^1_{et}(\bar{M}_I^2, \mathbb{Z}/l^r\mathbb{Z})$. If one does not appreciate the K-scheme $\bar{M}^2$, then one may view the expression $H^1_{et}(\bar{M}^2 \otimes K^{sep}, \overline{\mathbb{Q}}_l)$ as an abbreviation of the direct limit

$$\varinjlim H^1_{et}(\bar{M}_I^2 \otimes K^{sep}, \overline{\mathbb{Q}}_l).$$

3. A standard construction in étale cohomology yields a canonical action of the Galois group $Gal(K^{sep}/K)$ on each of the $\overline{\mathbb{Q}}_l$-vector spaces $H^1_{et}(\bar{M}_I^2 \otimes K^{sep}, \overline{\mathbb{Q}}_l)$. This provides $\underline{H}$ with a $Gal(K^{sep}/K)$-action. We note that the dimension of the vector space $H^1_{et}(\bar{M}_I^2 \otimes K^{sep}, \overline{\mathbb{Q}}_l)$ is known to be $2g$, where g is the genus of the (possibly not connected) curve $\bar{M}_I^2$.

1.2 *The action of* $G(\mathcal{A}_f) \times Gal(K^{sep}/K)$ *on* $\underline{H}$

First we have to introduce some notations.

(1.2.1) G denotes the group Gl_2 (over $\mathbb{Z}$). Its center is given the name Z. For any ring R one can consider the groups $G(R)$ and $Z(R)$.

(1.2.2) For any place (i.e., the equivalence class of a discrete valuation) v of K, the field K_v is the completion of K with respect to this place. The ring of integers of K_v is denoted by $\mathcal{O}_v$. Of special interest is the infinite place $v = \infty$. A generator of the maximal ideal of the valuation ring $\mathcal{O}_\infty$ is denoted by π_∞.

(1.2.3) The ring of adèles of the field K is denoted by $\mathcal{A}$. We recall its definition. $\mathcal{A}$ is the restricted product $\prod'_v K_v$, i.e., consists of the elements $(a_v)_v$ of the ordinary product $\prod_v K_v$ (v runs in the set of all places of K) such that for all but finitely many v one has $a_v \in \mathcal{O}_v$. The field K is diagonally embedded in $\mathcal{A}$.

(1.2.4) $\mathcal{O}$ denotes the product $\prod_v \mathcal{O}_v$.

(1.2.5) The places of K, different from ∞ are called the finite places. The ring $\mathcal{A}$ can be written as $\mathcal{A}_f \times K_\infty$, where $\mathcal{A}_f$ (the ring of finite adèles) is the restricted product $\prod'_{v\neq\infty} K_v$. Likewise $\mathcal{O} = \mathcal{O}_f \times \mathcal{O}_\infty$ with $\mathcal{O}_f = \prod_{v\neq\infty} \mathcal{O}_v$.

(1.2.6) We note that $\mathcal{O}_f$ is the completion $\hat{A}$ of A with respect to the set of all its non-zero ideals. It is easily seen that the obvious map $\hat{A} \otimes K \to \mathcal{A}_f$ is an isomorphism.

(1.2.7) We want to indicate how the group $G(\mathcal{A}_f)$ acts on $\underline{H}$. The procedure is somewhat delicate. For the classical modular curves, the situation is quite similar and we copy our construction from that case.
We want to find a right-action of the group $G(\mathcal{A}_f)$ on $\bar{M}^2$. This will induce a left $G(\mathcal{A}_f)$-action on $\underline{H}$. It suffices to find such an action on M^2.

Above M^2 there lives a universal Drinfeld module, which we will call D, of rank two with a level-I structure for every I. Let $D[I]$ denote the kernel of ϕ_I acting on D and let D_{tors} denote the direct limit of all $D[I]$. All the level structures can be combined into one isomorphism $\lambda : (\mathcal{A}_f/\hat{A})^2 \to D_{tors}$, where we note that K/A is canonically isomorphic to $\mathcal{A}_f/\hat{A}$. The object (M^2, λ) is universal in the category of the (equivalence classes of) pairs $(F/S, (\mathcal{A}_f/\hat{A})^2 \xrightarrow{\cong} F_{tors})$, where S is any K-scheme and F/S is a Drinfeld module of rank two over S with a level I-structure for every I.

The group of the automorphisms of the A-module $(\mathcal{A}_f/\hat{A})^2$ is easily seen to be $G(\hat{A})$. Using this one can define an action of $G(\hat{A})$ on M^2.
This is not quite what we want since the action of $G(\mathcal{A}_f)$ is not yet visible. One replaces an object above, i.e., some $(F/S, (\mathcal{A}_f/\hat{A})^2 \xrightarrow{\cong} F_{tors})$, by its isogeny class. The term F_{tors} is replaced by the Tate module $T(F)$, which is defined as the projective limit of the $F[I] := ker(\phi_I, F)$. Since we are working with isogeny classes, $T(F)$ is not well defined but $T(F) \otimes_A K$ is well defined. Thus we are looking at isogeny classes $(F/S, (\mathcal{A}_f)^2 \xrightarrow{\cong} T(F) \otimes K)$. This defines a new functor an K-schemes. It can be shown that this new functor is representable and that (the isogeny class of) the object $(D/M^2, \lambda : (\mathcal{A}_f)^2 \xrightarrow{\cong} T(D) \otimes K)$ is universal. The ordinary action of the group $G(\mathcal{A}_f)$ on $(\mathcal{A}_f)^2$ induces then a right action on M^2. Indeed, for any $g \in G(\mathcal{A}_f)$ one can change the level structure λ into $\lambda \circ g$. The universal property implies the existence of an automorphism $\tilde{g} : M^2 \to M^2$ such that $\tilde{g}$ transports the level structure λ into $\lambda \circ g$. In this way one finds an action of $G(\mathcal{A}_f)$ on M^2. This action extends to a right action on $\bar{M}^2$ and a left action on $\underline{H}$. For another description of the action of $G(\mathcal{A}_f)$ on M^2 we refer to [Ge 2].

We conclude that the $\overline{\mathbf{Q}}_l$-vector space $\underline{H}$ has a natural structure of $G(\mathcal{A}_f) \times Gal(K^{sep}/K)$-module. In other words $\underline{H}$ is a representation of the group $G(\mathcal{A}_f) \times Gal(K^{sep}/K)$. We note that the algebraic closure $\overline{\mathbf{Q}}_l$ of $\mathbf{Q}_l$ is needed later on, when one wants to decompose the representation $\underline{H}$.

At the moment the representation of the group $G(\mathcal{A}_f) \times Gal(K^{sep}/K)$ in the $\overline{\mathbf{Q}}_l$-vector space $\underline{H}$ is a rather "hazy" object. We have to introduce automorphic forms, analytic étale cohomology and harmonic cochains in order to make $\underline{H}$ more concrete.

2 Automorphic forms

2.1 Adèles, double cosets and the Bruhat-Tits tree

We introduce a number of the concepts, notations and properties that will play a rôle in the theorems.

(2.1.1) The groups $G(\mathcal{A})$, $G(\mathcal{O}_v)$, et cetera have a canonical topology. The ring $\mathcal{O}$ is compact and thus the group $G(\mathcal{O})$ is also compact. We will consider various spaces of double cosets. The first one is $G(K)\backslash G(\mathcal{A}_f)/G(\hat{A})$. In order to describe this we consider lattices.
An A-lattice $Y \subset K^2$ is a projective, rank two A-submodule of K^2. On this set the group $G(K)$ operates. The set of $G(K)$-orbits is equal to the set of isomor-

phy classes of projective rank two A-modules. Since A is a Dedekind ring, any projective rank two A-module Z can be written as a direct sum $A \oplus B$, with B a rank one projective A-module. Clearly $\Lambda^2 Z \cong B$. Using this one finds that the set of isomorphy classes of projective rank two modules is isomorphic to the Picard group $Pic(A)$ of A. This Picard group is known to be finite.

An $\hat{A}$-lattice Λ in $\mathcal{A}_f^2 = \hat{A} \otimes_A K^2$ is a $\hat{A}$-submodule such that for every $v \neq \infty$ one has that Λ_v is a free $\mathcal{O}_v$-module of rank two. Moreover one supposes that for all but finitely many $v \neq \infty$ one has $\Lambda_v = \mathcal{O}_v^2$. On the collection of the $\hat{A}$-lattices the group $G(\mathcal{A}_f)$ acts. The stabilizer of the (trivial) lattice $\hat{A}^2$ is the subgroup $G(\hat{A})$. It is obvious that $G(\mathcal{A}_f)$ acts transitively on this set of $\hat{A}$-lattices. And thus the set of $\hat{A}$-lattices can be identified with $G(\mathcal{A}_f)/G(\hat{A})$. For every A-lattice $Y \subset K^2$ one obtains the $\hat{A}$-lattice $\hat{A} \otimes_A Y \subset \hat{A} \otimes K^2 = \mathcal{A}_f^2$. This map can be seen to be bijective. Thus we find that the double coset space $G(K)\backslash G(\mathcal{A}_f)/G(\hat{A})$ can be identified with the finite set $Pic(A)$.

(2.1.2) We recall from lecture 6 that the Bruhat-Tits tree $\mathcal{T}$ associated to the group $G(K_\infty)$ is obtained from its action on the equivalence classes $[M]$ of $\mathcal{O}_\infty$-lattices in K_∞^2. The set of the vertices of this tree can be identified with the discrete set $G(K_\infty)/G(\mathcal{O}_\infty)Z(K_\infty)$. Indeed, the action of $G(K_\infty)$ on the equivalence classes of $\mathcal{O}_\infty$-lattices is transitive and the stabilizer of $[\mathcal{O}_\infty \oplus \mathcal{O}_\infty]$ is $G(\mathcal{O}_\infty)Z(K_\infty)$.

The set of the (oriented) edges of $\mathcal{T}$ can be identified with the discrete set $G(K_\infty)/I_\infty Z(K_\infty)$ with I_∞ the *Iwahori group*. Indeed, $G(K_\infty)$ acts transitively on the set of oriented edges and $I_\infty \subset G(\mathcal{O}_\infty)$ is defined as the stabilizer of the oriented edge $[\pi_\infty \mathcal{O}_\infty \oplus \mathcal{O}_\infty] \to [\mathcal{O}_\infty \oplus \mathcal{O}_\infty]$. It is easily seen that I_∞ is explicitly given as $I_\infty = \{\begin{pmatrix} a & b \\ c & d \end{pmatrix} \in G(\mathcal{O}_\infty) \mid c \equiv 0 \bmod \pi_\infty\}$.

(2.1.3) For any open subgroup $\mathcal{K}$ of $G(\mathcal{O})$ one is interested in the double coset space $G(K)\backslash G(\mathcal{A})/\mathcal{K}Z(K_\infty)$. For general $\mathcal{K}$ it is somewhat difficult to see what this double coset space is. However the group $\mathcal{K}$ is of finite index in $G(\mathcal{O})$ since $G(\mathcal{O})$ is compact. Thus all the groups $\mathcal{K}$ that we consider are commensurable. To get the picture we may suppose that $\mathcal{K}$ has the form $\mathcal{K}_f \times \mathcal{K}_\infty$, with $\mathcal{K}_f$ an open subgroup of $G(\mathcal{O}_f) = G(\hat{A})$ and $\mathcal{K}_\infty$ an open subgroup of $G(\mathcal{O}_\infty)$. We will underline elements of adelic nature and write pairs (,) to indicate the finite and the infinite part of an element in $G(\mathcal{A})$. We note that the group $G(K)$ is diagonally embedded in $G(\mathcal{A}) = G(\mathcal{A}_f) \times G(K_\infty)$. An element γ of $G(K)$, seen as element of the group $G(\mathcal{A}_f) \times G(K_\infty)$, will then be denoted by (γ, γ). In order to keep the notation readable we will make some abuses. Let S denote a set of representatives of the finite coset space $G(K)\backslash G(\mathcal{A}_f)/\mathcal{K}_f$. Every element $\underline{g}$ in $G(\mathcal{A})$ can then be written as

$$(\gamma, \gamma)(\underline{x}\underline{k}, g_\infty) \text{ with } \gamma \in G(K),\ \underline{x} \in S,\ \underline{k} \in \mathcal{K}_f,\ g_\infty \in G(K_\infty).$$

The element $\underline{x}$ is uniquely determined by $\underline{g}$. Suppose that we have another presentation of $\underline{g}$, now with $\tilde{\gamma}$, $\underline{\tilde{k}}$ and $\tilde{g}_\infty$. Then one sees that $\tilde{g}_\infty = \tilde{\gamma}\gamma^{-1}g_\infty$ and $\tilde{\gamma}\gamma^{-1} = \underline{x}\underline{k}\underline{\tilde{k}}^{-1}\underline{x}^{-1}$. Put $\Gamma_{\underline{x}} = G(K) \cap \underline{x}\mathcal{K}_f\underline{x}^{-1}$, where this intersection takes place in the group $G(\mathcal{A}_f)$. Thus $\tilde{\gamma}\gamma^{-1}$ lies in $\Gamma_{\underline{x}}$. In other words, $\underline{g}$ determines a unique coset $\Gamma_{\underline{x}}g_\infty$. The group $\Gamma_{\underline{x}} \subset G(K)$ can be seen to be a subgroup of finite index in the stabilizer of the A-lattice in K^2 corresponding with the representative $\underline{x}$. More or less by definition, this tells us that $\Gamma_{\underline{x}}$ is an *arithmetic subgroup* of $G(K)$. In this way one finds the following bijection

$$G(K)\backslash G(\mathcal{A})/\mathcal{K}Z(K_\infty) \xrightarrow{\sim} \bigsqcup_{\underline{x}\in S} \Gamma_{\underline{x}}\backslash G(K_\infty)/\mathcal{K}_\infty Z(K_\infty) \text{ (disjoint union)}.$$

For $\underline{g} = (\gamma,\gamma)(\underline{x}\underline{k}, g_\infty)$ the image of the double coset $G(K)\underline{g}\mathcal{K}Z(K_\infty)$ is the double coset $\Gamma_{\underline{x}}g_\infty\mathcal{K}_\infty Z(K_\infty)$, which is an element of the right hand side.

(2.1.4) In the special case with $\mathcal{K} = \mathcal{K}_f \times I_\infty$ we can identify $G(K_\infty)/I_\infty Z(K_\infty)$ with $edges(\mathcal{T})$, the set of the oriented edges of the Bruhat-Tits tree $\mathcal{T}$. We may suppose (under some mild condition concerning $\mathcal{K}_f$) that the group $\Gamma_{\underline{x}}$ acts on $\mathcal{T}$ without inversion. $\Gamma_{\underline{x}}$ has finite stabilizers on the edges and vertices of the tree $\mathcal{T}$ and thus $\Gamma_{\underline{x}}\backslash\mathcal{T}$ is a graph. Its set of oriented edges is denoted by $edges(\Gamma_{\underline{x}}\backslash\mathcal{T})$. This leads to a natural bijection

$$G(K)\backslash G(\mathcal{A})/\mathcal{K}Z(K_\infty) \xrightarrow{\sim} \bigsqcup_{\underline{x}\in S} edges(\Gamma_{\underline{x}}\backslash\mathcal{T}).$$

(2.1.5) Since any open subgroup $\mathcal{K}$ of $G(\mathcal{O})$ is commensurable with $G(\mathcal{O}_f) \times I_\infty$ we conclude that $G(K)\backslash G(\mathcal{A})/\mathcal{K}Z(K_\infty)$ is a countable infinite set with the discrete topology.

2.2 Automorphic Forms

Definition 2.1 *An* automorphic form *with respect to the open compact subgroup $\mathcal{K}$ of $G(\mathcal{O})$ is a (complex valued) function*

$$f : G(K)\backslash G(\mathcal{A})/\mathcal{K}Z(K_\infty) \to \mathbb{C}.$$

Equivalently, $f : G(\mathcal{A}) \to \mathbb{C}$ is a function such that, for all $\gamma \in G(K)$, $\underline{g} \in G(\mathcal{A}), \underline{k} \in \mathcal{K}Z(K_\infty)$, the equality $f(\gamma\underline{g}\underline{k})) = f(\underline{g})$ holds.

Remark 2.2

In the special case $\mathcal{K} = \mathcal{K}_f \times I_\infty$, we conclude that the automorphic forms are the functions on the union of the oriented edges of a finite set of graphs of the form $\Gamma\backslash\mathcal{T}$, where $\mathcal{T}$ is the Bruhat-Tits tree and Γ an arithmetic subgroup of $G(K)$ acting upon $\mathcal{T}$.

Definition 2.3 *A* cusp form *with respect to the compact open subgroup $\mathcal{K}$ of $G(\mathcal{O})$ is an automorphic form f with respect to $\mathcal{K}$, such that, for all $\underline{g} \in G(\mathcal{A})$*

$$\int_{K\backslash\mathcal{A}} f(\begin{pmatrix} 1 & \underline{v} \\ 0 & 1 \end{pmatrix} \underline{g})\, d\underline{v} = 0$$

($d\underline{v}$ is the normalized Haar measure on the compact group $K\backslash\mathcal{A}$).

Remarks 2.4

(2.4.1) The function $\tilde{f} : \underline{v} \mapsto f(\begin{pmatrix} 1 & \underline{v} \\ 0 & 1 \end{pmatrix} \underline{g})$ is seen as a function on the compact group $K\backslash\mathcal{A}$. The image of the compact set $K\backslash\mathcal{A}$ (for any choice of $\underline{g}$) in the double coset space $G(K)\backslash G(\mathcal{A})/\mathcal{K}Z(K_\infty)$ is finite and thus $\tilde{f}$ is a locally constant function on $K\backslash\mathcal{A}$. The integral is therefore a finite sum. Thus *the field of complex numbers does not play any role here.* We may replace $\mathbb{C}$ by any commutative ring L containing $\mathbb{Q}$ and f by a function with values in L. The integral is well defined since any $\tilde{f}$ is locally constant.
(2.4.2) In the literature ([Gel, Ha, J-L] et al.) one finds many definitions of the space of cusp forms. The most general one seems to be:
"a cusp form is a continuous function $f : G(K)\backslash G(\mathcal{A}) \to \mathbb{C}$ with a certain prescribed behaviour with respect to open compact subgroups of $G(\mathcal{A})$ and satisfying the condition of definition 2.3."
On this space of functions the center $Z(\mathcal{A})$ of $G(\mathcal{A})$ acts. The action of $Z(K)$ is, by definition, trivial. One can decompose the space of all cusps forms with respect to certain characters $\psi : Z(K)\backslash Z(\mathcal{A}) \to \mathbb{C}^*$, which are called grössencharacters. Our definition of cusp form is adapted to the space $\underline{H}$ that we want to study. In our situation we consider in fact the direct sum of the spaces of cusp forms with respect to the grössencharacters ψ, such that ψ is trivial on $Z(K_\infty)$.

Definition 2.5 *Let $L \supset \mathbb{Q}$ be a commutative ring and let $\mathcal{K}$ be an open compact subgroup of $G(\mathcal{O})$. Then $\mathcal{W}(\mathcal{K}, L)$ and $\mathcal{W}_o(\mathcal{K}, L)$ will denote the set of automorphic forms and the set of cusp forms with respect to $\mathcal{K}$ and with values in L.*
For an open compact subgroup $\mathcal{K}_f$ of $G(\mathcal{O}_f)$, one defines the following objects:

$$\mathcal{W}^{\mathcal{K}_f}(L) = \cup_{\mathcal{K}_\infty} \mathcal{W}(\mathcal{K}_f \times \mathcal{K}_\infty, L) \text{ and } \mathcal{W}_o^{\mathcal{K}_f}(L) = \cup_{\mathcal{K}_\infty} \mathcal{W}_o(\mathcal{K}_f \times \mathcal{K}_\infty, L),$$

where $\mathcal{K}_\infty$ runs through the set of the open compact subgroups of $G(\mathcal{O}_\infty)$.

$$\mathcal{W}(L) = \cup_{\mathcal{K}_f} \mathcal{W}^{\mathcal{K}_f}(L) \text{ and } \mathcal{W}_o(L) = \cup_{\mathcal{K}_f} \mathcal{W}_o^{\mathcal{K}_f}(L),$$

where $\mathcal{K}_f$ runs through the set of the open compact subgroups of $G(\mathcal{O}_f)$.

Remarks 2.6

(2.6.1) The "unramified Hecke operators" act on $\mathcal{W}^{\mathcal{K}_f}(L)$ and on $\mathcal{W}_o^{\mathcal{K}_f}(L)$. We will not explain this but refer to [Gel] §5B for an exhaustive treatment and to [Ge-Re] §4.9 for a review.

(2.6.2) The group $G(K_\infty)$ acts on $\mathcal{W}^{\mathcal{K}_f}(L)$ and $\mathcal{W}_o^{\mathcal{K}_f}(L)$. We will give more details. The elements of $\mathcal{W}^{\mathcal{K}_f}(L)$ are L-valued functions on the disjoint union $\bigsqcup_{\underline{x}\in S}\Gamma_{\underline{x}}\backslash G(K_\infty)$, such that the restriction of f to each $\Gamma_{\underline{x}}\backslash G(K_\infty)$ satisfies $f(ukz) = f(u)$ for $u \in \Gamma_{\underline{x}}\backslash G(K_\infty)$, $k \in \mathcal{K}_\infty$ for some open subgroup $\mathcal{K}_\infty$ of $G(\mathcal{O}_\infty)$ and $z \in Z(K_\infty)$. The action of $g \in G(K_\infty)$ on $\mathcal{W}^{\mathcal{K}_f}(L)$ transforms an element f into $\tilde{f}$, which is given by the formula $\tilde{f}(u) = f(ug)$. The function $\tilde{f}$ satisfies $\tilde{f}(ukz) = \tilde{f}(u)$ for $z \in Z(K_\infty)$ and $k \in (g\mathcal{K}_\infty g^{-1}) \cap G(\mathcal{O}_\infty)$. The last group is again an open subgroup of $G(\mathcal{O}_\infty)$ and thus $\tilde{f}$ belongs to $\mathcal{W}^{\mathcal{K}_f}(L)$. This defines the action of $G(K_\infty)$. It is clear that $\mathcal{W}_o^{\mathcal{K}_f}(L)$ is a $G(K_\infty)$-invariant subspace.

(2.6.3) By definition $\mathcal{W}(L)$ consists of the functions $f : G(K)\backslash G(\mathcal{A})/Z(K_\infty) \to L$ which are right-invariant under the action of some open compact subgroup $\mathcal{K}$ of $G(\mathcal{A})$. As in (2.6.2) above, the group $G(\mathcal{A})$ acts on $\mathcal{W}(L)$. For $f \in \mathcal{W}(L)$ and $g \in G(\mathcal{A})$, the function $\tilde{f} = gf$ is defined by $\tilde{f}(u) = f(ug)$. The subspace $\mathcal{W}_o(L)$ consists of those functions for which all the integrals of definition 2.3 are 0. It is clear that $\mathcal{W}_o(L)$ is stable under the action of $G(\mathcal{A})$.

(2.6.4) The condition in definition 2.3, which defines the subspaces $\mathcal{W}_o(\mathcal{K}, L)$ and $\mathcal{W}_o(L)$ of $\mathcal{W}(\mathcal{K}, L)$ and $\mathcal{W}(L)$, is somewhat difficult to handle. A nice translation is given in [Ha], (1.2.3). For $G = GL_2$ (and for more general reductive groups), Harder proves that $\mathcal{W}_o(\mathcal{K}, L)$ consists of the functions in $\mathcal{W}(\mathcal{K}, L)$ which have a finite support in $G(K)\backslash G(\mathcal{A})/\mathcal{K}Z(K_\infty)$. In fact he proves the stronger result that all the functions in $\mathcal{W}_o(\mathcal{K}, L)$ have their support in a fixed finite subset of $G(K)\backslash G(\mathcal{A})/\mathcal{K}Z(K_\infty)$. In particular, the space $\mathcal{W}_o(\mathcal{K}, L)$ is finite dimensional. We state the two results as a theorem and a corollary and we will not give a proof. We note however that, for the special case $A = \mathbb{F}_q[T]$ and $G = Gl_2$, it is possible to give an easy (but long) proof along the lines of [Go].

Theorem 2.7 *Let $\mathcal{K}$ be an open compact subgroup of $G(\mathcal{O})$ and L be a field of zero characteristic. Then $\mathcal{W}_o(\mathcal{K}, L)$ is a finite dimensional vector space over L.*

Corollary 2.8 *The subspaces $\mathcal{W}_o(\mathcal{K}, L) \subset \mathcal{W}(\mathcal{K}, L)$ and $\mathcal{W}_o(L) \subset \mathcal{W}(L)$ consist of the functions which have a compact support in $G(K)\backslash G(\mathcal{A})/Z(K_\infty)$.*

Another direct consequence of Theorem 2.7 is

Corollary 2.9 *Consider again a field $L \supset \mathbb{Q}$. The natural maps*

$$\mathcal{W}_o(\mathcal{K}, \mathbb{Q}) \otimes L \to \mathcal{W}_o(\mathcal{K}, L) \text{ and } \mathcal{W}_o(\mathbb{Q}) \otimes L \to \mathcal{W}_o(L)$$

are isomorphisms.

2.3 Harmonic cochains and measures

Let $\mathcal{T}$ be again the Bruhat-Tits tree and let Γ be an arithmetic group acting upon $\mathcal{T}$. We recall that $\Gamma \subset G(K)$ "arithmetic" means that the group is commensurable with $G(A)$. Let B be any additive group. Then $\underline{H}(\mathcal{T}, B)$ denotes the group of the ***harmonic cochains with values in*** B, i.e., the functions $h : edges(\mathcal{T}) \to B$ such that:
(1) $h(\bar{e}) = -h(e)$, where $\bar{e}$ is the opposite of the oriented edge e.
(2) For every vertex v of $\mathcal{T}$ the sum $\sum_e h(e)$ is 0, where the summation is taken over all oriented edges e starting at v.

Remark 2.10

We note that for any locally finite graph Gr, i.e., every vertex of Gr is the end point of only finitely many edges, one can extend the definition above to define the group $\underline{H}(Gr, B)$.

We continue with the Bruhat-Tits tree $\mathcal{T}$. The symbol $\underline{H}(\mathcal{T}, B)^{\Gamma}$ denotes the group of the Γ-invariant harmonic cochains and the group of the *cuspidal harmonic cochains for* Γ, denoted by $\underline{H}_!(\mathcal{T}, B)^{\Gamma}$, consists of the elements of $\underline{H}(\mathcal{T}, B)^{\Gamma}$ which have a finite support modulo the action of Γ.

One can identify the points of $\mathbb{P}^1(K_\infty)$ with the ends of the Bruhat-Tits tree $\mathcal{T}$. This identification may be described in the following way:
A point corresponds to a one dimensional linear subspace $K_\infty f_1 \subset K_\infty^2$, where f_1, f_2 is some basis of K_∞^2. One associates the end $\{[\pi_\infty^{-n}\mathcal{O}_\infty f_1 + \mathcal{O}_\infty f_2]\}_{n \geq 0}$ to $K_\infty f_1$.

Let $e \in edges(\mathcal{T})$ be an oriented edge. The set $U(e)$ is defined as the set of all half lines starting with e, interpreted as subset of $\mathbb{P}^1(K_\infty)$. One easily sees that every $U(e)$ is open and compact. Moreover for every open compact subset U of $\mathbb{P}^1(K_\infty)$ there is a finite set of oriented edges E such that U is equal to the disjoint union of the $\{U(e)\}_{e \in E}$.

A *finite additive B-valued measure* m, defined on the collection of the compact open subsets of $\mathbb{P}^1(K_\infty)$, with total mass 0, is defined by:

(1) A value $m(U) \in B$ for every open compact U.
(2) $m(U \cup V) = m(U) + m(V)$ if $U \cap V = \emptyset$.
(3) $m(\mathbb{P}^1(K_\infty)) = 0$.
For a given m one defines the harmonic cochain $\tilde{m}$ by $\tilde{m}(e) = m(U(e))$. From $U(e) \cup U(\bar{e}) = \mathbb{P}^1(K_\infty)$ it follows that $\tilde{m}(\bar{e}) = -\tilde{m}(e)$. Let v be a vertex of $\mathcal{T}$, then $\mathbb{P}^1(K_\infty)$ is the disjoint union of the $U(e)$ where the union is taken over the oriented edges starting at v. From (2) and (3) it follows that $\sum_e \tilde{m}(e) = 0$. We conclude that $\tilde{m}$ is indeed a harmonic cochain.
Conversely, for a harmonic cochain h one defines a finite additive measure $\tilde{h}$ by putting $\tilde{h}(U) = \sum_{e \in E} h(e)$, where $\{U(e)\}_{e \in E}$ is a disjoint covering of U.

2.4 The special representation

We suppose now that $L \supset \mathbb{Q}$ is a commutative ring with unit element and that B is a module over L. Let $V(L)$ denote the group of the locally constant functions on $\mathbb{P}^1(K_\infty)$ with values in L. Put $V_{sp}(L) = V(L)/L \cdot 1$. Then there is a natural isomorphism

$$\underline{H}(\mathcal{T}, B) \rightarrow Hom_L(V_{sp}(L), B).$$

To a $h \in \underline{H}(\mathcal{T}, B)$ we associate the measure $\tilde{h}$. One can integrate locally constant L-valued functions with respect to $\tilde{h}$. Since the total measure of $\tilde{h}$ is 0, one can consider $\tilde{h}$ as an element of $Hom_L(V_{sp}(L), B)$. It is clear that the described map is an isomorphism of L-modules.

The space $V_{sp}(L)$ has a natural $G(K_\infty)$-action , denoted by sp:
For $g \in G(K_\infty)$ and $f \in V_{sp}(L)$, $sp(g)f$ is the function $u \mapsto f(ug)$. This representation is the so called (*L-valued*) *special representation of* $G(K_\infty)$.
The group $G(K_\infty)$ acts on $\underline{H}(\mathcal{T}, B)$ by the formula $(gh)(e) := h(g^{-1}e)$. The isomorphisms above is $G(K_\infty)$-equivariant.

2.5 Automorphic forms and harmonic cochains

Proposition 2.11 *Let L be a commutative ring containing $\mathbb{Q}$, let $\mathcal{K}_f$ be an open compact subgroup of $G(\mathcal{O}_f)$ and let S, $\Gamma_{\underline{x}}$ et cetera be as in* (2.1.3).
There are L-linear isomorphisms

$$\oplus_{\underline{x} \in S} \underline{H}(\mathcal{T}, L)^{\Gamma_{\underline{x}}} \simeq Hom_{G(K_\infty)}(V_{sp}(L), \mathcal{W}^{\mathcal{K}_f}(L))$$

$$\oplus_{\underline{x} \in S} \underline{H}_!(\mathcal{T}, L)^{\Gamma_{\underline{x}}} \simeq Hom_{G(K_\infty)}(V_{sp}(L), \mathcal{W}_o^{\mathcal{K}_f}(L))$$

Proof. We start by proving *the first isomorphism*. The module $\mathcal{W}^{\mathcal{K}_f}(L))$ is, according to (2.1.3), a direct sum (over the $\underline{x} \in S$) of modules $Q_{\underline{x}}$. The module $Q_{\underline{x}}$ consists of the functions $f : G(K_\infty) \to L$ which have the property: $f(\gamma g h z) = f(g)$ for all $\gamma \in \Gamma_{\underline{x}}$, $g \in G(K_\infty)$, h lying in some open subgroup $\mathcal{K}_\infty$ of $G(\mathcal{O}_\infty)$ and $z \in Z(K_\infty)$.

Let Γ denote any of the $\Gamma_{\underline{x}}$ and write Q for $Q_{\underline{x}}$, then we have to prove that there is an isomorphism

$$s : \underline{H}(\mathcal{T}, L)^\Gamma \to Hom_{G(K_\infty)}(V_{sp}(L), Q).$$

An element of $\underline{H}(\mathcal{T}, L)^\Gamma$ is seen as an L-valued (finite additive) measure m on $\mathbb{P}^1(K_\infty)$, with total measure 0, and invariant under Γ. One defines $\psi := s(m)$ by the formula $\psi(f)(g) = m(sp(g)f)$, where $f \in V_{sp}(L)$, $g \in G(K_\infty)$ and $m(sp(g)f)$ denotes the integral of the function $sp(g)f$ with respect to the measure m. One has to verify that $\psi(f)$ belongs to Q, which means that we have to show that $\psi(f)(\gamma g k z) = \psi(f)(g)$ for $\gamma \in \Gamma$, $z \in Z(K_\infty)$ and k in some open subgroup of $G(\mathcal{O}_\infty)$.

The left invariance under $\gamma \in \Gamma$ follows from the Γ-invariance of m; the invariance under $z \in Z(K_\infty)$ follows from the trivial action of $Z(K_\infty)$ on $\mathbb{P}^1(K_\infty)$; the function f is locally constant and therefore invariant under $sp(k)$ for k in some open subgroup of $G(\mathcal{O}_\infty)$ (depending on f). Finally, it is clear the ψ defined above is L-linear and $G(K_\infty)$-equivariant.

On the other hand, let $\psi \in Hom_{G(K_\infty)}(V_{sp}(L), Q)$ be given. Then we define the measure m on $\mathbb{P}^1(K_\infty)$ by the formula $m(f) = \psi(f)(1)$, for any locally constant L-valued function f on $\mathbb{P}^1(K_\infty)$ and where 1 denotes the element 1 in $G(K_\infty)$. It is clear that m belongs to $\underline{H}(\mathcal{T}, L)^\Gamma$. It is also easy to see that the two maps defined in this way are each others inverses. This proves the first part of the theorem.

The second isomorphism

According to [Ha], Corollary 1.2.3. (see also corollary 2.8), $\mathcal{W}_o^{\mathcal{K}_f}(L)$ consists of the functions on $G(K)\backslash G(\mathcal{A})/\mathcal{K}_f Z(K_\infty)$ which have a compact support. Let $Q_o \subset Q$ denote the functions in Q such that their support in $\Gamma\backslash G(K_\infty)/Z(K_\infty)$ is compact. We have to show that the isomorphism

$$s : \underline{H}(\mathcal{T}, L)^\Gamma \to Hom_{G(K_\infty)}(V_{sp}(L), Q)$$

has the property $m \in \underline{H}_!(\mathcal{T}, L)^\Gamma$ if and only if $s(m) \in Hom_{G(K_\infty)}(V_{sp}(L), Q_o)$. Take $m \in \underline{H}_!(\mathcal{T}, L)^\Gamma$. It suffices to show that $\psi = s(m)$ satisfies $\psi(1_{U(e)}) \in Q_o$ for every edge $e \in edges(\mathcal{T})$. We recall that $U(e)$ is an open subset of $\mathbb{P}^1(K_\infty)$ associated to the oriented edge e and that $1_{U(e)}$ is the indicator function of this subset. The condition on m means that there are finitely many $e_1, \ldots, e_r \in edges(\mathcal{T})$ such that $m(1_{U(e)}) \neq 0$ if and only if $e \in \Gamma e_1 \cup \Gamma e_2 \cup \ldots \cup \Gamma e_r$. Thus $\psi(1_{U(e)})(g) = m(1_{U(g^{-1}e)}) \neq 0$ if and only if $g^{-1}e \in \Gamma e_1 \cup \Gamma e_2 \cup \ldots \cup \Gamma e_r$.

For any edge e' the set of the $g^{-1} \in G(K_\infty)$ with $ge = e'$ is compact modulo $Z(K_\infty)$. This proves that the function $\psi(1_{U(e)})(g)$ has a compact support in $\Gamma\backslash G(K_\infty)/Z(K_\infty)$.
The converse is proved in a similar way. □

The formulation of the proposition is somewhat clumsy because it involves seemingly unrelated groups $\Gamma_{\underline{x}}$. One can remedy this by introducing

$$\mathcal{T}_{\mathcal{K}_f} := \mathcal{T} \times G(\mathcal{A}_f)/\mathcal{K}_f.$$

The set $G(\mathcal{A}_f)/\mathcal{K}_f$ is discrete and we can see $\mathcal{T}_{\mathcal{K}_f}$ as a graph which is the disjoint union of trees isomorphic to $\mathcal{T}$. Thus $\underline{H}(\mathcal{T}_{\mathcal{K}_f}, L)$ is well defined. The group $G(K)$ acts on $\mathcal{T}_{\mathcal{K}_f}$ by the usual action on $\mathcal{T}$ and left-multiplication on $G(\mathcal{A}_f)/\mathcal{K}_f$. Using (2.1.3), one easily sees that the group $\underline{H}(\mathcal{T}_{\mathcal{K}_f}, L)^{G(K)}$ is identical with $\oplus_{\underline{x}\in S}\underline{H}(\mathcal{T}, L)^{\Gamma_{\underline{x}}}$. A similar thing holds for the $\underline{H}_!$ groups. Thus the second part of the proposition can be translated into

Corollary 2.12 *Let $\mathcal{K}_f$ be an open compact subgroup of $G(\mathcal{O}_f)$. Then there is a canonical L-linear isomorphism*

$$\underline{H}_!(\mathcal{T} \times G(\mathcal{A}_f)/\mathcal{K}_f, L)^{G(K)} \to Hom_{G(K_\infty)}(V_{sp}(L), \mathcal{W}_o^{\mathcal{K}_f}(L)).$$

The expression "canonical" in the corollary means that the constructed isomorphisms are compatible with respect to inclusions $\mathcal{K}'_f \subset \mathcal{K}_f$ of open compact subgroups of $G(\mathcal{O}_f)$. This invites one to consider the limit over all open compact subgroups $\mathcal{K}_f$. We consider the set $\tilde{\mathcal{T}} = \mathcal{T} \times G(\mathcal{A}_f)$ as a projective limit of the graphs $\mathcal{T} \times G(\mathcal{A}_f)/\mathcal{K}_f$. This projective limit is no longer a graph. We will not examine its structure but simply define $\underline{H}(\tilde{\mathcal{T}}, L)$ as the direct limit of the $\underline{H}(\mathcal{T} \times G(\mathcal{A}_f)/\mathcal{K}_f, L)$. We note that the group $G(\mathcal{A}_f)$ acts in an obvious way on $\underline{H}_!(\tilde{\mathcal{T}}, L)^{G(K)}$. The statement which will be used later on is

Theorem 2.13 *There is an L-linear and $G(\mathcal{A}_f)$-equivariant isomorphism*

$$\underline{H}_!(\tilde{\mathcal{T}}, L)^{G(K)} \to Hom_{G(K_\infty)}(V_{sp}(L), \mathcal{W}_o(L)).$$

Proof. The isomorphism is of course the direct limit of the isomorphisms in corollary 2.11. The verification of the $G(\mathcal{A}_f)$-invariance is straightforward. □

3 A review of rigid étale cohomology

Drinfeld's reciprocity law involves étale cohomology of analytic spaces. In [Dr] an ad hoc construction and some properties were given. The theme of analytic étale cohomology has been worked out by many authors. The main references are the book [Hu] of R. Huber and the articles [Be] of V. Berkovich and [dJ-vdP] of J. de Jong and M. van der Put. Of course, analytic étale cohomology is closely related to ordinary étale cohomology of algebraic varieties. For the latter one may consult the book [Ta] of G. Tamme.

In this section precise definitions of the étale cohomology groups will be given. Then we focus on the determination of the group $H^1_{étale}(Y\hat{\otimes}C, \mathbb{Z}/n\mathbb{Z})$ where Y is a Mumford curve. The structure of this Galois module is an important ingredient needed for the proof of Drinfeld's reciprocity law. The proofs given here are essentially complete. The details about étale cohomology that are omitted can easily be found in [dJ-vdP]. Omitted details about one-dimensional spaces, trees et cetera can be found in [vdP 1].

3.1 An outline and calculations for dimenson one

(3.1.1) Let L be a complete non-archimedean valued field. C will be the completion of the separable algebraic closure L^{sep} of L. It is well known that C is algebraically closed. The group $\mathcal{G}$ of all continuous automorphisms of C over L can be identified with $Gal(L^{sep}/L)$. We will consider (rigid) analytic spaces over L and define étale cohomology on those spaces. In the application to Drinfeld's reciprocity law one takes $L = K_\infty$ and $C = \mathbf{C}$.

(3.1.2) A morphism $f : Y \to X$ between analytic spaces over L is called *étale* if for every point $y \in Y$ the induced homomorphism of local rings $O_{X,f(y)} \to O_{Y,y}$ is flat and unramified. The term unramified means that $O_{Y,y}/mO_{Y,y}$ is a (finite) separable field extension of the field $O_{X,f(y)}/m$, where m denotes the maximal ideal of $O_{X,f(y)}$.

(3.1.3) *Example* ([dJ-vdP] 3.1.2):
A morphism of affinoid spaces $Y \to X$ is étale if and only if $O(Y)$ has as $O(X)$-algebra a presentation of the form

$$O(Y) = O(X) < T_1, \ldots, T_m > /(G_1, \ldots, G_m)$$

such that the determinant $\det(\frac{\partial G_i}{\partial T_j})$ is invertible in $O(Y)$.

(3.1.4) *The rigid site* X_{rigid} of an analytic space X is a category provided with a Grothendieck topology. The objects are the admissible open subsets Y of X. The morphisms $Y_1 \to Y_2$ are the inclusions $Y_1 \subset Y_2$ of admissible open subsets. A family $\{Y_i \subset Y\}_{i\in I}$ is a covering of the object Y in the Grothendieck topology if $\cup_i Y_i$ is an admissible covering. (See lecture 6 for the term "admissible").

The machinery of sheaves on sites produces for abelian sheaves S on X_{rigid} cohomology groups which are denoted by $H^*_{rigid}(X, S)$ or by $H^*(X_{rigid}, S)$.

(3.1.5) *The étale site $X_{étale}$* of X is a category with a Grothendieck topology. The objects are the étale morphisms $Y \to X$. A morphism from $Y_1 \xrightarrow{f_1} X$ to $Y_2 \xrightarrow{f_2} X$ is a morphism of analytic spaces $g : Y_1 \to Y_2$ such that $f_2 g = f_1$. The morphism g is seen to be étale as well. A family of étale morphisms $\{g_1 : Y_i \to Y\}_{i \in I}$ is an *étale covering* of the object Y if it has the following property:
For any (or some) choice of admissible affinoid coverings $Y_i = \cup_j Y_{i,j}$, we have $Y = \cup_{i,j} g_i(Y_{i,j})$ and this is an admissible covering of Y.
The cohomology groups of an abelian sheaf S on $X_{étale}$ are denoted by $H^*_{étale}(X, S)$ or by $H^*(X_{étale}, S)$.

(3.1.6) *The sheaves on $X_{étale}$ which interest us here are:*
(i) $\mathbb{G}_m$, defined by $\mathbb{G}_m(Y) = O(Y)^*$ for every object Y. We will also use the expression O^* for this sheaf.
(ii) Suppose that the integer $n > 1$ is not divisible by the residue characteristic. The morphism of étale sheaves $\mathbb{G}_m \xrightarrow{n} \mathbb{G}_m$ is surjective (which means that m induces surjective maps in all the stalks). The kernel of this morphism is by definition the étale sheaf μ_n.
(iii) Let A be an abelian group. Then $A = A_X$ is the sheaf on $X_{étale}$ given by: $A_X(Y)$ is the group of the A-valued functions on Y which are constant on every connected component of Y.

(3.1.7) If $L \neq C$ then the groups $H^*_{étale}(X, S)$ are in general too big and too complicated to work with. The problem is that the étale cohomology of the field L (this is in fact the Galois cohomology of the group $\mathcal{G}$) contributes to $H^*_{étale}(X, S)$. The remedy is to consider the groups $H^*_{étale}(X\hat{\otimes}C, S)$.
We will first explain the notation $X\hat{\otimes}C$. For an affinoid space Y over L the expression $Y\hat{\otimes}C$ denotes the affinoid space over C with affinoid algebra $O(Y)\hat{\otimes}_L C$. In general, let $\{X_i\}$ denote an admissible affinoid covering of X then $X\hat{\otimes}C$ is the analytic space over C obtained by gluing the affinoid spaces $X_i\hat{\otimes}C$. In particular, $\{X_i\hat{\otimes}C\}$ is an admissible affinoid covering of $X\hat{\otimes}C$.

Any $\sigma \in \mathcal{G}$ induces obvious automorphisms of the affinoid algebras $O(X_i)\hat{\otimes}_L C$. They glue to a "general morphism" $\phi_\sigma : X\hat{\otimes}C \to X\hat{\otimes}C$. This is not a morphism of C-analytic spaces unless $\sigma = id_C$. However ϕ_σ is an automorphism of the étale site $X_{étale}$. One has an obvious "general morphism" $\alpha : X\hat{\otimes}C \to X$, which can be interpreted as a morphism of the étale sites. Let S be a sheaf on $X_{étale}$. Since it is clear that $\alpha = \alpha \circ \phi_\sigma$, we obtain isomorphisms $\alpha^*(S) \cong (\alpha\phi_\sigma)^*(S) \cong$

$(\phi_\sigma)^*\alpha^*(S)$. Thus we find an isomorphism

$$\phi_\sigma^* : H^i_{\acute{e}tale}(X\hat{\otimes}C, \alpha^*(S)) \to H^i_{\acute{e}tale}(X\hat{\otimes}C, \alpha^*(S)).$$

This defines an action of $\mathcal{G}$ on the groups $H^i_{\acute{e}tale}(X\hat{\otimes}C, \alpha^*(S))$. In the sequel we will omit α^* in the notations.

(3.1.8) We will use the morphism of sites $r : X_{\acute{e}tale} \to X_{rigid}$ in the calculations of étale cohomology groups. We remark that r actually stands for a morphism of categories in the other direction $X_{rigid} \to X_{\acute{e}tale}$. With this notation we are following the usual one. This morphism yields a Leray spectral sequence $E_2^{a,b} := H^a_{rigid}(X, R^b r^* S)$ with limit $H^{a+b}_{\acute{e}tale}(X, S)$.

(3.1.9) *Let X denote an analytic space over C of dimension one.*
The sheaf $\mathbb{G}_m$ has the property that $H^1_{\acute{e}tale}(X, \mathbb{G}_m) = H^1_{rigid}(X, \mathbb{G}_m)$ = the group of the isomorphy classes of the line bundles on X. This has as consequence that $R^1 r^* \mathbb{G}_m = 0$. For $n > 1$, not divisible by the residue characteristic of C, one has an exact sequence

$$0 \to \mu_n \to \mathbb{G}_m \to \mathbb{G}_m \to 0$$

of sheaves on $X_{\acute{e}tale}$. Thus $O^* \xrightarrow{n} O^* \to R^1 r^* \mu_n \to 0$ is exact on X_{rigid}. Using Hilbert 90 one can show that $R^q r^* \mu_n = 0$ for $q \geq 2$. Further the cohomology groups $H^i_{rigid}(X, *)$ are zero for $i \geq 2$. As we work over the field C the sheaves μ_n and $\mathbb{Z}/n\mathbb{Z}$ are isomorphic. Using (3.1.8) one concludes that there is an exact sequence:

$$0 \to H^1_{rigid}(X, \mathbb{Z}/n\mathbb{Z}) \to H^1_{\acute{e}tale}(X, \mathbb{Z}/n\mathbb{Z}) \to H^0_{rigid}(X, R^1 r^* \mathbb{Z}/n\mathbb{Z}) \to 0.$$

(3.1.10) *The cohomology of a projective curve X over C*
We suppose that X is non-singular and connected and that n is not divisible by the characetristic of C. From the exact sequence

$$0 \to \mu_n \to \mathbb{G}_m \xrightarrow{n} \mathbb{G}_m \to 0$$

of sheaves on $X_{\acute{e}tale}$, one concludes that

$$0 \to H^1_{\acute{e}tale}(X, \mu_n) \to H^1_{\acute{e}tale}(X, \mathbb{G}_m) \xrightarrow{n} H^1_{\acute{e}tale}(X, \mathbb{G}_m)$$

is exact and thus $H^1_{\acute{e}tale}(X, \mu_n)$ is isomorphic to $Jac(X)[n]$= the group of the n-torsion points of the Jacobian variety of X. This group is isomorphic to $(\mathbb{Z}/n\mathbb{Z})^{2g}$ where g is the genus of X. The same result holds for the algebraic étale cohomology and we conclude *that the analytic étale cohomology and the algebraic étale cohomology coincide for non-singular projective curves.*

(3.1.11) *The cohomology of an open subspace of* $\mathbb{P}^1$.
The subspaces $U \subset \mathbb{P}^1_C$ that we are interested in are of the form $U = \mathbb{P}^1_C \setminus \mathcal{L}$, where $\mathcal{L}$ is a compact subset. We note that U has an admissible pure covering. There corresponds to this an analytic reduction and a tree $\mathcal{B}$, which is the intersection graph of the components of the analytic reduction. In the lectures 6 and 7, the details are given for the special case that $\mathcal{L} = \mathbb{P}^1(F)$, where F is some local field.

It is known that any constant sheaf on U has trivial cohomology for the rigid topology. Further any line bundle on U is trivial, in other words $H^1_{rigid}(U, O^*) = 0$. Another important fact about the function theory on U is the exact sequence (see [vdP 1] Theorem 2.1)

$$0 \to C^* \to O(U)^* \to \underline{H}(\mathcal{B}, \mathbb{Z}) \to 0.$$

Now $R^1 r^* \mu_n$ is the sheaf $O^*/(O^*)^n$ on U. This implies that $H^0_{rigid}(U, R^1 r^* \mu_n) = O(U)^*/(O(U)^*)^n \cong \underline{H}(\mathcal{B}, \mathbb{Z}/n\mathbb{Z})$. Using the exact sequence of (3.1.9), one finds that $H^1_{\acute{e}tale}(U, \mu_n)$ is canonically isomorphic to $\underline{H}(\mathcal{B}, \mathbb{Z}/n\mathbb{Z})$.

(3.1.12) *The cohomology of a Mumford curve*
We recall that a Mumford curve X over C is a complete irreducible smooth curve such that all the irreducible components of its (semi-)stable reduction are projective lines. For a Mumford curve there exists a discontinous subgroup Δ of $PGl(2, C)$ acting upon $\mathbb{P}^1_C$ such that:
(i) Δ is a free group on g (this is the genus of X) generators.
(ii) Let $\mathcal{L}$ denote the compact set of the limit points of Δ and put $U = \mathbb{P}^1_C \setminus \mathcal{L}$. Then Δ acts discontinuosly and without fixed points on U.
(iii) There is a morphism $u : U \to X$ which induces an analytic isomorphism $\Delta \backslash U \to X$.
One calls $u : U \to X$ *the analytic uniformization of* X.

For sheaves S on X_{rigid} one finds in the usual way a spectral sequence $E_2^{a,b} := H^a(\Delta, H^b_{rigid}(U, u^*S))$ with limit $H^{a+b}_{rigid}(X, S)$. The cohomology groups $H^a(\Delta, -)$ are zero for $a > 1$, since Δ is a free group. The rigid cohomology groups $H^b_{rigid}(U, -)$ are zero for $b > 1$. This leads to the exact sequence:

$$0 \to H^1(\Delta, H^0_{rigid}(U, u^*S)) \to H^1_{rigid}(X, S) \to H^1_{rigid}(U, u^*S)^{\Delta} \to 0.$$

Now we can identify the two extreme terms of the exact sequence of (3.1.9).

The group $H^1_{rigid}(X, \mathbb{Z}/n\mathbb{Z})$ is equal to $H^1(\Delta, \mathbb{Z}/n\mathbb{Z})$, since the constant sheaf $u^*\mathbb{Z}/n\mathbb{Z} = \mathbb{Z}/n\mathbb{Z}$ has trivial cohomology on U. The action of Δ on $\mathbb{Z}/n\mathbb{Z}$ is trivial and so $H^1(\Delta, \mathbb{Z}/n\mathbb{Z})$ is equal to $Hom(\Delta, \mathbb{Z}/n\mathbb{Z}) \cong (\mathbb{Z}/n\mathbb{Z})^g$.

One can give another description of this cohomology group. Let $\mathcal{B}$ denote the tree associated to U. The canonical map $\tilde{u} : \mathcal{B} \to \Delta\backslash\mathcal{B}$ is the universal covering of the finite graph $\Delta\backslash\mathcal{B}$. We note in passing that this finite graph is in fact the graph of the (semi-)stable reduction of X. This implies that $H^1(\Delta\backslash\mathcal{B}, \mathbb{Z}/n\mathbb{Z})$ is canonically isomorphic to $H^1(\Delta, \mathbb{Z}/n\mathbb{Z})$.

The next group to study is $H^0_{rigid}(X, R^1r^*\mu_n)$. Using the uniformization, that $R^1r^*\mu_n = O^*/(O^*)^n$ and that $O(U)^*/(O(U)^*)^n$ is isomorphic to $\underline{H}(\mathcal{B}, \mathbb{Z}/n\mathbb{Z})$, one finds a canonical isomorphism

$$H^0_{rigid}(X, R^1r^*\mu_n) \cong \underline{H}(\mathcal{B}, \mathbb{Z}/n\mathbb{Z})^{\Delta} = \underline{H}(\Delta\backslash\mathcal{B}, \mathbb{Z}/n\mathbb{Z})$$

The last group is again isomorphic to $(\mathbb{Z}/n\mathbb{Z})^g$. If we fix an isomorphism between μ_n and $\mathbb{Z}/n\mathbb{Z}$ then $H^0_{rigid}(X, R^1r^*\mathbb{Z}/n\mathbb{Z})$ is also canonically isomorphic to $\underline{H}(\mathcal{B}, \mathbb{Z}/n\mathbb{Z})^{\Delta}$.

(3.1.13) *The Galois action on* $H^1_{étale}(X, \mathbb{Z}/n\mathbb{Z})$
We will now assume that $X \cong Y\hat{\otimes}C$ such that Y is a Mumford curve over L, which has already a (semi-)stable reduction over L. We note that in this special case the analytic object $Y\hat{\otimes}C$ is in fact the rigid analytic space associated to the algebraic curve $Y \otimes C$. The group $\mathcal{G}$ acts on the cohomology and also on μ_n. Let us introduce some notations:
(a) $\mu_n = \mathbb{Z}/n\mathbb{Z}(1)$, which indicates the action of $\mathcal{G}$ on the group μ_n of the nth roots of unity in L^{sep}.
(b) For any integer k one defines $\mathbb{Z}/n\mathbb{Z}(k) := \mu_n^{\otimes k}$.
(c) For any $\mathcal{G}$-module M which is also a $\mathbb{Z}/n\mathbb{Z}$-module, one writes $M(k)$ for $M \otimes \mathbb{Z}/n\mathbb{Z}(k)$.

We note that $\mathcal{G}$ acts trivially on $H^1(\Delta\backslash\mathcal{B}, \mathbb{Z}/n\mathbb{Z})$ and on $\underline{H}(\Delta\backslash\mathcal{B}, \mathbb{Z}/n\mathbb{Z})$. This leads to an exact sequence of $\mathcal{G}$-modules

$$0 \to H^1(\Delta\backslash\mathcal{B}, \mathbb{Z}/n\mathbb{Z}) \to H^1_{étale}(X, \mathbb{Z}/n\mathbb{Z}) \to \underline{H}(\Delta\backslash\mathcal{B}, \mathbb{Z}/n\mathbb{Z})(-1) \to 0.$$

Unfortunately, this sequence does not completely determine the $\mathcal{G}$-action on $H^1_{étale}(X, \mathbb{Z}/n\mathbb{Z})$. We postpone the investigation of the precise Galois structure to subsection 3.3.

(3.1.14) *The cohomology groups* $H^*_{étale}(X, \mathbb{Q}_l)$
Let l be a prime, different from the residue characteristic p of L. Suppose, as before, that Y is an analytic space over L and write $X := Y\hat{\otimes}C$, where C is the completion of the separable closure of L. The groups $H^i_{étale}(X, \mathbb{Z}/l^m\mathbb{Z})$ form a projective system of $\mathcal{G}$-modules. The projective limit is denoted by $H^i_{étale}(X, \hat{\mathbb{Z}}_l)$. It is a module over the ring $\hat{\mathbb{Z}}_l$, which is the projective limit of

the rings $\mathbb{Z}/l^m\mathbb{Z}$. The ring $\hat{\mathbb{Z}}_l$ is also the ring of integers of the field of l-adic numbers $\mathbb{Q}_l$. The analytic l-adic cohomology groups of X, are defined by

$$H^i_{\acute{e}tale}(X, \mathbb{Q}_l) := H^i_{\acute{e}tale}(X, \hat{\mathbb{Z}}_l) \otimes_{\hat{\mathbb{Z}}_l} \mathbb{Q}_l.$$

Those objects are vector spaces over $\mathbb{Q}_l$ equipped with a $\mathcal{G}$-action.

3.2 Some representations of the absolute Galois group

(3.2.1) We will suppose that L is a local field. The residue field of L is supposed to have characteristic p and the cardinal number of the residue field is denoted by q. Let π denote a generator of the maximal ideal of the valuation ring of L. The special case which interests us is $L = K_\infty$, $\pi = \pi_\infty$ and $q = q_\infty$.

We will use the same notations as in (3.1.1), namely: L^{sep} is a separable algebraic closure of L. The absolute Galois group of L is the group $\mathcal{G} := Gal(L^{sep}/L)$. Further C is the completion of L^{sep}. Then C is algebraically closed and $\mathcal{G}$ is equal to the group of all continuous automorphisms of C/L. We will construct some representations of $\mathcal{G}$.

All the separable algebraic extensions of L, that we consider, are supposed to be subfields of L^{sep}. Let L^{nr} be the maximal unramified extension of L. Its residue field $\mathbb{F}_q^{alg}$ is the algebraic closure of $\mathbb{F}_q$, the residue field of L. One has

$$Gal(L^{nr}/L) \simeq Gal(\mathbb{F}_q^{alg}/\mathbb{F}_q) \cong \hat{\mathbb{Z}},$$

where $\hat{\mathbb{Z}}$ is the profinite completion of $\mathbb{Z}$ (i.e., the projective limit of all $\mathbb{Z}/n\mathbb{Z}$). Let F denote the element of $Gal(L^{nr}/L)$ which acts on the residue field as the relative Frobenius. The element F is a topological generator of $Gal(L^{nr}/L)$.

(3.2.2) Let $n > 1$ be an integer not divisible by p. Consider the extension $L^{nr}(\pi^{1/n}) \supset L$. It is a Galois extension. The Galois group is topologically generated by the two elements F, u given by:
(1) The restriction of F to L^{nr} is the F described above.
(2) u is the identity on L^{nr} and $u(\pi^{1/n}) = \zeta_n \pi^{1/n}$, where ζ_n is a chosen primitive nth root of unity in L^{nr}.
The only relations for F, u are $FuF^{-1} = u^q$ and $u^n = 1$. From this one obtains a Galois representation $\rho^{(n)}$ given by

$$\rho^{(n)} : \mathcal{G} \overset{rest}{\to} Gal(L^{nr}(\pi^{1/n})/L) \to GL_2(\mathbb{Z}/n\mathbb{Z})$$

with $\rho^{(n)}(F) = \begin{pmatrix} 1 & 0 \\ 0 & q^{-1} \end{pmatrix}$ and $\rho^{(n)}(u) = \begin{pmatrix} 1 & 1 \\ 0 & 1 \end{pmatrix}$. Let $U^{(n)}$ be the space $(\mathbb{Z}/n\mathbb{Z})^2$ equipped with this Galois action.

(3.2.3) Let l be a prime number different from p. We consider the projective limit of the representations $U^{(l^r)}$ for $r \to \infty$. One can make this explicit as follows. Let E_l be the field obtained by adjoining to L^{nr} all the l^r-roots of π with $r \in \mathbb{N}$. The extension E_l/L^{nr} is Galois. Its Galois group can be identified with $\hat{\mathbb{Z}}_l$, the group of the l-adic integers. A topological generator u can be described as follows.
Fix a sequence $\{\zeta_{l^r}\}_{r\geq 0}$ of primitive l^rth roots of unity in L^{nr} with the property $\zeta_{l^r}^l = \zeta_{l^{r-1}}$ for all $r \geq 1$ and $\zeta_0 = 1$. Fix also a sequence $\{d_r\}_{r\geq 0}$ in E_l such that $d_r^l = d_{r-1}$ for all $r \geq 1$ and $d_0 = \pi$. Then $u \in Gal(E_l/L^{nr})$ is defined by $u(d_r) = \zeta_{l^r} d_r$ for all $r \geq 1$.
The field extension E_l/L is again Galois and there is an exact sequence

$$1 \to Gal(E_l/L^{nr}) \to Gal(E_l/L) \to Gal(L^{nr}/L) \to 1$$

The sequence splits because the last non-trivial group is a free topological group. Let F denote now a lift of F to an element of $Gal(E_l/L)$. Then $Gal(E_l/L)$ is a semi-direct product $\hat{\mathbb{Z}} \propto \hat{\mathbb{Z}}_l$ with topological generators F, u having the only relation $FuF^{-1} = u^q$.

The following Galois representation is called the *special l-adic Galois representation*.

$$\rho_l \ : \mathcal{G} \overset{rest}{\to} Gal(E_l/L) \overset{\rho_l'}{\to} Gl_2(\hat{\mathbb{Z}}_l) \subset GL_2(\mathbb{Q}_l),$$

where the first map is the restriction and ρ_l' is given by $\rho_l'(F) = \begin{pmatrix} 1 & 0 \\ 0 & q^{-1} \end{pmatrix}$ and $\rho_l'(u) = \begin{pmatrix} 1 & 1 \\ 0 & 1 \end{pmatrix}$.
Let U_l denote the $\mathbb{Q}_l$-vector space $\mathbb{Q}_l^2$ equipped by this $\mathcal{G}$-action.

(3.2.4) We recall the notation $\mathbb{Z}/n\mathbb{Z}(1)$ for μ_n with its $\mathcal{G}$-action and $\mathbb{Z}/n\mathbb{Z}(k)$ for the $\mathcal{G}$-module μ_n^k, where $k \in \mathbb{Z}$ and $n \in \mathbb{Z}$ not divisible by the residue characteristic. This notation can be extended to projective limits, e.g. let l be a prime number (different from the residue characteristic) then the $\mathcal{G}$-module $\hat{\mathbb{Z}}_l(k)$ denotes the projective limt of the modules $\mathbb{Z}/l^r\mathbb{Z}(k)$. Moreover the $\mathcal{G}$-module $\mathbb{Q}_l(k)$ is defined as $\hat{\mathbb{Z}}_l(k) \otimes_{\hat{\mathbb{Z}}_l} \mathbb{Q}_l$.
The definition of the representation U_l of $\mathcal{G}$ implies the existence of an exact sequence of $\mathcal{G}$-modules

$$0 \to \mathbb{Q}_l \to U_l \to \mathbb{Q}_l(-1) \to 0,$$

since for a suitable basis e_1, e_2 of U_l the actions of F and u are given by $F(e_1) = e_1$, $F(e_2) = q^{-1}e_2$, $u(e_1) = e_1$, $u(e_2) = e_2 + e_1$. The action of u shows that the exact sequence does not split. It is easily seen that any non-split extension of $\mathbb{Q}_l(-1)$ by $\mathbb{Q}_l$ is isomorphic to U_l.

3.3 The étale cohomology of a Mumford curve

We make the same assumptions on L as in subsection 3.2. Let Y be a Mumford curve over L of genus g. The uniformization is denoted by $u : U \to Y$. The discontinuous subgroup $\Delta \subset PGl(2, L)$ with $\Delta\backslash U \cong Y$ is free on g generators. Let $\mathcal{B}$ denote the tree associated to the open analytic subspace $U \subset \mathbb{P}^1_L$.

Theorem 3.1 *The $\mathcal{G}$-module $H^1_{\acute{e}tale}(Y\hat{\otimes}C, \mathbb{Q}_l)$ is canonically isomorphic to $\underline{H}(\Delta\backslash\mathcal{B}, \mathbb{Q}_l) \otimes U_l$.*

Proof. The Jacobian variety $Jac(Y)$ of Y will play an essential rôle in the proof. Let $T_l(Jac(Y))$ denote the l-adic Tate-module of $Jac(Y)$. From the comparison statement of (3.1.10), by taking projective limits and by taking the tensor product with $\mathbb{Q}_l$ one obtains an isomorphism of $\mathcal{G}$-modules

$$H^1_{\acute{e}tale}(Y\hat{\otimes}C, \mathbb{Q}_l) \cong T_l(Jac(Y)) \otimes \mathbb{Q}_l(-1).$$

For a more detailed study of the l^r-torsion points of $Jac(Y)$ we need the uniformization of $Jac(Y)$.

Since Y is a Mumford curve over L, the Jacobian variety $Jac(Y)$ admits a uniformization of the form $\mathbb{T}_L/\Lambda \cong Jac(Y)$, where $\mathbb{T}_L$ is the split torus over L of dimension g and Λ is a multiplicative lattice of rank g in $\mathbb{T}_L(L) \cong (L^*)^g$. For any field F with $L \subset F \subset L^{sep}$, one has an exact sequence

$$0 \to \Lambda \to \mathbb{T}_L(F) \to Jac(Y)(F) \to 0.$$

One draws from this the conclusion that all the l^r-torsion points of $Jac(Y)$ lie in $Jac(Y)(E_l)$, where E_l is the field introduced in section 3.1.1. In particular the action of $\mathcal{G}$ on $T_l(Jac(Y)) \otimes \mathbb{Q}_l$ factors over the group $Gal(E_l/L)$ and we have to study the action of the two topological generators F and u on $T_l(Jac(Y))\otimes\mathbb{Q}_l$.

We recall that L^{nr} is the field of the u-invariant elements of E_l. Thus the u-invariant l^r-torsion elements (any r) of $Jac(Y)$ lie in $Jac(Y)(L^{nr})$. We consider again the exact sequence

$$0 \to \Lambda \to \mathbb{T}_L(L^{nr}) \to Jac(Y)(L^{nr}) \to 0.$$

The induces an injective homomorphism between the torsion subgroups $\mathbb{T}_L(L^{nr})_{tors} \to Jac(Y)(L^{nr})_{tors}$. We want to prove that the cokernel of this map is a finite group. For this purpose we introduce the additive valuation $v : L^* \to \mathbb{Z}$, which extends uniquely to $v : (L^{nr})^* \to \mathbb{Z}$. Further we will need the principal polarization of $Jac(Y)$.

This principal polarization has the following translation in terms of the algebraic torus $\mathbb{T}_L$ and the multiplicative lattice Λ:
Let $Char$ denote the group of the characters of $\mathbb{T}_L$. The elements χ of $Char$ are the homomorphism of algebraic groups $\chi : \mathbb{T}_L \to \mathbb{G}_{m,L}$. The principal polarization induces a bijective homomorphism $\sigma : \Lambda \to Char$, which has the properties:
(i) $\sigma(\lambda_1)(\lambda_2) = \sigma(\lambda_2)(\lambda_1)$.
(ii) The symmetric bilinear form $v(\sigma(\lambda_1)(\lambda_2))$ on Λ is positive definite.
Details and proofs can be found in [F-vdP].

The group $\mathbb{T}_L(L^{nr})$ can be identified with the group $Hom(Char, (L^{nr})^*)$. The map $v : (L^{nr})^* \to \mathbb{Z}$ induces a homomorphism

$$\tilde{v} : \mathbb{T}_L(L^{nr}) \to Hom(Char, \mathbb{Z}).$$

From (ii) one concludes that the restriction of $\tilde{v}$ to Λ is injective. Let $Jac(Y)(L^{nr})^+$ denote the kernel of the induced homomorphism from $Jac(Y)(L^{nr})$ to the finite group $Hom(Char, \mathbb{Z})/\tilde{v}(\Lambda)$. Let $x \in Jac(Y)(L^{nr})^+$ be a torsion element of order m. There is a $y \in \mathbb{T}_L(L^{nr})$ with image x and with image 0 in $Hom(Char, \mathbb{Z})$. Now y^m lies in Λ and still has image 0 in $Hom(Char, \mathbb{Z})$. Therefore $y^m = 1$. Thus we have proved that the torsion groups of $\mathbb{T}_L(L^{nr})$ and $Jac(Y)(L^{nr})^+$ coincide and as a consequence the cokernel of the map $\mathbb{T}_L(L^{nr})_{tors} \to Jac(Y)(L^{nr})_{tors}$ is finite. A direct consequence is that the set of u-invariant elements of $T_l(Jac(Y)) \otimes \mathbb{Q}_l(-1)$ is isomorphic to the g-dimensional $\mathbb{Q}_l$-vector space $T_l(\mathbb{T}_L) \otimes \mathbb{Q}_l$.

We return now to the group $H^1_{\acute{e}tale}(Y \hat{\otimes} C, \mathbb{Q}_l)$ and continue the calculations of (3.1.13). By taking the projective limit over $n = l^r$ with $r \to \infty$ and taking the tensor product with $\mathbb{Q}_l$ one finds an exact sequence of $\mathcal{G}$-modules

$$0 \to H^1(\Delta \backslash \mathcal{B}, \mathbb{Q}_l) \to H^1_{\acute{e}tale}(Y \hat{\otimes} C, \mathbb{Q}_l) \to \underline{H}(\Delta \backslash \mathcal{B}, \mathbb{Q}_l)(-1) \to 0.$$

We have shown above that the group $\mathcal{G}$ acts via the group $Gal(E_l/L)$. It suffices to calculate the actions of F and u on the terms of this sequence. We recall that $\mathcal{G}$ acts trivial on $H^1(\Delta \backslash \mathcal{B}, \mathbb{Q}_l)$ and on $\underline{H}(\Delta \backslash \mathcal{B}, \mathbb{Q}_l)$. Therefore F and u act trivially on $H^1(\Delta \backslash \mathcal{B}, \mathbb{Q}_l)$. Further F acts as $q^{-1}id$ and u acts trivially on $\underline{H}(\Delta \backslash \mathcal{B}, \mathbb{Q}_l)(-1)$. Thus the $2g$-dimensional $\mathbb{Q}_l$-vector space $H^1_{\acute{e}tale}(Y \hat{\otimes} C, \mathbb{Q}_l)$

has a basis $e_1, \ldots e_g, f_1, \ldots, f_g$ such that $e_1, \ldots, e_g$ is a basis of $H^1(\Delta\backslash\mathcal{B}, \mathbb{Q}_l)$ and such that the matrices of F and u have the block form

$$\begin{pmatrix} id & 0 \\ 0 & q^{-1}id \end{pmatrix} \text{ and } \begin{pmatrix} id & A \\ 0 & id \end{pmatrix},$$

where A is a $\mathbb{Q}_l$-linear map from $\underline{H}(\Delta\backslash\mathcal{B}, \mathbb{Q}_l)$ to $H^1(\Delta\backslash\mathcal{B}, \mathbb{Q}_l)$. Suppose that we know that A is a bijection, then we can choose the basis $e_1, \ldots, f_g$ such that $u(f_i) = f_i + e_i$ for $i = 1, \ldots, g$. Then clearly $H^1_{\acute{e}tale}(Y\hat{\otimes}C, \mathbb{Q}_l)$ is a direct sum of copies of the representation U_l and the statement of the theorem follows.

The $\mathbb{Q}_l$-vector space $H^1(\Delta\backslash\mathcal{B}, \mathbb{Q}_l)$ of dimension g consists of u-invariant elements. If we can show that this vector space is the set of *all* u-invariant elements of $H^1_{\acute{e}tale}(Y\hat{\otimes}C, \mathbb{Q}_l)$, then it follows that the map A is bijective. So we have to prove that the set of the u-invariant elements of $H^1_{\acute{e}tale}(Y\hat{\otimes}C, \mathbb{Q}_l)$ is a $\mathbb{Q}_l$-vector space of dimension g. This has already been shown above. □

Remark 3.2

The assertion in theorem 3.1 that the isomorphism is "canonical" has a precise meaning. Let $f : Y_2 \to Y_1$ be a non-constant morphism between Mumford curves over L. The Mumford uniformizations are denoted by $u_i : U_i \to Y_i$. The two free groups are denoted by Δ_i and the two trees by $\mathcal{B}_i$. The statement is that f gives rise to a natural map

$$f^* : \underline{H}(\Delta_1\backslash\mathcal{B}_1, \mathbb{Q}_l) \to \underline{H}(\Delta_2\backslash\mathcal{B}_2, \mathbb{Q}_2),$$

which is compatible with the canonical morphism of $\mathcal{G}$-modules

$$H^1_{\acute{e}tale}(Y_1\hat{\otimes}C, \mathbb{Q}_l) \to H^1_{\acute{e}tale}(Y_2\hat{\otimes}C, \mathbb{Q}_l).$$

The construction of f^* goes as follows. There is a lift $\tilde{f} : U_2 \to U_1$ of f. This yields homomorphisms $O(U_1)^* \to O(U_2)^*$ and $\underline{H}(\mathcal{B}_1, \mathbb{Z}) \to \underline{H}(\mathcal{B}_2, \mathbb{Z})$, compatible with the action of $\mathcal{G}$. The final result is the required homomorphism f^*.

3.3 *A variation on theorem 3.1*

As before L is a local field. Let $\Omega := \mathbb{P}^1_L \setminus \mathbb{P}^1(L)$ and let $\mathcal{T}$ denote the Bruhat-Tits tree associated to Ω (see the lectures 6 and 7). Suppose that $\Gamma \subset PGl(2, L)$ is a discrete subgroup such that the quotient $Y_0 := \Gamma\backslash\Omega$ is an *affine* curve over L. Let Y denote the compactification of Y_0. It is not difficult to see that Y is a Mumford curve over L (see [vdP 1]).
The Mumford uniformization of Y will be denoted by $u : U \to Y$. The corresponding group and tree by Δ and $\mathcal{B}$. We note that the restriction of u to the

morphism $u : u^{-1}(Y_0) \to Y_0$ is also a universal covering. Indeed, $u^{-1}(Y_0)$ is a connected open subset of U and has therefore only trivial analytic coverings. Moreover Δ acts discontinuous and without fixed points on $U^{-1}(Y_0)$.
Theorem 3.1 gives an expression for the cohomology group $H^1_{\acute{e}tale}(Y \hat{\otimes} C, \mathbb{Q}_l)$ involving $\underline{H}(\Delta \backslash \mathcal{B}, \mathbb{Q}_l)$. In the sequel we prefer to have an expression which uses the tree $\mathcal{T}$. Thus we have to find the relation between the two trees $\mathcal{B}$ and $\mathcal{T}$. This relation has been given in [vdP 1] and we will sketch the procedure.

Let Γ_{tors} denote the subgroup of Γ generated by the elements of finite order in Γ. Clearly $G := \Gamma_{tors}$ is a normal subgroup of Γ. We may assume that Γ acts without inversion on the Bruhat-Tits tree $\mathcal{T}$. (If this is not the case then we can refine the tree $\mathcal{T}$. This is a harmless operation).
We apply [Se 2], p77, cor.1, to the group G. Any element of finite order of G stabilizes a vertex a of $\mathcal{T}$. Thus G is generated by the union of all the stabilizers G_a of the vertices a of $\mathcal{T}$. Then the graph $G \backslash \mathcal{T}$ has a trivial fundamental group and is therefore a tree. We will denote this tree by $\mathcal{B}_1$. There is a canonical isomorphism $\underline{H}(\mathcal{T}, \mathbb{Q}_l)^{\Gamma_{tors}}$ with $\underline{H}(\mathcal{B}_1, \mathbb{Q}_l)$. This isomorphism sends the cochain f to the cochain $\tilde{f}$, given by $\tilde{f}(\tilde{e}) = \frac{\# Stab_v}{\# Stab_e} f(e)$, where the oriented edge $\tilde{e}$ of $\mathcal{B}_1$ is the image of the oriented edge e of $\mathcal{T}$ and where $\# Stab_e$ and $\# Stab_v$ denote the cardinal number of the stabilizers in Γ of the edge e and the vertex v.
The group $\Delta^* := \Gamma / \Gamma_{tors}$ has no torsion elements and acts on the tree $\mathcal{B}_1$. Therefore Δ^* is a free group (see [Se 2]). One can conclude that $\underline{H}(\mathcal{T}, \mathbb{Q}_l)^{\Gamma}$ is isomorphic to $\underline{H}(\mathcal{B}_1, \mathbb{Q}_l)^{\Delta^*}$. The same statement holds for cochains with compact support modulo Γ (resp. modulo Δ^*). That is, there is a canonical isomorphism

$$\underline{H}_!(\mathcal{T}, \mathbb{Q}_l)^{\Gamma} \cong \underline{H}_!(\mathcal{B}_1, \mathbb{Q}_l)^{\Delta^*}.$$

We consider now the one-dimensional analytic space $\Omega_1 := \Gamma_{tors} \backslash \Omega$. The standard covering of Ω (see lecture 7) by affinoids $\{O(e)\}_e$, where e runs in the set of the (non-oriented) edges of $\mathcal{T}$, is stable under the action of Γ_{tors}. The quotient Ω_1 is obtained by gluing the affinoid pieces $P(e) := Stab_e \backslash O(e)$. Each affinoid piece $P(e)$ is again an affinoid subset of $\mathbb{P}^1_L$, since $Stab_e \backslash \mathbb{P}^1_L$ is isomorphic to $\mathbb{P}^1_L$. The $\{P(e)\}$ form a pure (or formal) admissible covering of Ω_1 and thus define an analytic reduction of Ω_1. The intersection graph of this analytic reduction coincides with $\mathcal{B}_1$. Since this is a tree and since the valuation of L is discrete, one concludes (see [G-vdP], Chapter 4 Corollary (2.5)) that Ω_1 has the form $\mathbb{P}^1_L \setminus \mathcal{L}_1$ for some compact subset $\mathcal{L}_1$ of $\mathbb{P}^1(L)$.

Any element $\delta \in \Delta^*$ acts on Ω_1 and has a (unique) extension to an automorphism of $\mathbb{P}^1_L$ (see [G-vdP], Chapter 4 lemma (3.8)). In this way Δ^* is a subgroup of $PGl(2, L)$. This subgroup is discontinuous, since its action on Ω_1 is discontinuous. Let $\mathcal{L}_2 \subset \mathcal{L}_1$ denote the set of the limit points of Δ^*. Then

$\Omega_2 := \mathbb{P}^1_L \setminus \mathcal{L}_2$ is the open subspace of the ordinary points for Δ^*. The map $\Omega_1 \to \Delta^*\backslash\Omega_1 = Y_0$ is the universal analytic covering of Y_0 since Δ^* acts without fixed points on Ω_1 and Ω_1 is, in the analytic sense, simply connected. That implies that we may identify the covering $\Omega_1 \to \Delta^*\backslash\Omega_1 = Y_0$ and the group Δ^* with the covering $u : u^{-1}(Y_0) \to Y_0$ and the group Δ. Further Ω_2 can be identified with U.

Now all we are left with is proving that $\underline{H}_!(\mathcal{B}_1, \mathbb{Q}_l)^{\Delta}$ coincides with $\underline{H}(\mathcal{B}, \mathbb{Q}_l)^{\Delta}$. The first step is to observe that $\underline{H}(\mathcal{B}_1, \mathbb{Q}_l)^{\Delta}$ is equal to $\underline{H}(\Delta\backslash\mathcal{B}_1, \mathbb{Q}_l)$. This holds also for "compact supports", i.e. $\underline{H}_!(\mathcal{B}_1, \mathbb{Q}_l)^{\Delta}$ is isomorphic to $\underline{H}_!(\Delta\backslash\mathcal{B}_1, \mathbb{Q}_l)$. In the same way $\underline{H}(\mathcal{B}, \mathbb{Q}_l)^{\Delta}$ can be identified with $\underline{H}(\Delta\backslash\mathcal{B}, \mathbb{Q}_l)$. Thus we have to know the relation between the two graphs $Gr_1 := \Delta\backslash\mathcal{B}_1$ and $Gr := \Delta\backslash\mathcal{B}$. The graph Gr is the (finite) intersection graph of a (semi-)stable reduction of Y and the graph Gr_1 is the intersection graph of a (semi-)stable reduction of Y_0. Since $Y \setminus Y_0$ consists of finitely many points (which we will call cusps), one can easily see that Gr_1 is obtained from Gr by attaching for every cusp an end. From this $\underline{H}_!(Gr_1, \mathbb{Q}_l) \cong \underline{H}(Gr, \mathbb{Q}_l)$ easily follows. We have proved the following corollary.

Corollary 3.4 *Let Y, Γ and $\mathcal{T}$ be as above. Then there is a canonical isomorphism of $\mathcal{G}$-modules $\underline{H}_!(\mathcal{T}, \mathbb{Q}_l)^{\Gamma} \otimes U_l \to H^1_{\acute{e}tale}(Y \hat{\otimes} C, \mathbb{Q}_l)$.*

4 Comparing cohomologies

For an open compact subgroup $\mathcal{K}_f$ of $G(\mathcal{O}_f)$ one considers the analytic space $\Omega_{\mathcal{K}_f} := \Omega \times G(\mathcal{A}_f)/\mathcal{K}_f$. The set $G(\mathcal{A}_f)/\mathcal{K}_f$ is discrete and $\Omega_{\mathcal{K}_f}$ is seen as a disjoint union of copies of Ω. The group $G(K)$ acts on Ω in the usual way and acts on $G(\mathcal{A}_f)/\mathcal{K}_f$ by left-multiplication. The quotient $G(K)\backslash\Omega_{\mathcal{K}_f}$ can also be described as the disjoint union $\sqcup_{\underline{x} \in S} \Gamma_{\underline{x}} \backslash \Omega$, with the notations of (2.1.3). This is clearly an affine curve over K_∞. We will denote this curve by $M^2_{\mathcal{K}_f, K_\infty}$ and its compactification by $\bar{M}^2_{\mathcal{K}_f, K_\infty}$. The term K_∞ is added in the notation to stress the fact that the curves are considered as curves over this field.

Proposition 4.1 *There are canonical isomorphisms*

$$Hom_{G(K_\infty)}(V_{sp}(\mathbb{Q}_l), \mathcal{W}_o^{\mathcal{K}_f}(\mathbb{Q}_l)) \otimes U_l \to \underline{H}_!(\mathcal{T} \times G(\mathcal{A}_f)/\mathcal{K}_f, \mathbb{Q}_l)^{G(K)} \otimes U_l$$

$$\to H^1_{\acute{e}tale}(\bar{M}^2_{\mathcal{K}_f, K_\infty} \otimes \mathbf{C}, \mathbb{Q}_l).$$

Proof. The first isomorphism follows from corollary 2.11 and the second one from corollary 3.4. □

Again we would like to take the limit over all open compact subgroups $\mathcal{K}_f$ of $G(\mathcal{A}_f)$. The family of curves $\bar{M}^2_{\mathcal{K}_f,K_\infty}$ forms a projective system. The projective limit of this projective system is a scheme over K_∞, which we will denote by $\bar{M}^2_{K_\infty}$. This is a well defined object. There seems to be no reasonable rigid analytic space attached to this object. However we define, as a convenient formalism, the analytic étale cohomology group $H^1_{\acute{e}tale}(\bar{M}^2_{K_\infty} \otimes \mathbf{C}, \mathbb{Q}_l)$ as the direct limit of the groups $H^1_{\acute{e}tale}(\bar{M}^2_{\mathcal{K}_f,K_\infty} \otimes \mathbf{C}, \mathbb{Q}_l)$.
We let the elements $\underline{g} \in G(\mathcal{A}_f)$ act on $G(\mathcal{A}_f)$ by multiplication on the right. This induces a bijection $G(\mathcal{A}_f)/\mathcal{K}_f \to G(\mathcal{A}_f)/(g^{-1}\mathcal{K}_f g)$ and an isomorphism $(\Omega \times G(\mathcal{A}_f)/\mathcal{K}_f) \to (\Omega \times G(\mathcal{A}_f)/(g^{-1}\mathcal{K}_f g))$, which commutes with the left-action of $G(K)$. After dividing by the $G(K)$-actions, one obtains an isomorphism between the affine curves $M^2_{\mathcal{K}_f,K_\infty} \to M^2_{g^{-1}\mathcal{K}_f g,K_\infty}$ and also between their compactifications. There results a right-action of $G(\mathcal{A}_f)$ on $\bar{M}^2_{K_\infty}$ and a left-action of $G(\mathcal{A}_f)$ on $H^1_{\acute{e}tale}(\bar{M}^2_{K_\infty} \otimes \mathbf{C}, \mathbb{Q}_l)$. Proposition 4.1 and the definition of the action of $G(\mathcal{A}_f)$ on the space $\mathcal{W}_o(\mathbb{Q}_l)$ lead to the following result.

Theorem 4.2 *There is a $G(\mathcal{A}_f) \times Gal(K^{sep}_\infty/K_\infty)$- equivariant isomorphism*

$$Hom_{G(K_\infty)}(V_{sp}(\mathbb{Q}_l), \mathcal{W}_o(\mathbb{Q}_l)) \otimes U_l \to H^1_{\acute{e}tale}(\bar{M}^2_{K_\infty} \otimes \mathbf{C}, \mathbb{Q}_l).$$

Remark 4.3

Let $I \subset A$ be a nontrivial ideal. One associates with I is the open compact subgroup $\mathcal{K}_I$ of $G(\mathcal{O}_f) = G(\hat{A})$ which is the kernel of the canonical map $G(\hat{A}) \to G(A/I)$. The family $\{\mathcal{K}_I\}$ is cofinal in the set of all open compact subgroups of $G(\mathcal{A}_f)$. Thus we may restrict the limits above to the limits taken over the subgroups $\mathcal{K}_I$. The affine and the projective curve associated to I will be denoted by M^2_{I,K_∞} and $\bar{M}^2_{I,K_\infty}$. The family of curves $\{\bar{M}^2_{I,K_\infty}\}_I$ forms again projective system with limit $\bar{M}^2_{K_\infty}$. Moreover, $H^1_{\acute{e}tale}(\bar{M}^2_{K_\infty} \otimes \mathbf{C}, \mathbb{Q}_l)$ is the direct limit of the groups $H^1_{\acute{e}tale}(\bar{M}^2_{I,K_\infty} \otimes \mathbf{C}, \mathbb{Q}_l)$.

4.4 *Comparison with the $G(\mathcal{A}_f) \times Gal(K^{sep}/K)$-module $\underline{H}$.*

The story continues as follows. In the definition of the algebraic étale cohomology group $\underline{H}$ we may replace K^{sep} by any bigger field which is (separably) algebraicaly closed. We take for this bigger field $\mathbf{C}$. The curve $\bar{M}^2_I$ is replaced by $\bar{M}^2_I \otimes_K K_\infty$. One can then write $\underline{H}$ as the direct limit of the cohomology groups

$$H^1_{et}((\bar{M}^2_I \otimes_K K_\infty) \otimes_{K_\infty} \mathbf{C}, \overline{\mathbf{Q}}_l).$$

If one forgets that the curve $\bar{M}^2_I \otimes_K K_\infty$ over K_∞ comes from a curve over K, then one finds an action of the group $G(\mathcal{A}_f) \times Gal(K^{sep}_\infty/K_\infty)$ on $\underline{H}$. We recall that $Gal(K_\infty^{sep}/K_\infty)$ coincides with the group of the continuous automorphisms of $\mathbf{C}/K_\infty$. The field K^{sep} can be identified with the separable algebraic closure of K in $K_\infty^{sep} \subset \mathbf{C}$. Thus $Gal(K_\infty^{sep}/K_\infty)$ is a subgroup of $Gal(K^{sep}/K)$ and the action of $G(\mathcal{A}_f) \times Gal(K_\infty^{sep}/K_\infty)$ on $\underline{H}$ is the one obtained by restriction.

As a consequence of (3.1.10), the algebraic étale cohomology and the rigid étale cohomology coincide for the curve $\bar{M}_I^2 \otimes_K K_\infty$ over K_∞. By taking direct limits, one obtains an isomorphism between the $Gal(K_\infty^{sep}/K_\infty)$-modules $\underline{H}$ and $H^1_{étale}(\bar{M}^2_{K_\infty} \otimes \mathbf{C}, \overline{\mathbb{Q}}_l)$. This isomorphism respects the $G(\mathcal{A}_f)$-action on both objects. Indeed, the two actions are induced by a right action on $\bar{M}^2 \otimes K_\infty$, defined in (1.2.7), and a right action on $\bar{M}^2_{K_\infty}$, defined in the paragraph above Theorem 4.2. Those two actions coincide as one can verify. Thus Theorem 4.2 has the following corollary.

Corollary 4.5 *There is a $G(\mathcal{A}_f) \times Gal(K_\infty^{sep}/K_\infty)$-equivariant isomorphism*

$$Hom_{G(K_\infty)}(V_{sp}(\overline{\mathbb{Q}}_l), \mathcal{W}_o(\overline{\mathbb{Q}}_l)) \otimes U_l \to \underline{H}.$$

5 Automorphic representations

(5.1) We start by giving a description of the type of groups and representations that we will encounter. More details can be found in [Gel] and [J-L]. The topological groups H that we will work with have the properties:
(1) H is a Hausdorff space and any neighbourhood of $1 \in H$ contains a compact open subgroup H_0 of H.
(2) For one (or any) compact open subgroup H_0, the quotient H/H_0 is a discrete countable or finite set.
An *admissible representation* (see [J-L], (2.1) and (2.2)) of H is a complex vector space V equipped with a left $\mathbb{C}$-linear action of H such that:
(i) The stabilizer in H of any $v \in V$ is an open subgroup of H.
(ii) For any open compact subgroup H_0 the vector space V^{H_0}, consisting of the vectors which are invariant under H_0, is finite dimensional.
It is clear from this definition that the topology of the complex numbers does not play any rôle here. In the sequel we may replace the field complex numbers by any field L which is algebraically closed and has characteristic 0, such as $\overline{\mathbb{Q}}_l$.

Let V be an irreducible admissible representation of the group G. The following result, which is called *Schur's lemma*, holds:

Any linear $l : V \to V$, which commutes with the G-action, has the form $l = \lambda\, id$ for some $\lambda \in L$.

We will give the rather simple proof. For a suitable open compact subgroup H of G, the vector space V^H is finite-dimensional and nonzero. The space V^H is invariant under l and $lv = \lambda v$ holds for some $v \in V^H$, $v \neq 0$, $\lambda \in L$. Then $\ker(l - \lambda\, id, V)$ is a nonzero G-invariant subspace of V and thus equal to V itself.

Let V be again an admissible irreducible representation of G. According to Schur's lemma, the center Z of G acts on V by scalar multiplication. In other words, there is a homomorphism $\psi : Z \to L^*$, which is called the *central character of* V, such that Z acts on V as ψ.

(5.2) *Examples.* The groups that interest us here are $G(\mathcal{A})$, $G(\mathcal{A}_f)$ and $G(K_v)$ for every place v of K.

The space $\mathcal{W}_o(L)$ is an admissible representations of $G(\mathcal{A})$.

Indeed, property (i) is part of the definition of $\mathcal{W}_o(L)$. Property (ii) is in fact Theorem 2.7.

$V_{sp}(L)$ is an admissible irreducible representation of $G(K_\infty)$

Indeed, any locally constant L-valued function on $\mathbb{P}^1(K_\infty)$ is invariant under some open compact subgroup of $G(K_\infty)$. Further (ii) holds, since the quotient of $\mathbb{P}^1(K_\infty)$ by any open compact subgroup of $G(K_\infty)$ is finite. The irreducibility of the representation is proved in [J-L], (2.9.1).

$\underline{H}_!(\tilde{\mathcal{T}}, L)^{G(K)}$ is an admissible representation of $G(\mathcal{A}_f)$

This space is defined as the direct limit of spaces $\underline{H}_!(\mathcal{T} \times \mathcal{A}_f/\mathcal{K}_f, L)^{G(K)}$. Thus property (i) is obvious. Using corollary 3.4, one deducts that each term of this direct limit is finite dimensional. This proves (ii).

$\underline{H}$ is an admissible representation of $G(\mathcal{A}_f)$

This follows from the previous example, corollary 4.4 and theorem 2.12.

(5.3) *The irreducible admissible representations of $G(K_v)$*
We will first summarize the classification of the *complex* admissible irreducible representations, as given in [Gel], §4 and [J-L], §2.
Any admissible irreducible representation of finite dimension is in fact one-dimensional (See [J-L], Proposition 2.7). Those representations do not interest us here and will be neglected.
An irreducible admissible representation V is called *"class 1" (or spherical or*

unramified) if the subspace $V^{G(\mathcal{O}_v)}$ of the $G(\mathcal{O}_v)$-invariant vectors has dimension one. One constructs all class 1 representations as follows.
An *unramified (quasi-) character* is a homomorphism $\chi : K_v^* \to \mathbb{C}^*$ such that χ maps $\mathcal{O}_v^*$ to 1. Thus an unramified character is determined by the value $\chi(\pi_v)$, where π_v is a generator of the maximal ideal of $\mathcal{O}_v$. One takes two unramified characters χ_1 and χ_2 such that:
The fraction $\frac{\chi_1(\pi_v)}{\chi_2(\pi_v)}$ is different from $q_v^{\pm 1}$, where q_v is the cardinal of the residue field of $\mathcal{O}_v$.
The pair χ_1, χ_2 is used to define a one-dimensional representation $B(K_v) \to \mathbb{C}^*$ of the Borel group

$$B(K_v) := \{\begin{pmatrix} a & b \\ 0 & d \end{pmatrix} \mid a, b, d \in K_v;\ ad \in K_v^*\},$$

by mapping the matrix $\begin{pmatrix} a & b \\ 0 & d \end{pmatrix}$ to $\chi_1(a)\chi_2(d) \in \mathbb{C}^*$. This one-dimensional representation of $B(K_v)$ is induced to an admissible representation $\rho(\chi_1, \chi_2)$ of $G(K_v)$. The proces of induction is described in [Gel], p 66. For the convenience of the reader we will give some details. The vector space V of this representation consists of the functions $f : G(K_v) \to \mathbb{C}$, which are locally constant and satisfy the relation

$$f(\begin{pmatrix} t_1 & x \\ 0 & t_2 \end{pmatrix} g) = \chi_1(t_1)\chi_2(t_2)|t_1/t_2|^{1/2} f(g),$$

for all $t_1, t_2 \in K_v^*$, $x \in K_v$ and $g \in G(K_v)$. The action of $G(K_v)$ on V is given by $(gf)(h) = f(hg)$ for all $g, h \in G(K_v)$. The representation $\rho(\chi_1, \chi_2)$ is a class one representation. Moreover, $\rho(\chi_1, \chi_2)$ is a (pre)-unitary representation, i.e., there is a $G(K_v)$-invariant inner product on V, if and only if $|\chi_1(\pi_v)| = |\chi_2(\pi_v)| = 1$. Every class one representation has the form $\rho(\chi_1, \chi_2)$. Two class one representations $\rho(\chi_1, \chi_2)$ and $\rho(\chi_1', \chi_2')$ are isomorphic if and only if $\{\chi_1, \chi_2\} = \{\chi_1', \chi_2'\}$.

This is in fact all that one has to know about the irreducible representations of $G(K_v)$ for understanding Drinfeld's reciprocity law. In order to get a more complete picture we have added some more information about those irreducible representations.

If the unramified characters χ_1 and χ_2 do not satisfy the condition above, then one can define $\rho(\chi_1, \chi_2)$ in the same way. However, this representation is reducible. If the fraction $\frac{\chi_1(\pi_v)}{\chi_2(\pi_v)}$ is equal to q_v, then it has a codimension one irreducible admissible subrepresentation. If the fraction $\frac{\chi_1(\pi_v)}{\chi_2(\pi_v)}$ is equal to q_v^{-1}, then there is a one-dimensional invariant subspace, such that the quotient by this subspace is an admissible irreducible representation. The two types of representations that we have just described are called *special* representations.

There are other irreducible representations. First of all the representations obtained by choosing ramified characters $\chi_{,1}, \chi_2$, i.e., the kernels of the continuous homomorphisms $\chi_i : K_v^* \to L^*$ do not both contain $\mathcal{O}_v^*$. This gives rise to the ramified principal series and ramified special representations. Further the so-called ***supercuspidal representations***. We will not need their description, but mention an important property ([Gel], Theorem 4.24):

For any admissible irreducible representation V, with central character ψ, there is a minimal integer $n \geq 0$ and a nonzero vector $v \in V$ such that

$$\begin{pmatrix} a & b \\ c & d \end{pmatrix} v = \psi(a)v$$

holds for all the elements of the subgroup

$$\{\begin{pmatrix} a & b \\ c & d \end{pmatrix} \mid a, b, c, d \in \mathcal{O}_v,\ ad - bc \in \mathcal{O}_v^*,\ c \in \pi_v^n \mathcal{O}_v\} \text{ of } G(\mathcal{O}_v).$$

The vector v is unique up to scalar multiplication.

The ideal $\pi_v^n \mathcal{O}_v$ is called the *conductor* of the representation. According to [Gel], Remark 4.25, the class 1 representations are the ones with $n = 0$. Further for a class 1 representation V, the central character ψ is easily seen to be $\chi_1\chi_2$. Since ψ is trivial on $\mathcal{O}_v^*$, the space of invariants $V^{G(\mathcal{O}_v)}$ has dimension one.

Let L be an algebraically closed field of characteristic 0. We will need to know the admissible irreducible representations in vector spaces V over L. Let such a V be given and let $v \in V$, $v \neq 0$. The stabilizer of v is some open subgroup of $G(K_v)$ and the orbit $G(K_v)v$ is (at most) countable. This orbit spans V since V is irreducible, and so V has a basis, which is at most countable. The representation is determined by the coefficients of the gv, with $g \in G(K_v)$, with respect to this basis. We conclude that V depends on a countable set of data and we may replace L by a subfield $\tilde{L}$ which is algebraically closed and (at most) countably generated over $\mathbb{Q}$. In other words, $V = L \otimes_{\tilde{L}} W$ with W an admissible irreducible representation over the field $\tilde{L}$. The field $\tilde{L}$ can be embedded into $\mathbb{C}$ and $\mathbb{C} \otimes_{\tilde{L}} W$ is an admissible irreducible complex representation. The conclusion is that the admissible irreducible representations of class 1, in vector spaces over L, have the same description as in the complex case. This conclusion is easily extended to the special representations. The same conclusion is also valid for the remaining representations, the ramified principal series, the ramified special representations and the supercuspidal ones. We will not give a proof of this since it involves a more detailed description of these representations.

(5.4) *Restricted tensor products of admissible representations*
We will now construct an admissible representation of $G(\mathcal{A})$ from a family of representations $\{H_v\}$ for every place v of K. The construction works over any algebraically closed field $L \supset \mathbb{Q}$. The family $\{H_v\}$ is supposed to satisfy the condition that for all but finitely many v the representation H_v of $G(K_v)$ is class 1. The finite set of places v, for which H_v is not class 1, is denoted by S_0. For the other places v we fix a basis e_v of the one dimensional vector space $H_v^{G(\mathcal{O}_v)}$.
For any finite set of places S, containing S_0, one defines the object
$H_S = \otimes_{v\in S} H_v$, which is a vector space equipped with an obvious action of
$\prod_{v\in S} G(K_v)$. We extend this action to an action of the subgroup $\prod_{v\in S} G(K_v)\times \prod_{v\notin S} G(\mathcal{O}_v)$ of $G(\mathcal{A})$, by imposing the trivial action of the factor $\prod_{v\notin S} G(\mathcal{O}_v)$ on H_S. For $S \subset S'$, with S' again a finite set of places, one maps H_S into $H_{S'}$ be sending a vector $w \in H_S$ to $w \otimes e_{v_1} \otimes e_{v_2} \otimes \ldots \otimes e_{v_n} \in H_{S'}$, where $S' \setminus S = \{v_1, \ldots, v_n\}$.
The *restricted tensor product*, in notation $\otimes_v' H_v$, is the direct limit $\lim_{\rightarrow} H_S$, equipped with the obvious action of $G(\mathcal{A})$. The representations H_v are called the *local factors* of the restricted tensor product.
The following nice result permits to classify the admissible irreducible representations of $G(\mathcal{A})$:

Every admissible irreducible representation π of $G(\mathcal{A})$ is in a unique way a restricted tensor product $\otimes_v' \pi_v$. The π_v are called the local factors of π. They are (of course) irreducible and class 1 for all but finitely many places v.

This result is stated in [Gel], §4, for the case of complex representations. The reasoning of (5.3) implies its validity for any algebraically closed field $L \supset \mathbb{Q}$.

(5.5) *Decomposition of the representation $\mathcal{W}_o(\mathbb{C})$*
Statements and proofs can be found in [Gel], Thm 5.7, [J-L], Proposition 11.1.1, [De 2], Théorème 2.5.6 (for the number field case) and [P]. The important result about this representation is: $\mathcal{W}_o(\mathbb{C})$ is a direct sum $\oplus_{\pi\in\Pi} V_\pi$ of a certain family Π of unitary, irreducible admissible representations V_π. Moreover:
For $\pi \neq \pi'$ the representations V_π and $V_{\pi'}$ are not isomorphic. This property is called "multiplicity one". An even stronger statement, with the name "strong multiplicity one", holds:
If the local factors of V_π and $V_{\pi'}$ are isomorphic for all but finitely many places v, then $\pi = \pi'$.

We want to deduce a similar statement with $\mathbb{C}$ replaced by any algebraically

closed field $L \supset \mathbb{Q}$. We start by choosing an embedding $\overline{\mathbb{Q}} \subset \mathbb{C}$. According to corollary 2.9, one has $\mathcal{W}_o(\mathbb{C}) = \mathbb{C} \otimes \mathcal{W}_o(\overline{\mathbb{Q}})$. Every automorphism of $\mathbb{C}/\overline{\mathbb{Q}}$ acts on $\mathcal{W}_o(\mathbb{C})$ and permutes the V_π, because of "multiplicity one". We claim that the V_π are in fact invariant under the action of the automorphisms of $\mathbb{C}/\overline{\mathbb{Q}}$. Fix a V_π and let $V_{\pi,v}$ be a local factor which is class 1 and is defined by the two characters $\{\chi_1, \chi_2\}$. If the two values $\{\chi_1(\pi_v), \chi_2(\pi_v)\}$ are not both in $\overline{\mathbb{Q}}$, then the orbit of $V_{\pi,v}$, and thus also the orbit of V_π, under the group of all automorphisms of $\mathbb{C}/\overline{\mathbb{Q}}$ is uncountable. This is a contradiction and we have proved the claim.
It follows that $\mathcal{W}_o(\overline{\mathbb{Q}})$ is a direct sum (over $\pi \in \Pi$) of admissible irreducible representations W_π, such that $\mathbb{C} \otimes W_\pi = V_\pi$. Moreover for every π and every place v such that $V_{\pi,v}$ is class 1, the algebraic numbers $\{\chi_1(\pi_v), \chi_2(\pi_v)\}$ have absolute value 1. The Galois group $Gal(\overline{\mathbb{Q}}/\mathbb{Q})$ acts in an obvious way on $\mathcal{W}_o(\overline{\mathbb{Q}}) = \overline{\mathbb{Q}} \otimes \mathcal{W}_o(\mathbb{Q})$. Thus $Gal(\overline{\mathbb{Q}}/\mathbb{Q})$ acts as a permutation group on the collection $\{W_\pi\}$ and thus on Π. Fix a $\pi \in \Pi$, an unramified place v of W_π corresponding to the pair $\{\chi_1, \chi_2\}$ and let $\gamma \in Gal(\overline{\mathbb{Q}}/\mathbb{Q})$. Then $\gamma(W_\pi) = W_{\gamma(\pi)}$ and v is again unramified for $W_{\gamma(\pi)}$ and corresponds to the pair $\{\gamma\chi_1, \gamma\chi_2\}$. We conclude from this that the algebraic numbers $\{\chi_1(\pi_v), \chi_2(\pi_v)\}$ have absolute value 1 for every embedding of $\overline{\mathbb{Q}}$ in $\mathbb{C}$.

For any algebraically closed field $L \supset \overline{\mathbb{Q}}$ one has $\mathcal{W}_o(L) = L \otimes \mathcal{W}_o(\overline{\mathbb{Q}})$. The conclusion is that $\mathcal{W}_o(L)$ decomposes as a direct sum of irreducible admissible factors $L \otimes W_\pi$ with $\pi \in \Pi$. We will adopt the notation $W_\pi(L) := L \otimes W_\pi$. For any unramified place v of $W_\pi(L)$ the corresponding characters $\{\chi_1, \chi_2\}$ have the property that the $\chi_i(\pi_v) \in L^*$ are algebraic numbers and have absolute value 1 for every embedding of $\overline{\mathbb{Q}}$ in $\mathbb{C}$.

(5.6) *Decomposition of the representation* $\underline{H}$
We fix an embedding of $\overline{\mathbb{Q}}$ in $\overline{\mathbb{Q}}_l$ and start by determining

$$Hom_{G(K_\infty)}(V_{sp}(\overline{\mathbb{Q}}_l), \mathcal{W}_o(\overline{\mathbb{Q}}_l)).$$

It is a direct sum (over $\pi \in \Pi$) of the terms $Hom_{G(K_\infty)}(V_{sp}(\overline{\mathbb{Q}}_l), W_\pi(\overline{\mathbb{Q}}_l))$. For each π we have only to consider the $G(K_\infty)$-action on $W_\pi(\overline{\mathbb{Q}}_l)$. We recall that the special representation $V_{sp}(\overline{\mathbb{Q}}_l)$ of $G(K_\infty)$ is an admissible irreducible representation. It follows that the Hom, considered here, is zero if the infinite component $W_{\pi,\infty}$ of W_π is not isomorphic to V_{sp}. Otherwise this Hom is equal to the restricted tensor product $\otimes_{v\neq\infty} W_{\pi,v}(\overline{\mathbb{Q}}_l)$. Let Π_∞ be the subset of the elements $\pi \in \Pi$ such that $W_{\pi,\infty} \cong V_{sp}$. Then

$Hom_{G(K_\infty)}(V_{sp}(\overline{\mathbb{Q}}_l), \mathcal{W}_o(\overline{\mathbb{Q}}_l))$ is isomorphic, as $G(\mathcal{A}_f)$ – module, to

$$\oplus_{\pi\in\Pi_\infty} \otimes'_{v\neq\infty} W_{\pi,v}(\overline{\mathbb{Q}}_l).$$

As a consequence $\underline{H}$ is, as $G(\mathcal{A}_f) \times Gal(K_\infty^{sep}/K_\infty)$-module, isomorphic to

$$\oplus_{\pi\in\Pi_\infty} \otimes'_{v\neq\infty} W_{\pi,v}(\overline{\mathbb{Q}}_l) \otimes U_l.$$

By construction the group $Gal(K_\infty^{sep}/K_\infty)$ acts on $\otimes'_{v\neq\infty} W_{\pi,v}(\overline{\mathbb{Q}}_l) \otimes U_l$ by the trivial action on all the components of the tensor product except for the last one U_l. We remark that every admissible irreducible $G(\mathcal{A}_f)$-representation $\otimes'_{v\neq\infty} W_{\pi,v}(\overline{\mathbb{Q}}_l)$ (with $\pi \in \Pi_\infty$) occurs twice in $\underline{H}$. The group $Gal(K^{sep}/K)$ acts also on $\underline{H}$ and this action commutes with the action of $G(\mathcal{A}_f)$. For every $\gamma \in Gal(K^{sep}/K)$ and every $\pi \in \Pi_\infty$ the space $\otimes'_{v\neq\infty} W_{\pi,v}(\overline{\mathbb{Q}}_l) \otimes U_l$ is therefore invariant under γ. Consider the obvious homomorpism of $\overline{\mathbb{Q}}_l$-algebras

$$End(U_l) \to End(\otimes'_{v\neq\infty} W_{\pi,v}(\overline{\mathbb{Q}}_l) \otimes U_l),$$

where the first *End* means the $\overline{\mathbb{Q}}_l$-linear endomorphisms and the second *End* is defined as the $\overline{\mathbb{Q}}_l$-linear endomorphisms, which commute with the $G(\mathcal{A}_f)$-action. This homomorphism is bijective by Schur's lemma and the irreducibility of the representation $\otimes'_{v\neq\infty} W_{\pi,v}(\overline{\mathbb{Q}}_l)$. Thus γ acts on the two-dimensional space U_l. This action depends on the choices of l and $\pi \in \Pi_\infty$. It will be denoted by $\sigma(\pi)_l(\gamma)$. In other words $\underline{H}$ is as $G(\mathcal{A}_f) \times Gal(K^{sep}/K)$-module a direct sum of irreducible representations

$$\oplus_{\pi\in\Pi_\infty} \otimes'_{v\neq\infty} W_{\pi,v}(\overline{\mathbb{Q}}_l) \otimes \sigma(\pi)_l.$$

At this point it is good to remark that the two-dimensional representation $\sigma(\pi)_l$ of $Gal(K^{sep}/K)$ is a continuous representation and uses, unlike the $W_\pi(\overline{\mathbb{Q}}_l)$, the topology of the field $\overline{\mathbb{Q}}_l$.

6 The reciprocity law

The following theorem gives the crucial link between π and $\sigma(\pi)_l$.

Theorem 6.1 *Let $\pi \in \Pi_\infty$ and let $v \neq \infty$ be an unramified place of W_π corresponding to the pair of characters $\{\chi_1, \chi_2\}$. Then v is an unramified place for the representation $\sigma(\pi)_l : Gal(K^{sep}/K) \to Gl(2, \overline{\mathbb{Q}}_l)$. Let $Fr_v \in Gal(K^{sep}/K)$ be a choice for the Frobenius element at the place v. Then the eigenvalues of $\sigma(\pi)_l(Fr_v)$ are $\{q_v^{1/2}\chi_1(\pi_v), q_v^{1/2}\chi_2(\pi_v)\}$.*

We will give no proof, but refer to the expended version of the proof in [Dr], given in [De-Hu], Chapter 5, §3. We recall that q_v is the cardinal number of the residue field at the place v. A direct consequence of theorem 6.1 is: $\sigma(\pi)_l \cong \sigma(\pi')_l$ implies $\pi = \pi'$.

Indeed, for every place v, which is unramified for both W_π and $W_{\pi'}$, the local factors $W_{\pi,v}$ and $W_{\pi',v}$ are isomorphic. The "strong multiplicity one" statement of (5.3) proves that $\pi = \pi'$.

We choose for every prime $l \neq p$ an embedding of $\overline{\mathbb{Q}}$ into $\overline{\mathbb{Q}}_l$. *Drinfeld's reciprocity map* is the map which associates to $\pi \in \Pi_\infty$ the family $\{\sigma(\pi)_l\}$ of continuous two-dimensional representations of $Gal(K^{sep}/K)$.
Let us define the set Σ_∞ as follows:
Every element is a family $\{\rho_l\}_{l\neq p}$ such that $\rho_l : Gal(K^{sep}/K) \to Gl(2, \overline{\mathbb{Q}}_l)$ is (an isomorphy class of) a continuous representation with respect to the ordinary topology on $\overline{\mathbb{Q}}_l$. This family is supposed to have the following properties:

1. For every l the restriction of ρ_l to the subgroup $Gal(K_\infty^{sep}/K_\infty)$ of $Gal(K^{sep}/K)$ is isomorphic to $\overline{\mathbb{Q}}_l \otimes U_l$.

2. There is a finite subset S_0 of places, containing ∞, such that every place $v \notin S_0$ is unramified for all ρ_l.

3. For any $v \notin S_0$, the characteristic polynomials of $\rho_l(Fr_v)$ do not depend on l. In other words, there is a polynomial $T^2 - a_v T + b_v \in \overline{\mathbb{Q}}[T]$ which coincides with the characteristic polynomial of $\rho_l(Fr_v)$ for every l.

4. The zeroes of $T^2 - a_v T + b_v \in \overline{\mathbb{Q}}[T]$ are algebraic numbers which have absolute value $q^{1/2}$ for every embedding of $\overline{\mathbb{Q}}$ in $\mathbb{C}$.

Drinfeld's reciprocity law, which coincides with a special case of the Langlands conjecture for $Gl(2)$ over function fields, is the following statement

Theorem 6.2 *The map $\Pi_\infty \to \Sigma_\infty$, given by $\pi \mapsto \{\sigma(\pi)_l\}$, is bijective.*

Proof. For $\pi \in \Pi_\infty$, the family $\{\sigma(\pi)_l\}$ belongs to Σ_∞ by theorem 6.1 and (5.5). We have already proven that the map is injective. Deligne has observed that the map is surjective. *The idea is the following*:
Let $\{\rho_l\} \in \Sigma_\infty$ be given. In [De 1], §9 and in particular in proposition 9.9, it is shown that the L-function of the representation ρ_l, which is known to be a rational function, satisfies a certain functional equation and is actually a polynomial. Then [J-L], theorem 12.2 asserts the existence of a $\pi \in \Pi_\infty$ with $\sigma(\pi) = \{\rho_l\}$. □

We note that an elliptic curve E over the field K, such that $E \otimes K_\infty$ is a Tate curve (in other words, E has a split multiplicative reduction at ∞), gives rise to an element $\{\rho_l\}$ of Σ_∞. The representation ρ_l is just the $Gal(K^{sep}/K)$-action on the Tate-module $T_l(E) \otimes \overline{\mathbb{Q}}_l$. This collection is known to satisfy the properties 2, 3 and 4 of the definition of Σ_∞. That the restriction of ρ_l to $Gal(K_\infty^{sep}/K_\infty)$ is isomorphic to $U_l \otimes \overline{\mathbb{Q}}_l$ can easily be deduced from the Tate-uniformization of

$E \otimes K_\infty$. The latter is an analytic representation of the curve $E \otimes K_\infty$ as a quotient $\mathbf{G}_{m,K_\infty}/a^{\mathbf{Z}}$, where $a \in K_\infty$ is some element with $0 < |a| < 1$. Property 1 follows from theorem 3.1 applied with $Y = E \otimes K_\infty$ and $L = K_\infty$.

Theorem 6.2 associates to the elliptic curve E above, an irreducible component W_π of $\mathcal{W}_0(\overline{\mathbb{Q}})$ with $W_{\pi,\infty} \cong V_{sp}(\overline{\mathbb{Q}})$.
In the classical case, i.e., the theory of the ordinary modular curves, one can associate to an admissible irreducible, with the correct behaviour at the infinite place, a certain modular form of weight two. This modular form induces a surjective morphism of the Jacobian of a suitable modular curve to the elliptic curve over $\mathbb{Q}$ in question.
This procedure fails here since modular forms of weight two for the Drinfeld modular curves are objects in characteristic p. In the Drinfeld situation, the rôle of modular forms of weight two is taken over by Theta functions. Those Theta functions are still objects in characteristic p. However the absolute value of a Theta function is an object in characteristic 0. In lecture 12 these matters will be explained.

Remark 6.3

We note that the Drinfeld reciprocity map depends on the choices of the embeddings of $\overline{\mathbb{Q}}$ in $\overline{\mathbb{Q}}_l$ for every $l \neq p$. In the special case of an elliptic curve E/K with mutiplicative reduction at ∞, the associated element $\{\rho_l\} \in \Sigma_\infty$ has the property that the characteristic polynomials of $\rho_l(Fr_v)$ (for unramified v and any l) lie in $\mathbb{Q}[T]$. As a consequence the element $\pi \in \Pi_\infty$, with $\{\sigma(\pi)_l\} = \{\rho_l\}$, is stable under the action of $Gal(\overline{\mathbb{Q}}/\mathbb{Q})$. In other words, π is already defined over $\mathbb{Q}$ and the image $\{\sigma(\pi)_l\}$ does not depend on the choices of the embeddings $\overline{\mathbb{Q}} \subset \overline{\mathbb{Q}}_l$.

References

[Be] V. Berkovich - *Étale cohomology for non-archimedean analytic spaces* - Publ. Math. I.H.E.S. 78 (1993).

[Co] G. Cornelissen - *A survey of Drinfeld modular forms* - This volume.

[De 1] P. Deligne - *Les constantes des équations fonctionelles des fonctions L* - Lecture Notes in Math. 349, 143-316 (1973).

[De 2] P. Deligne - *Formes modulaires et représentations de $GL(2)$* - Lecture Notes in Math. 349, 55-105 (1973).

[De-Hu] P. Deligne and D. Husemöller - *Survey of Drinfel'd Modules* - Contemporary Mathematices, Volume 67,p. 25- 91 (1987)

[Dr] V.G. Drinfeld - *Elliptic Modules* - (russian). Math. Sbornik 94, 594-627 (1974). English translation: Math. USSR-Sbornik 23, 561-592 (1976).

[F-vdP] J. Fresnel and M. van der Put - *Géométrie analytique rigide et applications* - P.M.18, Birkhaüser (1981).

[Ge-Re] E.-U. Gekeler and M. Reversat - *Jacobians of Drinfeld modular Curves* - J. reine angew. Math. 476, 27-93 (1996).

[Ge] E.-U. Gekeler - *Jacquet-Langlands theory over K and relations with elliptic curves* - This volume.

[Ge 2] E.-U. Gekeler -*Drinfeld modular curves* - Lect. Notes in Math. 1231, 1986, Springer Verlag.

[Gel] S. Gelbart - *Automorphic forms on adele groups* - Annals of Math. Studies 83, Princeton Univ. Press (1975).

[Go] R. Godement - *Domaines fondamentaux des groupes arithmétiques* - Sém. Bourbaki n 257 (1962/63).

[G- vdP] L. Gerritzen and M. van der Put - *Schottky groups and Mumford curves* - Lect. Notes in Math. 817, Springer Verlag (1980).

[Ha] G. Harder - *Chevalley groups over function fields and automorphic forms* - Ann. of Math. 100, 249-306 (1974).

[Hu] R. Huber - *Étale cohomology of rigid analytic varieties and adic spaces* - Aspects of Math., Vieweg (1996).

[J-L] H. Jacquet and R.P. Langlands - *Automorphic forms on $GL(2)$* - Lecture Notes in Math. vol 114, (1970)

[dJ-vdP] J. de Jong and M. van der Put - *Étale cohomology of rigid analytic spaces* - Documenta Mathematica, Journal der Deutschen Mathematiker-Vereinigung, Band 1, 1996

[P] I. Piatetski-Shapiro - *Multiplicity One Theorems* - Proceedings of Symposia in Pure Mathematics, Vol 33 (1979), part 1, pp 209-212.

[vdP 1] M. van der Put - *Discrete Groups, Mumford Curves and Theta Functions* - Ann. Fac. Sc. Toulouse sér. 6 vol. 1 , 399-435 (1992).

[vdP 2] M. van der Put - *The structure of* Ω *and its quotients* - This volume.

[Se 1] J.-P. Serre - *Corps locaux* - Hermann (1962/68).

[Se 2] J.-P. Serre - *Arbres, Amalgames,* SL_2 - Astérisque 46 , Soc. Math. France (1977).

[Ta] G. Tamme - *Introduction to Étale cohomology* - Springer-Verlag (1994).

Lecture 12

Jacquet-Langlands theory over K and relations with elliptic curves

Ernst-Ulrich Gekeler
Fachbereich 9 Mathematik
Universität des Saarlandes, Postfach 15 11 50
D-66041 Saarbrücken, gekeler@math.uni-sb.de

The aim of the present article is to introduce the reader into the relationship between the theory of automorphic forms over a global function field K and Drinfeld modular curves over K. The automorphic forms considered here (forms on GL(2) of a rather special type) may be regarded as functions on oriented edges of a certain Bruhat-Tits tree $\mathcal{T}$. They translate geometrical properties of Drinfeld modular curves into representation theoretic terms, in such a way that e.g. *Weil curves* (elliptic curves covered through Drinfeld modular curves) may be found as very special functions (integral valued Hecke eigenforms) on $\mathcal{T}$. The connection between automorphic forms, galois representations and modular curves is made explicit in Drinfeld's reciprocity law [4]. Unfortunately, Drinfeld's paper is accessible only to those with a very strong background in algebraic geometry and automorphic forms, and is certainly not suitable to enter the subject. Therefore, a down-to-earth presentation of some aspects of Drinfeld's results has been undertaken in [I.11].

As an outflow of the reciprocity law, there results a Shimura-Taniyama-Weil correspondence over K which, in contrast with the number theoretic case, can be proved to hold in full generality. The link between *automorphic forms* (characteristic-zero-valued data) on the one hand side and Drinfeld modular curves and *modular forms* (characteristic-p-valued data) on the other side is provided by the construction of Jacobians of Drinfeld modular curves through theta functions. These theta functions, being themselves rigid analytic functions on the Drinfeld upper half-plane, via logarithmic derivation give rise to differentials on modular curves, while the logarithmic difference operator applied to their absolute values (a discrete variant of logarithmic derivation) produces automorphic forms. Matters are made transparent by our diagram (5.14) along with the modular interpretation of the various groups it contains. The study of theta functions is crucial both for the constructions alluded to and for understanding which features of the classical theory carry over the function field case (and vice versa), and which do not. This is why their investigation occupies the largest part of this work.

With few exceptions, the material here is taken from [7], [10], and [8]. We are not ambitious to present new results, but rather try to provide motivation, background, explanations for these and other original articles. In some cases, we supply (almost) complete proofs (basic properties of theta functions, construction of the Jacobian), in others, we restrict to offering guidance for a more extended study (general theory of automorphic forms, Hecke operators). Since we don't want to obscure the geometric ideas through accidental technical complications, we formulated and discussed the Shimura-Taniyama-Weil correspondence for a polynomial groundring $A = \mathbb{F}_q[T]$ only, and refer to [10] for the general case.

The reader is assumed to dispose of some prerequisites from algebraic number theory as well as from algebraic and rigid geometry, to the extent as circumscribed through the other articles of this volume. Also, some knowledge about classical modular/automorphic forms and elliptic curves will be helpful. The books of Gelbart [11] and Silverman [27], [28] serve as our basic references for these fields.

We use the standard notation of these introductory articles: A, K, K_∞, C denote the basic arithmetic rings of characteristic p that are analogous with $\mathbb{Z}$, $\mathbb{Q}$, $\mathbb{R}$, $\mathbb{C}$. $\mathbb{F}_q$ is always the exact field of constants of A and K. The *standard example* is provided by a polynomial ring $A = \mathbb{F}_q[T]$ with its field of fractions $K = \mathbb{F}_q[T]$. K_∞ is the completion of K at a place ∞, with ring of integers O_∞, a uniformizer π_∞, absolute value "$|?|$" extended to C and normalized such that $|\pi_\infty|^{-1} = q_\infty = \sharp$ (residue class field at ∞). The "log" function will be with respect to base q_∞. $\Omega = \mathbb{P}^1(C) - \mathbb{P}^1(K_\infty)$ is Drinfeld's upper half-plane, $\mathcal{T}$ the Bruhat-Tits tree of $\mathrm{GL}(2, K_\infty)$, with sets

$$\begin{aligned} X(\mathcal{T}) &= \mathrm{GL}(2,K_\infty)/Z(K_\infty)\cdot\mathcal{K}_\infty \quad \text{of vertices, and} \\ Y(\mathcal{T}) &= \mathrm{GL}(2,K_\infty)/Z(K_\infty)\cdot\mathcal{I}_\infty \quad \text{of oriented edges,} \end{aligned}$$

respectively. Here $Z(K_\infty)$ is the group of scalar matrices, and $\mathcal{K}_\infty = \mathrm{GL}(2, O_\infty)$, $\mathcal{I}_\infty = \{\left(\begin{smallmatrix} a & b \\ c & d \end{smallmatrix}\right) \in \mathcal{K}_\infty \mid c \equiv 0 \ (\bmod\ \pi_\infty)\}$. We do not distinguish between positive divisors on Spec A and non-zero ideals of A, labelled by $\mathfrak{n}, \mathfrak{m}, \ldots$ Acccordingly, *primes* $\mathfrak{p}, \mathfrak{q}, \ldots$ and *places* have the same meaning, where $\mathbb{F}_\mathfrak{p} = A/\mathfrak{p}$ is the residue class field. If the group G acts on the set X, $\gamma \in G$, $x \in X$, we denote by γx, G_x, Gx, $G \setminus X$ the image $\gamma(x)$, the stabilizer group, the orbit, the set of all orbits, respectively.

1. The classical Shimura-Taniyama-Weil correspondence.

Let for the moment E be an elliptic curve defined over the rational numbers $\mathbb{Q}$. Hence E is given through an affine equation

$$Y^2 + a_1XY + a_3Y = X^3 + a_2X^2 + a_4X + a_6 \tag{1.1}$$

with $a_i \in \mathbb{Q}$, whose discriminant $\Delta(a_1, \ldots, a_6)$, a certain isobaric polynomial of weight 12 in the a_i, doesn't vanish. The *conductor* of E is a natural number $\mathrm{cond}(E)$ divisible by precisely those prime numbers p modulo which equation (1.1) (or more precisely, an equation $(1.1)'_p$ equivalent with (1.1) with p-integral coefficients a'_i, and whose discriminant $\Delta(a'_1, \ldots, a'_6)$ has minimal p-adic valuation) becomes singular. If p divides $\mathrm{cond}(E)$, the reduced curve $\tilde{E}/\mathbb{F}_p$ has either an ordinary double point or a cusp as its unique singularities, in which case we speak of *multiplicative* or *additive* reduction at p, respectively. The exponents $f_p(E)$ of $\mathrm{cond}(E) = \prod_{p \text{ prime}} p^{f_p(E)}$ satisfy

$$(1.2) \qquad f_p(E) = \left\{ \begin{array}{c} 0 \\ 1 \\ \geq 2 \end{array} \right\} \text{ if } E \text{ has } \left\{ \begin{array}{c} \text{good} \\ \text{multiplicative} \\ \text{additive} \end{array} \right\} \text{ reduction at } p.$$

Moreover, $f_p(E) > 2$ can occur for $p = 2$ and 3 only.

(1.3) Let now $X_0(N)$ be the elliptic modular curve of Hecke type with conductor $N \in \mathbb{N}$. If $\Gamma_0(N)$ denotes the Hecke congruence subgroup $\{\left(\begin{smallmatrix} a & b \\ c & d \end{smallmatrix}\right) \in \mathrm{SL}(2, \mathbb{Z}) \mid c \equiv 0 \pmod{N}\}$, which acts through linear transformations on the complex upper half-plane H, $X_0(N)$ results from compactifying the open Riemann surface $\Gamma_0(N) \setminus H$ by attaching a finite number of "*cusps*":

$$X_0(N) = \Gamma_0(N) \setminus H \cup \Gamma_0(N) \setminus \mathbb{P}^1(\mathbb{Q}).$$

It is well-known that the compact Riemann surface $X_0(N)$ is the set of complex points of a (uniquely determined) algebraic curve defined over $\mathbb{Q}$, labelled also by $X_0(N)$ or $X_0(N)/\mathbb{Q}$. We have canonical isomorphisms of the following complex vector spaces of dimension g = genus of $X_0(N)$:

$$\text{(a)} \quad \left\{ \begin{array}{c} \text{holomorphic differential forms on} \\ X_0(N)/\mathbb{Q} \text{ in the algebraic sense} \end{array} \right\} \otimes_{\mathbb{Q}} \mathbb{C}$$

$$\text{(b)} \quad \left\{ \begin{array}{c} \text{holomorphic differential forms on} \\ X_0(N)/\mathbb{C} \text{ in the algebraic sense} \end{array} \right\}$$

$$\text{(c)} \quad \left\{ \begin{array}{c} \text{holomorphic differential forms on} \\ X_0(N) \text{ in the analytic sense} \end{array} \right\}$$

$$\text{(d)} \quad \{\text{cusp forms of weight two for } \Gamma_0(N)\} =: S_2(N).$$

Here (a) $\xrightarrow{\cong}$ (b) comes from base extension, (b) $\xrightarrow{\cong}$ (c) is the GAGA isomorphism between algebraic and analytic data, and (d) $\xrightarrow{\cong}$ (c) is given by $f(z) \longmapsto f(z)dz$. (The transformation rule of $f(z)$ asserts the $\Gamma_0(N)$-invariance of $f(z)dz$, the cusp condition for $f(z)$ gives the holomorphy of $f(z)dz$ at cusps.) Similar assertions hold for other modular curves. For each natural number $m \in \mathbb{N}$ (whether or not m and N are coprime), we have a Hecke operator T_m acting on (a) ... (d), and the isomorphisms between these spaces are compatible with the respective Hecke actions. We therefore do not further distinguish between (a) ... (d). The reader may consult e.g. [18] Ch. VII for the precise definition of the T_m.

(1.4) The "*classical*" conjecture of Shimura, Taniyama and Weil states:

> $\boxed{\text{STW}}$ If the elliptic curve $E/\mathbb{Q}$ has conductor N, there exists a dominant morphism, defined over $\mathbb{Q}$, of algebraic curves
>
> $$\pi : X_0(N)/\mathbb{Q} \longrightarrow E/\mathbb{Q}.$$

Such a π is called a *Weil uniformization* of E; it is a *strong Weil uniformization* if π cannot be properly factored through another Weil uniformization π' : $X_0(N)/\mathbb{Q} \xrightarrow{\pi'} E'/\mathbb{Q} \longrightarrow E/\mathbb{Q}$. For each Weil uniformization π, there exists an isogeneous curve E', the *strong Weil curve*, provided with a strong Weil uniformization π', and π' is uniquely determined up to automorphisms of E'.

(1.5) Now if the conjecture is true, well-known classical arguments show that we have bijections between the following three sets of

(a) $\mathbb{Q}$-isogeny classes of elliptic curves $E/\mathbb{Q}$ with $\mathrm{cond}(E) = N$ (the conductor of an elliptic curve is an isogeny invariant!);

(b) isogeny factors of dimension one of $J_0^{\mathrm{new}}(N)$, the new part of the Jacobian $J_0(N)$ of $X_0(N)$;

(c) classes (up to scaling) of Hecke eigenforms with rational eigenvalues in $S_2^{\mathrm{new}}(N)$, the new part of $S_2(N)$.

The new parts in (b) and (c) are those parts that genuinely belong to the conductor N and not to proper divisors N' of N (for precise definitions, see e.g. [18] Ch. VIII). The map from (b) to (a) is the tautological one; its well-definedness and bijectivity depends on $\boxed{\text{STW}}$. The map from (b) to (c) is defined as follows. Let $X_0(N) \hookrightarrow J_0(N)$ be the Abel-Jacobi map with respect to a $\mathbb{Q}$-rational point of $X_0(N)$ (for example, the cusp represented by ∞), and let $J_0(N) \longrightarrow J_0^{\mathrm{new}}(N) \longrightarrow E$ be the quotient map that belongs to $E \in$ (b). Together, this yields a uniformization $\pi : X_0(N) \longrightarrow E$. Let further ω be a nowhere vanishing holomorphic differential form on E. Then $\pi^*(\omega)$ is a holomorphic differential form on $X_0(N)$, hence $\pi^*(\omega) = f(z)dz$ with some

$f(z) \in S_2(N)$. Actually $f(z) \in S_2^{\rm new}(N)$, and it is a Hecke eigenform with rational (hence integral) eigenvalues that are related to the reduction properties of E. The well-definedness and bijectivity of the map (class of E) $\longmapsto$ (class of $f(z)$) from (b) to (c) is much easier than the one above, and is fully established without assuming $\boxed{\text{STW}}$. Taking $\boxed{\text{STW}}$ for granted yields an effective method to determine all the elliptic curves $E/\mathbb{Q}$ with a given conductor N, in that the classification is reduced to linear algebra in the vector space $S_2(N)$. For tables and numerical results, see [2]. Now A. Wiles has established the truth of $\boxed{\text{STW}}$ for a large class of conductors N, including all the square-free ones [31], [29]. Due to work of Ribet [22] based on ideas of Frey and Serre, it is known that this special case of $\boxed{\text{STW}}$ implies for instance Fermat's last theorem.

2. The function field setting.

Let us now come back to the function field case with our usual objects K, ∞, A, K_∞, C etc. We dispose of

- elliptic curves over K,
- (Drinfeld) modular curves over K,
- (Drinfeld) modular forms (see [I.8], [I.10]),

all of which behave largely similarly to their counterparts over number fields (although there are also some important differences). It is therefore natural to ask for a drinfeldian analogue of $\boxed{\text{STW}}$. In describing such an analogue, we postpone precise definitions and statements to later sections.

(2.1) Let $\Gamma \subset \mathrm{GL}(2, K)$ be an arithmetic subgroup, and let M_Γ be the associated affine algebraic curve with C-points $M_\Gamma(C) = \Gamma \setminus \Omega$ and smooth compactification $\overline{M}_\Gamma$. Then M_Γ and $\overline{M}_\Gamma$ may be defined over K_∞ (in fact, over a finite extension K' of K with $K' \subset K_\infty$). Let J_Γ/K_∞ be the Jacobian. As follows from the analytic description of M_Γ and J_Γ, the abelian variety J_Γ/K_∞, and hence each of its isogeny factors, has split multiplicative reduction. This means that an elliptic curve E/K that admits a "*Weil uniformization*" through a Drinfeld modular curve necessarily has split multiplicative reduction at ∞.

(2.2) Recall that E has *split multiplicative reduction* at ∞ if it has multiplicative reduction and the two tangents at the double point of the reduction are rational over the residue class field of ∞. Equivalently ([28] V), E regarded as a curve over K_∞ is a Tate curve, i.e., may be described as the multiplicative group $\mathbb{G}_m/K_\infty$ divided by some lattice $t^{\mathbb{Z}}$, where $|t| < 1$.

(2.3) While looking out for a Shimura-Taniyama-Weil correspondence over (K,∞), it is therefore necessary to restrict to elliptic curves E/K with split multiplicative reduction at ∞. That restriction is however not very serious: Suppose that the j-invariant $j(E)$ of E satisfies $v_\infty(j(E)) < 0$. Then there exists a finite separable extension K' of K and an extension ∞' of ∞ to K' such that $E/K'_{\infty'}$ has split multiplicative reduction. Hence a given curve E/K either has a constant j-invariant $j(E) \in \mathbb{F}_q$ (these E are easily classified), or, after some base extension, there exists a place ∞ at which E acquires split multiplicative reduction, in which case E is covered by the considerations below.

(2.4) We can now formulate a K-version of $\boxed{\text{STW}}$ in the given framework. For technical reasons, we restrict to the case of our standard example $(K,\infty,A) = (\mathbb{F}_q(T),\infty,\mathbb{F}_q[T])$, and refer to [10] for the general case. We put $\Gamma(1) = \mathrm{GL}(2,A)$
$= \mathrm{GL}(2,\mathbb{F}_q[T])$, and for some ideal $0 \neq \mathfrak{n} \subset A$, $\Gamma_0(\mathfrak{n}) = \{\left(\begin{smallmatrix} a & b \\ c & d\end{smallmatrix}\right) \in \Gamma(1) \mid c \in \mathfrak{n}\}$. We further denote by $M_0(\mathfrak{n})$ and $\overline{M}_0(\mathfrak{n})$ the affine and projective Drinfeld modular curve given by $\Gamma_0(\mathfrak{n})$, respectively. I.e., $M_0(\mathfrak{n})(C) = \Gamma_0(\mathfrak{n}) \setminus \Omega$, $\overline{M}_0(\mathfrak{n})(C) = \Gamma_0(\mathfrak{n}) \setminus \Omega \cup \Gamma_0(\mathfrak{n}) \setminus \mathbb{P}^1(K)$ with Drinfeld's upper half-plane $\Omega = \mathbb{P}^1(C) - \mathbb{P}^1(K_\infty)$. Then $\overline{M}_0(\mathfrak{n})$ and its Jacobian $J_0(\mathfrak{n})$ are defined over K. If an elliptic curve E/K has split multiplicative reduction at ∞, its conductor $\mathrm{cond}(E)$ (a positive divisor of K) contains ∞ with multiplicity one, and has therefore the form $\mathfrak{n}\cdot(\infty)$ with some positive divisor $\mathfrak{n}$ coprime with ∞, which we regard as an ideal of A.

(2.5) The K-version of the Shimura-Taniyama-Weil conjecture (here it is a theorem!) is:

> $\boxed{\text{STW/K}}$ If the elliptic curve E/K with split multiplicative reduction at ∞ has conductor $\mathrm{cond}(E) = \mathfrak{n}\cdot(\infty)$, there exists a dominant morphism, defined over K, of algebraic curves
>
> $$\pi : \overline{M}_0(\mathfrak{n})/K \longrightarrow E/K.$$

Such a π is called a *Weil uniformization.* Accordingly, we define a strong Weil uniformization and the strong Weil curve in a given isogeny class.

In the remainder of this article, we will develop some of the necessary tools, sketch the proof, and derive bijections analogous to those given in (1.5).

3. Arithmetic groups and harmonic cochains.

(3.1) An *arithmetic subgroup* Γ of $\mathrm{GL}(2,K)$ is a subgroup commensurable with

$\Gamma(1) = GL(2, A)$, i.e., such that $\Gamma \cap \Gamma(1)$ has finite index in both Γ and $\Gamma(1)$. The main examples are (a) principal congruence subgroups $\Gamma(\mathfrak{n}) = \{\gamma \in \Gamma(1) \mid \gamma \equiv \left(\begin{smallmatrix}1&0\\0&1\end{smallmatrix}\right) \pmod{\mathfrak{n}}\}$ for an ideal $0 \neq \mathfrak{n}$ of A and (b) $GL(Y)$, where $Y \subset K^2$ is a projective A-submodule of rank two.

Recall ([I.6], [I.7]) that $\Omega = \mathbb{P}^1(C) - \mathbb{P}^1(K_\infty)$ is Drinfeld's upper half-plane and $\mathcal{T}$ the Bruhat-Tits tree of $GL(2, K_\infty)$. The sets of vertices and oriented edges of $\mathcal{T}$ are given by

$$\begin{aligned} X(\mathcal{T}) &= GL(2, K_\infty)/Z(K_\infty) \cdot \mathcal{K}_\infty \quad \text{and} \\ Y(\mathcal{T}) &= GL(2, K_\infty)/Z(K_\infty) \cdot \mathcal{I}_\infty, \end{aligned}$$

respectively. The group $GL(2, K_\infty)$ acts from the left on both Ω and $\mathcal{T}$, which are related by a $GL(2, K_\infty)$-equivariant map $\lambda : \Omega \longrightarrow \mathcal{T}(\mathbb{R})$.

Most of the following properties of arithmetic subgroups Γ are proved for $\Gamma = \Gamma(1)$ in [25] or in [7], and are easily seen to be shared by general arithmetic groups as defined above. See also [I.7] and [I.8].

(3.2) Γ acts with finite stabilizers on Ω and $\mathcal{T}$. Hence we can form the quotient $\Gamma \setminus \Omega$ as an analytic space (defined over C and even over K_∞) and the quotient graph $\Gamma \setminus \mathcal{T}$. An admissible covering of $\Gamma \setminus \Omega$ may be described by means of $\Gamma \setminus \mathcal{T}$ ([I.7] 2.2).

(3.3) $\Gamma \setminus \mathcal{T}$ is the union of a finite graph and a finite number of infinite half-lines • – – – • – – – • – – – ···, the *ends* of $\Gamma \setminus \mathcal{T}$.

(3.4) There exists a smooth connected affine algebraic curve M_Γ / C (which may even be defined over a finite subextension K' of K_∞ / K) whose set $M_\Gamma(C)$ of C-valued points agrees with $\Gamma \setminus \Omega$.

We let $\overline{M}_\Gamma$ be the canonical smooth compactification of M_Γ.

(3.5) There are canonical bijections between the sets of

(a) ends of $\Gamma \setminus \mathcal{T}$;

(b) cusps of M_Γ (i.e., $\overline{M}_\Gamma(C) - M_\Gamma(C)$);

(c) orbits of Γ on $\mathbb{P}^1(K)$ (i.e., $\Gamma \setminus \mathbb{P}^1(K)$).

We identify these sets and label them by $cusps(\Gamma)$.

(3.6) We have equality between the three numbers

(a) $\text{rank}_{\mathbb{Z}} H_1(\Gamma \setminus \mathcal{T}, \mathbb{Z})$

(b) $g(\Gamma)$ = genus of $\overline{M}_\Gamma$

(c) $\mathrm{rank}_{\mathbb{Z}}(\Gamma^{ab})$, where Γ^{ab} is the maximal abelian quotient of Γ.

Here (a) = (c) is an easy group-theoretical fact, explained in [25]. The equality (a) = (b) is deeper, it is a consequence of Drinfeld's reciprocity law [I.11].

(3.7) We define some other groups attached to Γ. First, let

$$\tilde{\Gamma} = \Gamma/\Gamma \cap Z(K_\infty).$$

Note that $\Gamma \cap Z(K_\infty)$ (which is isomorphic with a subgroup of $\mathbb{F}_q^*$) is the kernel of the actions of Γ on Ω and on $\mathcal{T}$. Hence $\tilde{\Gamma}$ acts effectively. Next, let

$$\overline{\Gamma} = \Gamma^{ab}/\mathrm{tor}(\Gamma^{ab})$$

be the maximal torsion-free abelian quotient of Γ and Γ_f the (normal) subgroup generated by the elements of finite order.

3.8 Proposition. (i) Γ/Γ_f *is free in* $g = g(\Gamma)$ *generators.*
(ii) *The canonical map* $\overline{\Gamma} \longrightarrow (\Gamma/\Gamma_f)^{ab}$ *is an isomorphism.*
(iii) $\mathrm{tor}(\Gamma^{ab})$ *equals the image of* Γ_f *in* Γ^{ab}.

Proof. Note that an element of Γ has finite order if and only if it stabilizes a vertex of $\mathcal{T}$. Therefore, by [25] I, Thm. 13, Cor. 1, Γ/Γ_f is the fundamental group of $\Gamma \setminus \mathcal{T}$, which by (3.6) is free in g generators, thus (i). The map in (ii) is well-defined by (i) and obviously surjective; it is therefore bijective since both groups are free abelian in g generators. (iii) is an easy consequence of (ii). □

(3.9) For each abelian group B we let $\underline{H}(\mathcal{T}, B)$ be the group (actually, a right $\mathrm{GL}(2, K_\infty)$-module) of maps $\varphi : Y(\mathcal{T}) \longrightarrow B$ such that the conditions (i) and (ii) hold. (Such φ are called B-valued *harmonic cochains.*)

(i) $\varphi(\overline{e}) = -\varphi(e) \quad \forall\, e \in Y(\mathcal{T})$ ($\overline{e}$ = edge e with inverse orientation)

(ii) $\sum_{o(e)=v} \varphi(e) = 0 \quad \forall v \in X(\mathcal{T})$ ($o(e)$ = origin of the edge e).

Let $\underline{H}(\mathcal{T}, B)^\Gamma \supset \underline{H}_!(\mathcal{T}, B)^\Gamma \supset \underline{H}_{!!}(\mathcal{T}, B)^\Gamma$ be the subgroups of $\underline{H}(\mathcal{T}, B)$ that moreover satisfy (iii), (iii)+(iv), (iii)+(iv)', respectivley, where

(iii) $\varphi(\gamma e) = \varphi(e) \quad \forall e \in Y(\mathcal{T}), \forall \gamma \in \Gamma$;

(iv) φ has finite support mod Γ;

(iv)' φ lies in the image of the canonical map

$$\underline{H}_!(\mathcal{T},\mathbb{Z})^\Gamma \otimes B \longrightarrow \underline{H}(\mathcal{T},B)^\Gamma.$$

Note that elements of $\underline{H}(\mathcal{T},B)^\Gamma$ may be considered as functions on $Y(\Gamma\setminus\mathcal{T})$ that satisfy (i) and (ii)$_\Gamma$, with

$$\text{(ii)}_\Gamma \quad \sum_{o(e)=v} m(e)\varphi(e) = 0 \quad \forall \text{ vertices } v \text{ of } \Gamma\setminus\mathcal{T}.$$

Here the $m(e)$ are certain multiplicities that count how many edges of $\mathcal{T}$ are identified modulo Γ. The precise formula is $m(e) = [\tilde{\Gamma}_{o(e)} : \tilde{\Gamma}_e]$, the index of respective stabilizers in $\tilde{\Gamma}$. Elements of $\underline{H}_!(\mathcal{T},B)^\Gamma$ ($\underline{H}_{!!}(\mathcal{T},B)^\Gamma$) are called *cuspidal* (*double cuspidal*) B-valued *harmonic cochains* for Γ. In practice, B will be one of $\mathbb{Z}$, $\mathbb{Q}$, $\mathbb{C}$, $\mathbb{F}_p$, or C.

3.10 Proposition. (i) *If B is torsion-free, any $\varphi \in \underline{H}_!(\mathcal{T},B)^\Gamma$ vanishes on the ends of $\Gamma\setminus\mathcal{T}$.*
(ii) *If $pB = 0$ then $\underline{H}(\mathcal{T},B)^\Gamma = \underline{H}_!(\mathcal{T},B)^\Gamma$.*

Proof. (i) In view of (3.3), $\varphi \in \underline{H}_!(\mathcal{T},B)^\Gamma$ vanishes eventually on a given end of $\Gamma\setminus\mathcal{T}$, and therefore on the full end, using (3.9) (ii)$_\Gamma$.
(ii) Let two edges e, e' of $\Gamma\setminus\mathcal{T}$ lie on an end h as indicated in

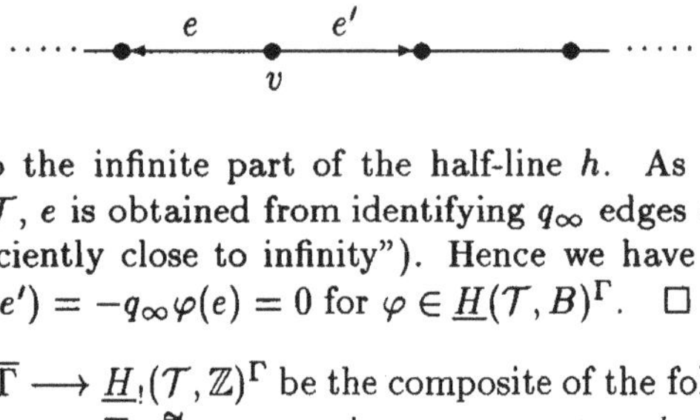

where e' points to the infinite part of the half-line h. As results from the description of $\Gamma\setminus\mathcal{T}$, e is obtained from identifying q_∞ edges of $\mathcal{T}$ under Γ (at least, if v is "sufficiently close to infinity"). Hence we have $m(e) = q_\infty$ and $m(e') = 1$, thus $\varphi(e') = -q_\infty\varphi(e) = 0$ for $\varphi \in \underline{H}(\mathcal{T},B)^\Gamma$. □

(3.11) Let now $j : \overline{\Gamma} \longrightarrow \underline{H}_!(\mathcal{T},\mathbb{Z})^\Gamma$ be the composite of the following canonical maps: the isomorphism $\overline{\Gamma} \xrightarrow{\cong} (\Gamma/\Gamma_f)^{ab}$ of (3.8), $(\Gamma/\Gamma_f)^{ab} \xrightarrow{\cong} H_1(\Gamma\setminus\mathcal{T},\mathbb{Z})$ (since Γ/Γ_f is the fundamental group of $\Gamma\setminus\mathcal{T}$), and

$$\begin{array}{ccc} H_1(\Gamma\setminus\mathcal{T},\mathbb{Z}) & \longrightarrow & \underline{H}_!(\mathcal{T},\mathbb{Z})^\Gamma, \\ \varphi & \longmapsto & \varphi^* \end{array}$$

where $\varphi^*(e) = \sharp(\tilde{\Gamma}_e)\varphi(e)$ (writing e both for vertices of $\mathcal{T}$ and their class mod Γ). It is easily checked that j is well-defined (i.e., φ^* lies in $\underline{H}_!(\mathcal{T},\mathbb{Z})^\Gamma$), injective, and has finite cokernel. The finite cokernel property follows from $H_1(\Gamma\setminus\mathcal{T},\mathbb{Q}) \xrightarrow{\cong} \underline{H}_!(\mathcal{T},\mathbb{Q})^\Gamma$, which is obvious since we can write down the inverse map $\varphi^* \longmapsto \varphi$ over $\mathbb{Q}$. A more explicit description of j is given in [10] 3.3. We know of no example of group Γ where j fails to be bijective. What is known about j is summarized in the next result.

3.12 Theorem. (i) *The cokernel of j has always an order not divisible by p.*

(ii) *j is bijective at least if one of* (a) *or* (b) *hold:*

(a) *Γ has no non-p torsion (i.e., every torsion element of Γ has order a power of p).*

(b) *A is as in the standard example (i.e., a polynomial ring), and Γ is one of the standard congruence subgroups* $\Gamma_0(\mathfrak{n})$, $\Gamma_1^*(\mathfrak{n})$, $\Gamma_1(\mathfrak{n})$, $\Gamma(\mathfrak{n}) \cdots$ of $\Gamma(1) = \mathrm{GL}(2, A)$.

Proof. (i) results from comparing the maps $j = j_\Gamma$ and j_Δ for Γ and a carefully chosen subgroup $\Delta \subset \Gamma$ via the transfer map. Details are carried out in [10] VI, notably prop. 6.4.4.
(ii) follows from a close inspection of the quotient graph $\Gamma \setminus \mathcal{T}$, which can be found in *loc. cit.* 3.4 for case (a) and in [9] for case (b). □

By means of j, we will regard $\overline{\Gamma}$ as a subgroup of $\underline{H}_!(\mathcal{T}, \mathbb{Z})^\Gamma \hookrightarrow \underline{H}_!(\mathcal{T}, \mathbb{C})^\Gamma$, a space of automorphic forms. Assertion(i) will be crucial for the construction of the Jacobian J_Γ of $\overline{M}_\Gamma$, while (ii) is important for the calculation of examples.

4. Theta functions for arithmetic groups.

(4.1) We fix an arithmetic subgroup Γ of $\mathrm{GL}(2, K)$. A *meromorphic theta function* for Γ is a meromorphic function $0 \neq f : \Omega \longrightarrow C$ that

(i) for each $\gamma \in \Gamma$ satisfies a functional equation

$$f(\gamma z) = c_f(\gamma) f(z)$$

with some constant $c_f(\gamma)$ independent of $z \in \Omega$, and

(ii) is invertible around cusps of $M_\Gamma = \Gamma \setminus \Omega$.

For a *holomorphic theta function*, we require that f be holomorphic and nowhere zero on Ω. (The "nowhere zero" condition will turn out later to be superfluous, see (6.9).) Obviously, the map $c_f : \gamma \longmapsto c_f(\gamma)$ is a homomorphism from Γ to C^*, called the *multiplier* of f.

We briefly explain the meaning of conditon (ii). Let $s \in \mathbb{P}^1(K)$ represent the cusp $[s]$, t_s the associated uniformizer (see [I.10]), and let $f : \Omega \longrightarrow C$ satisfy (i). Since C has no p-th roots of unity, the multiplier c_f vanishes on $(\Gamma_s)^u$, the unipotent radical of the stabilizer Γ_s of s. In other words, f is invariant under $(\Gamma_s)^u$, and has therefore a Laurent expansion $\sum_{k\in\mathbb{Z}} a_k t_s^k$ around $t_s = 0$. Condition (ii) for $[s]$ now requires that $a_k = 0$ if $k < 0$ and $a_0 \neq 0$. More details can be found in [7] V sect. 2.

We put $\Theta_h(\Gamma)$ and $\Theta_m(\Gamma)$ for the multiplicative groups of holomorphic and meromorphic theta functions for Γ, respectively. The main examples of theta functions are given through the following construction. Let ω and η be two fixed elements of Ω, and put

$$\theta(\omega,\eta,z) = \prod_{\gamma\in\tilde{\Gamma}} \frac{z-\gamma\omega}{z-\gamma\eta}.$$

Here the factors are rational functions in $z \in \mathbb{P}^1(C)$, and the product is evaluated according to the usual rules for infinite products of analytic functions. Note that the product is not over Γ but over $\tilde{\Gamma} = \Gamma/Z(K_\infty)\cap\Gamma$.

4.3 Theorem. (i) *The product* (4.2) *converges locally uniformly on* Ω. *The resulting meromorphic function* $\theta(\omega,\eta,\cdot)$ *has zeroes of order* $\sharp\tilde{\Gamma}_\omega$ *at the points of* $\Gamma\omega$ *and poles of order* $\sharp\tilde{\Gamma}_\eta$ *at* $\Gamma\eta$, *if* $\Gamma\omega \neq \Gamma\eta$, *and has neither zeroes nor poles on* Ω *if* $\Gamma\omega = \Gamma\eta$.

(ii) *For* $\alpha \in \Gamma$ *there exists a constant* $c(\omega,\eta,\alpha)$ *such that*

$$\theta(\omega,\eta,\alpha z) = c(\omega,\eta,\alpha)\theta(\omega,\eta,z)$$

independently of z.

(iii) $\theta(\omega,\eta,z)$ *is invertible around cusps of* Γ.

(iv) $c(\omega,\eta,\alpha)$ *depends only on the class* $\overline{\alpha}$ *of* α *in* $\overline{\Gamma}$. *Thus* $\alpha \longmapsto c(\omega,\eta,\alpha)$ *defines a group homomorphism* $c(\omega,\eta,\cdot) : \Gamma \longrightarrow \overline{\Gamma} \longrightarrow C^*$.

(v) *The holomorphic function* $u_\alpha(z) := c(\omega,\alpha\omega,z)$ *is independent of the choice of* $\omega \in \Omega$ *and depends only on* $\overline{\alpha} \in \overline{\Gamma}$.

(vi) *For* $\alpha,\beta \in \Gamma$ *we have* $u_{\alpha,\beta} = u_\alpha u_\beta$.

(vii) $c(\omega,\eta,\alpha) = u_\alpha(\eta)/u_\alpha(\omega)$ *and, in particular, is holomorphic in* ω *and* η.

(viii) *Let* $c_\alpha = c(\omega,\alpha\omega,\cdot) : \Gamma \longrightarrow C^*$ *be the multiplier of* u_α. *Then* $(\alpha,\beta) \longmapsto c_\alpha(\beta)$ *defines a symmetric bilinear map from* $\overline{\Gamma}\times\overline{\Gamma}$ *to* C^*.

Proof. (i) The locally uniform convergence of the product means that there exists an admissible covering (U_n) of the analytic space Ω such that, given U_n and $\epsilon > 0$, almost all of the factors of (4.2) satisfy $\left|\frac{z-\gamma\omega}{z-\gamma\eta}\right| < \epsilon$. Such a covering is specified, and the required estimates are carried out, in [10] 5.3. This also gives the assertion on zeroes/poles of $\theta(\omega, \eta, z)$.
(ii) First observe that we have the identity of rational functions in z:

$$\frac{\alpha z - \gamma\omega}{\alpha z - \gamma\eta} = h_{\alpha,\gamma}\frac{z - \alpha^{-1}\gamma\omega}{z - \alpha^{-1}\gamma\eta} \text{ with } h_{\alpha,\gamma} = \frac{\alpha\infty - \gamma\omega}{\alpha\infty - \gamma\eta} \in C^*.$$

(The two rational functions on $\mathbb{P}^1(C)$ have the same divisor. Evaluating on $z = \infty$ yields the value for the comparison factor $h_{\alpha,\gamma}$.) Therefore,

$$\theta(\omega, \eta, \alpha z) = \prod_{\gamma\in\tilde{\Gamma}} (h_{\alpha,\gamma}\frac{z - \alpha^{-1}\gamma\omega}{z - \alpha^{-1}\gamma\eta}) = c(\omega, \eta, \alpha)\theta(\omega, \eta, z),$$

where the convergence of $c(\omega, \eta, \alpha) = \prod h_{\alpha,\gamma}$ follows from that of $\theta(\omega, \eta, z)$.
(iii) Let $s \in \mathbb{P}^1(K)$ represent a cusp of Γ. Possibly replacing the arithmetic group Γ by $\gamma^{-1}\Gamma\gamma$, where $\gamma \in \mathrm{GL}(2, K)$ is such that $\gamma(\infty) = s$, we can assume that $s = \infty$. It results from the way how cusps are attached to $\Gamma \setminus \Omega$ (see [I.9], 3.2.2) that

$$\lim_{z\to\infty} \left(\frac{z - \gamma\omega}{z - \gamma\eta}\right) = 1$$

uniformly in the $\gamma \in \tilde{\Gamma}$. Here the lim has to be taken according to the topology on $\overline{M}_\Gamma$. Hence, if t_∞ is the uniformizer at ∞, $\lim_{z\to\infty}$ means $\lim_{t_\infty\to 0}$, and therefore $\theta(\omega, \eta, z) = 1 + o(t_\infty)$.
(v) For two points ω, η in Ω, we calculate

$$\theta(\omega, \alpha\omega, z)\theta(\eta, \alpha\eta, z)^{-1} = \prod_\gamma \left(\frac{z - \gamma\omega}{z - \gamma\alpha\omega}\right) \prod_\gamma \left(\frac{z - \gamma\eta}{z - \gamma\alpha\eta}\right)^{-1}$$

$$= \prod_\gamma \left(\frac{z - \gamma\omega}{z - \gamma\eta}\right) \prod_\gamma \left(\frac{z - \gamma\alpha\eta}{z - \gamma\alpha\omega}\right) = \theta(\omega, \eta, z)\theta(\eta, \omega, z) = 1,$$

hence n_α is independent of the base point ω.
(vi) $u_{\alpha,\beta} = \theta(\omega, \alpha\beta\omega, z) = \theta(\omega, \beta\omega, z)\theta(\beta\omega, \alpha\beta\omega, z) = u_\beta(z)u_\alpha(z)$.
(v) (continued) Still we have to show that u_α only depends on $\overline{\alpha}$. Let α have order m in Γ^{ab}. In view of (vi), $u_\alpha^m = 1$, hence u_α is constant equal to an m-th root of unity. Evaluating at $z = \infty$ (see (iii)) yields $u_\alpha = 1$.
(vii) Rearrangement of factors similar to the one in (v) gives the stated formula.
(iv) Clearly $(c(\omega, \eta, \cdot)$ factors over Γ^{ab}. Let $\alpha \in \Gamma$ have order m in Γ^{ab}. Fixing

ω, the holomorphic function $\eta \longmapsto c(\omega, \eta, \alpha)$ takes its values in the m-th roots of unity, hence is constant. Evaluating at $\eta = \omega$ gives $c(\omega, \eta, \alpha) = 1$.
(viii) $c_\alpha(\beta)$ is defined on $\overline{\Gamma} \times \overline{\Gamma}$ and bimultiplicative. By (vii), we have $c_\alpha(\beta) = c(\omega, \alpha\omega, \beta) = u_\beta(\alpha\omega)/u_\beta(\omega) = c_\beta(\alpha)$. □

4.4 Remark. The first three properties of (4.3) show that $\theta(\omega, \eta, z)$ is a meromorphic theta function as defined in (4.1), and even a holomorphic theta function if $\Gamma\omega = \Gamma\eta$. As we will see later, the group $\Theta'_m(\Gamma)$ generated by the $\theta(\omega, \eta, z)$ and the constants equals the subgroup of those theta functions whose multipliers factor over $\overline{\Gamma}$. The embedding

$$\begin{array}{ccc} \Theta_m(\Gamma)/\Theta'_m(\Gamma) & \hookrightarrow & \mathrm{Hom}(\mathrm{tor}(\Gamma^{ab}), C^*) \\ f & \longmapsto & c_f|_{\mathrm{tor}(\Gamma^{ab})} \end{array}$$

then shows that $\Theta'_m(\Gamma)$ has finite index in $\Theta_m(\Gamma)$, since the prime-to-p part of $\mathrm{tor}(\Gamma^{ab})$ is always finite ([25] p. 130, Ex. 2).

We later need the

4.5 Corollary. *The constants $c_\alpha(\beta)$ lie in $K^*_\infty \subset C^*$.*

Proof. Let $\omega \in \Omega$ be algebraic over K_∞, and write $u_\alpha = \theta(\omega, \alpha\omega, \cdot)$. Then by the very definition of $c_\alpha(\beta)$, it lies in $K_\infty(\omega)$. Since u_α and $c_\alpha(\beta)$ are independent of the choice of ω, $c_\alpha(\beta)$ lies in each algebraic extension of K_∞ and thus in K_∞. □

5. Theta functions and harmonic cochains.

In this section, we relate theta functions for an arithmetic group Γ with Γ-invariant $\mathbb{Z}$-valued harmonic cochains, by means of some sort of "*logarithmic derivative*" mapping. A more extensive discussion of the concepts introduced below, containing most of the proofs, can be found in [10] I.

Let $\lambda : \Omega \longrightarrow \mathcal{T}(\mathbb{R})$ be the building map. Recall that it takes values in $\mathcal{T}(\mathbb{Q})$, and is onto $\mathcal{T}(\mathbb{Q})$. The inverse image of a vertex v (an edge e) of $\mathcal{T}$ is an analytic space U_v isomorphic with $\mathbb{P}^1(C) - (q_\infty + 1)$ disjoint balls (an annulus U_e), respectively. These are glued together according to the incidence relations on $\mathcal{T}$. Note that U_e depends only on the underlying non-oriented edge; choosing an orientation on e is the same as choosing an orientation on U_e, which allows to distinguish its *inner* from its *outer* boundary. The next observation is crucial.

5.1 Lemma. *Let $f : \Omega \longrightarrow C$ be an invertible holomorphic function and $| \,.\, | : C \longrightarrow \mathbb{R}_{>0}$ the normalized absolute value on C. Then $|f(z)|$ depends only on $\lambda(z) \in \mathcal{T}(\mathbb{Q})$.*

Idea of proof. The occurrence of two diffent absolute values on e.g. U_v yields a critical radius for the Laurent expansion of f around one of its *holes.* By a well-known principle of non-archimedean function theory, this implies the existence of a zero of f on U_v. □

We put $\mathcal{O}_\Omega$ for the sheaf of holomorphic functions on Ω and $\mathcal{O}_\Omega(\Omega)^*$ for its invertible sections. For $f \in \mathcal{O}_\Omega(\Omega)^*$ and an edge e with origin $o(e)$ and terminus $t(e)$, we can therefore define

$$r(f)(e) := \log \left| \frac{f(w)}{f(z)} \right|, \tag{5.2}$$

where $z, w \in \Omega$ with $\lambda(z) = o(e)$, $\lambda(w) = t(e)$, and "log" is the logarithm to base q_∞. The next result is due to M. van der Put. Proofs are given in [5] and [21].

5.3 Proposition. (i) $r(f)(e) \in \mathbb{Z}$.

(ii) $r(f) : e \longmapsto r(f)(e)$ *lies in* $\underline{H}(\mathcal{T}, \mathbb{Z})$.

(iii) $r : f \longmapsto r(f)$ *gives rise to an exact sequence of* $\mathrm{GL}(2, K_\infty)$*-modules*

$$0 \longrightarrow C^* \longrightarrow \mathcal{O}_\Omega(\Omega)^* \xrightarrow{r} \underline{H}(\mathcal{T}, \mathbb{Z}) \longrightarrow 0. \tag{5.4}$$

Intuitively, the harmonic cochain $r(f)$ describes the growth (or decay) of $f(z)$ along $U_e = \lambda^{-1}(e)$. Similar to (5.2), we can associate a residue $\mathrm{res}(\omega)(e) \in C$ to a holomorphic differential form ω defined on U_e. Namely, writing $\omega = f(z)dz$ in a coordinate z (which we assume oriented such that the interior of U_e w.r.t. z corresponds to the vertex $o(e)$) and developing $f(z) = \sum_{n \in \mathbb{Z}} a_n z^n$, we put

$$\mathrm{res}(\omega)(e) = a_{-1}. \tag{5.5}$$

Its main properties are consequences of the non-archimedean residue theorem ([13] p. 95), and are summarized as follows:

5.6 Proposition. (i) $\mathrm{res}(\omega)(e)$ *is well-defined independently of the choice of coordinate* z *for the oriented annulus* U_e.

(ii) *Let* ω *be a holomorphic differential on* Ω. *Then* $\mathrm{res}(\omega) : e \longmapsto \mathrm{res}(\omega)(e)$ *lies in* $\underline{H}(\mathcal{T}, C)$.

The two maps r and res are related by the commutative diagram (straightforward from (5.3) and (5.6))

$$\begin{array}{ccc} \mathcal{O}_\Omega(\Omega)^* & \xrightarrow{r} & \underline{H}(\mathcal{T}, \mathbb{Z}) \\ \mathrm{dlog} \downarrow & & \downarrow \mathrm{red} \\ \mathcal{O}_\Omega(\Omega) & \xrightarrow{\mathrm{res}} & \underline{H}(\mathcal{T}, C), \end{array} \tag{5.7}$$

where dlog is logarithmic derivation $f \longmapsto f'/f = \frac{df}{dz}/f$, red is reduction mod p, and $\mathrm{res}(g) = \mathrm{res}(g(z)dz)$. In view of (5.4) and (5.7), we regard the map r as a legitimate substitute for the classical logarithmic derivative map.

So far, no group Γ was in the game. We now fix an arithmetic subgroup Γ of $\mathrm{GL}(2, K)$. As is easily seen, the functional equation $f(\gamma z) = c_f(\gamma) f(z)$ for a holomorphic theta function f for Γ simply translates to Γ-invariance of $r(f) \in \underline{H}(\mathcal{T}, \mathbb{Z})$. Hence (5.4) yields a map

$$\overline{r} : \Theta_h(\Gamma)/C^* \longrightarrow \underline{H}(\mathcal{T}, \mathbb{Z})^\Gamma.$$

Let further $\overline{u} : \overline{\Gamma} \longrightarrow \Theta_h(\Gamma)/C^*$ be the map induced from $\alpha \longmapsto u_\alpha$.

5.8 Theorem. *For $\alpha \in \Gamma$ with class $\overline{\alpha}$ in $\overline{\Gamma}$ we have $r(u_\alpha) = j(\overline{\alpha})$. In other words, the diagram*

$$\begin{array}{ccc} & \overline{\Gamma} & \\ {\scriptstyle \overline{u}} \swarrow & & \searrow {\scriptstyle j} \\ \Theta_h(\Gamma)/C^* & \xrightarrow{\ \overline{r}\ } & \underline{H}(\mathcal{T}, \mathbb{Z})^\Gamma \end{array}$$

is commutative.

The proof, given in [10] 5.6.1 is straightforward but technical, and will therefore be omitted. Since j takes values in $\underline{H}_!(\mathcal{T}, \mathbb{Z})^\Gamma$, we have

5.9 Corollary. $r(u_\alpha) \in \underline{H}_!(\mathcal{T}, \mathbb{Z})^\Gamma$.

Remember that $\underline{H}_!(\mathcal{T}, \mathbb{C})^\Gamma$ is a space of functions on the set

$$Y(\Gamma \setminus \mathcal{T}) := \Gamma \setminus \mathrm{GL}(2, K_\infty)/Z(K_\infty) \cdot \mathcal{I}_\infty,$$

and as such has a natural interpretation as a space of automorphic forms in the sense of Jacquet-Langlands [14] (see [I.11] 1.3). The Haar measure on the locally compact group $\mathrm{GL}(2, K_\infty)$ induces a measure μ on $Y(\Gamma \setminus \mathcal{T})$, which can be normalized such that the values of all the double classes in $Y(\Gamma \setminus \mathcal{T})$ are rational numbers. (Note that, as $\mathcal{J}_\infty$ is open in $\mathrm{GL}(2, K_\infty)$, $Y(\Gamma \setminus \mathcal{T})$ is in fact a discrete set, and neither of the considerations below really involve analysis.) The formula is:

(5.10) If $x \in \mathrm{GL}(2, K_\infty)$ corresponds to the edge $e \in Y(\mathcal{T}) = \mathrm{GL}(2, K_\infty)/ Z(K_\infty)\mathcal{J}_\infty$ then

$$\mathrm{vol}_\mu(\text{class of } x \text{ in} Y(\Gamma \setminus \mathcal{T})) = \frac{1}{2\#\tilde{\Gamma}_e}.$$

Since elements of $\underline{H}_!(\mathcal{T},\mathbb{C})^\Gamma$ have finite support as functions on $Y(\Gamma\setminus\mathcal{T})$, $\underline{H}_!(\mathcal{T},\mathbb{C})^\Gamma \hookrightarrow L^2(Y(\Gamma\setminus\mathcal{T}))$ inherits a scalar product $(.,.)_\mu$, the *Petersson product*, which takes rational values on the lattice $\underline{H}_!(\mathcal{T},\mathbb{Z})^\Gamma \hookrightarrow \underline{H}_!(\mathcal{T},\mathbb{C})^\Gamma$. On the other hand, due to (4.3) (viii) and (4.5), we have a bilinear map

$$(.,.) : \overline{\Gamma}\times\overline{\Gamma}\longrightarrow\mathbb{Z}$$

that to the classes $\overline{\alpha},\overline{\beta}$ of $\alpha,\beta\in\Gamma$ associates $\log|c_\alpha(\beta)|$. These two agree by the next result, whose proof again will be omitted.

5.11 Theorem ([10] 5.7.1). *For* $\overline{\alpha},\overline{\beta}\in\overline{\Gamma}$ *we have* $(\overline{\alpha},\overline{\beta}) = (j(\overline{\alpha}),j(\overline{\beta}))_\mu$.

We will mainly use its following immediate consequence:

5.12 Corollary. *The bilinear form* $(.,.)$ *on* $\overline{\Gamma}$ *is positive definite.*

5.13 Remarks. (i) Since $\mathrm{GL}(2,K_\infty)$ is reductive, Haar measures are bi-invariant, which exempts us from choosing between left or right Haar measures. (ii) The present normalization of μ differs by the factor 1/2 from the one chosen in [10]. That factor compensates the fact that each geometric edge $\{e,\overline{e}\}$ of $\Gamma\setminus\mathcal{T}$ contributes twice to

$$(\overline{\alpha},\beta) = (j(\overline{\alpha}),j(\overline{\beta}))_\mu = \sum_{e\in Y(\Gamma\setminus\mathcal{T})} j(\overline{\alpha})(e)j(\overline{\beta})(e)\mathrm{vol}_\mu(e).$$

(iii) Although we do not need this fact, we note that $\mathrm{vol}_\mu(Y(\Gamma\setminus\mathcal{T}))$ is finite for each arithmetic group Γ. The precise value is related with special values of the zeta function of K through a Gauss-Bonnet formula. See [25] II, exercises to 2.3.

We can extend the diagram of theorem 5.8 in taking into account the effect of the mappings dlog and res of (5.7). First, as is obvious from differentiating the functional equation of a holomorphic theta function f, $f'/f\in M_{2,1}(\Gamma)$, the space of modular froms of weight two and type one for Γ (see [I.10] 1.4). Moreover, in view of $t_s'/t_s = -t_s$ for the uniformizer t_s at $s\in\mathbb{P}^1(K)$ (*loc. cit.* 1.5), f'/f vanishes of order at least two at cusps, and therefore lies in $M_{2,1}^2(\Gamma)$, the space of *double cuspidal forms* in $M_{2,1}(\Gamma)$. Second, as follows from (5.7), $\mathrm{res}(f) = \mathrm{res}(f(z)dz)$ is Γ-invariant for $f\in M_{2,1}(\Gamma)$. We finally get the following commutative diagram which encorporates the relations between arithmetical, geometrical and combinatorial data.

(5.14)

$$\begin{array}{ccccc}
 & & \overline{\Gamma} & & \\
 & \swarrow^{\bar{u}} & & \searrow^{j} & \\
 & \Theta_h(\Gamma)/C^* & \xrightarrow[\bar{r}]{\cong} & \underline{H}_!(\mathcal{T},\mathbb{Z})^{\Gamma} & \\
 \mathrm{f} & \big\downarrow & & \big\downarrow \mathrm{red} & \\
 \downarrow & & & & \\
 \mathrm{f}'/\mathrm{f} & M^2_{2,1}(\Gamma,\mathbb{F}_p) & \xrightarrow[\mathrm{res}]{\cong} & \underline{H}_{!!}(\mathcal{T},\mathbb{F}_p)^{\Gamma} &
\end{array}$$

Here,

- the commutativity of the upper triangle is (5.8);
- injectivity of $\bar{r}$ follows from (5.4);
- surjectivity of $\bar{r}$ also comes from (5.4), combined with the observation that f is invertible around cusps if $r(f) \in \underline{H}_!(\mathcal{T},\mathbb{Z})^{\Gamma}$;
- "red" is reduction mod p, which by definition is onto $\underline{H}_{!!}(\mathcal{T},\mathbb{F}_p)^{\Gamma} = \underline{H}_!(\mathcal{T},\mathbb{Z})^{\Gamma} \otimes \mathbb{F}_p$.

From (5.7) and the above remarks, we get a commutative square analogous to the lower square of (5.14), but where the bottom line has been replaced by

$$M^2_{2,1}(\Gamma) \xrightarrow{\mathrm{res}} \underline{H}_{!!}(\mathcal{T},C)^{\Gamma} = \underline{H}_{!!}(\mathcal{T},\mathbb{F}_p)^{\Gamma} \otimes_{\mathbb{F}_p} C.$$

We *define* $M^2_{2,1}(\Gamma,\mathbb{F}_p)$ as the pre-image in $M^2_{2,1}(\Gamma)$ of $\underline{H}_{!!}(\mathcal{T},\mathbb{F}_p)^{\Gamma}$. Then the map res of (5.14) is surjective. On the other hand, $M^2_{2,1}(\Gamma)$ is canonically isomorphic, through $f(z) \longmapsto f(z)dz$, with the space of holomorphic differentials on $\overline{M}_{\Gamma}$, and therefore has dimension $g = \mathrm{genus}(\overline{M}_{\Gamma}) = \mathrm{rank}_{\mathbb{Z}}\underline{H}_!(\mathcal{T},\mathbb{Z})^{\Gamma} = \dim_{\mathbb{F}_p} \underline{H}_{!!}(\mathcal{T},\mathbb{F}_p)^{\Gamma}$. By dimension reasons, res is also injective, which completes the discussion of (5.14).

We note the following observations and comments.

(5.15) $M^2_{2,1}(\Gamma,\mathbb{F}_p)$ is an $\mathbb{F}_p$-structure on the C-vector space $M^2_{2,1}(\Gamma)$.

(5.16) Let $\alpha_1,\ldots,\alpha_g$ be elements of Γ whose classes in $\overline{\Gamma}$ form a $\mathbb{Z}$-basis. Then the $u'_{\alpha_i}/u_{\alpha_i}$ form an $\mathbb{F}_p$-basis of $M^2_{2,1}(\Gamma,\mathbb{F}_p)$ and thus a C-basis of $M^2_{2,1}(\Gamma)$. Equivalently, the $(u'_{\alpha_i}/u_{\alpha_i})dz$ form a basis of the C-space of holomorphic differential forms on $\overline{M}_{\Gamma}$. (Here we use that, by virtue of (3.12) (i), $\overline{\Gamma}$ maps surjectively to $\underline{H}_{!!}(\mathcal{T},\mathbb{F}_p)^{\Gamma}$.)

(5.17) Taking the horizontal arrows in (5.14) as identifications, the rigid analytic modular forms in $M^2_{2,1}(\Gamma, \mathbb{F}_p)$ are just the reductions mod p of the integral-valued automorphic forms in $\underline{H}_!(\mathcal{T}, \mathbb{Z})^\Gamma$.

(5.18) A cuspidal $\mathbb{F}_p$-valued harmonic cochain $\mathrm{res}(f(z)dz) \in \underline{H}_!(\mathcal{T}, \mathbb{F}_p)^\Gamma$ can be lifted to a cuspidal $\mathbb{Z}$-valued harmonic cochain if and only if f is double cuspidal. The lifting of $\mathrm{res}(f(z)dz)$ amounts to solving $u'/u = f$ for a holomorphic theta function u for Γ.

6. Construction of the Jacobian.

(6.1) Recall that a split torus over a field L is an algebraic group T/L isomorphic with a certain power $(\mathbb{G}_m)^g$ of the multiplicative group $\mathbb{G}_m/L$. Its *character group* $X(T)$ is $\mathrm{Hom}(T, \mathbb{G}_m) \cong \mathbb{Z}^g$. We can recover T from $X(T)$: If a finitely generated free abelian group M is given then $T_M := \mathrm{Hom}(M, \mathbb{G}_m)$ (where for an extension L'/L we have $T_M(L') = \mathrm{Hom}(M, \mathbb{G}_m(L')) = \mathrm{Hom}(M, L'^*)$) is a split torus with character group M. In fact, $T_M \longmapsto X(T)$ is an anti-equivalence of categories (see e.g. [1] III, Cor. 8.3).

(6.2) We now specialize to tori over the algebraically closed valued field C. With some additional efforts, the requirement of algebraic closedness for the base field can be suppressed, and we will have similar considerations and results notably over the field K_∞.

Let T be a split torus of dimension g over C with character group M, and fix a $\mathbb{Z}$-basis $\underline{\chi} = \{\chi_1, \ldots, \chi_g\}$ of M. Consider the map

$$\begin{aligned} l_{\underline{\chi}} : \ & T(C) \longrightarrow \mathbb{R}^g \text{ defined by} \\ & t \longmapsto (\log|\chi_1(t)|, \ldots, \log|\chi_g(t)|). \end{aligned}$$

A *lattice* in T/C is a subgrup $\Lambda \cong \mathbb{Z}^g$ of $T(C)$ whose image $l_{\underline{\chi}}(\Lambda)$ in $\mathbb{R}^g$ is a lattice in the usual sense, i.e., is discrete and cocompact. A moment's thought shows that the lattice property does not depend on the choice of $\underline{\chi}$. Given a lattice Λ in T/C, we can form the quotient T/Λ in the category of rigid analytic groups over C, and the question arises whether or not T/Λ is *algebraic*, i.e., the analytic space associated with an abelian variety A/C. (If such an A exists, it is uniquely determined by the GAGA principle.) The answer for the analogous problem over the complex numbers $\mathbb{C}$ is given in terms of the existence of a Riemann form on Λ, see [23], Thm. A. In our case, there exists a similar criterion, due to Gerritzen.

6.3 Theorem ([12]). T/Λ is algebraic in the sense defined above if and only if there exists a homomorphism σ from Λ to $M = X(T)$ such that the bilinear map

$$\begin{array}{ccc} \Lambda \times \Lambda & \longrightarrow & C^* \\ \alpha \ \beta & \longmapsto & \sigma(\alpha)(\beta) \end{array}$$

is symmetric and positive definite. (The last condition means that $\log|\sigma(\alpha)(\alpha)| > 0$ for $1 \neq \alpha \in \Lambda$.)

Let now $\Gamma \subset \mathrm{GL}(2, K)$ be an arithmetic group as in the preceding sections. By abuse of language, we write T_Γ for the torus $T_{\overline{\Gamma}} = \mathrm{Hom}(\overline{\Gamma}, \mathbb{G}_m)$ with character group $\overline{\Gamma}$. We regard $\overline{\Gamma}$ as a subgroup of $T_\Gamma(C)$ via the embedding $\bar{c} : \overline{\Gamma} \longrightarrow \mathrm{Hom}(\overline{\Gamma}, C^*)$ induced from $\alpha \longmapsto c_\alpha$. Theorems 4.3 and 5.11 together with their corollaries show that $\bar{c}$ is well-defined and injective, that $\overline{\Gamma}$ is a lattice in $T_\Gamma(C)$, and that the requirements of Theorem 6.3 are fulfilled (take $\sigma = \mathrm{id}$). Hence there exists an abelian variety A_Γ/C characterized through the short exact sequence of groups of C-valued points

$$1 \longrightarrow \overline{\Gamma} \xrightarrow{\bar{c}} T_\Gamma(C) \longrightarrow A_\Gamma(C) \longrightarrow 0. \tag{6.4}$$

It turns out that A_Γ actually agrees with the Jacobian of $\overline{M}_\Gamma$.

6.5 Theorem. *The quotient A_Γ of T_Γ by the lattice $\bar{c}(\overline{\Gamma})$ is canonically isomorphic with the Jacobian J_Γ of $\overline{M}_\Gamma$.*

Proof (see [10]). The proof is divided into two steps.
(a) We first construct, by means of theta functions, a morphism $\varphi : J_\Gamma \longrightarrow A_\Gamma$ which is injective on C-valued points.
(b) We then show, by inspecting the effect of φ on differentials, that φ is separable.
In view of $\dim J_\Gamma = g(\Gamma) = \dim A_\Gamma$, φ will then be surjective, hence an isomorphism, and the theorem is proved.

(a) Fix $\omega_0 \in \Omega$ and let $\psi : \Omega \longrightarrow A_\Gamma(C)$ be the map that to each $\omega \in \Omega$ associates the class in $A_\Gamma(C) = \mathrm{Hom}(\overline{\Gamma}, C^*)/\bar{c}(\overline{\Gamma})$ of $c(\omega_0, \omega, \cdot)$. It follows from (4.3) (vii) that ψ is analytic. We claim that ψ is Γ-invariant. For let $\alpha, \beta \in \Gamma$, then

$$\frac{c(\omega_0, \alpha\omega, \beta)}{c(\omega_0, \omega, \beta)} = \frac{u_\beta(\alpha\omega)}{u_\beta(\omega)} = c_\beta(\alpha) = c_\alpha(\beta),$$

hence $\psi(\alpha\omega) = \psi(\omega)$.

Now ψ factors through the natural projection $p_\Gamma : \Omega \longrightarrow \Gamma \backslash \Omega$, which yields an analytic map $\psi_\Gamma : M_\Gamma(C) = \Gamma \backslash \Omega \longrightarrow A_\Gamma(C)$. As $c(\omega_0, \omega, \alpha) = u_\alpha(\omega)/u_\alpha(\omega_0)$ and the u_α are holomorphic and invertible around cusps, ψ_Γ extends to an analytic map $\overline{\psi} : \overline{M}_\Gamma \longrightarrow A_\Gamma(C)$, which by the GAGA-theorems [15], [16], [17] is a morphism of algebraic varietes.

(6.6)

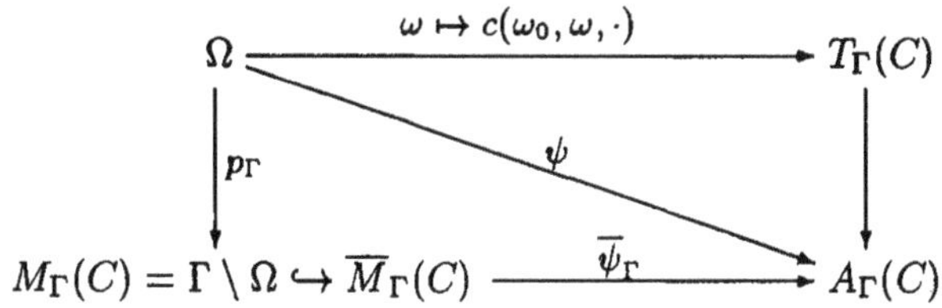

Let $P_0 = p_\Gamma(\omega_0)$ and $\kappa : \overline{M}_\Gamma \longrightarrow J_\Gamma$ be the Abel-Jacobi map, which to each $P \in \overline{M}_\Gamma(C)$ associates the divisor class of $P - P_0$. From the universal property of Jacobians, we get a uniquely determined morphism $\varphi : J_\Gamma \longrightarrow A_\Gamma$ of abelian varieties that fits into the diagram:

(6.7)

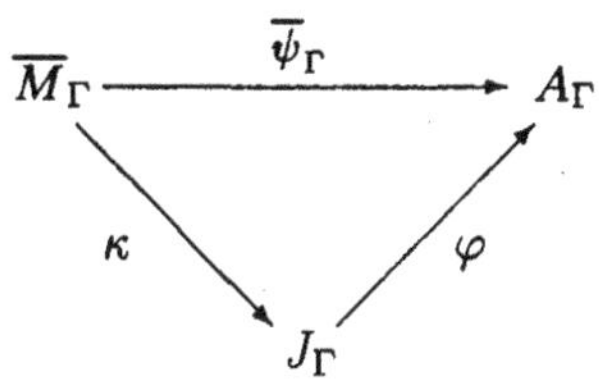

We show that φ is injective. Let ω_i $(i = 1, \ldots, n)$ be points of Ω, $P_i = p_\Gamma(\omega_i)$ their images in M_Γ, and let $[D]$ be the class of the divisor $D = P_1 + \cdots + P_n - nP_0$. Each element of $J_\Gamma(C)$ can be represented as such a $[D]$. Suppose that $\varphi([D]) = 0$. Then there exists $\alpha \in \Gamma$ such that

$$c_\alpha(\cdot) = \prod_{1 \leq i \leq n} c(\omega_i, \omega_0, \cdot).$$

The function

$$u_\alpha^{-1} \prod_{1 \leq i \leq n} \theta(\omega_i, \omega_0, z)$$

is invariant under Γ and holomorphic invertible around cusps. It is therefore a meromorphic function on $\overline{M}_\Gamma$, whose divisor equals D (see Thm. (4.3) (i)). Hence already $[D] = 0$, and φ is injective on C-valued points.

(b) We want to show that φ is separable. Let for the moment Ω^1 be the respective sheaf of differential 1-forms on the varieties $\overline{M}_\Gamma$, J_Γ, A_Γ. Then, looking at (6.7),

$$\begin{array}{llllll} \varphi \text{ is separable} & \Leftrightarrow & \varphi^* : H^0(A_\Gamma, \Omega^1) & \longrightarrow & H^0(J_\Gamma, \Omega^1) & \text{is bijective} \\ & \Leftrightarrow & \overline{\psi}_\Gamma^* : H^0(A_\Gamma, \Omega^1) & \longrightarrow & H^0(\overline{M}_\Gamma, \Omega^1) & \text{is bijective,} \end{array}$$

since $\kappa^* : H^0(J_\Gamma, \Omega^1) \longrightarrow H^0(\overline{M}_\Gamma, \Omega^1)$ is always bijective ([19] Prop. 5.3). Choose $\alpha_1, \ldots, \alpha_g \in \Gamma$ whose classes $\overline{\alpha}_i$ in $\overline{\Gamma}$ form a $\mathbb{Z}$-basis, and identify $T_\Gamma(C) \xrightarrow{\cong} (C^*)^g$ by means of the α_i. Then $\psi : \Omega \longrightarrow A_\Gamma(C)$ (see (6.6)) is given by $\omega \longmapsto$ class of $(c(\omega_0, \omega, \alpha_1), \ldots, c(\omega_0, \omega, \alpha_g))$. Let $x_1, \ldots, x_g$ be the coordinates on $(C^*)^g$. Then $\{\frac{dx_i}{x_i} \mid 1 \leq i \leq g\}$ is a basis for $H^0(A_\Gamma, \Omega^1)$. Putting $u_i := u_{\alpha_i}$ and $u_i' = \frac{du_i}{d\omega}$, the relation $c(\omega_0, \omega, \alpha) = u_i(\omega)/u_i(\omega_0)$ yields

$$\frac{dx_i}{d\omega} = \frac{u_i'(\omega)}{u_i(\omega_0)}.$$

Finally, $\psi^*(\frac{dx_i}{x_i}) = \frac{u_i'(\omega)}{u_i(\omega)} d\omega = \mathrm{dlog}\ u_i$, which is a Γ-invariant holomorphic differential form on Ω that extends to $\overline{M}_\Gamma$. By (5.16), the dlog u_i form a basis of $H^0(\overline{M}_\Gamma, \Omega^1)$. Hence $\overline{\psi}_\Gamma^*$ is bijective, φ is separable, and we are done. □

Among numerous immediate consequences of the theorem, we restrict to mention the three following.

6.8 Corollary. *For each $\chi \in \mathrm{Hom}(\overline{\Gamma}, C^*)$ there exists a $2n$-tuple $\omega_1, \ldots, \omega_n$, $\eta_1 \ldots, \eta_n$ of elements of Ω such that $g(z) = \Pi\theta(\omega_i, \eta_i, z)$ has multiplier $c_g = \chi$.*

Since the divisor of a meromorphic theta function is Γ-stable, we may regard it as a divisor on $\Gamma \setminus \Omega = M_\Gamma(C)$. For example, div $\theta(\omega, \eta, z) = P - Q$ with $P = \Gamma\omega$, $Q = \Gamma\eta$.

6.9 Corollary. *Let $f \in \Theta_m(\Gamma)$ have divisor D on $\overline{M}_\Gamma$. Then* deg $D = 0$. *In particular, each f without poles is a holomorphic theta function as defined in* (4.1).

Proof. Since the non-p-part of tor(Γ^{ab}) is finite, after possibly replacing f by a suitable power, its multiplier c_f is trivial on tor(Γ^{ab}). By (6.8), there exists $g(z) = \Pi\theta(\omega_i, \eta_i, z)$ with $c_f = c_g$. Then f/g is a meromorphic function on $\overline{M}_\Gamma$, hence deg (div f) = deg(div g) = 0. □

6.10 Corollary. *Let $f \in \Theta_m(\Gamma)$ be such that c_f restricted to* tor(Γ^{ab}) *is trivial. Then, up to constant, $f(z) = \Pi\theta(\omega_i, \eta_i, z)$ for suitable ω_i, η_i in Ω. If moreover $f \in \Theta_h(\Gamma)$, we even have $f(z) =$* const. *$u_\alpha(z)$ for some $\alpha \in \Gamma$.*

Proof. Dividing $f(z)$ by factors of type $\theta(\omega_i, \eta_i, z)$ if necessary, there results a holomorphic theta function. It therefore suffices to show the second assertion. Thus let $f \in \Theta_h(\Gamma)$, and let $g = \Pi\theta(\omega_i, \eta_i, \cdot)$ have the same multiplier as f. Then g/f is meromorphic on $\overline{M}_\Gamma$, so its divisor div(g/f) = div(g) is principal, which means that $\Pi c(\omega_i, \eta_i, \cdot)$ equals c_α for some $\alpha \in \Gamma$. Now f and u_α are both holomorphic with the same multiplier c_α, thus, by (5.4), they agree up to a constant. □

6.11 Remark. Recall that the Drinfeld modular curve $\overline{M}_\Gamma$ as well as its Jacobian is defined over a finite subextension K'/K of K_∞. It therefore makes sense to ask for an analytical description of J_Γ already over the field K_∞. With suitable modifications, we can indeed carry out all the constructions of this section (including that underlying Gerritzen's theorem 6.3) over K_∞. E.g., if L is a complete subextension of C/K_∞, there results an exact sequence analogous to (6.4):

$$1 \longrightarrow \overline{\Gamma} \longrightarrow T_\Gamma(L) \longrightarrow J_\Gamma(L) \longrightarrow 0 \tag{6.12}$$

7. Hecke operators.

(7.1) Consider diagram (5.14). All the five groups that appear are equipped with action of certain Hecke operators, and all the maps are compatible with these operators. They are defined

- on $\overline{\Gamma}$ by group theoretical means, involving some sort of "*Verlagerung*";
- on $\Theta_h(\Gamma)/C^*$ and on $M_{2,1}^2(\Gamma)$ (like on other spaces of modular forms) through a certain averaging of the canonical action of $\mathrm{GL}(2, K_\infty)$ on Ω, which results in a correspondence on $\Gamma \setminus \Omega$. On modular forms, one finds the action given by the familiar formulas;
- on spaces of type $\underline{H}_?(\mathcal{T}, ?)^\Gamma$ through averaging of the action of $\mathrm{GL}(2, K_\infty)$ on $\mathcal{T}$, which results in a correspondence of $\Gamma \setminus \mathcal{T}$.

Compatibility of these operators with the lower square of (5.14) will then be pretty obvious from the $\mathrm{GL}(2, K_\infty)$-equivariance of $\lambda : \Omega \longrightarrow \mathcal{T}(\mathbb{R})$. Compatibility with the upper triangle of (5.14) is somewhat deeper. All these operators are defined and compatibilities are checked in [10] 4.9 and 9.3, see also [7] VIII. Here we restrict to giving the definition on $\underline{H}(\mathcal{T}, \mathbb{Z})^\Gamma$, which suffices to state and discuss results. In order to keep technical and terminological ramifications under control, we make for the remainder of this article the

7.2 Assumptions. (a) $(K, \infty, A) = (\mathbb{F}_q(T), \infty, \mathbb{F}_q[T])$ as in the standard example;
(b) $\Gamma = \Gamma_0(\mathfrak{n})$, with notations as in (2.4).
Then the class number of A is one, $\overline{M}_\Gamma = \overline{M}_0(\mathfrak{n})$ is defined over K, and the map $j : \overline{\Gamma} \longrightarrow \underline{H}_!(\mathcal{T}, \mathbb{Z})^\Gamma$ is bijective, all of which considerably simplifies the presentation.

(7.3) In the following, we let $\mathfrak{m}, \mathfrak{n}, \mathfrak{n}'$, ... be ideals with monic generators m, n, n', ..., $\mathfrak{p}, \mathfrak{q}$, ... prime ideals, and "deg" the degree function on $A = \mathbb{F}_q[T]$. Recall that $Y(\mathcal{T}) = \mathrm{GL}(2, K_\infty)/Z(K_\infty) \cdot \mathcal{J}_\infty$ and hence functions on $Y(\mathcal{T})$ are functions on $\mathrm{GL}(2, K_\infty)$ right invariant under $Z(K_\infty) \cdot \mathcal{J}_\infty$.

For a function φ on $\mathrm{GL}(2, K_\infty)$ we put

$$(T_\mathfrak{m}\varphi)(g) = \sum \varphi(\begin{pmatrix} a & b \\ 0 & d \end{pmatrix} g),$$

where the sum is over triples (a, b, c) of A such that a, d are monic, $(ad) = \mathfrak{m}$, $(a, \mathfrak{n}) = 1$, and deg $b <$ deg d. We are confident that there results no confusion between the *Hecke* operator $T_\mathfrak{m}$ and $T \in A$. The following properties are standard and easily verified (e.g. [18], [30], [26]):

(7.4) (i) For each coefficient group B and $? =$ empty, !, !!, $T_\mathfrak{m} : \varphi \longmapsto T_\mathfrak{m}\varphi$ maps $\underline{H}_?(\mathcal{T}, B)^\Gamma$ into itself.
We henceforth consider $T_\mathfrak{m}$ as an operator on $\underline{H}_!(\mathcal{T}, B)^\Gamma$.

(ii) All the $T_\mathfrak{m}$ commute.

(iii) If $\mathfrak{m}$ and $\mathfrak{m}'$ are coprime, $T_{\mathfrak{m}\mathfrak{m}'} = T_\mathfrak{m} \circ T_{\mathfrak{m}'}$.

(iv) If $(\mathfrak{p}, \mathfrak{n}) = 1$ then
$T_{\mathfrak{p}^{n+1}} = T_{\mathfrak{p}^n} \circ T_\mathfrak{p} - q^{\deg \mathfrak{p}} T_{\mathfrak{p}^{n-1}}$.

(v) If $(\mathfrak{m}, \mathfrak{n}) = 1$ then $T_\mathfrak{m}$ is hermitian w.r.t. the Petersson product $(.,.)_\mu$ on $\underline{H}_!(\mathcal{T}, \mathbb{C})^\Gamma$.

Hecke operators $T_\mathfrak{m}$ with $(\mathfrak{m}, \mathfrak{n}) = 1$ are called *unramified.*

(7.5) For abbreviation, we put

$$H(\mathfrak{n}) := \underline{H}_!(\mathcal{T}, \mathbb{Z})^{\Gamma_0(\mathfrak{n})}.$$

We also let $\mathcal{H}(\mathfrak{n})$ be the $\mathbb{Z}$-algebra generated by the operators $T_\mathfrak{m}$ on $H(\mathfrak{n})$. Suppose that $\mathfrak{n}'$ divides $\mathfrak{n}$. For each monic divisor of n/n' we have an embedding $i_{a,\mathfrak{n}'} : H(\mathfrak{n}') \longrightarrow H(\mathfrak{n})$ given by

$$i_{a,\mathfrak{n}'}(\varphi)(g) = \varphi(\begin{pmatrix} a & 0 \\ 0 & 1 \end{pmatrix} g).$$

We let $(H(\mathfrak{n}) \otimes \mathbb{Q})^{\mathrm{new}}$ be the orthogonal complement in $H(\mathfrak{n}) \otimes \mathbb{Q}$ with respect to $(.,.)_\mu$ of the images of all the $i_{a,\mathfrak{n}'} \otimes \mathbb{Q}$, where $\mathfrak{n}'$ runs through the proper divisors of $\mathfrak{n}$. We further put $H^{\mathrm{new}}(\mathfrak{n}) = H(\mathfrak{n}) \cap (H(\mathfrak{n}) \otimes \mathbb{Q})^{\mathrm{new}}$. Then clearly $H^{\mathrm{new}}(\mathfrak{n}) \otimes \mathbb{Q} = (H(\mathfrak{n}) \otimes \mathbb{Q})^{\mathrm{new}}$. Intuitively, $H^{\mathrm{new}}(\mathfrak{n})$ is that part of $H(\mathfrak{n})$ that genuinely belongs to the level $\mathfrak{n}$.

The next two facts are well-known from the theory of automorphic forms, although it is difficult to find convenient references. We advice the reader to consult [11] for an overview.

(vi) $T_\mathfrak{m}$ maps $H^{\text{new}}(\mathfrak{n})$ into itself. Putting $\mathcal{H}^{\text{new}}(\mathfrak{n})$ for the $\mathbb{Z}$-algebra generated by the $T_\mathfrak{m}|_{H^{\text{new}}(\mathfrak{n})}$, the $\mathbb{Q}$-algebra $\mathcal{H}^{\text{new}}(\mathfrak{n})\otimes\mathbb{Q}$ is commutative semisimple, at fortiori a product of totally real number fields.

In view of (vi), $H^{\text{new}}(\mathfrak{n})\otimes\mathbb{C}$ has a basis of simultaneous eigenforms φ for the action of $\mathcal{H}^{\text{new}}(\mathfrak{n})$. Let $\lambda(\varphi,\mathfrak{p})$ be the eigenvalue of φ under $T_\mathfrak{p}$.

(vii) If $\varphi,\varphi'\in H^{\text{new}}(\mathfrak{u})\otimes\mathbb{C}$ are eigenforms for $\mathcal{H}^{\text{new}}(\mathfrak{u})$ with $\lambda(\varphi,\mathfrak{p})=\lambda(\varphi',\mathfrak{p})$ for all but a finite number of primes $\mathfrak{p}$ of A, then $\varphi'=\text{const.}\ \varphi$ ("strong multiplicity one principle").

We mention the next result for the sake of completeness, although we don't really use it. It follows from combining Drinfeld's reciprocity law with the Weil conjecture for the l-adic cohomology of $\overline{M}_0(\mathfrak{n})$.

(viii) For each eigenform φ in $H^{\text{new}}(\mathfrak{n})\otimes\mathbb{C}$ and each prime $\mathfrak{p}$ of A, the estimate

$$|\lambda(\varphi,\mathfrak{p})|\leq 2q^{1/2\ \deg\ \mathfrak{p}}$$

holds.

(7.6) We give another interpretation of the $\mathbb{Z}$-algebra $\mathcal{H}(\mathfrak{n})$. Let for the moment $\tilde{T}_\mathfrak{m}$ denote the $\mathfrak{m}$-th Hecke correspondence on $\overline{M}_0(\mathfrak{n})$. Recall that points of $\overline{M}_0(\mathfrak{n})$ correspond to rank-two Drinfeld modules ϕ (degenerate, if the point is a cusp) provided with a "*Hecke* $\mathfrak{n}$-*structure*" H. Now $\tilde{T}_\mathfrak{m}$ assigns to (ϕ,H) the collection of all (ϕ',H') that are related to (ϕ,H) by a cyclic $\mathfrak{m}$-isogeny. An alternative definition for $\tilde{T}_\mathfrak{m}$ was as a certain algebraic curve on the product $\overline{M}_0(\mathfrak{n})\times\overline{M}_0(\mathfrak{n})$, such that $((\phi,H),(\phi',H'))$ belongs to that curve if and only if (ϕ',H') occurs in $T_\mathfrak{m}(\phi,H)$. Now such a correspondence gives rise to an endomorphism, equally denoted by $\tilde{T}_\mathfrak{m}$, of $J_0(\mathfrak{n}):=J_{\Gamma_0(\mathfrak{n})}$. Let $\tilde{\mathcal{H}}(\mathfrak{n})$ be the $\mathbb{Z}$-algebra generated by the $\tilde{T}_\mathfrak{m}$. Taking into account that:

(a) every endomorphism of $J_0(\mathfrak{n})$ may be lifted to $T_{\Gamma_0(\mathfrak{n})}$ (see (6.12)); and

(b) $\overline{\Gamma}\longmapsto T_\Gamma=\text{Hom}(\overline{\Gamma},\mathbb{G}_m)$ comes from an anti-equivalence of categories;

we get a faithful anti-representation $\text{End}(J_0(\mathfrak{n}))\hookrightarrow\text{End}(\overline{\Gamma})$ (i.e., the order of multiplication is reversed). In view of

(c) $j:\overline{\Gamma}\xrightarrow{\cong}H(\mathfrak{n})$, there results a diagram, clearly commutative

(7.7)

$$\begin{array}{ccccc} \tilde{T}_{\mathfrak{m}} & & \tilde{\mathcal{H}}(\mathfrak{n}) & \hookrightarrow & \operatorname{End}(J_0(\mathfrak{n})) \\ \updownarrow & & \downarrow \wr & & \downarrow \\ T_{\mathfrak{m}} & & \mathcal{H}(\mathfrak{n}) & \hookrightarrow & \operatorname{End}(H(\mathfrak{n})) , \end{array}$$

which entitles us to identify $\tilde{T}_{\mathfrak{m}}$ with $T_{\mathfrak{m}}$ and $\tilde{\mathcal{H}}(\mathfrak{n})$ with $\mathcal{H}(\mathfrak{n})$. (Recall that $\mathcal{H}(\mathfrak{n})$ is commutative, so the "anti"-identification of $\tilde{\mathcal{H}}(\mathfrak{n})$ with $\mathcal{H}(\mathfrak{n})$ doesn't matter!) In the sequel, we omit the tildes "$\sim$".

(7.8) Now (7.7) implies that each $\mathcal{H}(\mathfrak{n})$-stable splitting $\oplus H_i$ of $H(\mathfrak{n}) \otimes \mathbb{Q}$ gives rise to an $\mathcal{H}(\mathfrak{n})$-splitting up to isogeny $J_0(\mathfrak{n}) \sim \Pi A_i$ of $J_0(\mathfrak{n})$ into abelian varieties A_i. Moreover, the correspondence $\oplus H_i \longmapsto \Pi A_i$ is dimension-preserving, as results from (6.1). In particular, we can define the "new part" $J_0^{\text{new}}(\mathfrak{n})$ of $J_0(\mathfrak{n})$ as the quotient of $J_0(\mathfrak{n})$ that corresponds to $H^{\text{new}}(\mathfrak{n}) \otimes \mathbb{Q}$. (There is an intrinsic definition of $J_0^{\text{new}}(\mathfrak{n})$ avoiding the use of Hecke operators, which of course yields the same result.) Since all the $T_{\mathfrak{m}}$ are K-rational endomorphisms of $J_0(\mathfrak{n})$, all the pieces A_i are K-rational. Restricting to the new parts, we get from (7.7)

(7.9)

$$\begin{array}{ccc} \tilde{\mathcal{H}}^{\text{new}}(\mathfrak{n}) & \hookrightarrow & \operatorname{End}({J_0}^{\text{new}}(\mathfrak{n})) \\ \| & & \downarrow \\ \mathcal{H}^{\text{new}}(\mathfrak{n}) & \hookrightarrow & \operatorname{End}(H^{\text{new}}(\mathfrak{n})). \end{array}$$

By (7.5) (vii), $\mathcal{H}^{\text{new}}(\mathfrak{n}) \otimes \mathbb{Q}$ is a commutative semisimple $\mathbb{Q}$-algebra of dimension $g^{\text{new}} := \dim\ J_0^{\text{new}}(\mathfrak{n}) = \operatorname{rank}_{\mathbb{Z}} H^{\text{new}}(\mathfrak{n})$, product of totally real number fields F_i:

$$(*) \qquad \mathcal{H}^{\text{new}}(\mathfrak{n}) \otimes \mathbb{Q} = \Pi F_i .$$

From the classification of possible endomorphism rings of abelian varieties [20] it follows that $(*)$ produces the complete splitting up to isogeny of $J_0^{\text{new}}(\mathfrak{n})$ over K. Let $\varphi \in H^{\text{new}}(\mathfrak{n}) \otimes \mathbb{C}$ be an eigenform for $\mathcal{H}^{\text{new}}(\mathfrak{n})$ with rational eigenvalues $\lambda(\varphi, \mathfrak{p})$. (In fact, the $\lambda(\varphi, \mathfrak{p})$ are then integral, and up to scaling, φ lies even in $H^{\text{new}}(\mathfrak{n})$.) The relation between φ and the corresponding one-dimensional factor E of $J_0^{\text{new}}(\mathfrak{n})$ (hence E is an elliptic curve, well-defined up to isogeny) is given by

$$(7.10) \qquad \lambda(\varphi, \mathfrak{p}) = q^{\deg\, \mathfrak{p}} + 1 - \sharp E(A/\mathfrak{p}) \quad \text{if } \mathfrak{p} \nmid \mathfrak{n}.$$

Recall that $\overline{M}_0(\mathfrak{n})$, hence $J_0(\mathfrak{n})$ and also E have good reduction at $\mathfrak{p}$ if $\mathfrak{p} \nmid \mathfrak{n}$, with $\sharp E(A/\mathfrak{p})$ rational points over $A/\mathfrak{p}$. Therefore:

(7.11) One-dimensional eigenspaces in $H^{\text{new}}(\mathfrak{n}) \otimes \mathbb{Q}$ (or in $H^{\text{new}}(\mathfrak{n}) \otimes \mathbb{C}$, what amounts to the same) with rational (thus integral) eigenvalues under the $T_{\mathfrak{m}}$ correspond bijectively to isogeny classes of elliptic curves E over K that are factors of $J_0^{\text{new}}(\mathfrak{n})$. (The injectivity of this map results from the strong multiplicity one principle.)

Now, as is shown in [10], sect. 8, 9, using the apparatus of automorphic forms, and combining deep results of Grothendieck, Jacquet-Langlands, Deligne, and Drinfeld:

(7.12) The curves E that appear in (7.11) have Tate reduction at ∞ (this is easy) and conductor $\mathfrak{n} \cdot (\infty)$ (this is less easy).

(7.13) All the curves E with Tate reduction at ∞ and conductor $\mathfrak{n} \cdot (\infty)$ are isogeny factors of $J_0(\mathfrak{n})$, or, what is the same, possess a Weil uniformization

$$\pi : \overline{M}_0(\mathfrak{n})/K \longrightarrow E.$$

(7.14) One finally arrives at canonical bijections between the three sets of

(a) K-isogeny classes of elliptic curves E/K with Tate reduction at ∞ and $\text{cond}(E) = \mathfrak{n} \cdot (\infty)$;

(b) isogeny factors of dimension one of $J_0^{\text{new}}(\mathfrak{n})$;

(c) classes (up to scaling) of Hecke eigenforms with rational eigenvalues in $H^{\text{new}}(\mathfrak{n}) \otimes \mathbb{C}$;

which are analogous to the (still partially conjectural) bijections in (1.5).

7.15 Remarks. (i) The reader might ask for the reason why we presently can prove more in the function field case compared to the "*classical*" case outlined in section one. That question is related to the construction of the map from (a) to (b) or (c) in (7.14). Due to work of Grothendieck and Deligne [3], for each $E \in (a)$, the zeta function ζ_E has the "*right*" analytical properties such that Jacquet-Langlands' theory may be applied to ensure the existence of a Hecke eigenform $\varphi = \varphi_E$ as in (c). Perhaps the reason for these properties of ζ_E (if there is reason) lies in the existence of a global Frobenius endomorphism of the $\mathbb{F}_p$-scheme E, a structure which is not available for elliptic curves over $\mathbb{Q}$. The link that is missing in the classical case (and that is bridged by Grothendieck's results in our function field case) is what traditionally has been known as Hasse's conjecture, which is about the meromorphic continuation of the zeta functions of elliptic curves over $\mathbb{Q}$. Of course, we didn't even sketch here how to find φ_E from E, see [10] for more explanation. In what follows, we take the existence of φ_E for granted, and show how E can be recovered.

(ii) We should point out that our present Hecke algebras $\mathcal{H}(\mathfrak{n})$ and $\mathcal{H}^{\text{new}}(\mathfrak{n})$ slightly differ from those appearing in [10], due to the fact that ramified Hecke operators are admitted.

8. The strong Weil uniformization. In this section, we give an explicit construction of the Weil uniformization of an elliptic curve E/K as in (7.14) (a), starting with the eigenform $\varphi = \varphi_E$ as in (7.14) (c).

Consider the short exact sequence

$$1 \longrightarrow \overline{\Gamma} \xrightarrow{\bar{c}} \text{Hom}(\overline{\Gamma}, C^*) \longrightarrow J_0(\mathfrak{n})(C) \longrightarrow 0, \tag{8.1}$$

which is a special case of (6.12). Note that $\overline{\Gamma} \xrightarrow{\cong} \underline{H}_!(\mathcal{T}, \mathbb{Z})^{\Gamma} = H(\mathfrak{n})$ as Hecke modules. Also, $T_\Gamma(C) = \text{Hom}(\overline{\Gamma}, C^*)$ inherits from $\overline{\Gamma}$ a Hecke action.

8.2 Proposition. *Let* $(\mathfrak{m}, \mathfrak{n}) = 1$. *The Hecke operator* $T_\mathfrak{m}$ *on* $\overline{\Gamma}$ *is formally self-adjoint w.r.t. the bilinear map from* $\overline{\Gamma}\times\overline{\Gamma}$ *to* K_∞^* *induced from* $(\alpha, \beta) \longmapsto c_\alpha(\beta)$.

Proof. [10] 9.3.3.

The proposition implies that $\bar{c}$ is compatible with unramified Hecke operators. This may also be checked for the quotient map from T_Γ to J_Γ. Therefore, 8.1 is an exact sequence of modules under the unramified Hecke operators. Recall that, by the strong multiplicity one principle, the eigenvalues $\lambda(\varphi, \mathfrak{p})$ for primes $\mathfrak{p} \nmid \mathfrak{n}$ characterize an eigenform up to constants.

Now fix an eigenform $\varphi \in H^{\text{new}}(\mathfrak{n}) \otimes \mathbb{C}$ with rational eigenvalues. Scaling φ, we may assume that $\varphi \in H^{\text{new}}(\mathfrak{n})$ and is primitive (not divisible by integers $n \neq \pm 1$). Consider the evaluation map $\text{ev}_\varphi : \text{Hom}(\overline{\Gamma}, C^*) \longrightarrow C^*$ given by $f \longmapsto f(\varphi)$, where φ is regarded as an element of $\overline{\Gamma} \xrightarrow{\cong} H(\mathfrak{n})$.

8.3 Theorem ([10], [8]). *Under our assumptions, the image* Λ *of* ev_φ *is a lattice in* C^*, *generated by some* $t \in K_\infty^*$ *with* $0 < |t| < 1$.

The fact that Λ contains a subgroup of finite index which has the indicated shape $t^{\mathbb{Z}}$ is [10] 9.5.1. Its proof uses techniques from the theory of automorphic forms. The assertion that Λ itself has the form $t^{\mathbb{Z}}$ is a non-trivial extension, and requires supplementary ingredients, e.g. Fourier coefficients of elements of $H(\mathfrak{n}) \otimes \mathbb{C}$. It is shown in [8] Thm. 3.2. Combining the preceding, we get a commutative diagram

$$
(8.4)\qquad
\begin{array}{ccccccccc}
1 & \longrightarrow & \overline{\Gamma} & \stackrel{\bar{c}}{\longrightarrow} & \mathrm{Hom}(\overline{\Gamma}, C^*) & \longrightarrow & J_0(\mathfrak{n})(C) & \longrightarrow & 0 \\
 & & \downarrow & & \downarrow \mathrm{ev}_\varphi & & \downarrow \mathrm{pr}_\varphi(C) & & \\
1 & \longrightarrow & \Lambda & \longrightarrow & C^* & \longrightarrow & C^*/\Lambda & \longrightarrow & 0
\end{array}
$$

(where each occurrence of C may be replaced by $L =$ any complete subextension of C/K_∞). Now

- C^*/Λ is the group of C-points of an elliptic curve E_φ/K_∞;
- pr_φ is an analytic map of abelian varieties over K_∞, hence algebraic;
- The Hecke operator $T_\mathfrak{m}$ with $(\mathfrak{m}, \mathfrak{n}) = 1$ induces on E_φ the endomorphism "multiplication by $\lambda(\varphi, \mathfrak{m})$".

Thus E_φ is the isogeny factor of $J_0(\mathfrak{n})$ associated with the eigenspace $\mathbb{C}\varphi$, and E_φ as well as pr_φ are K-rational. Through the choice of a K-rational point (for example, the cusp represented by ∞), we embed $\overline{M}(\mathfrak{n})$ into its Jacobian $J_0(\mathfrak{n})$. Put π_φ for the composite map from $\overline{M}_0(\mathfrak{n})$ to E_φ, which is a Weil uniformization for E_φ, and is maximal by its definition. If as usual, u_φ denotes the theta function associated to $\varphi \in \overline{\Gamma}$, we have the commutative square

$$
(8.5)\qquad
\begin{array}{ccc}
\Omega & \xrightarrow{\quad u_\varphi \quad} & C^* \\
\downarrow & & \downarrow \\
\Gamma \backslash \Omega & & C^*/\Lambda \\
\downarrow & & \| \\
M_0(\mathfrak{n})(C) \hookrightarrow \overline{M}_0(\mathfrak{n})(C) & \xrightarrow{\pi_\varphi} & E_\varphi(C)
\end{array} .
$$

For the coordinate w on C^* and the associated invariant differential $\omega = \frac{dw}{w}$ on E_φ, the relation

$$
(8.6)\qquad \pi_\varphi^* = f(z)dz
$$

holds, where $f(z) \in M_{2,1}^2(\Gamma)$ is the modular form $u_\varphi'(z)/u_\varphi(z)$, i.e., "$\varphi$ reduced mod p" (see (5.14)).

8.7 Remark. Unlike the classical case, knowledge of $f(z)$ does not suffice to determine E_φ, due to the loss of information caused by reduction mod p. See example 9.4.

(8.8) As we have seen, the splitting of $J_0(\mathfrak{n})$ and in particular its factors of dimension one are governed by the $\mathbb{Z}$-module $\overline{\Gamma} \xrightarrow{\cong} H(\mathfrak{n})$, together with its Hecke action. We can read off some more properties of π_φ and E_φ from $H(\mathfrak{n})$. Namely, for $\varphi \in H(\mathfrak{n})$ as above (primitive new eigenform with rational eigenvalues), we put

$$\begin{aligned} m(\varphi) &:= \min\{(\varphi, \alpha) > 0 \mid \alpha \in H(\mathfrak{n})\} \\ r(\varphi) &:= (\varphi, \varphi)/m(\varphi). \end{aligned}$$

Here $(.,.)$ is the scalar product on $H(\mathfrak{n})$. The next two results are proved in [8].

8.9 Theorem. *The number $m(\varphi)$ equals the pole order $-v_\infty(j(E_\varphi))$ of the j-invariant $j(E_\varphi)$ of the strong Weil curve E_φ attached to φ, or, what is the same, the number of irreducible components of the Néron model of E_φ at ∞.*

8.10 Theorem. *The number $r(\varphi)$ equals the degree* deg π_φ *of the strong Weil uniformization $\pi_\varphi : \overline{M}_0(\mathfrak{n}) \longrightarrow E_\varphi$.*

9. Examples.

We keep notations of the last sections and present some examples on the respective structures of $\Gamma \setminus \mathcal{T}$, $H(\mathfrak{n})$, $J_0(\mathfrak{n})$ and related Weil uniformizations. These are compiled from different sources, mainly from [6], [8] and [24].

9.1 Example [6]. Let $\mathfrak{p}$ be a prime of degree 3. Then $\overline{M}_0(\mathfrak{p})$ has genus q and $H^{\text{new}}(\mathfrak{p}) = H(\mathfrak{p})$. The graph $\Gamma_0(\mathfrak{p}) \setminus \mathcal{T}$ has the shape

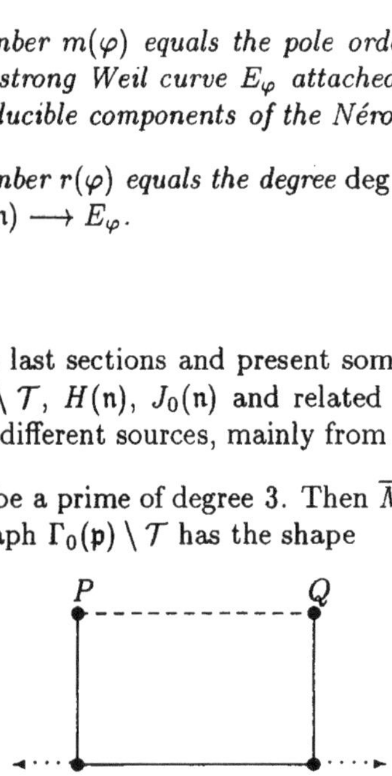

where $\cdots >$ stands for an end, i.e., an infinite half-line, and $\bullet - - - - - - - - \bullet$ for q edges e_x joining P to Q, indexed by the elements x of $\mathbb{F}_q$. We further put e_∞ for the unique edge e with $o(e) = P$, $t(e) \neq Q$. Hence

$$\varphi \longmapsto (\varphi(e_0), \ldots, \varphi(e_\infty))$$

identifies $H(\mathfrak{p})$ with the space $C_0(\mathbb{P}^1(\mathbb{F}_q))$ of $\mathbb{Z}$-valued functions on $\mathbb{P}^1(\mathbb{F}_q)$ with total degree zero. Under this identification, the scalar product $(.,.)$ on $H(\mathfrak{p})$ corresponds to the product on $C_0(\mathbb{P}^1(\mathbb{F}_q))$ with $\mathrm{vol}(x) = 1$ for $x \neq \infty$, $\mathrm{vol}(\infty) = q + 1$. The matrices of the Hecke operators $T_\mathfrak{q}$ ($\mathfrak{q}$ a prime of degree one) w.r.t. the obvious basis of $H(\mathfrak{p})$ are calculated in [6] Satz 6.8. It results that e.g. $H(\mathfrak{p})$ contains no eigenforms with rational eigenvalues if $q = 2, 3, 5$ and $\mathfrak{p}$ is an arbitrary prime of degree 3 (*loc. cit.*, 10.2). If however $q = 7$ and $\mathfrak{p}$ is the prime $(T^3 - 2)$ then there exists a unique (up to scaling) eigenform φ with rational eigenvalues, given by

$$\varphi = (a_0, \ldots, a_\infty) = (4, 1, 1, -2, 1, -2, -2, -1).$$

We have $(\varphi, \varphi) = 39$ and $m(\varphi) = 3$. Hence we can predict $v_\infty(j) = -3$ and $\deg \pi = 13$ for the invariant $j = j(E_\varphi)$ and the strong Weil uniformization $\pi : \overline{M}_0(\mathfrak{p}) \longrightarrow E_\varphi$ attached to φ. Further, E_φ will have good reduction at $\mathfrak{q} = (T - x)$ with $8 + a_x$ rational points over $A/(T - x) \cong \mathbb{F}_7$. It can be worked out that

$$Y^2 = X^3 + aX + b$$

with $a = -3T(T^3 + 2)$, $b = -2T^6 + 3T^3 + 1$ is an equation for E_φ.

(9.2) For the remainder, we assume that $q = 2$. Then the scalar products on the spaces $H(\mathfrak{n}) = \underline{H}_!(\mathcal{T}, \mathbb{Z})^{\Gamma_0(\mathfrak{n})}$ are the obvious ones, where each geometric edge of $\Gamma_0(\mathfrak{n}) \setminus \mathcal{T}$ has volume one.

9.3 Example [6]. Let $\mathfrak{n}$ be the ideal (T^3) of $A = \mathbb{F}_2[T]$. Then $\Gamma \setminus \mathcal{T}$ looks

.

We have $H(\mathfrak{n}) = \mathbb{Z}\psi$ with $(\psi, \psi) = 4$. $\overline{M}_0(\mathfrak{n}) = J_0(\mathfrak{n})$ is itself elliptic, with a j-invariant that satisfies $v(j) = -4$. An equation is

$$Y^2 + TXY = X^3 + T^2.$$

9.4 Example [6], [8]. $\mathfrak{n} = T(T^2 + T + 1)$. Then $\Gamma \setminus \mathcal{T}$ looks

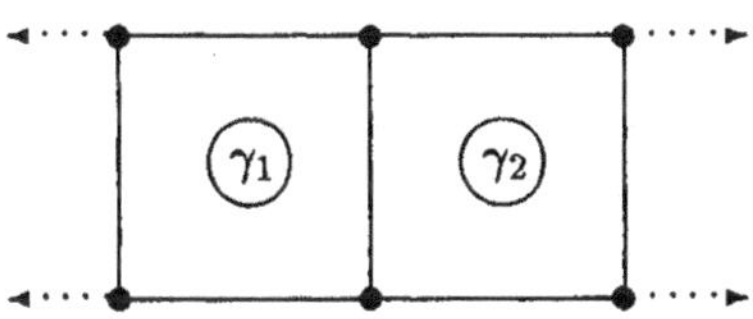

.

Let γ_1 and γ_2 be the two cycles of length 4, oriented counterclockwise. Then $\varphi_1 = \gamma_1 + \gamma_2$, $\varphi_2 = - - \gamma_1 + \gamma_2$ are eigenforms with rational eigenvalues.

Moreover, if w_i ($i = 1, 2$) are the Atkin-Lehner involutions associated to the prime divisors $\mathfrak{p}_1 = (T)$, $\mathfrak{p}_2 = (T^2 + T + 1)$ of $\mathfrak{n}$ then

$$\begin{array}{rcccl} w_1(\varphi_1) & = & \varphi_1 & = & -w_2(\varphi_1) \\ w_2(\varphi_2) & = & \varphi_2 & = & -w_1(\varphi_2). \end{array}$$

Hence $J_0(\mathfrak{n})$ is isogeneous with $E_1 \times E_2$, where E_i corresponds to φ_i. We have $m(\varphi_1) = (\varphi_1, \gamma_1) = 3$, $m(\varphi_2) = (\varphi_2, \gamma_2) = 5$, $(\varphi_1, \varphi_1) = 6$, $(\varphi_2, \varphi_2) = 10$, hence $r(\varphi_1) = r(\varphi_2) = 2$, which corresponds to the fact that the strong Weil uniformization for the φ_i are the quotient maps $\overline{M}_0(\mathfrak{n}) \longrightarrow \overline{M}_0(\mathfrak{n})/w_i$ of degree two. Equations are given by

$$\begin{array}{rrcl} E_1: & Y^2 + (T+1)XY + Y & = & X^3 + T(T^2 + T + 1) \\ E_2: & Y^2 + (T+1)XY + Y & = & X^3 + X^2 + T + 1. \end{array}$$

Although E_1 and E_2 are non-isogeneous, the two modular forms u_i'/u_i ($u_i := u_{\varphi_i}$) agree since $\varphi_1 \equiv \varphi_2 \pmod 2$.

I owe the next example to A. Schweizer (unpublished; see however [24] for related material).

9.5 Example. Let now $\mathfrak{n} = (T^4)$. Here the graph $\Gamma_0(\mathfrak{n}) \backslash \mathcal{T}$ is the one-skeleton of a tetrahedron:

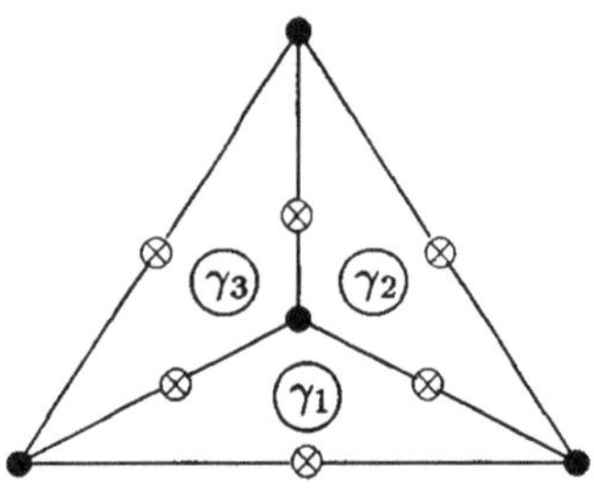

where at each vertex of type $\otimes$, an end is added. (Despite the symmetry of $\Gamma \backslash \mathcal{T}$, the modular interpretation distinguishes the different vertices and edges.) Let $\gamma_1, \gamma_2, \gamma_3$ be the three cycles of length 6, oriented counterclockwise. They form a basis of $H(\mathfrak{n})$. The Atkin-Lehner involution w acts on $\Gamma \backslash \mathcal{T}$ as the reflexion at the middle axis, i.e., as the matrix M (see below) w.r.t the γ_i. Furthermore, the canonical embedding i_{1,T^3} (see (7.5)) of $H((T^3))$ into $H((T^4))$ maps a generator ψ of $H((T^3))$ to $(\gamma_1 + \gamma_2)$. Since $H^{\text{new}}(\mathfrak{n})$ is w-stable, we easily see that $H^{\text{new}}(\mathfrak{n}) = \mathbb{Z}\varphi_1$ with $\varphi_1 = \gamma_2 + \gamma_3$. Then $(\varphi, \varphi) = 8$, $m(\varphi) = (\varphi, \gamma_1) = 4$, $r(\varphi) = 2$. Hence the strong Weil curve E_φ with conductor $(T^4)(\infty)$ is the quotient of $\overline{M}_0((T^4))$ modulo an involution w' different from w, given by the matrix M'.

$$M = \begin{pmatrix} -1 & 0 & 0 \\ 0 & 0 & -1 \\ 0 & -1 & 0 \end{pmatrix} \quad M' = \begin{pmatrix} -1 & 0 & 0 \\ 0 & 0 & 1 \\ 0 & 1 & 0 \end{pmatrix}.$$

Summarizing, $J_0((T^4))$ splits up to isogeny into three elliptic curves E_i corresponding to the eigenforms (for unramified Hecke operators) $\varphi_1 = \gamma_2 + \gamma_3$, $\varphi_2 = \gamma_1 + \gamma_3$, $\varphi_3 = \gamma_1 + \gamma_2$. We have $J_0^{\text{new}}((T^4)) \sim E_1$, $E_2 \sim E_3 \sim J_0((T^3))$. All of these are quotients of $\overline{M}_0((T^4))$ w.r.t. certain involutions. The equation of $J_0^{\text{new}}((T^4)) = J_0((T^4))/w'$ is

$$Y^2 + TXY = X^3 + TX^2 + T^2.$$

A more extended discussion of this circle of ideas can be found in [24], where, among others, all the elliptic and hyperelliptic modular curves $\overline{M}_0(\mathfrak{n})$ and moreover the "involutory" elliptic curves E/K are classified, allowing $K = \mathbb{F}_q(T)$ with q an arbitrary prime power.

References

[1] A. Borel: Linear algebraic groups. Springer-Verlag 1991.

[2] J.E. Cremona: Algorithms for modular elliptic curves. Cambridge University Press 1992.

[3] P. Deligne: Les constantes des équations fonctionelles des fonction L. Lect. Notes Math. **349**, 501-597, Springer-Verlag 1973.

[4] V.G. Drinfeld: Elliptic modules. Math. USSR-Sbornik **23**, 561-592 (1976).

[5] J. Fresnel and M. van der Put: Géométrie analytique et applications. Birkhäuser-Verlag 1981.

[6] E.-U. Gekeler: Automorphe Formen über $\mathbb{F}_q(T)$ mit kleinem Führer. Abh. Math. Sem. Univ. Hamburg **55**, 111-146 (1985).

[7] E.-U. Gekeler: Drinfeld modular curves. Lect. Notes Math. **1231**, Springer-Verlag 1986.

[8] E.-U. Gekeler: Analytical construction of Weil curves over function fields. J. Th. Nomb. Bordeaux **7**, 27-49 (1995).

[9] E.-U. Gekeler and U. Nonnengardt: Fundamental domains of some arithmetic groups over function fields. Int. J. Math. **6**, 689-708 (1995).

[10] E.-U. Gekeler and M. Reversat: Jacobians of Drinfeld modular curves. J. reine angew. Math. **476**, 27-93 (1996).

[11] S. Gelbart: Automorphic forms on adele groups. Ann. Math. Stud. **83**, Princeton University Press 1975.

[12] L. Gerritzen: On non-archimedean representations of abelian varieties. Math. Ann. **196**, 323-346 (1972).

[13] L. Gerritzen and M. van der Put: Schottky groups and Mumford curves. Lect. Notes Math. **817**, Springer-Verlag 1980.

[14] H. Jacquet and R.P. Langlands: Automorphic forms on GL(2). Lect. Notes Math. **114**, Springer-Verlag 1970.

[15] R. Kiehl: Der Endlichkeitssatz für eigentliche Abbildungen in der nichtarchimedischen Funktionentheorie. Invent. Math. **2**, 191-214 (1967).

[16] R. Kiehl: Theorem A und Theorem B in der nichtarchimedischen Funktionentheorie. Invent. Math. **2**, 256-273 (1967).

[17] U. Köpf: Über eigentliche Familien algebraischer Varietäten über affinoiden Räumen. Schriftenr. Math. Inst. Univ. Münster **7** (1974).

[18] S. Lang: Introduction to modular forms. Springer-Verlag 1976.

[19] J.S. Milne: Abelian varieties. In: G.C. Cornell and J.H. Silverman (eds.), Arithmetic geometry. Springer-Verlag 1986.

[20] D. Mumford: Abelian varieties. Oxford University Press 1970.

[21] M. van der Put: Les fonctions thêta d'une courbe de Mumford. Groupe d'étude d'analyse ultramétrique 1981/82, Paris 1982.

[22] K. Ribet: On modular representations of $\mathrm{Gal}(\overline{\mathbb{Q}})/\mathbb{Q})$ arising from modular forms. Invent. Math. **100**, 431-476 (1990).

[23] M. Rosen: Abelian varieties over $\mathbb{C}$. In: G.C. Cornell and J.H. Silverman (eds.), Arithmetic geometry. Springer-Verlag 1986.

[24] A. Schweizer: Zur Arithmetik der Drinfeld'schen Modulkurven $X_0(\mathfrak{n})$. Thesis Saarbrücken 1996.

[25] J.-P. Serre: Trees. Springer-Verlag 1980.

[26] G. Shimura: Introduction to the arithmetic theory of automorphic functions. Publ. Math. Soc. Japan 1971.

[27] J.H. Silverman: The arithmetic of elliptic curves. Springer-Verlag 1986.

[28] J.H. Silverman: Advanced topics in the arithmetic of elliptic curves. Springer-Verlag 1994.

[29] R. Taylor and A. Wiles: Ring-theoretic properties of certain Hecke algebras. Ann. Math. **141**, 553-572 (1995).

[30] A. Weil: Dirichlet series and automorphic forms. Lect. Notes Math. **189**. Springer-Verlag 1971.

[31] A. Wiles: Modular elliptic curves and Fermat's last theorem. Ann. Math. **141**, 443-551 (1995).

References to instructional papers in this volume are labelled by [I.number]. Hence e.g. [I.6] refers to the sixth instructional paper.

Research Papers

On Hilbert class field towers of global function fields

Bruno Angles

Laboratoire de Mathématiques Emile Picard,
C.N.R.S. U.M.R. 5580, U.F.R. M.I.G. Université Toulouse III,
118 route de Narbonne 31062 Toulouse Cedex 4 France

Introduction

Let $\mathbb{F}_q$ be a finite field with q elements. Set:

$$A(q) = \limsup_X \frac{\# X(\mathbb{F}_q)}{g_X},$$

where the limit is taken over smooth projective absolutely irreducible curves X over $\mathbb{F}_q$ of genus g_X tending to ∞. In 1981, V. G. Drinfeld and S. G. Vlâdut ([D&V]) found the following bound:

$$A(q) \leq \sqrt{q} - 1.$$

When q is a square, by a result of Ihara ([IH]), it follows that $A(q) = \sqrt{q} - 1$.

In 1983, J.-P. Serre ([SE1]) proved that there exists a positive constant c such that, for all q, we have:

$$A(q) \geq c \log q.$$

In 1992, R. Schoof ([SC], Theorem 1.1) found the following bound for $A(2)$:

$$A(2) \geq \frac{2}{9}.$$

In 1993, C.-P. Xing ([XI], Theorem 9) found:

$$A(3) \geq \frac{1}{3} \text{ and } A(5) \geq \frac{1}{2}.$$

The determination of the value of $A(q)$, when q is not a square, remains an open problem.

Most of the lower bounds of $A(q)$ have been obtained by showing that an "Hilbert class field" tower of some global function field is infinite. In this paper, we study abelian extensions K of a fixed global function field k and we give some results about the infinitude of the Hilbert class field tower of K.

Let k be a global function field and let K/k be a finite separable extension. The fact that the Hilbert class field tower of K is infinite comes from the Theorem of Golod-Shafarevitch for function fields (see for example [SC], Theorem 2.3) which we recall in section 1 and which states that if the ℓ-rank of the ideal class group of the "ring of integers" of K is "sufficiently large" for some prime number ℓ, then the Hilbert class field tower of K must be infinite.

Let K/k be an abelian extension and let ℓ be a prime number that divides $[K : k]$, we give, in section 2, a lower bound of the ℓ-rank of $\operatorname{Pic} O_K$ (see Theorem 2.1); furthermore if the Galois group of K/k is cyclic, we show (Theorem 2.2) that, if σ is a generator of $\operatorname{Gal}(K/k)$, $\operatorname{Pic} O_K/(\operatorname{Pic} O_K)^{\sigma-1}$ is isomorphic to $\operatorname{Gal}(\Gamma/K)$, where O_K is the ring of integers of K, $\operatorname{Pic} O_K$ is the ideal class group of O_K and Γ is the maximal abelian extension of k contained in the Hilbert class field of O_K (see the notations below).

In the last section, we take $k = \mathbb{F}_q(T)$ where $\mathbb{F}_q(T)$ is the field of rational functions over a finite field $\mathbb{F}_q$ and let $M \in \mathbb{F}_q[T]$, $M \neq 0$. Let t be the number of distinct monic irreducible divisors of M, let $k(M)$ be the Mth cyclotomic function field (see for example [GAL& ROS], section 1) and let ℓ be a prime divisor of $q-1$. We will apply the results obtained in section 2 to show that, if t is "sufficiently large", then the Hilbert ℓ-class field towers of $k(M)$ and $F(M)$ are infinite (Proposition 3.3), where $F(M)$ is the "maximal real subfield" of $k(M)$. In particular, if q is odd and if $t \geq 9$, then the Hilbert 2-class field towers of $k(M)$ and $F(M)$ are infinite.

The author thanks M. Reversat for helpfull comments and suggestions.

Notations

We fix a global function field k and S, $S \neq \emptyset$, a finite set of places of k. Let $\mathbb{F}_q$ be the constant field of k and let O_k be the ring of "S-integers" of k, i.e. O_k is the set of elements of k which are

regular outside every place in S. Let k^{sep} be a fixed separable closure of k. Let p be the characteristic of k.

Let $K \subset k^{sep}$ be a finite extension of k. We have the following notations:

- S_K: the set of all places of K above the places in S;
- O_K: the integral closure of O_k in K;
- J_K: the group of non-zero fractional ideals of O_K;
- P_K: the group of non-zero principal ideals of O_K;
- $\operatorname{Pic} O_K = J_K/P_K$: the ideal class group of O_K;
- E_K: the group of units of O_K;
- I_K: the group of idèles of K;

• $U_K = \prod_{v\in S_K} K_v^* \prod_{v\notin S_K} U_v$, where, for every place v of K, K_v is the completion of K at v and U_v is the group of units of the valuation ring of K_v.

A "finite prime" (or a finite place) will be a prime of K which does not lie in S_K. K^*U_K is an open subgroup of I_K of finite index (this comes from the fact that $\operatorname{Pic} O_K$ is a finite group), we denote the abelian extension of K in k^{sep} which corresponds to K^*U_K by H_K. The field H_K is called the "Hilbert class field" of O_K.

By class field theory, H_K is the maximal unramified abelian extension of K contained in k^{sep} such that every prime in S_K splits completely in H_K. For further properties of H_K, we refer the reader to [ROS1].

1. Hilbert ℓ-class field towers

In this section, we fix a prime number ℓ. Let G be any group. We denote the maximal abelian factor group of G of exponent ℓ by G/ℓ. The ℓ-rank of G is:

$$d_\ell(G) = \dim_{\mathbb{F}_\ell} G/\ell,$$

where $\mathbb{F}_\ell = \mathbb{Z}/\ell\mathbb{Z}$.

Let $K \subset k^{sep}$ be a finite extension of k. Let K_1 be the maximal ℓ-extension of K contained in H_K. The field K_1 is called the "Hilbert ℓ-class field " of O_K and the Galois group $\operatorname{Gal}(K_1/K)$ is

isomorphic to the ℓ-Sylow subgroup of $\operatorname{Pic} O_K$ (see [ROQ]). Let K_2 be the maximal ℓ-extension of K_1 contained in H_{K_1}. Repeating this process, we obtain a ℓ-class field tower:

$$K \subset K_1 \subset K_2 \subset \cdots$$

Set $K^{(\ell)} = \bigcup_{n\geq 1} K_n$. Set $\delta(K,\ell) = 1$ if $(\ell,p) = 1$ and if K contains the ℓth roots of unity, and $\delta(K,\ell) = 0$ in the other cases. As in the classical case, we have the Theorem of Golod-Shafarevitch:

1.1. Theorem. *If $K^{(\ell)}/K$ is a finite extension, we have:*

$$d_\ell(\operatorname{Pic} O_K) < 2 + 2\sqrt{\#(S_K) + \delta(K,\ell)}\,.$$

Proof. See [SC], Theorem 2.3. ◇

We have the following lower bound of $d_\ell(\operatorname{Pic} O_K)$:

1.2. Proposition. *We assume that the constant field of K is $\mathbb{F}_q$. Let t_ℓ be the number of primes of k for which each of the ramification indices in K is divisible by ℓ. Let $m = [K : k]$ and let $w_\ell(m)$ be the exponent of ℓ occuring in m. Then:*

$$d_\ell(\operatorname{Pic} O_K) \geq t_\ell - 1 - w_\ell(m) - \#(S_K).$$

Proof. Let D_S be the set of divisors of K having their supports in S_K, and let $D_{0,S}$ be the set of divisors of D_S of degree zero. Let P_S be the set of principal divisors of K having their supports in S_K, and let $\operatorname{Cl} K$ be the class group of divisors of degree zero of K.

We have the following exact sequence (see [ROS1], Lemma 4.1):

$$0 \to D_{0,S}/P_S \to \operatorname{Cl} K \to \operatorname{Pic} O_K \to \mathbb{Z}/r\mathbb{Z} \to 0,$$

where r is the g.c.d. of the degrees of the places in S_K. This implies:

$$d_\ell(\operatorname{Pic} O_K) \geq d_\ell(\operatorname{Cl} K) - d_\ell(D_{0,S}/P_S).$$

Now, we have:

$$d_\ell(D_{0,S}/P_S) \leq d_\ell(D_{0,S}) \leq d_\ell(D_S) = \#(S_K),$$

and by [MAD], Theorem 2, we have:

$$d_\ell(\operatorname{Cl} K) \geq t_\ell - 1 - w_\ell(m).$$

The Proposition follows. ◇

2. On abelian extensions of k

In this section, we improve Proposition 1.2 when K/k is abelian (Theorem 2.1). We have been inspired by the genus field theory for function fields developed by R. Clement in [CL].

Let $K \subset k^{\text{sep}}$ be a finite abelian extension of k of degree m and let $G = \operatorname{Gal}(K/k)$. Let ℓ be a prime number such that ℓ divides m.

Let $Q_1, \cdots, Q_t$ be the finite primes of k which are ramified in K. Let e_i be the ramification index of Q_i in K for $i = 1, \cdots, t$. Let u_ℓ be the number of Q_i for which ℓ divides e_i.

For each place $w \in S$, let m_w be the local degree of K/k at w, i.e. let v be a place in S_K above w, then $m_w = [K_v : k_w]$. Let ε_ℓ be the number of places $w \in S$ for which ℓ divides m_w.

Let $N : I_K \to I_k$ be the norm map. Let Γ be the maximal abelian extension of k contained in H_K. We have obtained the following results:

2.1. Theorem. $d_\ell(\operatorname{Pic} O_K) \geq u_\ell - \delta(k,\ell) + \varepsilon_\ell - \#(S) + 1 - d_\ell(G)$.

2.2. Theorem. *Suppose that G is cyclic. If σ is a generator of G, then:*

$$\frac{\operatorname{Pic} O_K}{(\operatorname{Pic} O_K)^{\sigma-1}} \simeq \operatorname{Gal}(\Gamma/K),$$

where $(\operatorname{Pic} O_K)^{\sigma-1} = \{\sigma(x)/x,\ x \in \operatorname{Pic} O_K\}$.

In order to prove these two Theorems, we need some auxiliary Lemmas:

2.3. Lemma.

$$[\Gamma : k] = \#(\operatorname{Pic} O_k) \prod_{w \in S} m_w \prod_{i=1}^{t} e_i \,(E_k : E_k \cap N(U_K))^{-1},$$

and if G is cyclic:

$$E_k \cap N(U_K) = E_k \cap N(K^*).$$

Proof. By [CL], Lemma 2.2, the subgroup of the idèles of k which corresponds to Γ is $k^* N(U_K)$. Thus:

$$\operatorname{Gal}(\Gamma/k) \simeq \frac{I_k}{k^* N(U_K)}.$$

Therefore:

$$\begin{aligned}[\Gamma : k] &= (I_k : k^* N(U_K)) \\ &= (I_k : k^* U_k)\,(k^* U_k : k^* N(U_K)) \\ &= (I_k : k^* U_k)\,(U_k : N(U_K))\,(k^* \cap U_k : k^* \cap N(U_K))^{-1}.\end{aligned}$$

Now, it is clear that $k^* \cap U_k = E_k$. If α is in $E_k \cap N(K^*)$, then α is a local norm at every place of k, thus, since $\alpha \in E_k$, we have that α is in $k^* \cap N(U_K)$. Reciprocally, if G is cyclic, we have: $k^* \cap N(U_K) \subset E_k \cap N(K^*)$. Thus: $k^* \cap N(U_K) = E_k \cap N(K^*)$ when G is cyclic.

By [TA], Proposition 7.2 and [SE2], section 4, Theorem 1, we have:

$$(U_k : N(U_K)) = \prod_{w \in S} m_w \prod_{i=1}^{t} e_i.$$

Finally, note that:

$$(I_k : k^*\, U_k) = \#(\operatorname{Pic} O_k). \ \diamond$$

2.4. Lemma.

$$d_\ell(\operatorname{Gal}(\Gamma/KH_k)) \geq u_\ell - \delta(k,\ell) + \varepsilon_\ell - \#(S) + 1 - d_\ell(G).$$

Proof. We have the following exact sequence:

$$1 \to \frac{E_k}{E_k \cap N(U_K)} \to \frac{U_k}{N(U_K)} \to \frac{k^*\, U_k}{k^*\, N(U_K)} \to 1.$$

It is clear that:

$$d_\ell(\frac{U_k}{N(U_K)}) \geq u_\ell + \varepsilon_\ell.$$

Now:

$$d_\ell\left(\frac{E_k}{E_k \cap N(U_K)}\right) \leq d_\ell(E_k) = \#(S) + \delta(k,\ell) - 1.$$

Thus:

$$d_\ell\left(\frac{k^*\, U_k}{k^*\, N(U_K)}\right) \geq u_\ell + \varepsilon_\ell + 1 - \delta(k,\ell) - \#(S).$$

But, $\operatorname{Gal}(\Gamma/H_k) \simeq k^*\, U_k/(k^*\, N(U_K))$, and we have the following exact sequence:

$$1 \to \operatorname{Gal}(\Gamma/KH_k) \to \operatorname{Gal}(\Gamma/H_k) \to \operatorname{Gal}(K/K \cap H_k) \to 1.$$

Recall that $\operatorname{Gal}(K/K \cap H_k)$ is a subgroup of G, thus:

$$d_\ell(\operatorname{Gal}(\Gamma/KH_k)) \geq d_\ell(\operatorname{Gal}(\Gamma/H_k)) - d_\ell(G). \ \diamond$$

Proof of Theorem 2.1. We have: $d_\ell(\operatorname{Gal}(\Gamma/K)) \geq d_\ell(\operatorname{Gal}(\Gamma/KH_k)$. But $\operatorname{Pic} O_K \simeq \operatorname{Gal}(H_K/K)$ and $\Gamma \subset H_K$. Now the Theorem follows from Lemma 2.4 . $\diamond$

Proof of Theorem 2.2. Set $P_k = \{xO_K \mid x \in k^*\}$. For each finite prime Q of k set $\Phi(Q) = \prod P$ where the product is over all the finite primes of K that divide Q. Let J_K^G be the subgroup of elements of J_K which are invariant under the action of G. Then J_K^G is the free abelian group generated by the $\Phi(Q)$, where Q runs through the set of finite primes of k. Thus:

$$\#\left(\frac{J_K^G}{P_k}\right) = \#(\operatorname{Pic} O_k) \prod_{i=1}^{t} e_i.$$

From the proof of Theorem 8.3 in [TA], we have:

$$\frac{\#(H^2(G,E_K))}{\#(H^1(G,E_K))} = \frac{\prod_{w\in S} m_w}{m}.$$

Thus:

$$\#(H^1(G,E_K)) = \frac{m\,(E_k : N(E_K))}{\prod_{w\in S} m_w}.$$

We have the following exact sequence:

$$1 \to P_K \to J_K \to \operatorname{Pic} O_K \to 1,$$

and recall that $H^1(G,J_K) = \{1\}$, therefore:

$$1 \to \frac{P_K^G}{P_k} \to \frac{J_K^G}{P_k} \to (\operatorname{Pic} O_K)^G \to H^1(G,P_K) \to 1.$$

But, we have also:

$$1 \to E_K \to K^* \to P_K \to 1,$$

thus:

$$1 \to E_k \to k^* \to P_K^G \to H^1(G,E_K) \to 1.$$

This leads to:

$$H^1(G,E_K) \simeq \frac{P_K^G}{P_k}.$$

Finally, we have:

$$1 \to H^1(G,P_K) \to H^2(G,E_K) \to H^2(G,K^*),$$

this implies that:

$$\#(H^1(G,P_K)) = (E_k \cap N(K^*) : N(E_K)).$$

Therefore, by Lemma 2.3, we have:

$$\begin{aligned}\#((\operatorname{Pic} O_K)^G) &= \frac{\#(H^1(G,P_K))\,\#(J_K^G/P_k)}{\#(H^1(G,E_K))}\\ &= \#(\operatorname{Pic} O_k) \prod_{i=1}^{t} e_i \prod_{w\in S} m_w\; m^{-1}\,(E_k : E_k \cap N(K^*))^{-1}\\ &= [\Gamma : K].\end{aligned}$$

It is not difficult to show that H_K/k is a Galois extension. Set $\mathcal{G} = \mathrm{Gal}(\mathcal{H}_K/||)$. Let σ be a generator of G and let $\bar{\sigma} \in \mathcal{G}$ such that the restriction of $\bar{\sigma}$ to K is equal to σ.

Let $\phi : \operatorname{Pic} O_K \to \mathrm{Gal}(H_K/K)$ be the isomorphism given by the Artin symbol, and let $\Omega = \phi((\operatorname{Pic} O_K)^{\sigma-1})$.

Let $[a] \in \operatorname{Pic} O_K$, then:

$$\begin{aligned}\phi([a]^{\sigma-1}) &= \phi([a]^{\sigma})\,\phi([a]^{-1})\\ &= (\bar{\sigma})^{-1}\,\phi([a])\,\bar{\sigma}\,\phi([a])^{-1}.\end{aligned}$$

Thus Ω lies in the derived subgroup of $\mathcal{G}$. This implies:

$$\Gamma \subset H_K^{\Omega},$$

where H_K^{Ω} is the subfield of H_K fixed by Ω. But:

$$\mathrm{Gal}(H_K^{\Omega}/K) \simeq \frac{\operatorname{Pic} O_K}{(\operatorname{Pic} O_K)^{\sigma-1}},$$

and:

$$\frac{\operatorname{Pic} O_K}{\operatorname{Pic} O_K^G} \simeq (\operatorname{Pic} O_K)^{\sigma-1}.$$

Thus: $H_K^{\Omega} = \Gamma$ and the proposition follows. $\diamondsuit$

3. Some applications.

(3.1) In this paragraph $k = \mathbb{F}_q(T), S = \{1/T\}$, thus $O_k = \mathbb{F}_q[T]$. Let ℓ be a prime number and let K/k be a finite cyclic extension of degree ℓ. We keep the notations of the precedent sections.

It is clear that $\mathrm{Gal}(\Gamma/k) \simeq k^*\,U_k/(k^*\,N(U_K))$ is an elementary abelian ℓ-group. Let $Q_1(T), \cdots, Q_t(T)$ be the monic irreducible elements of $\mathbb{F}_q[T]$ which are ramified in K. Set $n_i = \deg_T Q_i(T)$ for $i = 1, \cdots, t$. We have by Lemma 2.3:

$$[\Gamma : K] = \ell_\infty\,\ell^{t-1}\,(\mathbb{F}_q^* : \mathbb{F}_q^* \cap N(K^*))^{-1},$$

where ℓ_∞ is the local degree of K/k at $1/T$.

Set:

$m(K/k) = q - 1$ if ℓ does not divide $q - 1$,

$m(K/k) = (q-1)/\ell$ if ℓ divides $q - 1$ and $1/T$ is ramified in K,

$m(K/k) = (q-1)\,\ell^{\epsilon-1}$, where $\ell^\epsilon = \text{g.c.d.}\,(\text{g.c.d.}\,(n_1, \ell), \cdots, \text{g.c.d.}\,(n_t, \ell))$, if ℓ divides $q-1$ and $1/T$ is unramified in K.

Then by [ROS2], Proposition 2:

$$(\mathbb{F}_q^* : \mathbb{F}_q^* \cap N(K^*))^{-1} = \frac{m(K/k)}{q-1}.$$

If we apply Theorem 2.2, we get the following five cases:
(3.1.1) K/k is totally imaginary (i.e. $\ell_\infty = \ell$) and ℓ does not divide $q-1$. We have $[\Gamma : k] = \ell^t$. Thus, ℓ divides $\#(\mathrm{Pic}\, O_K)$ if and only if $t \geq 1$. Moreover ℓ^t divides $\#(\mathrm{Pic}\, O_K)$. Furthermore, if $t \geq 4$ then $K^{(\ell)}/K$ is infinite.
(3.1.2) K/k is totally imaginary, ℓ divides $q-1$ and $1/T$ is ramified in K. We have: $[\Gamma : K] = \ell^{t-1}$. Thus ℓ divides $\#(\mathrm{Pic}\, O_K)$ if and only if $t \geq 2$. If $t \geq 6$ then $K^{(\ell)}/K$ is infinite.
(3.1.3) K/k is totally imaginary, ℓ divides $q-1$ and $1/T$ is unramified. We have: $[\Gamma : k] = \ell^{t-1+\varepsilon}$. Thus, ℓ divides $\#(\mathrm{Pic}\, O_K)$ if and only if $t \geq 2-\varepsilon$. If $t \geq 6-\varepsilon$ then $K^{(\ell)}/K$ is infinite.
(3.1.4) K/k is totally real (i.e. $\ell_\infty = 1$) and ℓ does not divide $q-1$. We have $[\Gamma : K] = \ell^{t-1}$. Thus ℓ divide $\#(\mathrm{Pic}\, O_K)$ if and only if $t \geq 2$. If $t \geq 3 + 2\sqrt{\ell}$ then $K^{(\ell)}/K$ is infinite.
(3.1.5) K/k is totally real and ℓ divides $q-1$. We have: $[\Gamma : K] = \ell^{t-2+\varepsilon}$. Thus ℓ divides $\#(\mathrm{Pic}\, O_K)$ if and only if $t \geq 3-\varepsilon$. If $t \geq 4-\varepsilon+2\sqrt{\ell+1}$ then $K^{(\ell)}/K$ is infinite.

Let again $k = \mathbb{F}_q(T), S = \{1/T\}, O_k = \mathbb{F}_q[T]$ and $q \geq 3$. For each non-zero element $M \in \mathbb{F}_q[T]$, let $k(M)$ be the Mth cyclotomic function field and let $F(M)$ be its maximal real subfield (see [GAL& ROS], section 1).

3.2. Lemma. *Let $P(T)$ be a monic irreducible polynomial in $\mathbb{F}_q[T]$. Let $d = \deg P$ and let ℓ be a prime number that divides $q-1$.*
i) If ℓ divides d then $k(\sqrt[\ell]{P(T)}) \subset F(P)$.
ii) If ℓ does not divide d, let n, $1 \leq n \leq \ell - 1$, be such that ℓ divides $nd-1$. Then $k(\sqrt[\ell]{-P(T)^n}) \subset k(P)$.

Proof.
i) Recall that $F(P)$ is the ray class field of $\mathbb{F}_q[T]$ associated to the ideal $P(T)\mathbb{F}_q[T]$ ([HA], Theorem 9.7). Thus, the subgroup of the idèles of k which corresponds to $F(P)$ is:

$$k^*\, k_\infty^*\, U_P^1 \prod_{Q \neq P} U_Q,$$

where k_∞^* is the completion of k at $1/T$, for each finite prime Q, U_Q is the group of Q-adic units in the completion k_Q of k at Q, and where U_P^1 is the group of units in k_P which are congruent to 1 modulo P.

Set $K = k(\sqrt[\ell]{P(T)})$. Then $P(T)$ is totally and tamely ramified in K, $1/T$ splits completely in K and $P(T)$ is the only prime of k that ramifies in K. Thus:

$$k^* \, k_\infty^* \, U_P^1 \prod_{Q \neq P} U_Q \subset k^* \, N(I_K).$$

Therefore, by class field theory, we have: $K \subset F(P)$.
ii) Let $\mathcal{P}$ be a prime of $k(P)$ above $1/T$, then by [GAL& ROS], Lemma 1.1:

$$k(P)_{\mathcal{P}} = \mathbb{F}_q((\sqrt[q-1]{-1/T})).$$

The group of idèles of k which corresponds to $k(P)$ is:

$$k^* \, N(k(P)_{\mathcal{P}}^*) \, U_P^1 \prod_{Q \neq P} U_Q.$$

Thus, by class field theory: $k(\sqrt[\ell]{-P(T)^n}) \subset k_P$. ◇

3.3. Proposition. *Let $M \neq 0$, $M \in \mathbb{F}_q[T]$, and let $P_1(T), \cdots P_t(T)$ be the distinct monic irreducible divisors of M. Let ℓ be a prime number that divides $q-1$.*

For each i, $1 \leq i \leq t$, let $n_i = 1$ if ℓ divides $\deg P_i$, and if ℓ does not divide $\deg P_i$, let n_i be such that: $1 \leq n_i \leq \ell - 1$ and ℓ divides $n_i \deg P_i - 1$. Let μ_ℓ be the number of P_i such that ℓ does not divide $\deg P_i$.

Set: $N_1(T) = \prod_{i=1}^t P_i(T)$ and $N_2(T) = (-1)^{\mu_\ell} \prod_{i=1}^t P_i(T)^{n_i}$. Then:
A) Case ℓ divides $\deg N_1$ or ℓ divides $\deg N_2$.

A.1) Case $\mu_\ell = 0$. If $t \geq 3 + 2\sqrt{\ell+1}$ then $F(M)^{(\ell)}/F(M)$ is an infinite extension.

A.2) Case $\mu_\ell \neq 0$. If $t \geq 4 + 2\sqrt{\ell+1}$ then $F(M)^{(\ell)}/F(M)$ is infinite.

B) Case ℓ does not divide $\deg N_1$ and ℓ does not divide $\deg N_2$.

B.1) Case $\ell \neq 2$. If $t - \mu_\ell \geq 3 + 2\sqrt{\ell+1}$ then $F(M)^{(\ell)}/F(M)$ is infinite.

B.2) Case $\ell = 2$. If $t \geq 9$ then $F(M)^{(2)}/F(M)$ is infinite.

C) We have the following cases for the infinitude of $k(M)^{(\ell)}/k(M)$:

C.1) Case ℓ does not divide $\deg N_2$. If $t \geq 6$ the $k(M)^{(\ell)}/k(M)$ is infinite.

C.2) Case ℓ divides $\deg N_2$ and $\mu_\ell \neq 0$. If $t \geq 7$, then $k(M)^{(\ell)}/k(M)$ is infinite.

C.3) Case ℓ divides $\deg N_2$ and $\mu_\ell = 0$. If $t \geq 3 + 2\sqrt{\ell+1}$ then $k(M)^{(\ell)}/k(M)$ is infinite.

Proof. The proposition comes from (3.1) and the following facts:
- if K and L are two finite separable extensions of k such that $K \subset L$ and if $K^{(\ell)}/K$ is infinite then $L^{(\ell)}/L$ is infinite too.

- If $\deg N_1$ is divisible by ℓ, by the same arguments than that given in part i) of the proof of Lemma 3.2, $k(\sqrt[\ell]{N_1(T)}) \subset F(M)$.
- By Lemma 3.2, $k(\sqrt[\ell]{N_2(T)}) \subset k(M)$, and if ℓ divides $\deg N_2$ then $k(\sqrt[\ell]{N_2(T)}) \subset F(M)$.
- If $\ell = 2$ and if 2 does not divide $\deg N_2$ then there exists j such that 2 does not divide $\deg P_j(T)$. Set $N_3(T) = \prod_{i \neq j} P_i(T)$. Then 2 divides $\deg N_3$ and $k(\sqrt{N_3(T)}) \subset F(M)$. $\diamond$

References

[CL] R. CLEMENT, The genus field of an algebraic function field, Journal of Number Theory **40**, 359-375, 1992.

[D&V] V. G. DRINFELD and S. G. VLADUT, Number of points of an algebraic curve, Funct. Anal. **17**, 53-54, 1983.

[GAL& ROS] S. GALOVICH and M. ROSEN, Units and class groups in cyclotomic function fields, Journal of Number Theory **14**, 156-184, 1982.

[HA] D. R. HAYES, Explicit class field theory in global function fields, "Studies in Algebra and Number Theory", Advances in Mathematisc Supplementary Studies, Vol. 6, 173-217, Academic Press, New York, 1979.

[IH] Y. IHARA, Some remarks on the number of rational points of algebraic curves over finite fields, J. Fac. Sci. Tokyo **28**, 721-724, 1981.

[MAD] M. L. MADAN, Class number and ramification in fields of algebraic functions, Arch. Math. Vol. XIX, 121-124, 1968.

[ROQ] P. ROQUETTE, Chapter 9 in "Algebraic Number Theory", Edited by J. W. S. Cassels and A. Fröhlich, Academic Press, 1967.

[ROS1] M. ROSEN, The Hilbert class field in function fields, Expo. Math. 5 , 365-378, 1987.

[ROS2] M. ROSEN, Ambiguous divisor classes in function fields, Journal of Number Theory **9**, 160-174, 1977.

[SC] R. SCHOOF, Algebraic curves over $\mathbb{F}_2$ with many rational points, Journal of Number Theory **41**, 6-14, 1992.

[SE1] J.-P. SERRE, Sur le nombre des points rationnels d'une courbe algébrique sur un corps fini, C. R. Acad. Sc. Paris, t. 296, Série I, 397-402, 1983.

[SE2] J.-P. SERRE, Chapter 6 in "Algebraic Number Theory", Edited by J. W. S. Cassels and A. Fröhlich, Academic Press, 1967.

[TA] J. T. TATE, Chapter 7 in "Algebraic Number Theory", Edited by J. W. S. Cassels and A. Fröhlich, Academic Press, 1967.

[XI] C.-P. XING, Multiple Kummer extension and the number of prime divisors of degree one in function fields, Journal of Pure and Applied Algebra 84, 85-93, 1993.

Drinfeld modular forms of level T

Gunther Cornelissen [1]
University of Gent
Dept. of Pure Mathematics and Computer Algebra
Galglaan 2, B-9000 Gent, Belgium
e-mail gc@cage.rug.ac.be

This paper should be seen as a companion to [2], giving an introduction to the results contained therein by working out an easy example: we will give a presentation for the ring of Drinfeld modular forms for the principal congruence group of level T, $\Gamma(T)$ over $\mathbf{F}_q[T]$ by generators and relations. In our example, the modular curve $\bar{M}_{\Gamma(T)}$ is of genus zero, which greatly simplifies the arguments. It will be our policy to always indicate what becomes more difficult or remains unproven for modular curves of general genus. Our notations will be compatible with [3] in this volume. Apart from this, the paper is written in a self-contained way, in particular, independent of [2].

To simplify notations, we will fix once and for all a level $N \in \mathbf{F}_q[T]$, and let $\Gamma := \Gamma(N)$.

(1.1) Example By $\bar{M}_{\Gamma(T)}$ we will mean the C-projective curve which is the compactification of $\Gamma(T)\backslash\Omega$. It has $q+1$ cusps and genus zero. This curve (and its K-structure) is also studied in the instructional lectures 8 and 9. We will represent the cusps by $\{\infty = (1:0), (\alpha:1), \alpha \in \mathbf{F}_q\} \in \Gamma(T)\backslash\mathbf{P}^1(\mathbf{F}_q(T)) = \mathbf{P}^1(\mathbf{F}_q)$. As prototypical modular forms of weight one for $\Gamma(T)$ we have the Eisenstein series ([3], (1.5.5)):

$$E_u(\omega) := \sum_{(a_1,a_2)\equiv u \bmod T} \frac{1}{a_1\omega + a_2},$$

for different $u \in \mathbf{F}_q[T]^2 - \{(0,0)\}$. There are q^2-1 different such functions, one for each class modulo T in $\mathbf{F}_q[T]^2$; choose e.g. as representatives $u \in \mathbf{F}_q^* \times U$ with $U := \{\infty = (0,1), (1,\beta), \beta \in \mathbf{F}_q\}$. (Note that one can make a different choice of representatives for the cusps and the indices U. We have chosen them in the above "dual" way, because this simplifies some formulae.) There exist C-linear relations between these functions, namely $E_{\lambda u} = \lambda^{-1}E_u$ for all $\lambda \in \mathbf{F}_q^*$, but these are the only ones, *i.e.* if we choose only representatives in U (having "sign one"), we get linearly independent functions, as can be seen by calculating the order of these functions at the cusps: it is clear that the function E_u can only be non-zero at the cusp ∞ if $a_1 = 0$ for all a_1 occuring in the sum, so $u = (0,1)$. If one is at a different cusp $s = (s_1, s_2) = \gamma \cdot \infty$,and $\gamma \in \Gamma(1)$, then $\mathrm{ord}_s E_u = \mathrm{ord}_\infty E_{u\cdot\gamma} = 0$ only if $u_1 s_1 + u_2 s_2 = 0$. So we

[1] supported as "aspirant" by the Belgian National Fund for Scientific Research (NFWO)

see that there is a bijection between the cusps and U, given by associating to any cusp the unique Eisenstein series that doesn't vanish at that cusp. This suffices to see that they are linearly independent: any relation $\sum a_u E_u$ can be evaluated at all cusps to find $a_u = 0, \forall u$. To simplify future notations, we will write E_β for $E_{(1,\beta)}$.

On the other hand, If $\mathcal{M}$ is the divisor of germs of modular forms of weight one on $\bar{M}_{\Gamma(T)}$, we have $\deg \mathcal{M} = q$, and application of the Riemann-Roch theorem gives

$$\dim M_1(\Gamma(T)) = \dim H^0(\bar{M}_{\Gamma(T)}, \mathcal{M}) = \deg(\mathcal{M}) + 1 = q + 1,$$

using the fact that $\bar{M}_{\Gamma(T)}$ has genus zero. Hence $\{E_u : u \in U\}$ form a basis for $M_1(\Gamma(T))$.

Note that since $\bar{M}_{\Gamma(T)} = \mathbf{P}^1$, a bundle is uniquely determined up to isomorphism by its degree, so $\mathcal{M} \cong \mathcal{O}(q)$, *cf.* the notation of lecture 9.

(1.2) Generalizations For a level N with $\deg N > 1$, the modular curve $\bar{M}_\Gamma$ is no longer of genus zero. Also, the linear relations between the different Eisenstein series are not so easy to find, but there is a trick due to Hecke [8], which roughly goes as follows: choose only *primitive* representatives $u = (u_1, u_2)$, which means that u_1, u_2, N have no common divisor. Change the definition of E_u by adding the condition in the sum that a_1, a_2 have no common divisor. Then there is again a one-one correspondence between cusps and primitive altered series

$$\sum_{\substack{(a_1,a_2)\equiv u \bmod T \\ (a_1;a_2)=1}} \frac{1}{a_1\omega + a_2},$$

not vanishing at that cusp, so the different altered functions span a space of dimension $= \#S$, the number of cusps. One then uses Moebius-inversion to show that the original and altered functions can be linearly expressed in one another, and that the non-primitive functions can be expressed by primitive ones. In conclusion, the $\{E_u\}$ span a space of dimension $\#S$, and a linearly independent set is given by choosing only primitive u.

For the dimension of the space of modular forms of weight one, the Riemann-Roch theorem only gives

$$\dim M_1(\Gamma) = \#S + H^1(\bar{M}_\Gamma, \mathcal{M}),$$

but E.-U. Gekeler has shown that the H^1 vanishes (*see* [3], (6.9.1) — use duality to interpret H^1 as a space of double cusp forms of weight one, and calculate, following Teitelbaum [12], the "periods" of its q-th power). In conclusion, we still find that the primitive Eisenstein series span $M_1(\Gamma)$.

Please note that this conclusion is very different from the case of classical modular forms, where e.g. the square of the Dedekind eta-function $\eta^2 = e^{\frac{\pi i z}{6}} \prod(1 - e^{2\pi i n z})^2$ is a cusp form of weight one for $\Gamma(12)$.

(2.1) Example The next question that arises is whether the Eisenstein series of weight one generate the whole algebra of modular forms, *i.e.* whether any modular form of level T is a polynomial in the $\{E_u\}$. One has:

$$\dim M_k(\Gamma(T)) = kq + 1.$$

For instance, $M_2(\Gamma(T))$ has dimension $2q+1$; for any fixed $v \in U$, $\{E_u^2 : u \in U\} \cup \{E_v.E_u : u \in U - \{v\}\}$ are linearly independent (same argument as before), hence form a basis for $M_2(\Gamma(T))$. Similarly, the $3q+1$-dimensional $M_3(\Gamma(T))$ is spanned, for fixed $v, w \in U$ by the union of the sets $\{E_u^3 : u \in U\}$,

$$\{E_v^2 E_u : u \in U - \{v\}\}, \{E_v E_u^2 \text{ some } u \neq v\}, \{E_u E_v E_w : u \in U - \{v, w\}\}.$$

This follows by a refinement of the independence argument in (1.1), now not only locating the zeroes, but also calculating their order. A calculation (e.g. following the product expansion by Gekeler in [5], (2.1)) shows that the series development of E_β in the parameter t at the cusp ∞ is

(2.1.1) $\quad (T\pi)^{-1} E_\beta(\infty) = t + a_\beta t^2 + a_\beta^2 t^3 + \ldots - (T-1)T^{q-1} t^q + o(t^{q+1}),$

where $a_\beta = -T e_{\pi \mathbf{F}_q[T]}(T^{-1}\pi\beta)$ is a $q-1$-th root of $-T$.

(2.1.2) *Remark* A suitable multiple of E_β admits a series development defined over K, namely if we let ζ be such that $\zeta^{q-1} = -T$, and we choose as parameter at ∞, $s := \zeta T t$, then

$$\frac{\beta\zeta}{\pi} E_\beta(\infty) = \beta s + \beta^2 s^2 + \beta^3 s^3 + \ldots + (1 - \frac{1}{T})\beta s^q + o(s^{q+1}) \in K[[s]].$$

From the point of view of rationality questions (*i.e.* the K-structure on $M_{\Gamma(T)}$), this approach seems to be more suitable.

Since $\bar{M}_{\Gamma(T)} \cong \mathbf{P}^1$, we have $\mathcal{M} \cong \mathcal{O}(q)$, and, one can see immediately that $M(\Gamma(T)) \cong \Gamma_*(\mathcal{O}(q))$ is generated by the Eisenstein series of weight one $M_1 = H^0(\bar{M}_{\Gamma(T)}, \mathcal{O}(1))$.

(2.2) Generalizations Since $M(\Gamma)$ is a finitely generated algebra, there must be some bound b, such that the elements of weight $\leq b$ generate $M(\Gamma)$. It is no longer immediate that $b = 1$ (as for positive bundles on the projective line). There is a general theory attacking this question, due to Castelnuovo and Mumford (C.I.M.E.-course [10]). They study conditions under which the natural map $M_k(\Gamma) \otimes M_l(\Gamma) \to M_{k+l}(\Gamma)$, reinterpreted as

$$H^0(\bar{M}_\Gamma, \mathcal{M}^{\otimes k}) \otimes H^0(\bar{M}_\Gamma, \mathcal{M}^{\otimes l}) \to H^0(\bar{M}_\Gamma, \mathcal{M}^{\otimes k+l})$$

is surjective. One of their results says that this map is surjective if $\min\{k,l\}\cdot \deg \mathcal{M} > 2g+1$. It has been remarked by D. Goss that this implies $M(\Gamma)$ is generated in weight ≤ 3. Actually, a refinement shows that weight ≤ 2 suffices ([2]). I don't see how the proof can be adapted to show that weight one suffices; the bundle $\mathcal{M}$ is in general approximately half-canonical and non-special.

If $M(\Gamma)$ is generated in weight one, then certainly $\mathcal{M}$ has to be very ample (see *loc. cit.*). We can first address the weaker question of whether this holds in general. So consider the map induced by $\mathcal{M}$

$$i : \bar{M}_\Gamma \to \mathbf{P}^n : \omega \mapsto (E_i(\omega)),$$

where we have chosen an ordering $E_1, ..., E_n$ on the *different* Eisenstein series E_u. (Note that although the values of E_u are not well defined for $z \in \bar{M}_{\Gamma(T)}$, their ratios are). It turns out that this is an embedding in general, so $\mathcal{M}$ *is* very ample. The homogeneous coordinate ring of $\mathrm{im}(i)$ is exactly the ring $E(\Gamma)$, defined as the subring of $M(\Gamma)$ generated by the Eisenstein series of weight one. Hence the closed points of $\mathrm{Proj}(E(\Gamma))$ are isomorphic to $\bar{M}_\Gamma$.

If R is a graded domain, let $R_{((0))}$ denote the ring consisting of quotients of homogeneous elements of the same degree in R. Since the function field of $\bar{M}_\Gamma$ is the field of modular functions $M(\Gamma)_{((0))}$, and the function field of $\mathrm{Proj}(E(\Gamma))$ is $E(\Gamma)_{((0))}$, we find that these are equal. But let $f \in M_k(\Gamma)$ be a modular form of weight k, then fE_u^{-k} is a modular function, hence belongs to $E(\Gamma)_{((0))}$ by the above. It follows that the quotient fields of $E(\Gamma)$ and $M(\Gamma)$ are equal. But what is more, all rings of modular forms for arithmetic groups are normal (*i.e.* integrally closed in their quotient field); this is classically due to Igusa ([9] III.5 – basically, one puts a grading on the integral closure, and then shows that homogeneous integral elements satisfy the correct transformation equation for a modular form. Finally, holomorphy is preserved for integral elements everywhere locally, as follows from Weierstrass' lemma). In conclusion, we see that $M(\Gamma)$ is the integral closure of $E(\Gamma)$.

In this way, the question whether $E(\Gamma) = M(\Gamma)$ turns out to be equivalent to the normality of $E(\Gamma)$, or combining the fact that Proj $E(\Gamma)$ is smooth with Serre's criterium for normality ([4], 11.5), it is equivalent to the fact that $E(\Gamma)$ is Cohen-Macaulay. The latter can be interpreted by saying that $E(\Gamma)$ is free as a $C[g,\Delta]$-module. An affirmative answer to this question is not known, except when $\deg N = 1$ or $(q, \deg N) = (2,2)$ (the latter since the embedding via $\mathcal{M}$ turns out to be an extremal curve w.r.t. Castelnuovo's bound — see e.g. [1], p. 116).

(3.1) Example So far, we know that $M(\Gamma(T))$ is generated by $\{E_u\}$, so the next question is to find the algebraic relations between these functions.

The degree of the embedding

$$i : \bar{M}_{\Gamma(T)} \to \mathbf{P}^q : \omega \mapsto (E_u(\omega) : u \in U)$$

(the number of points of intersection of im(i) with a general plane in $\mathbf{P}^q$) equals the degree of $\mathcal{M}$ as a line bundle $= q$. We also have that im(i) is non-degenerate, by which we mean that it doesn't belong to any hyperplane: the basis points $(1 : ... : 0), ..., (0 : ... : 1)$ belong to it (they are the images of the cusps of $\bar{M}_{\Gamma(T)}$). But a smooth non-degenerate curve of degree q in $\mathbf{P}^q$ is projectively isomorphic to the rational normal curve $\mathcal{R}_q$ given by $\mathbf{P}^1 \to \mathbf{P}^q$: $(X : Y) \mapsto (X^q : X^{q-1}Y : ... : XY^{q-1} : Y^q)$ ([7], 1.14).

From the fact that i embeds the modular curve as a rational normal curve, we know it is the intersection of quadrics. Let $x_\beta : \beta \in \mathbf{F}_q \cup \{\infty\}$ be $q+1$ variables, and define the "ideal of relations" I as the kernel of

$$R := C[x_\beta : \beta \in \mathbf{F}_q \cup \{\infty\}] \to E(\Gamma) : x_\beta \to E_\beta.$$

Then I can be generated by elements of degree two. We will now try to determine these.

(3.1.1) Here is an ad-hoc way of finding I: use the series expansions (2.1.1) to produce quadratic polynomials in E_β that have a zero of order ≥ 2 at all cusps, hence identically zero since there are no cusp forms of weight two. Make as many relations as required to let R/I have the same Hilbert function as $E(\Gamma)$.

(3.1.2) Second method: use the fact that im(i) is projectively isomorphic to the standard rational normal curve $\mathcal{R}_q$ in $\mathbf{P}^q$, which has equations

(3.1.3) $\quad Z_i Z_j - Z_{i-1} Z_{j+1}, \quad (1 \leq i \leq j \leq q-1),$

in terms of the coordinates $Z_0, ..., Z_q$ on $\mathbf{P}^q$ ([7], 5.4). The curve $\mathcal{R}_q$ passes through the $q+1$ points $(0 : ... : 1), (1 : \alpha : \alpha^2 : ... : \alpha^q), \alpha \in \mathbf{F}_q$, which are linearly independent (their determinant is Vandermonde). On the other hand, evaluating i at the different cusps of $\bar{M}_{\Gamma(T)}$, we see that im(i) passes through the $q+1$ "base"-points $(0 : ... : 1 : ... : 0)$. Suppose we transform coordinates in $\mathbf{P}^q$ via the transformation of these two bases of the underlying affine space $\mathbf{A}^{q+1}$:

$$T\pi \cdot Z_j = \sum_{\alpha \in \mathbf{F}_q} \alpha^j x_\alpha + \delta_{j,q} x_\infty, \quad j = 0, ..., q \quad \text{(convention: } 0^0 = 1),$$

where δ is a Kronecker symbol. There is in general a unique rational normal curve through $q+3$ points, of which no $q+1$ lie in a hyperplane; so there is no immediate reason why the image of $\mathcal{R}_q$ under this transformation would have to be im(i). But we are lucky, because we will now show that the relations (3.1.3) are identically satisfied when substituting $x_\beta = E_\beta(\omega), \forall \omega \in \Omega$.

Proof. We will prove this by showing that the series expansions vanish at all cusps of order two. This will suffice to show that the relations hold identically, since setting $x_\beta = E_\beta$ will produce a modular form in $M_2^2(\Gamma(T)) = 0$. Let us therefore first of all expand the result from (2.1.1) to include all cusps. We will skip the calculation, and just state that for a non-zero constant $\zeta = \in C^*$:

$$\begin{cases} (T\pi)^{-1} \cdot E_\infty(\infty) = \zeta + o(t^2) \\ (T\pi)^{-1} \cdot E_\beta(\infty) = t + o(t^2),\ \beta \neq \infty \\ (T\pi)^{-1} \cdot E_\infty((\alpha:1)) = t + o(t^2),\ \alpha \neq \infty \\ (T\pi)^{-1} \cdot E_\beta((\alpha:1)) = (\alpha+\beta)^{-1}t + o(t^2),\ 0 \neq \alpha+\beta < \infty \\ (T\pi)^{-1} \cdot E_\beta(\alpha:1)) = \zeta + o(t^2),\ \alpha+\beta = 0, \end{cases}$$

A little calculation using the transformation equations shows that

$$\begin{cases} Z_j(\infty) = o(t^2), j < q-1 \\ Z_{q-1}(\infty) = -t + o(t^2) \\ Z_q(\infty) = \zeta + o(t^2) \\ Z_j((\alpha:1)) = -\binom{q-2}{j-1}\alpha^{j-1}t + (-\alpha)^j\zeta + o(t^2),\ \alpha \neq \infty, j = 0...q. \end{cases}$$

From this, it is immediately clear that (3.1.3) is satisfied at ∞, up to t^2. For the other cusps, one finds:

$$\begin{aligned} &(Z_iZ_j - Z_{i-1}Z_{j+1})((\alpha:1)) \\ &= \alpha^{i+j-1}[(-1)^{j+1}(\tbinom{q-2}{i-1} + \tbinom{q-2}{i-2}) + (-1)^{i+1}(\tbinom{q-2}{j-1} + \tbinom{q-2}{j})] + o(t^2) \\ &= \alpha^{i+j-1}[(-1)^{j+1}\tbinom{q-1}{i-1} + (-1)^{i+1}\tbinom{q-1}{j}] + o(t^2) \\ &= \alpha^{i+j-1}[(-1)^{j+1}(-1)^{i-1} + (-1)^{i+1}(-1)^j] + o(t^2) \\ &= o(t^2), \end{aligned}$$

where we have used the fact that $\binom{q-1}{m} = (-1)^m \in \mathbf{F}_q$ (this follows inductively from $\binom{q}{m} = \binom{q-1}{m-1} + \binom{q-1}{m} = \delta_{m,0}$ for $m < q$). □

We can put these results together in the following theorem:

Theorem *The ring of (Drinfeld) modular forms of level T is isomorphic to*

$$C[x_\beta : \beta \in \mathbf{F}_q \cup \{\infty\}]/\langle f_{i,j} : 1 \leq i \leq j \leq q-1\rangle \cong M(\Gamma(T)),$$

where the isomorphism is given by sending $x_\beta \to E_\beta$, and the "relations" $f_{i,j}$ are given by

$$f_{i,j} := \sum_{\alpha,\beta \in \mathbf{F}_q} \alpha^{i-1}\beta^j(\alpha-\beta)x_\alpha x_\beta - \delta_{j,q-1} \cdot \sum_{\alpha \in \mathbf{F}_q} \alpha^{i-1}x_\alpha x_\infty,$$

where we again have adopted the convention that $0^0 = 1$.

(3.1.4) Here is one more way of finding the ideal of relations I: Since a general lattice function $e_\Lambda(\omega)$ is additive, its derivative $e'_\Lambda(\omega) = 1$. Hence by taking the logarithmic derivative of $e_\Lambda(\omega)$, we find that

$$\frac{1}{e_\Lambda(\omega)} = \sum_{\lambda\in\Lambda} \frac{1}{\omega - \lambda}, \quad \omega \notin \Lambda.$$

For our particular case, this implies that

$$E_u^{-1}(\omega) = T^{-1} e_{\Lambda_\omega}\left(\frac{u_1\omega + u_2}{T}\right), \quad \Lambda_\omega = \omega A \oplus A$$

is almost the T-torsion of the Drinfeld module ϕ^{Λ_ω} = the roots of $\phi_T^{\Lambda_\omega}$ ([3], (1.5.5)). This means that

$$\phi_T^{\Lambda_\omega} = TX + g(\omega)X^q + \Delta(\omega)X^{q^2} = TX \prod_{\beta\in U} (X^{q-1}T^{1-q}E_\beta^{q-1}(\omega) - 1)$$

holds as a formal identity in X for all $\omega \in \Omega$.

The idea is that any set of functions satisfying such an identity cannot be far away from $\{E_u\}$. This is formalised as follows: let $J' \subseteq C[x_1, \ldots, x_{q+1}, \gamma, \delta]$ be the ideal generated by the coefficients in the formal variable X of the identity

$$TX + \gamma X^q + \delta X^{q^2} = TX \prod_{\beta\in U} (X^{q-1}T^{1-q}x_\beta^{q-1} - 1).$$

Let $J \subseteq R$ be the ideal generated by the elements of J', but where the variables γ, δ are eliminated. Then

$$J = \langle \sigma_2(x_\beta^{q-1}), \ldots, \sigma_q(x_\beta^{q-1}) \rangle,$$

where σ_i is the symmetric function of weight i. This is certainly an ideal of relations between $E_\alpha(\omega), \forall\omega \in \Omega$, and it is not far away from I. As a matter of fact, for any permutation $\sigma \in Sym\{\mathbf{F}_q \cup \{\infty\}\}$ and any vector $\lambda = (\lambda_i) \in (\mathbf{F}_q^*)^{q+1}$, let $I^{\sigma,\lambda}$ be the kernel of

$$R \to E(\Gamma(T)) : x_\beta \mapsto \lambda_\beta E_{\sigma(\beta)}.$$

Then:

$$\bigcap I^{\sigma,\lambda} = J.$$

Proof. The proof is given in [2]. Let Z indicates the zero set of a homogeneous ideal in $\mathbf{P}^q$. First of all, one checks that the zero sets of both ideals are equal.

But the left ideal is radical, since $E(\Gamma(T))$ is reduced. On the other hand, R/J is a complete intersection (# equations = # variables - # relations). Hence it is reduced $\Longleftrightarrow$ the singularities of its zero locus have dimension zero ([4], 18.15). But one can find these singularities by calculating the Jacobian ideal of J. Via the formula for the Vandermonde determinant, one can even factor the Jacobian ideal, and one finds that all singularities belong to planes of the form $Z(x_\beta - \lambda x_{\beta'}), \beta \neq \beta', \lambda \in \mathbf{F}_q^*$. Hence there are only a finite number of singular points if the components of $Z(J)$ don't belong to such hyperplanes. But these components are $Z(I^{\sigma,\lambda})$, parametrized by $\bar{M}_{\Gamma(T)} \to \mathbf{P}^q : \omega \mapsto (\lambda_\beta E_{\sigma(\beta)}(\omega))$, and this never belongs to such a hyperplane, since otherwise $E_\beta(\omega) \equiv \lambda\lambda_\beta^{-1}\lambda_{\beta'}E_{\beta'}(\omega)$ for all ω, a contradiction since we have constructed the E_β as linearly independent functions. □

So to find I, it suffices to find one factor of the ideal J, which is explicitely given as a complete intersection. There exist algorithms for finding such factors, and one of them (Gianni, Trager, Zacharias [6]) was implemented by Hans Decker at Saarbrücken. Unfortunately, the complexity of the algorithm is rather bad, so we had to stick to the case $q = 3$, where a factor

$$I = \langle x_0x_1 + x_0x_2 + x_1x_2, x_0x_1 - x_0x_2 - x_1x_2, x_1x_2 + x_1x_\infty - x_2x_\infty \rangle$$

was found. This is compatible with the construction of I from (3.1.2).

(3.2) Generalizations The ad-hoc computation of (3.1.1) can of course always be made, within the limits of computation. The first case $q = 2, N = T(T+1)$ has 9 indeterminates, the ideal of relations will have approximately 15 generators, and these numbers grow rapidly. The calculation in (3.1.2) was based specifically on the rational normal curve, and this definitely doesn't hold any more if the level increases. Finally, calculation (3.1.3) continues to hold in general. This then produces the relations between $\{E_u\}$ up to permutation.

(4.1) Example We can calculate the number M of irreducible components of $Z(J)$: it equals the quotient of the degree of $Z(J)$ by the degree of $Z(I)$. The latter is q, the former can be calculated using Stanley's formula for the Hilbert function of a complete intersection ([11], (3.3)). One finds:

$$M = q^{-1}\deg(Z(J)) = q^{-1}\lim_{t\to 1}\prod_{i=2}^{q}\frac{1-t^{i(q-1)}}{1-t} = (q-1)! \cdot (q-1)^{q-1}.$$

This is a very big number, which might explain the complexity of factoring.

In the example $q = 3$, the group D_8 of permutations of the variables generated by a 4-cycle and a transposition, acts simply transitive on the 8 components of $Z(J)$. This implies that the orbit space $D_8\backslash Z(J)$ will be birational to one of the components $Z(I)$; in particular, the normalization of the cone over the orbit space will be isomorphic to Spec $M(\Gamma(T))$.

(4.2) Generalizations The calculation of the number of components immediately generalizes. If the genus of the modular curve is > 1, this calculation has an application to estimating its number of automorphisms. Namely, any element $\{\sigma, \lambda\}$ acting on the components of an ideal similar to J, which stabilizes such a component, induces an automorphism on $\bar{M}_\Gamma$.

5. *Remark* It is clear that all results in this paper hold, not only for level T, but for any level $N = aT + b, a \in \mathbf{F}_q^*, b \in \mathbf{F}_q$ of degree one.

6. *Addendum* D. Zagier has observed that the relations $\{f_{ij} = 0\}$ of the main theorem in section (3.1.3) are equivalent to

$$\{(\alpha - \beta)x_\alpha x_\beta + (x_\alpha - x_\beta)x_\infty = 0, \ \forall \alpha, \beta \in \mathbf{F}_q\},$$

and that these can be shown to hold by the following observation: for all $\beta \in \mathbf{F}_q$, the $\mathbf{F}_q$-additivity of e_{Λ_ω} gives

$$E_\beta^{-1}(\omega) = e_{\Lambda_\omega}(\frac{\omega + \beta}{T}) = e_{\Lambda_\omega}(\frac{\omega}{T}) + \beta e_{\Lambda_\omega}(\frac{1}{T}) = E_0^{-1}(\omega) + \beta E_\infty^{-1}(\omega),$$

hence $E_\beta^{-1} - \beta E_\infty^{-1}$ is independent of β.

References

[1] E. Arbarello, M. Cornalba, P.A. Griffiths, and J Harris, *Geometry of algebraic curves, volume 1*, Grundlehren der Math. Wiss., vol. 267, Springer, Berlin - Heidelberg - New York, 1985.

[2] G. Cornelissen, *Drinfeld modular forms of weight one*, `http://cage.rug.ac.be/~gc/`, preprint 1996.

[3] ———, *A survey of Drinfeld modular forms*, this volume.

[4] D. Eisenbud, *Commutative algebra*, Graduate Texts in Math., vol. 150, Springer, Berlin - Heidelberg - New York, 1995.

[5] E.-U. Gekeler, *Modulare Einheiten für Funktionenkörper*, J. Reine Angew. Math. **348** (1984), 94–115.

[6] P. Gianni, B. Trager, and G. Zacharias, *Gröbner bases and primary decomposition of polynomial ideals*, Computational aspects of Commutative Algebra (Robbiano, ed.), Academic Press, London - New York, 1988, pp. 15–33.

[7] J. Harris, *Algebraic geometry - a first course*, Graduate Texts in Math., vol. 133, Springer, Berlin - Heidelberg - New York, 1992.

[8] E. Hecke, *Theorie der Eisensteinschen Reihen höherer Stufe und ihre Anwendung auf Funktionentheorie und Arithmetik*, Abh. Math. Sem. Univ. Hamburg **5** (1927), 199–224, = Werke, 461-486.

[9] J. Igusa, *Theta functions*, Grundlehren der Math. Wiss., vol. 194, Springer, Berlin - Heidelberg - New York, 1972.

[10] D. Mumford, *Varieties defined by quadratic equations*, Proceedings of the Centro Internationale Matematico Estivo "Questions on algebraic varieties", Edizioni Cremonese, Roma, 1969, pp. 31–100.

[11] R. Stanley, *Hilbert functions of graded algebras*, Adv. in Math. **28** (1978), 57–83.

[12] J. T. Teitelbaum, *The Poisson kernel for Drinfeld modular curves*, J. Amer. Math. Soc. **4** (1991), no. 3, 491–511.

Explicit Reciprocity Law for a Rank One Drinfeld $\mathbf{F}_q[t]$-Module

Véronique MAUDUIT
Dépt. de Mathématiques, Université de Caen
14032 Caen Cedex, France
mauduit@matin.math.unicaen.fr

Introduction

We are looking for an analogue of the quadratic symbol and the reciprocity law for a rank one Drinfeld $\mathbf{F}_q[t]$-module. The most studied of all rank one Drinfeld modules is the "Carlitz module" $\gamma : \mathbf{F}_q[t] \to \mathbf{C}\{\tau\}$ defined by $\gamma_t = t + \tau$, where τ denotes the Frobenius automorphism. But, as Marius van der Put noticed, it is more natural to consider the rank one module ϕ defined by $\phi_t = t - t\,\tau$ for the following reason.

We will adopt the standard notations $K, K_\infty, \mathbf{C}$, i.e. K is the field of fractions of $\mathbf{F}_q[t]$; the completion of K at the valuation at ∞ is K_∞ and $\mathbf{C}$ denotes the completion of the algebraic closure of K_∞.

Using the Carlitz module γ, one can obtain the maximal abelian extension, called $K(\gamma)$ here, of K which is totally split at ∞ in the following way. Let $ker\,\gamma_f$ denote the kernel of the action of γ_f on $\mathbf{C}$ and let $K(ker\,\gamma_f)$ denote the field extension of K generated by $ker\,\gamma_f$. Then, the union $\cup_{f\in\mathbf{F}_q[t], f\neq 0} K(ker\,\gamma_f)$ is a subfield of $K_\infty(u)$, where u satisfies the equation $u^{q-1} = -t$. In fact u lies in the field $K(ker\,\gamma_t)$. Let σ denote a generator of the Galois group of $K_\infty(u)/K_\infty$. Then $K(\gamma)$ is equal to $(\cup_{f\in\mathbf{F}_q[t], f\neq 0} K(ker\,\gamma_f))^{<\sigma>}$. Moreover the Galois group of $K(\gamma)/K$ is equal to $\widehat{\mathbf{F}_q[t]}^{*}/\mathbf{F}_q^{*}$. (See [Ha] for more details).

Using similar notations for the Drinfeld module ϕ one finds that $K(\phi) := \cup_{f\in\mathbf{F}_q[t], f\neq 0} K(ker\,\phi_f)$ is contained in K_∞. This follows in fact from the analytic theory of Drinfeld modules. The Drinfeld module ϕ corresponds to a rank one lattice $\mathbf{F}_q[t]\eta \subset \mathbf{C}$ with $\eta \in K_\infty$. This implies that $ker\,\phi_f$ belongs to K_∞ for every $f \in \mathbf{F}_q[t], f \neq 0$. Thus $K(\phi)$ is an abelian extension of K which is completely split at ∞. Over the field $K(u)$ the Drinfeld modules γ and ϕ are equivalent since $u^{-1}\gamma_t u = \phi_t$. Therefore $K(\gamma)(u)$ and $K(\phi)(u)$ are equal. Taking invariants under σ one finds that $K(\gamma)$ equals $K(\phi)$.

We note that $ker\,\phi_t$ is equal to $\mathbf{F}_q$. As a consequence the subgroup $\mathbf{F}_q^*$ of $\widehat{\mathbf{F}_q[t]}^*$ acts trivially on $K(\phi)$. And again the Galois group of $K(\phi)/K$ is isomorphic to $\widehat{\mathbf{F}_q[t]}^*/\mathbf{F}_q^*$.

Another reason why the Drinfeld module ϕ seems more natural is that the rank one Drinfeld module corresponding to a cusp of the Drinfeld modular curve is over K_∞ isomorphic to ϕ and not to γ.

It seems at first that the price that one has to pay for using ϕ instead of γ is that ϕ has a bad reduction for the prime ideal (t). But a t-adic study of the situation (Newton polygon) shows that the anomaly of the t-residue symbol for the prime (t) is not greater for ϕ than for γ. In this paper we will give explicit formulas for the t-residue symbol. The relation of this symbol with a "Newton symbol" which enable us to give an elementary proof of the reciprocity law (as in case of the quadratic reciprocity law) is given. The latter turns out to be a residue of a certain rational functions at ∞. We note that the use of ϕ simplifies the expression of the reciprocity law.

1. The t-residue symbol.

We first recall some properties of the Carlitz module which will be used in the proofs.

Let $P \in \mathbf{F}_q[t]$ be a polynomial of degree n. We denote by $D^{(i)}(P)$ its i-th Galois derivative in t (cf.[He1] or [He2]). In the particular case of the Carlitz module, $D^{(i)}(P)$ can be defined by the following relation (1.1).

Proposition 1.1 *For any monic polynomial P we have the following equalities.*

$$\gamma_P(X) = \sum_{i=0}^{\infty} D^{(i)}(P) X^{q^i} \quad \textit{with } D^{(i)}(P) = 0 \textit{ if } i > \deg(P) \tag{1.1}$$

$$D^{(0)}(P) = P \quad \textit{and} \quad D^{(\deg(P))}(P) = 1$$

$$D^{(i)}(t\,P) = t\,D^{(i)}(P) + \left(D^{(i-1)}(P)\right)^q. \quad (\textit{Leibniz's formula})$$

If P is irreductible then P divides $D^{(i)}(P)$ for $0 \le i < \deg(P)$ (cf. [Ha] prop.2.4).

The divisibility of the $D^{(i)}$'s by P, when P is irreducible, can be seen as an analog of Fermat's little theorem. Indeed we then have:

$$\gamma_P(A) \equiv A^{q^{\deg(P)}} \equiv A \pmod{P}.$$

Let us now consider the module ϕ. As noted before, ϕ is isomorphic to γ over the field $K(u)$ with $u^{q-1} = -t$. Indeed, $u^{-1}\gamma_t u = \phi_t$ and thus for any polynomial $P \in \mathbf{F}_q[t]$ we have

$$(-t)^{\frac{1}{q-1}}\phi_P = \gamma_P\,(-t)^{\frac{1}{q-1}},$$

where we have adopted the notation $(-t)^{\frac{1}{q-1}}$ for u. Let us express ϕ in terms of Galois derivatives.

Proposition 1.2 *Let $P = \sum_0^n p_{n-i}t^i$ be a monic polynomial of degree n of $\mathbf{F}_q[t]$ and $A \in \mathbf{F}_q[t]$. We have:*

$$\phi_P = \sum_{i=0}^{n}(-t)^{\frac{q^i-1}{q-1}} D^{(i)}(P)\tau^i$$

moreover if P is irreducible we have

$$\phi_P(A) \equiv p_n\, A \pmod{P}.$$

Proof. For the equality, using the isomorphism between ϕ_P and γ_P we get:

$$\begin{aligned}
\phi_P &= (-t)^{\frac{-1}{q-1}}\gamma_P(-t)^{\frac{1}{q-1}}\\
&= (-t)^{\frac{-1}{q-1}}\sum_{i=0}^{n} D^{(i)}(P)\tau^i((-t)^{\frac{1}{q-1}})\\
&= (-t)^{\frac{-1}{q-1}}\sum_{i=0}^{n} D^{(i)}(P)(-t)^{\frac{q^i}{q-1}}\tau^i\\
&= \sum_{i=0}^{n}(-t)^{\frac{q^i-1}{q-1}} D^{(i)}(P)\tau^i.
\end{aligned}$$

When P is a monic prime and $i < n$ all the $D^{(i)}(P)$'s are divisible by P and $D^{(n)}(P) = 1$ thus:

$$\begin{aligned}
\phi_P &\equiv (-t)^{\frac{q^n-1}{q-1}}\tau^n \equiv (-t)^{q^{n-1}+\dots+q+1}\tau^n \pmod{P}\\
&\equiv (-1)^n \mathrm{N}_{\mathbf{F}_{q^n}/\mathbf{F}_q}(t)\tau^n \equiv (-1)^{2n}p_n\tau^n \pmod{P}\\
&\equiv p_n\tau^n \pmod{P}
\end{aligned}$$

and $$\phi_P(A) \equiv p_n\, A^{q^n} \equiv p_n\, A \pmod{P}$$ □

The last formula is an analogue of Fermat's little theorem for the module ϕ.

After this brief introduction of ϕ we are looking for a definition of an analogue of the quadratic residue symbol. In the classical case of the quadratic reciprocity law, the quadratic residue symbol can be defined in two different ways. The historical definition is Euler's criterion. A more general one, involving the Artin symbol, is given by Tate and Serre in [CF] (ex.1,p.348). Let us recall them in the classical setting.

Definition 1.3 Let a be a non zero rational integer and p a positive odd prime such that $(a,p) = 1$. The quadratic symbol of a at p denoted by $\left(\frac{a}{p}\right)$ is the element of $\{-1, 1\}$ such that:

$$\left(\frac{a}{p}\right) \equiv a^{\frac{p-1}{2}} \pmod{p}.$$

Definition 1.4 Let a be a non-zero rational integer and p a positive odd rational integer such that $(a,p) = 1$. The quadratic residue symbol $\left(\frac{a}{p}\right)$ is defined by:

$$(\sqrt{a})^{F_{\mathbf{Q}(\sqrt{a})/\mathbf{Q}}(p)} = \left(\frac{a}{p}\right)\sqrt{a}$$

where $F_{\mathbf{Q}(\sqrt{a})/\mathbf{Q}}(p)$ is the Artin symbol of p in $\mathbf{Q}(\sqrt{a})/\mathbf{Q}$.

So for two polynomials A and P of $\mathbf{F}_q[t]$, satisfying certain conditions, we want to define a symbol $\left(\frac{A}{P}\right)$ which can be seen as an analog of these two previous definitions.

We have seen in Proposition 1.2 that the analog of Fermat's little theorem is, for a monic prime, $\phi_P(A) \equiv p_n A \pmod{P}$ which is equivalent to $\phi_{P-p_n}(A) \equiv 0 \pmod{P}$. Comparing with definition 1.3 we see that, in our context, t will play the part of 2.

And now to follow definition 1.4, we have to check that the Artin symbol is well defined in the extension k_A/k where k_A denotes the splitting field of $f(t) := \phi_t(X) - A = -tX^q + tX - A$ over $k := \mathbf{F}_q(t)$. Let us recall the definition of the Artin symbol.

Definition 1.5 (cf.[S]) Let L/k be a Galois extension, A a Dedekind ring with quotient field k and B its integral closure in L. Let $\mathfrak{P}$ be a non zero prime ideal of B and $\mathfrak{p}=\mathfrak{P}\cap A$. If $\mathfrak{P}$ is unramified in L/k and if $A/\mathfrak{p}$ is a finite field with r elements, then the decomposition group $D_{\mathfrak{P}}$ can be identified with the Galois group $G(\bar{L}/\bar{k})$ of the residue extension $\bar{L}/\bar{k}$ which is cyclic with generator $x \mapsto x^r$. The element $F(\mathfrak{P})$ which corresponds to this generator satisfies

$$b^{F(\mathfrak{P})} \equiv b^r \pmod{\mathfrak{P}} \qquad \text{for any } b \in B$$

and is called the Frobenius automorphism.

If the extension L/k is Abelian, $F(\mathfrak{P})$ does not depend on the choice of $\mathfrak{P}$ over $\mathfrak{p}$. The Frobenius is then denoted by $F(\mathfrak{p})$ and called the Artin symbol. The Artin symbol is extended by linearity to any ideal $\mathfrak{a}$ of A such that $\mathfrak{a}$ contains no ramified prime.

As the extension k_A/k is a generalized Artin-Schreier extension, it is thus Galois and Abelian.

If the t-valuation of A is strictly positive, the discriminant equals 1 and the extension ramifies only at infinity. The extension ramifies also at zero if the t-valuation of A equals zero.

Thus the Artin symbol is defined for any monic non zero polynomial $P \in \mathbf{F}_q[t]$ different from t and also for $P = t$ if $v_t(A) > 0$. We will denote it by $F_A(P)$ to recall that it depends on the choice of A for the extension k_A/k.

Let us describe the roots α of f, i.e. the analog of the set $\{\sqrt{a}, -\sqrt{a}\}$. In the sequel we will adopt the notation X^P for the polynomial $\phi_P(X)$.

Lemma 1.6 *The set of the roots of* $X^t = -t\,X^q + t\,X$ *is* $\mathbf{F}_q$.

Lemma 1.7 *Let* α *be a root of* $f(X) = X^t - A = -t\,X^q + t\,X - A$, *then the other roots of this polynomial are the distinct points of the affine line through* α *parallel to the line of the roots of* X^t.

Proof. The points of this affine line are roots of f. As f has only q roots, all its roots are points of the affine line. □

Keeping in mind that we want an analogue of Definition 1.4 we now show how the Artin symbol acts on integral elements of k_A/k.

Lemma 1.8 *Let* A *be a polynomial in* $\mathbf{F}_q[t]$ *and* P *be a monic irreducible polynomial of* $\mathbf{F}_q[t]$ *different from* t. *Then for any integral* β *in* k_A/k, *we have:*

$$\beta^P \equiv p_n\, \beta^{F_A(P)} \pmod{P}.$$

Proof. As the cardinality of the residue class field of $\mathbf{F}_q(t)$ with respect to P is equal to $q^{\deg(P)}$, we have by definition:

$$\beta^{F_A(P)} \equiv \beta^{q^{\deg(P)}} \pmod P.$$

But by Proposition 1.2, we have:

$$\beta^P \equiv p_n\, \beta^{q^{\deg(P)}} \pmod P$$

Thus $$\beta^P \equiv p_n\, \beta^{F_A(P)} \pmod P. \qquad \square$$

Proposition and Definition 1.9 *Let $A \in \mathbf{F}_q[t]$ and let $P \in \mathbf{F}_q[t]$ be a monic irreducible polynomial different from t. Then there exists $\epsilon \in \mathbf{F}_q$ such that the following conditions are equivalent:*

a) $$A^{\frac{P-p_n}{t}} \equiv p_n\, \epsilon \pmod P$$

b) For any root α of $f(X) = X^t - A = -t\, X^q + t\, X - A$, we have:

$$\alpha^{F_A(P)} - \alpha = \epsilon \pmod P.$$

Accordingly, we define the t-residue symbol $\left(\frac{A}{P}\right)$ for irreducible P to be the common value $\epsilon \in \mathbf{F}_q$ of these two previous expressions.

If $P = t$ and $v_t(A) > 0$, we let $\left(\frac{A}{t}\right) := \alpha^{F_A(t)} - \alpha$.

If $P = P_1^{\nu_1} \ldots P_s^{\nu_s}$ is the product of irreducible polynomials P_i we define

$$\left(\frac{A}{P}\right) := \sum_{i=1}^{s} \nu_i \left(\frac{A}{P_i}\right).$$

Remarks 1.10 1) Condition a) is an analogue of the definition of the quadratic symbol by means of the Euler criterion, whereas b) is an analogue of the definition of the quadratic symbol by means of the Artin symbol.

2) Let $A \in \mathbf{F}_q[t]$ and P be a monic irreducible polynomial of $\mathbf{F}_q[t]$ different from t. Then $\left(\frac{A}{P}\right) = 0$ if and only if the congruence $A \equiv X^t \pmod P$ has a solution B with $B \in \mathbf{F}_q[t]$.

Indeed for any $B \in \mathbf{F}_q[t]$, we have $(B^t)^{(P-p_n)/t} = B^{P-p_n} \equiv 0 \pmod{P}$. If we consider the reduced module $\bar{\phi}_t : \mathbf{F}_q[t]/(P) \to \mathbf{F}_q[t]/(P)$. We have

$$\#\mathrm{Im}\,\bar{\phi}_t = \#\mathbf{F}_q[t]/(P) \Big/ \#\mathrm{Ker}\,\bar{\phi}_t = q^{\deg(P)-1}$$

by lemma 1.6. As the equation $\bar{\phi}_{(P-p_n)/t}(X) = 0$ has $q^{\deg(P)-1}$ solutions in $\mathbf{F}_q[t]/(P)$ all the solutions are the elements of $\mathrm{Im}\,\bar{\phi}_t$. Thus $A^{(P-p_n)/t} \equiv 0 \pmod{P}$ if and only if there exists $B \in \mathbf{F}_q[t]$ such that $A \equiv B^t \pmod{P}$.

3) The definition of the t-residue symbol when P is reducible is the natural one. Indeed if P and Q denote two monic primes of $\mathbf{F}_q[t]$ different from t then

$$\begin{aligned}\alpha^{F_A(PQ)} - \alpha &= \alpha^{F_A(P)F_A(Q)} - \alpha \\ &= \left(\alpha^{F_A(P)} - \alpha\right)^{F_A(Q)} + \alpha^{F_A(Q)} - \alpha \\ &= \alpha^{F_A(P)} - \alpha + \alpha^{F_A(Q)} - \alpha = \left(\frac{A}{P}\right) + \left(\frac{A}{Q}\right)\end{aligned}$$

as $\alpha^{F_A(P)} - \alpha \in \mathbf{F}_q$.

Proof of Proposition 1.9 As $P \neq t$, $p_n \neq 0$. By Lemma 1.8

$$\alpha^{P-p_n} \equiv \alpha^P - p_n\alpha \equiv p_n(\alpha^{F_A(P)} - \alpha) \pmod{P}.$$

Thus, as $v_t(P - p_n) \geq 1$ and as α is a root of $X^t - A$, we have:

$$A^{\frac{P-p_n}{t}} = (\alpha^t)^{\frac{P-p_n}{t}} = \alpha^{P-p_n} \equiv p_n(\alpha^{F_A(P)} - \alpha) \pmod{P}.$$

It remains to prove that this common value is in $\mathbf{F}_q$.

As the value of the Artin symbol is an element of the Galois group of the extension k_A/k, the image of α by $F_A(P)$ gives rise to another root of f. Thus

$$(\alpha^{F_A(P)} - \alpha)^t = A - A = 0$$

and $(\alpha^{F_A(P)} - \alpha)$ is a root of X^t and then, by Lemma 1.6, is in $\mathbf{F}_q$.

If $P = t$, $\alpha^{F_A(t)} - \alpha \in \mathbf{F}_q$ for the same reason. □

We now give some properties of the t-residue symbol. The first one refers to the notion of "bilinearity" and the others exhibit similarities between this symbol and the usual quadratic residue symbol.

Proposition 1.11 *The t-residue symbol is "bilinear", i.e. for A, $B \in \mathbf{F}_q[t]$, P, Q, R three monic polynomials of $\mathbf{F}_q[t]$, we have:*

$$\left(\frac{A}{RQ}\right) = \left(\frac{A}{R}\right) + \left(\frac{A}{Q}\right) \tag{1.2}$$

$$\left(\frac{A+B}{P}\right) = \left(\frac{A}{P}\right) + \left(\frac{B}{P}\right) \tag{1.3}$$

Proof. The first formula is the definition of the t-residue symbol when $P = RQ$ is reducible.

It suffices to prove the second formula when P is irreducible by (1.2). If $P \neq t$, this is another way of writing $(A+B)^{(P-p_n)/t} = A^{(P-p_n)/t} + B^{(P-p_n)/t}$.

If $P = t$ we let α (resp. β) be a root of $X^t - A$ (resp. $X^t - B$). Then $\alpha+\beta$ is a root of $X^t - (A+B)$. As k_A/k (resp. k_B/k, resp. k_{A+B}/k) is Galois, the Artin symbol $F(P)$ of P in $k(\alpha,\beta)/k$ restricted to k_A (resp. k_B, resp. k_{A+B}) is $F_A(P)$ (resp. $F_B(P)$, resp. $F_{A+B}(P)$) and the equality follows.

Corollary 1.12 *Let A, $B \in \mathbf{F}_q[t]$ and let $P \in \mathbf{F}_q[t]$ be a monic polynomial such that $v_t(P) = 0$. For $P = \prod P_i^{\nu_i}$ with P_i a monic prime we set* $\mathrm{rad}(P) = \prod P_i$. *If $A \equiv B \pmod{\mathrm{rad}(P)}$ then*

$$\left(\frac{A}{P}\right) = \left(\frac{B}{P}\right).$$

Proof. Suppose P irreducible and write $B = A + CP$. Then by (1.3) and Definition 1.9

$$\left(\frac{B}{P}\right) = \left(\frac{A+CP}{P}\right) = \left(\frac{A}{P}\right) + \left(\frac{CP}{P}\right)$$
$$\equiv p_n^{-1} A^{\frac{(P-p_n)}{t}} + p_n^{-1}(CP)^{\frac{(P-p_n)}{t}} \equiv p_n^{-1} A^{\frac{(P-p_n)}{t}} \equiv \left(\frac{A}{P}\right) \pmod{P}.$$

The result then follows by (1.2).

Remark 1.13 If $P = t$ we have $\left(\frac{0}{t}\right) = 0$ and $\left(\frac{t}{t}\right) \neq 0$. Indeed in the latter case k_t/k is a scalar extension as k_t is the splitting field of $X^q - X + 1$ and thus $F_t(t) \neq \mathrm{id}$.

Corollary 1.14 *For any integer m and any monic polynomial $P \in \mathbf{F}_q[t]$ we have:*

$$\left(\frac{t^{qm+1}}{P}\right) = \left(\frac{t^{m+1}}{P}\right)$$

Proof. As $-t^{qm+1} + t^{m+1} = (t^m)^t$ we have the result by remark 1.10 2). □

It is rather difficult and long to calculate the t-residue symbol by means of its definition. In order to get rid of this problem we will introduce in the next section the Newton symbol. The latter will be easy computable and, as we will see in section 3, the t-residue symbol can be expressed in an elementary way in terms of this symbol.

2. The Newton symbol.

In this part the results will be given for any commutative *ring* F. But in the following sections, they will be used only in the particular case $F = \mathbf{F}_q$. Most of the results will not be necessary in section 3 but they show how the Newton symbol works.

We denote by $F[t]$ the ring of polynomials with coefficients in F and by $M(F) \subset F[t]$ the multiplicative monoid of monic polynomials of $F[t]$. Given $P = t^n + p_1 t^{n-1} + \ldots + p_n \in M(F) \setminus \{1\}$, we consider the sequence $(s_0, s_1, s_2, \ldots)$ of the Newton sums for P with the usual convention $s_0 = \deg(P)$. We recall that this sequence can be defined inductively via the following formulas :

$$(N)\quad \begin{cases} s_0 = \deg(P) = n & \\ s_i + p_1 s_{i-1} + \ldots + p_{i-1} s_1 + i p_i = 0 & \text{if} \quad i \le \deg(P) = n \\ s_i + p_1 s_{i-1} + \ldots + p_{n-1} s_{i-n+1} + p_n s_{i-n} = 0 & \text{if} \quad i > \deg(P) = n \end{cases}$$

If $P = 1$, we take the sequence $(0, 0, \ldots)$.

Definition 2.1 Let $P \in M(F)$ and $A = \sum_0^d a_{d-i} t^i \in F[t]$. We say that the Newton symbol of A at P is:

$$\left(\frac{A}{P}\right)_N := a_0 s_d + a_1 s_{d-1} + \ldots + a_d s_0 \in F.$$

Remark 2.2 If $A = t^d$ and F is a field, then

$$\left(\frac{t^d}{P}\right)_N = \sum_{\rho \in \text{roots of } P} \rho^d = s_d$$

the d-th Newton sum of the roots of P in an algebraic closure of F.

Proposition 2.3 *For $P \in M(F)$, the F-linear form $(\frac{\cdot}{P})_N$ factors through $F[t]/(P)$.*

Proof. Thanks to the induction formulas we see that for any integer i we have $(\frac{t^i P}{P})_N = 0$. □

Proposition 2.4 *The Newton map: $P \mapsto (\frac{\cdot}{P})_N$ is a morphism from the monoid $(M(F), \times)$ to the dual of the F-module $F[t]$ considered as an abelian group.*

Proof. By linearity it suffices to see that $(\frac{t^i}{P_1 P_2})_N = (\frac{t^i}{P_1})_N + (\frac{t^i}{P_2})_N$. □

Remark 2.5 Don Zagier noticed that if the coefficients are in a field, these two results are immediate consequences of the equality

$$\left(\frac{A}{P}\right)_N = \mathrm{Res}(-A(t)P'(t)/P(t), \infty).$$

Corollary 2.6 *Let $P = P_1^{\nu_1} \ldots P_s^{\nu_s}$ and set $\mathrm{rad}(P) = P_1 \ldots P_s$ where $P_i \in M(F)$ are distinct irreducible primes for $1 \leq i \leq s$. Then we have:*

$$\left(\frac{\cdot}{P}\right)_N = \nu_1 \left(\frac{\cdot}{P_1}\right)_N + \ldots + \nu_s \left(\frac{\cdot}{P_s}\right)_N$$

$$(\, A \equiv B \pmod{\mathrm{rad}(P)}\,) \Longrightarrow (\, \left(\frac{A}{P}\right)_N = \left(\frac{B}{P}\right)_N \,).$$

The following proposition implies that the Newton symbol satisfies a kind of Artin reciprocity law. A corollary will be an analogue of the quadratic reciprocity law.

Proposition 2.7 *Let $A = \sum a_{d-i} t^i \in F[t]$ be a polynomial of degree d, and let $P, Q \in M(F)$. If P and Q are such that*

$$\frac{P}{Q} - 1 \equiv 0 \pmod{\frac{1}{t^{d+1}}}$$

then

$$\left(\frac{A}{P}\right)_N - \left(\frac{A}{Q}\right)_N = 0.$$

Proof. The congruence $(P/Q) - 1 \equiv 0 \pmod{t^{-(d+1)}}$ is equivalent to

$$P = t^n + p_1 t^{n-1} + \ldots + p_d t^{n-d} + p_{d+1} t^{n-d-1} + \ldots + p_n$$

and
$$Q = t^n + p_1 t^{n-1} + \ldots + p_d t^{n-d} + q_{d+1} t^{n-d-1} + \ldots + q_n$$

and implies then by formulas (N) that $(\frac{t^i}{P})_N = (\frac{t^i}{Q})_N$ for $0 \leq i \leq d$. We then have:

$$\left(\frac{A}{P}\right)_N = \sum_{i=0}^{d} a_{d-i} \left(\frac{t^i}{P}\right)_N = \sum_{i=0}^{d} a_{d-i} \left(\frac{t^i}{Q}\right)_N = \left(\frac{A}{Q}\right)_N . \qquad \square$$

Remark 2.8 If $\deg(P) = m > \deg(Q) = n$ Proposition 2.7 states that $(P/t^{m-n}Q) - 1 \equiv 0 \pmod{t^{-(d+1)}}$ implies

$$\left(\frac{A}{P}\right)_N = \left(\frac{A}{t^{m-n}Q}\right)_N$$

i.e.
$$\left(\frac{A}{P}\right)_N = \left(\frac{A}{Q}\right)_N + (m-n)\, a_d$$

by Proposition 2.4.

Corollary 2.9 *Let $P, Q \in M(F)$ and $r \in \mathbf{N}$ such that* $\deg(P) = \deg(Q) = 2r$ *or* $2r + 1$ *and* $p_1 = q_1, \ldots, p_r = q_r$ *then*

$$\left(\frac{P}{Q}\right)_N + \left(\frac{Q}{P}\right)_N = 0.$$

Proof. Take $A = P - Q$ in Proposition 2.7. $\square$

Remark 2.10 All these previous results except proposition 2.4 and its corollary are still valid if we consider the Wronski sums, defined by the formulas :

$$(W) \begin{cases} s_0 = 1 & \\ s_i + p_1 s_{i-1} + \ldots + p_{i-1} s_1 + p_i = 0 & \text{if} \quad i \leq \deg(P) \\ s_i + p_1 s_{i-1} + \ldots + p_{n-1} s_{i-n+1} + p_n s_{i-n} = 0 & \text{if} \quad i > \deg(P) = n \end{cases}$$

or any sequence of integers s_i defined by formulas of type

$$\begin{cases} s_0 = \lambda_n & \\ s_i + p_1 s_{i-1} + \ldots + p_{i-1} s_1 + \lambda_i p_i = 0 & \text{if} \quad i \leq \deg(P) \text{ and } \lambda_i \in F \\ s_i + p_1 s_{i-1} + \ldots + p_{n-1} s_{i-n+1} + p_n s_{i-n} = 0 & \text{if} \quad i > \deg(P) = n . \end{cases}$$

3. Expression of the t-residue symbol in terms of the Newton symbol.

We now come back to the t-residue symbol to give its expression in terms of the Newton symbol and the reciprocity law it satisfies.

Proposition 3.1 *For any monic and irreducible $P = \sum_0^n p_{n-i} t^i \in \mathbf{F}_q[t]$ different from t we have:*

$$\left(\frac{1}{P}\right) = p_n^{-1}\, p_{n-1}.$$

Proof. As $1^{t^i} = 0$ for $i \geq 1$,

$$\left(\frac{1}{P}\right) = p_n^{-1}\, 1^{\frac{P-p_n}{t}} = p_n^{-1}\, 1^{p_{n-1}} = p_n^{-1} p_{n-1}. \qquad \square$$

If the degree of A is greater than or equal to 1 the result is less easy. We will need the following lemma to prove the proposition in which the link between the t-residue symbol and the Newton symbol is established.

Lemma 3.2 *Let P be a monic irreducible polynomial of $\mathbf{F}_q[t]$ different from t. For $m \in \mathbf{N}$ and $i < \deg(P)$, we have :*

$$D^{(i)}\left(\tfrac{P-p_n}{t}\right) t^{q^i m} \equiv (-1)^{i+1}\; p_n\; t^{q^i m - \frac{q^{i+1}-1}{q-1}} \pmod{P}\;.$$

Proof. We proceed by induction on i. For the case $i = 0$ we have:

$$D^{(0)}\left(\tfrac{P-p_n}{t}\right) t^m = (P - p_n)\, t^{m-1} \equiv -p_n\, t^{m-1} \pmod{P}.$$

Assuming the formula holds for $i - 1$, we will prove it for $1 \leq i < \deg(P)$. By Leibniz's formula we have:

$$\begin{aligned} D^{(i)}\left(\tfrac{P-p_n}{t}\right) t^{q^i m} &= \Big(D^{(i)}(P) - D^{(i)}(p_n)\Big)\; t^{q^i m-1} - \Big(D^{(i-1)}\left(\tfrac{P-p_n}{t}\right)\Big)^q t^{q^i m-1} \\ &\equiv -\, \Big(D^{(i-1)}\left(\tfrac{P-p_n}{t}\right) t^{q^{i-1}m}\Big)^q\; t^{-1} \pmod{P} \end{aligned}$$

as $D^{(i)}(P) \equiv 0 \pmod{P}$ for $0 \leq i < \deg(P)$ and $D^{(i)}(p_n) = p_n\, D^{(i)}(1) = 0$ for $i \geq 1$. So, by induction hypothesis:

$$\begin{aligned} D^{(i)}\left(\tfrac{P-p_n}{t}\right) t^{q^i m} &\equiv -\, \Big((-1)^i\; p_n\; t^{q^{i-1}m - \frac{q^i-1}{q-1}}\Big)^q t^{-1} \pmod{P} \\ &\equiv (-1)^{i+1}\; p_n\; t^{q^i m - (q^i + q^{i-1} \ldots + 1)} \pmod{P} \\ &\equiv (-1)^{i+1}\; p_n\; t^{q^i m - \frac{q^{i+1}-1}{q-1}} \pmod{P} \qquad \square \end{aligned}$$

Proposition 3.3 *Let $P \in \mathbf{F}_q[t]$ be a monic irreducible polynomial, then for any $m \in \mathbf{N}^*$ we have:*

$$\left(\frac{t^m}{P}\right) = -\left(\frac{t^{m-1}}{P}\right)_N$$

in particular $\left(\frac{t}{P}\right) = -\deg(P)$ *and* $\left(\frac{t^2}{P}\right) = p_1$.

Proof. Suppose $P \neq t$. As P is irreducible, Definition 1.9 gives:

$$\begin{aligned}
\left(\frac{t^m}{P}\right) &\equiv p_n^{-1}(t^m)^{\frac{P-p_n}{t}} \pmod P \\
&\equiv p_n^{-1}\sum_{i=0}^{n-1}(-t)^{\frac{q^i-1}{q-1}}\, D^{(i)}\left(\tfrac{P-p_n}{t}\right) t^{q^i m} \pmod P \quad \text{by Proposition 1.2} \\
&\equiv p_n^{-1}\sum_{i=0}^{n-1}(-t)^{\frac{q^i-1}{q-1}}\, (-1)^{i+1}\, p_n\, t^{q^i m - \frac{q^{i+1}-1}{q-1}} \pmod P \quad \text{by Lemma 3.2} \\
&\equiv \sum_{i=1}^{n}(-1)^i\,(-1)^{i+1}\, t^{q^i m}\, t^{\frac{q^i-1}{q-1}-\frac{q^{i+1}-1}{q-1}} \pmod P \\
&\equiv -\sum_{i=0}^{n-1} t^{q^i m - q^i} \equiv -\sum_{i=0}^{n-1}(t^{m-1})^{q^i} \pmod P \\
&\equiv -\left(\frac{t^{m-1}}{P}\right)_N \pmod P.
\end{aligned}$$

As the elements in each side of the congruence are in $\mathbf{F}_q$ we obtain an equality.

If $P = t$, we have to distinguish two different cases.

$*$ If $m \geq 2$ then t splits completely in k_{t^m}/k as $X^q - X + t^{m-1} \equiv X^q - X \pmod t$. Thus $\left(\frac{t^m}{t}\right) = 0$. In the other side $\left(\frac{t^l}{t}\right)_N = 0$ for $l \geq 1$. Thus $\left(\frac{t^m}{t}\right) = \left(\frac{t^{m-1}}{t}\right)_N = 0$.

$*$ If $m = 1$, then k_A/k is the splitting field of $-tX^q + tX - t$ i.e. of $X^q - X + 1$. By definition of the Frobenius automorphism, $\alpha^{F_t(t)} \equiv \alpha^q \pmod t$. We deduce that $\alpha^{F_t(t)} - \alpha \equiv \alpha^q - \alpha \equiv -1 \pmod t$. Thus $\left(\frac{t}{t}\right) = -\left(\frac{1}{t}\right)_N = -1$. □

This proposition is a justification a posteriori of our definition of the t-residue symbol. Since $\left(\frac{1}{t}\right)$ cannot be defined in a natural way we arbitrarily let $\left(\frac{1}{t}\right) = 0$. The expression of $\left(\frac{A}{t}\right)$ for any A, then follows by bilinearity.

If we extend the t-residue symbol to any quotient of monic polynomials letting $\left(\frac{A}{PQ^{-1}}\right) = \left(\frac{A}{P}\right) - \left(\frac{A}{Q}\right)$ and if t divides A, the admissibility of the Artin symbol is a consequence of the following proposition (for the definition of admissible cf.[CF] chap.VII Def.3.7).

Proposition 3.4 *Let $A \in \mathbf{F}_q[t]$ and P, Q be two monic polynomials of $\mathbf{F}_q[t]$ such that:*

$$A = \sum_{i=0}^{d} a_{d-i} t^i \quad \textit{with} \quad \deg(A) = d$$

$$P = \sum_{i=0}^{m} p_{m-i} t^i \quad \textit{with} \quad \deg(P) = m \quad \textit{and} \quad v_t(P) = g$$

$$Q = \sum_{i=0}^{n} q_{n-i} t^i \quad \textit{with} \quad \deg(Q) = n \quad \textit{and} \quad v_t(Q) = h$$

$$p_i = q_i \ \textit{for} \ 0 \le i \le d-1$$

with the convention $p_i = 0$ if $i > m$ and $q_i = 0$ if $i > n$. Then

$$\left(\frac{A}{P}\right) - \left(\frac{A}{Q}\right) = a_{d-1}\,(n-m) + a_d\,(p_{m-g}^{-1}\, p_{m-g-1} - q_{n-h}^{-1}\, q_{n-h-1}) \ .$$

Proof. Suppose first that $v_t(P) = v_t(Q) = 0$. By "bilinearity" of the symbol and Propositions 3.3 and 3.1, we have the following equalities:

$$\left(\frac{A}{P}\right) - \left(\frac{A}{Q}\right) = \sum_{i=0}^{d} a_{d-i} \left[\left(\frac{t^i}{P}\right) - \left(\frac{t^i}{Q}\right)\right]$$

$$= \sum_{i=1}^{d} a_{d-i} \left[-\left(\frac{t^{i-1}}{P}\right)_N + \left(\frac{t^{i-1}}{Q}\right)_N\right] + a_d\,(p_m^{-1}\, p_{m-1} - q_n^{-1}\, q_{n-1})$$

$$(*) \qquad = a_{d-1}\,(n-m) + a_d(p_m^{-1}\, p_{m-1} - q_n^{-1}\, q_{n-1})$$

by Remark 2.8.

If $v_t(P) = g$ and $v_t(Q) = h$ we let $P = t^g P'$ and $Q = t^h Q'$

$$\left(\frac{A}{P}\right) - \left(\frac{A}{Q}\right) = \left(\frac{A}{t^g}\right) + \left(\frac{A}{P'}\right) - \left(\frac{A}{t^h}\right) - \left(\frac{A}{Q'}\right)$$

$$= (g-h)\left(\frac{A}{t}\right) + \left(\frac{A}{P'}\right) - \left(\frac{A}{Q'}\right)$$

$$= a_{d-1}(h-g) + a_{d-1}((n-h) - (m-g)) + a_d(p_{m-g}^{-1}\, p_{m-g-1} - q_{n-h}^{-1}\, q_{n-h-1})$$

$$= a_{d-1}(n-m) + a_d(p_{m-g}^{-1}\, p_{m-g-1} - q_{n-h}^{-1}\, q_{n-h-1})$$

using $(*)$ as $v_t(P') = v_t(Q') = 0$.

Remark 3.5 We can always suppose $\deg(A) < \mathrm{Max}(\deg(P), \deg(Q))$ by Corollary 1.12 and the result can be improved if $\deg(A) = mq+1$ by Corollary 1.14.

Corollary 3.6 (Reciprocity Law) *Let P, Q be two monic polynomials of $\mathbf{F}_q[t]$ and $r \in \mathbf{N}$ such that:*

$$P = \sum_{i=0}^{m} p_{m-i}t^i \quad with \quad \deg(P) = m \quad and \quad v_t(P) = g$$

$$Q = \sum_{i=0}^{n} q_{n-i}t^i \quad with \quad \deg(Q) = n \quad and \quad v_t(Q) = h$$

$$\deg(P - Q) \leq r + 1$$

$$p_i = q_i \ for \ 0 \leq i \leq r$$

with the convention $p_i = 0$ if $i > m$ and $q_i = 0$ if $i > n$. Then

$$\left(\frac{P}{Q}\right) + \left(\frac{Q}{P}\right) = (q_{n-1} - p_{m-1})\,(n-m) + (q_n - p_m)\,(p_{m-g}^{-1}\, p_{m-g-1} - q_{n-h}^{-1}\, q_{n-h-1}) \,.$$

Proof. Take $A = Q - P$ in Proposition 3.4. □

Remarks 3.7 1) If $\deg(P) = \deg(Q)$ and $v_t(P) = v_t(Q) = 0$ the reciprocity law takes exactly the form of the usual one.

2) If $\deg(P) = \deg(Q) = 2r + 2$, then the conditions $\deg(P + Q) \leq r + 1$ and $p_i = q_i$ for $0 \leq i \leq r$ are equivalent. If $\deg(P) = \deg(Q) > 2r + 2$ then $\deg(P + Q) \leq r + 1$ implies $p_i = q_i$ for $0 \leq i \leq r$ and, in the contrary, if $\deg(P) = \deg(Q) < 2r+2$ then $p_i = q_i$ for $0 \leq i \leq r$ implies $\deg(P+Q) \leq r+1$.

What would have happened if we had taken the Carlitz module γ ? The results would not have been very different for $\mathbf{F}_2$ or more generally for a finite field of characteristic 2. However in the case of an odd characteristic the results are much more complicated and can be found in F. Shultheis' thesis (cf.[Sch1] or [Sch2]) .

References:

[CF] J.W.S. Cassels - A. Fröhlich, *Algebraic Number Theory*, Academic Press, 1967.

[Ha] D.R. HAYES, *Explicit class field theory for rational function fields*, Trans. Amer. Math. soc., 189, 1974, p.77-91.

[He1] Y. HELLEGOUARCH, *Galois calculus and Carlitz exponentials*, The Arithmetic of Function Fields, D. Goss, D.R. Hayes, M.I. Rosen Editors, W. de Gruyter, 1992, p.33-50.

[He2] Y. HELLEGOUARCH, *Un analogue d'un théorème d'Euler*, C. R. Acad. Sci. Paris, Sér. I Math., 313, 1991, p.155-158.

[Sch1] F. SCHULTHEIS, *Explicit reciprocity laws in Algebraic Function Fields*, Ph. D. Thesis, University of Michigan,1987.

[Sch2] F. SCHULTHEIS, *Local and Global Residue Symbols for Algebraic Function Fields*, J. Number Theory, 52, n°1, 1995, 119-124.

[Se] J.- P. SERRE, *Corps locaux*, Hermann, 1968.

Groups of elliptic units and torsion points of Drinfeld modules

Hassan Oukhaba
Laboratoire de Mathématiques et Mécanique Théorique
16, route de Gray
25030 Besançon Cedex France e-mail: hassan@math.univ-fcomte.fr

1. Introduction

Let k be a global function field and ${}_q$ be its field of constants. Fix a place ∞ of k. Let $\mathcal{O}_k$ be the Dedekind ring of functions of k regulars outside of ∞, k_∞ the completion of k at ∞ and let C be the completion of the algebraic closure of k_∞.

For each finite extension $F \subseteq \mathrm{C}$ of k we let $\mathcal{O}_F$ be the integral closure of $\mathcal{O}_k$ in F. We know that $\mathcal{O}_F$ is a Dedekind ring with a finite ideal class number $h(\mathcal{O}_F)$. As usual we denote by $\mathcal{O}_F^\times$ the group of units of $\mathcal{O}_F$, and by $\mu(F) \subseteq \mathcal{O}_F^\times$ the finite multiplicative group of non-zero constants of F. We have $\mu(k) = {}^\times_q = \mathcal{O}_k^\times$, and in general the quotient group $\mathcal{O}_F^\times/\mu(F)$ is a free abelian group of rank $r_F - 1$, where r_F is the exact number of places of F sitting over ∞.

Suppose in addition that F is a finite abelian extension of k, and let $\mathcal{T}_{F/k}$ be the set of prime divisors of k which ramify in F/k. Then for each fractional ideal $\mathfrak{b}$ of $\mathcal{O}_k$ prime to $\mathcal{T}_{F/k}$, we denote by $(\mathfrak{b},\ F/k)$ the automorphism of F/k associated to $\mathfrak{b}$ by the so-called Artin map. We note $\mathcal{A}_{F/k}$ the group of all $\mathfrak{b}$ such that $(\mathfrak{b},\ F/k)$ is the identity of F. Let $\mathfrak{m}$ be an integral ideal of $\mathcal{O}_k$. Then there exists a unique finite abelian extension $H_\mathfrak{m} \subseteq k_\infty$ of k such that $\mathcal{A}_{H_\mathfrak{m}/k} = \{x\mathcal{O}_k,\quad x \in k^\times \text{ and } x \equiv 1 \mod .\mathfrak{m}\}$. $H_\mathfrak{m}$ is the ray class field of $\mathcal{O}_k$ modulo $\mathfrak{m}$. It is the maximal finite abelian extension of k contained in k_∞ and whose conductor divides $\mathfrak{m}$. In particular, $\mathrm{Gal}(H_\mathfrak{m}/k)$ is naturally isomorphic to the ideal class group $\mathcal{I}_\mathfrak{m}/\mathcal{A}_{H_\mathfrak{m}/k}$, where $\mathcal{I}_\mathfrak{m}$ is the group of the fractional ideals of $\mathcal{O}_k$ prime to $\mathfrak{m}$. Therefore we have $[H_\mathfrak{m} : k] = \frac{\varphi(\mathfrak{m})}{q-1} h(\mathcal{O}_k)$.

Let ϕ be a Drinfeld $\mathcal{O}_k$-module of rank one over C, normalized with respect to some sign function s. Let $K_\mathfrak{e}$ be the normalizing field of ϕ (see the definition below), and let $H := H_{(1)}$ be the maximal unramified abelian extension of k such that $H \subseteq k_\infty$. Then we associate to each $\tau \in \mathrm{Gal}(H/k)$ a subgroup $\bar{E}_\tau$ of $\mathcal{O}_{K_\mathfrak{e}}^\times$, and a subgroup $\mathcal{E}_{H,\tau}$ of $\mathcal{O}_H^\times$ defined as follows: Let $\mathfrak{p}$ be a prime ideal of $\mathcal{O}_k$ such that $(\mathfrak{p},\ H/k) = \tau$, $\bar{H}_\mathfrak{p} := H_\mathfrak{p} K_\mathfrak{e}$, and let λ be a non zero root of the polynomial $\phi_\mathfrak{p} \in K_\mathfrak{e}[t]$. Then $\bar{E}_\tau$ is generated by $\mu(K_\mathfrak{e}) = \mu(H)$ and by the norms $N_{\bar{H}_\mathfrak{p}/K_\mathfrak{e}}(\phi_\mathfrak{a}(\lambda)/\lambda)$, where $\mathfrak{a}$ is any integral ideal of $\mathcal{O}_k$ prime to $\mathfrak{p}$, while $\mathcal{E}_{H,\tau}$ is the intersection with $\mathcal{O}_H^\times$ of the group generated by the norms $N_{H_\mathfrak{p}/H}(\phi_\mathfrak{a}(\lambda)/\lambda^{N\mathfrak{a}})$.

Let Γ be a subgroup of $G :=\mathrm{Gal}(H/k)$. Then define $\bar{E}_\Gamma$ to be the subgroup of $\mathcal{O}_{K_\mathfrak{e}}^\times$ generated by $\bar{E}_\tau, \tau \in \Gamma$. In this paper we compute the index $[\mathcal{O}_{K_\mathfrak{e}}^\times : \bar{E}_\Gamma \mathcal{O}_L^\times]$, where L is the fixed field of Γ. This generalizes the index formula $[\mathcal{O}_{K_\mathfrak{e}}^\times : \bar{E}_G] = h(\mathcal{O}_H)$ obtained by L.Yin in [10]. We also consider the group $\bar{E}_{H,\Gamma} := \bar{E}_\Gamma \cap H$. In particular we give the index of $\bar{E}_{H,\Gamma}\mathcal{O}_L^\times$ in $\mathcal{O}_H^\times$. Recall that all these indices were already calculated by D.Hayes in the case when the degree of ∞ is one, cf. [3]. We have amply taken our inspiration from Hayes's work to make the computation in the general case. In an other request we prove that $\bar{E} \cap H = \mathcal{E}_H$, where $\mathcal{E}_H$ is the subgroup of $\mathcal{O}_H^\times$ generated by $\mathcal{E}_{H,\tau}, \tau \in G$. We proved in [6] that the elements of $\mathcal{E}_H$ can be obtained as finite products of special values of certain elliptic functions associated to the $\mathcal{O}_k$-lattices of C, and that $[\mathcal{O}_L^\times : \mu(L) N_{H/L}(\mathcal{E}_H)] = h(\mathcal{O}_L)/[H : L]$.

2. The unramified units

Let $K_\mathfrak{e}$ be the normalizing field of ϕ. By definition, cf. [5], $K_\mathfrak{e}$ is generated over k by the coefficients of ϕ_x, $x \in \mathcal{O}_k$. We know that the extension $K_\mathfrak{e}/k$ is finite, abelian and unramified outside of ∞. Moreover $K_\mathfrak{e}/H$ is totally ramified at ∞ and $[K_\mathfrak{e} : H] = \frac{q^{d_\infty}-1}{q-1}$, where d_∞ is the degree of ∞. Let $\mathfrak{a}$ be an ideal of $\mathcal{O}_k$. Then the automorphism $(\mathfrak{a}, K_\mathfrak{e}/k)$ is the identity of $K_\mathfrak{e}$ if, and only if, $\mathfrak{a}$ is principal generated by some element $x \in k^\times$ such that $s(x) = 1$; such an element will be called positive.

Now let $\mathfrak{m}$ be an integral ideal of $\mathcal{O}_k$ and $K_\mathfrak{m} := K_\mathfrak{e}(\Lambda_\mathfrak{m})$ be the field obtained by adjoining to $K_\mathfrak{e}$ the set of roots of the linear polynomial $\phi_\mathfrak{m} \in K_\mathfrak{e}[t]$. The arithmetical properties of $K_\mathfrak{m}$ are well known, cf. loc. cit. In particular $K_\mathfrak{m}$ is a finite abelian extension of k containing $H_\mathfrak{m}$. The ramification number in $K_\mathfrak{m}/k$ at each maximal ideal $\mathfrak{p}$ of $\mathcal{O}_k$ is $\varphi(\mathfrak{p}^e)$, where $\mathfrak{p}^e$ is the highest power of $\mathfrak{p}$ dividing $\mathfrak{m}$. The extension $K_\mathfrak{m}/H_\mathfrak{m}$ is totally ramified at ∞, with ramification number equal to $q^{d_\infty} - 1$. Let $\mathfrak{a}$ be an integral ideal of $\mathcal{O}_k$ prime to $\mathfrak{m}$. Then we have $\lambda^{(\mathfrak{a},\, K_\mathfrak{m}/k)} = \phi_\mathfrak{a}(\lambda)$, for all $\lambda \in \Lambda_\mathfrak{m}$; moreover, the automorphism $(\mathfrak{a}, K_\mathfrak{m}/k)$ is the identity of $K_\mathfrak{m}$ if, and only if, $\mathfrak{a}$ is principal generated by some positive element x of $\mathcal{O}_k$ such that $x \equiv 1$ mod.m. Hence $K_\mathfrak{m}$ depends only on $\mathfrak{m}$ and s.

In the sequel we fix a subgroup Γ of $\mathrm{Gal}(H/k)$. We will denote by n its cardinal and by L its fixed field. The cardinal of $\mu(L)$ will be written as w_L. Let $w_\infty := q^{d_\infty} - 1$. It is well known that $w_H = w_{K_\mathfrak{e}} = w_\infty$.

Lemma 1. *Let $\mathfrak{a}$ be an integral ideal of $\mathcal{O}_k$ prime to $\mathfrak{m}$. Let λ be a generator of $\Lambda_\mathfrak{m}$ considered as an $\mathcal{O}_k$-submodule of C, where the action of $\mathcal{O}_k$ on C is given by ϕ. Then $\lambda^{N\mathfrak{a}}/\phi_\mathfrak{a}(\lambda) \in H_\mathfrak{m}$. Moreover, suppose that $\mathfrak{m} = \mathfrak{q}^e$, where $e \geq 1$ and $\mathfrak{q}$ is a maximal ideal of $\mathcal{O}_k$, and let $\mathfrak{q}_{H_\mathfrak{m}}$ be the product of the maximal ideals of $\mathcal{O}_{H_\mathfrak{m}}$ which divide $\mathfrak{q}$. Then $\lambda^{N\mathfrak{a}}/\phi_\mathfrak{a}(\lambda)$ generates in $\mathcal{O}_{H_\mathfrak{m}}$*

the $\frac{N\mathfrak{a}-1}{q-1}$*-st power of the ideal* $\mathfrak{q}_{H_\mathfrak{m}}$*; otherwise it is a unit of* $\mathcal{O}_{H_\mathfrak{m}}$.

Proof. This is a direct consequence of Corollary 4.14 and Theorem 4.17 of [4]. ♡

Definition 1. *Let* $\mathfrak{p}$ *be a maximal ideal of* $\mathcal{O}_k$ *and* λ *be a non-zero element, and hence a generator, of* $\Lambda_\mathfrak{p}$. *Then we let* $R_{L,\phi,\mathfrak{p}}$ *be the subgroup of* $L^\times$ *generated by all the norms*

$$N_{H_\mathfrak{p}/L}\left(\frac{\lambda^{N\mathfrak{a}}}{\phi_\mathfrak{a}(\lambda)}\right), \tag{2.1}$$

where $\mathfrak{a}$ *runs over the set of integral ideals of* $\mathcal{O}_k$ *which are prime to* $\mathfrak{p}$.

The group $R_{L,\phi,\mathfrak{p}}$ does not depend on the generator λ of $\Lambda_\mathfrak{p}$. For, if λ' is an other generator of $\Lambda_\mathfrak{p}$, then $\lambda' = \phi_x(\lambda) = \lambda^{(x\mathcal{O}_k,\, K_\mathfrak{p}/k)}$ for some positive $x \in \mathcal{O}_k - \mathfrak{p}$. But since the automorphism $(x\mathcal{O}_k,\, K_\mathfrak{p}/k)$ is the identity on H, the quotient (2.1) does not change if we replace λ by λ'.

On the other hand, if $\mathfrak{a}$ and $\mathfrak{b}$ are integral ideals of $\mathcal{O}_k$ prime to $\mathfrak{p}$, then the following identity

$$\left(\frac{\lambda^{N\mathfrak{a}}}{\phi_\mathfrak{a}(\lambda)}\right)^{(\mathfrak{b},\, K_\mathfrak{p}/k)} = \left(\frac{\lambda^{N\mathfrak{a}\mathfrak{b}}}{\phi_{\mathfrak{a}\mathfrak{b}}(\lambda)}\right)\left(\frac{\lambda^{N\mathfrak{b}}}{\phi_\mathfrak{b}(\lambda)}\right)^{-N\mathfrak{a}}$$

shows that $R_{L,\phi,\mathfrak{p}}$ is stable under the natural action of the Galois group of L/k.

Lemma 2. *Let* ϕ' *be a Drinfeld* $\mathcal{O}_k$*-module of rank one over* C*, normalized with respect to some other sign function, say* s'. *Then we have*

$$\mu(L)R_{L,\phi,\mathfrak{p}} = \mu(L)R_{L,\phi',\mathfrak{p}}.$$

Actually, when $s' = s$*, the groups* $R_{L,\phi,\mathfrak{p}}$ *and* $R_{L,\phi',\mathfrak{p}}$ *are equal.*

Proof. It is well known that one can find an integral ideal $\mathfrak{a}$ of $\mathcal{O}_k$, which we may choose prime to $\mathfrak{p}$, and an element $\alpha \in C^\times$ such that $\alpha \cdot \phi' \cdot \alpha^{-1} = \mathfrak{a} * \phi$. Therefore the set $\Lambda'_\mathfrak{p}$ of roots of $\phi'_\mathfrak{p}$ satisfies

$$\Lambda'_\mathfrak{p} = \alpha \cdot \phi_\mathfrak{a}(\Lambda_\mathfrak{p}).$$

But since ϕ' is supposed to be normalized with respect to s' (resp. s), the element $\alpha^{q^{d_\infty}-1}$ (resp. α) must be a constant of H. It is now easy to complete the proof of the Lemma. ♡

Using the analytic theory of Drinfeld $\mathcal{O}_k$-modules of rank one over C, D.R.Hayes proved that for every fractional ideal $\mathfrak{b}$ of $\mathcal{O}_k$ there exists $\xi(\mathfrak{b}) \in C^\times$

such that the lattice $\xi(\mathfrak{b})\mathfrak{b}$ is special for s, that is to say, the Drinfeld $\mathcal{O}_k$-module associated to this lattice is s-normalized. The element $\xi(\mathfrak{b})$ is determined up to a non-zero constant in $\mu(H)$. In particular one can find a fractional ideal $\mathfrak{c}$ of $\mathcal{O}_k$ such that the lattice $\Gamma := \xi(\mathfrak{c})\mathfrak{c}$ is the one associated to ϕ. Let us fix both $\mathfrak{c}$ and $\xi(\mathfrak{c})$, and let $\mathfrak{a}$ be an integral ideal of $\mathcal{O}_k$. Then the Drinfeld $\mathcal{O}_k$-module $\phi^{(\mathfrak{a},\, K_\mathfrak{e}/k)} = \mathfrak{a} * \phi$ is s-normalized and associated to $D(\phi_\mathfrak{a})\mathfrak{a}^{-1}\Gamma$. Hence we can put $\xi(\mathfrak{a}^{-1}\mathfrak{c}) := D(\phi_\mathfrak{a})\xi(\mathfrak{c})$. This fixes a value of $\xi(\mathfrak{a}^{-1}\mathfrak{c})$. More generally, a fractional ideal $\mathfrak{d}$ of $\mathcal{O}_k$ has the form $\mathfrak{d}=\mathfrak{a}\mathfrak{b}^{-1}\mathfrak{c}$, where $\mathfrak{a}$ and $\mathfrak{b}$ are integral ideals of $\mathcal{O}_k$. Then one can verify that the definition

$$\xi(\mathfrak{d}) := \frac{D(\phi_\mathfrak{b})}{D(\phi_\mathfrak{a})^\tau}\xi(\mathfrak{c}) \tag{2.2}$$

where $\tau = (\mathfrak{d}^{-1}\mathfrak{c}, K_\mathfrak{e}/k)$, fixes a value of $\xi(\mathfrak{d})$. In particular we have the formula

$$\xi(x\mathfrak{d}) = \frac{\eta(x)^\tau}{x}\xi(\mathfrak{d}) \tag{2.3}$$

for all $x \in k^\times$, where η is the multiplicative function defined on $k^\times$ such that $\eta(x)$ is the leading coefficient of the polynomial ϕ_x if $x \in \mathcal{O}_k - \{0\}$. The formula (2.3) allows us to define a map ∂ :Gal$(H/k) \longrightarrow \mathbb{C}^\times$. Indeed, let $\sigma := (\mathfrak{d}^{-1}, H/k)$ and choose $\alpha \in k^\times$ such that $\mathfrak{d}^h = \alpha\mathcal{O}_k$, where $h := h(\mathcal{O}_k)$. Then we put

$$\partial(\sigma) := \left(\alpha\xi(\mathfrak{d})^h\right)^{w_\infty}.$$

Lemma 3. *Let $\mathfrak{d}_1$, $\mathfrak{d}_2$ and $\mathfrak{d}$ be fractional ideals of $\mathcal{O}_k$. Then the ideal $\mathfrak{d}_2\mathfrak{d}_1^{-1}\mathcal{O}_{K_\mathfrak{e}}$ is principal generated by $\xi(\mathfrak{d}_1)/\xi(\mathfrak{d}_2)$. Moreover, we have*

$$\left(\frac{\xi(\mathfrak{d}_1)}{\xi(\mathfrak{d}_2)}\right)^{(\mathfrak{d},K_\mathfrak{e}/k)} = \frac{\xi(\mathfrak{d}_1\mathfrak{d}^{-1})}{\xi(\mathfrak{d}_2\mathfrak{d}^{-1})}. \tag{2.4}$$

Proof. See [10]. ♡

Remark 1. One can prove, using Lemma 3, that the quotient $\partial(\sigma_1)/\partial(\sigma_2)$ is a unit of $\mathcal{O}_H$, and that we have

$$\left(\frac{\partial(\sigma_1)}{\partial(\sigma_2)}\right)^\sigma = \frac{\partial(\sigma_1\sigma)}{\partial(\sigma_2\sigma)} \quad \text{for all} \quad \sigma_1, \sigma_2, \sigma \in \mathrm{Gal}(H/k).$$

Since the polynomial $\phi_\mathfrak{p}(t)/t \in \mathcal{O}_{K_\mathfrak{e}}[t]$ is Eisenstein at $\mathfrak{p}$, we have $N_{K_\mathfrak{p}/K_\mathfrak{e}}(\lambda) = D(\phi_\mathfrak{p}) = \xi(\mathfrak{p}^{-1}\mathfrak{c})/\xi(\mathfrak{c})$, for all $\lambda \in \Lambda_\mathfrak{p} - \{0\}$. On the other hand, if we recall that

$$N_{H_\mathfrak{p}/H}\left(\frac{\lambda^{N\mathfrak{a}}}{\phi_\mathfrak{a}(\lambda)}\right)^{q-1} = N_{K_\mathfrak{p}/K_\mathfrak{e}}\left(\frac{\lambda^{N\mathfrak{a}}}{\phi_\mathfrak{a}(\lambda)}\right)$$

whenever $\mathfrak{a}$ is an integral ideal of $\mathcal{O}_k$ prime to $\mathfrak{p}$, we get

Proposition 1. *The subgroup* $(R_{L,\phi,\mathfrak{p}})^{q-1}$ *of* $L^\times$ *is generated by the following elements*

$$N_{H/L}\left(\left(\frac{\xi(\mathfrak{p}^{-1})}{\xi(\mathcal{O}_k)}\right)^{N\mathfrak{a}}\frac{\xi(\mathfrak{a}^{-1})}{\xi(\mathfrak{a}^{-1}\mathfrak{p}^{-1})}\right),$$

where $\mathfrak{a}$ *is any integral ideal of* $\mathcal{O}_k$ *prime to* $\mathfrak{p}$.

Proposition 2. *Put* $\mathcal{E}_{L,\phi,\mathfrak{p}} := R_{L,\phi,\mathfrak{p}} \cap \mathcal{O}_L^\times$. *Let* $x \in L^\times$. *Then* $x \in (\mathcal{E}_{L,\phi,\mathfrak{p}})^{q-1}$ *if, and only if,* x *can be written as*

$$N_{H/L}\left(\prod_{\mathfrak{a}\in\mathfrak{A}}\left(\frac{\xi(\mathcal{O}_k)}{\xi(\mathfrak{p}^{-1})}\frac{\xi(\mathfrak{a}^{-1}\mathfrak{p}^{-1})}{\xi(\mathfrak{a}^{-1})}\right)^{n_\mathfrak{a}}\right), \tag{2.5}$$

where $\mathfrak{A}$ *is a finite set of integral ideals of* $\mathcal{O}_k$; *and* $n_\mathfrak{a}$, $\mathfrak{a}\in\mathfrak{A}$, *are rational integers such that*

$$\sum_{\mathfrak{a}\in\mathfrak{A}} n_\mathfrak{a}(N\mathfrak{a}-1) \equiv 0 \bmod. w_\infty\ . \tag{2.6}$$

Proof. It is clear that an element of $(\mathcal{E}_{L,\phi,\mathfrak{p}})^{q-1}$ has the required form. Conversely, let us prove that the product (2.5) belongs to $(\mathcal{E}_{L,\phi,\mathfrak{p}})^{q-1}$. First, we remark that we may assume that the ideals $\mathfrak{a}\in\mathfrak{A}$ are all prime to $\mathfrak{p}$. Indeed, it suffices to replace each ideal $\mathfrak{a}\in\mathfrak{A}$ by an integral ideal $\mathfrak{a}'$ of $\mathcal{O}_k$ prime to $\mathfrak{p}$ and such that $(\mathfrak{a}', K_\mathfrak{e}/k) = (\mathfrak{a}, K_\mathfrak{e}/k)$. Now (2.6) implies that we can find a finite set $Z \subseteq \mathcal{O}_k - \{\mathfrak{p}\}$ of positive elements, and rational integers $n_z, z \in Z$ such that

$$\sum_{\mathfrak{a}\in\mathfrak{A}} n_\mathfrak{a}(N\mathfrak{a}-1) - \sum_{z\in Z} n_z(Nz\mathcal{O}_k - 1) = 0.$$

This allows us to write the product (2.5) as an element of $(\mathcal{E}_{L,\phi,\mathfrak{p}})^{q-1}$. ♡

Corollary 1. *Let* $\mathfrak{p}$ *and* $\mathfrak{p}'$ *be two maximal ideals of* $\mathcal{O}_k$ *such that* $(\mathfrak{p}, L/k) = (\mathfrak{p}', L/k)$. *Then*

$$(\mathcal{E}_{L,\phi,\mathfrak{p}})^{q-1} = (\mathcal{E}_{L,\phi,\mathfrak{p}'})^{q-1}.$$

In particular the group $\mathcal{E}_{L,\sigma} := \mu(L)\mathcal{E}_{L,\phi,\mathfrak{p}}$ *depends only on* $\sigma := (\mathfrak{p}, L/k)$.

Let $\mathcal{E}_{L,\phi}$ be the subgroup of $\mathcal{O}_L^\times$ generated by all $\mathcal{E}_{L,\phi,\mathfrak{p}}$, where $\mathfrak{p}$ describes the maximal ideals of $\mathcal{O}_k$. Since $\mathcal{E}_{L,\phi,\mathfrak{p}} = N_{H/L}(\mathcal{E}_{H,\phi,\mathfrak{p}})$, we also have

$$\mathcal{E}_{L,\phi} = N_{H/L}(\mathcal{E}_{H,\phi}).$$

The group $\mathcal{E}_L := \mu(L)\mathcal{E}_{L,\phi}$ does not depend on ϕ, as follows from Proposition 2. It is generated by $\mathcal{E}_{L,\sigma}, \sigma \in \mathrm{Gal}(L/k)$. Hence we have

$$\mathcal{E}_L := \prod_{\sigma \in \mathrm{Gal(L/k)}} \mathcal{E}_{L,\sigma}.$$

Remark 2. Recall that we have a group morphism $\psi : \mathrm{Gal}(H/k) \longrightarrow (/w_{\infty})^{\times}$ defined as follows. Let $\tau \in \mathrm{Gal}(H/k)$ and $\mathfrak{a}$ be an integral ideal of $\mathcal{O}_k$ such that $(\mathfrak{a},\, H/k) = \tau$, then $N\mathfrak{a} \bmod . w_{\infty}$ depends only on τ. By definition we put $\psi(\tau) := N\mathfrak{a} \bmod . w_{\infty}$.

In particular, let $\kappa := w_{\infty}/w_k$, $G :=\mathrm{Gal}(H/k)$ and I_G be the augmentation ideal of $[G]$. Then we obtain a homomorphism $\Psi^{(1)} : I_G \longrightarrow /\kappa$ defined so that the image of the sum $\sum_{\sigma \in G} n_{\sigma}(\sigma - 1)$ is the element $\sum_{\sigma \in G} n_{\sigma}(\frac{\psi(\sigma)-1}{q-1})$.

Theorem 1. *We have a well defined homomorphism* $\Psi^{(2)} : I_G^2 \longrightarrow /\kappa$ *given by*

$$\sum_{\tau,\tau' \in G} n_{\tau,\tau'}(\tau - 1)(\tau' - 1) \longmapsto \sum_{\tau,\tau' \in G} n_{\tau,\tau'}\Big(\frac{\psi(\tau) - 1}{q-1}\Big)\Big(\frac{\psi(\tau') - 1}{q-1}\Big).$$

Moreover, let $\Delta^{(i)}$ *be the kernel of* $\Psi^{(i)}, i \in \{1, 2\}$. *Then we have* $\Delta^{(2)} = I_G\Delta^{(1)}$.

Proof. See section 4 below. ♡

Theorem 2. *The group* $(\mathcal{E}_L)^{w_{\infty}}$ *is the set of all the following norms of* $\mathcal{O}_L^{\times}$

$$N_{H/L}\left(\prod_{(\mathfrak{a},\mathfrak{b}) \in \mathfrak{A}\times\mathfrak{B}} \left(\frac{\xi(\mathcal{O}_k)}{\xi(\mathfrak{b}^{-1})}\frac{\xi(\mathfrak{a}^{-1}\mathfrak{b}^{-1})}{\xi(\mathfrak{a}^{-1})}\right)^{n_{\mathfrak{a},\mathfrak{b}}}\right)^{\frac{w_{\infty}}{q-1}}, \tag{2.7}$$

where $\mathfrak{A}$ *and* $\mathfrak{B}$ *are finite sets of integral ideals of* $\mathcal{O}_k$; *and* $n_{\mathfrak{a},\mathfrak{b}}$, $(\mathfrak{a},\mathfrak{b}) \in \mathfrak{A}\times\mathfrak{B}$, *are rational integers such that*

$$\sum n_{\mathfrak{a},\mathfrak{b}}\Big(\frac{N\mathfrak{a} - 1}{q-1}\Big)\Big(\frac{N\mathfrak{b} - 1}{q-1}\Big) \equiv 0 \bmod . \frac{w_{\infty}}{q-1}. \tag{2.8}$$

Proof. It suffices to give the proof when $L = H$. On the other hand, if we denote by $\mathcal{F}$ the group of the elements (2.7) which satisfy (2.8), we have only to prove that

$$\mathcal{F}^{w_k h} = \mathcal{E}_H^{w_k w_{\infty} h},$$

because both $\mathcal{E}_H^{w_\infty}$ and $\mathcal{F}$ consist of positive elements. For this we use the logarithm map $l : H^\times \longrightarrow [G]$, defined by

$$l(x) = \sum_{\tau \in G} v_\infty(x^\tau)\tau^{-1},$$

where v_∞ is the extension to C of the normalized valuation of k_∞. In particular we have

$$l\left(\frac{\partial(\sigma)}{\partial(1)}\right) = \omega(\sigma - 1), \quad \text{for all} \quad \sigma \in G,$$

where ω is the element of [G] given by $\omega := \sum_{\tau \in G} v_\infty\left(\frac{\partial(\tau)}{\partial(1)}\right)\tau^{-1}$. Now Proposition 2 describes a set of generators of $\mathcal{E}_H^{w_k}$. We deduce from this description

$$l\left(\mathcal{E}_H^{w_k w_\infty h}\right) = \omega I_G \Delta^{(1)}.$$

On the other hand, we have $l(\mathcal{F}^{w_k h}) = \omega\Delta^{(2)}$. But since $I_G\Delta^{(1)} = \Delta^{(2)}$ by Theorem 1 above, and $\ker l \cap \mathcal{O}_H^\times = \mu(H)$, we get $\mathcal{F}^{w_k h} = \mathcal{E}_H^{w_k w_\infty h}$. This completes the proof of Theorem 2. ♡

Let $\mathfrak{p}$ be a maximal ideal of $\mathcal{O}_k$ and let λ be a non zero element of $\Lambda_\mathfrak{p}$. Then put $\bar{H}_\mathfrak{p} := H_\mathfrak{p} K_\mathfrak{e}$. It is not difficult to verify that $\lambda \in \bar{H}_\mathfrak{p}$ and that

$$N_{\bar{H}_\mathfrak{p}/K_\mathfrak{e}}\left(\frac{\phi_\mathfrak{a}(\lambda)}{\lambda}\right)^{q-1} = N_{K_\mathfrak{p}/K_\mathfrak{e}}\left(\frac{\phi_\mathfrak{a}(\lambda)}{\lambda}\right) = \left(\frac{\xi(\mathfrak{p}^{-1})}{\xi(\mathcal{O}_k)}\frac{\xi(\mathfrak{a}^{-1})}{\xi(\mathfrak{a}^{-1}\mathfrak{p}^{-1})}\right)^\tau \tag{2.9}$$

for any integral ideal $\mathfrak{a}$ of $\mathcal{O}_k$ prime to $\mathfrak{p}$, where $\tau := (\mathfrak{c}, K_\mathfrak{e}/k)$. This leads us to consider the subgroup $\bar{E}_{\phi,\mathfrak{p}}$ of $O_{K_\mathfrak{e}}^\times$ generated by

$$N_{\bar{H}_\mathfrak{p}/K_\mathfrak{e}}\left(\frac{\phi_\mathfrak{a}(\lambda)}{\lambda}\right),$$

where $\mathfrak{a}$ describes all the integral ideals of $\mathcal{O}_k$ prime to $\mathfrak{p}$. It is obvious that $\bar{E}_{\phi,\mathfrak{p}}$ is stable under the action of $\mathrm{Gal}(K_\mathfrak{e}/k)$; and since $\mu(K_\mathfrak{e})\bar{E}_{\phi,\mathfrak{p}}$ depends only on $\tau := (\mathfrak{p}, H/k)$, we put $\bar{E}_\tau := \mu(K_\mathfrak{e})\bar{E}_{\phi,\mathfrak{p}}$. Then we define $\bar{E}$ to be the subgroup of $\mathcal{O}_{K_\mathfrak{e}}^\times$ generated by $\bar{E}_\tau, \tau \in \mathrm{Gal}(H/k)$, i.e.,

$$\bar{E} := \prod_{\tau \in \mathrm{Gal}(H/k)} \bar{E}_\tau .$$

Corollary 2. *The groups $\mathcal{E}_H$ and $\bar{E} \cap H$ are equal, i.e., we have $\mathcal{E}_H = \bar{E} \cap H$.*

Proof. It is clear that $\mathcal{E}_H \subseteq \bar{E}$. Conversely, let $t \in \bar{E} \cap H$. By definition, there exists a finite set T of maximal ideals of $\mathcal{O}_k$, and for each $\mathfrak{p} \in T$ an element $t_\mathfrak{p} \in \bar{E}_{\phi,\mathfrak{p}}$ such that $t^{w_\infty} = \left(\prod_{\mathfrak{p} \in T} t_\mathfrak{p}\right)^{w_\infty}$.

Since $t_{\mathfrak{p}} \in \bar{E}_{\phi,\mathfrak{p}}$ for $\mathfrak{p} \in T$, we have

$$t_{\mathfrak{p}} = \prod_{\mathfrak{a} \in T_{\mathfrak{p}}} N_{\bar{H}_{\mathfrak{p}}/K_{\mathfrak{e}}} \left(\frac{\phi_{\mathfrak{a}}(\lambda_{\mathfrak{p}})}{\lambda_{\mathfrak{p}}} \right)^{n_{\mathfrak{p},\mathfrak{a}}},$$

where $T_{\mathfrak{p}}$ is a finite set of integral ideals of $\mathcal{O}_k$ prime to $\mathfrak{p}$, and $n_{\mathfrak{p},\mathfrak{a}}$ are rational integers defined for every $\mathfrak{a} \in T_{\mathfrak{p}}$. The element $\lambda_{\mathfrak{p}}$ is a non-zero root of the polynomial $\phi_{\mathfrak{p}}$.

Now let $x \in \mathcal{O}_k$ be such that $x \equiv 1$ mod. $\mathfrak{p}$, for all $\mathfrak{p} \in T$. Then, by [4] Corollary 4.14, we get

$$(t_{\mathfrak{p}})^{\tau_x} = \eta(x)^{-r_{\mathfrak{p}}} t_{\mathfrak{p}},$$

where $\tau_x := (x\mathcal{O}_k, K_{\mathfrak{e}}/k)$ and $r_{\mathfrak{p}} := \frac{N\mathfrak{p}-1}{q-1} \sum_{\mathfrak{a} \in T_{\mathfrak{p}}} n_{\mathfrak{p},\mathfrak{a}}(N\mathfrak{a}-1)$. But we want $t \in H$. Hence we must have

$$\sum_{\mathfrak{p} \in T} \frac{N\mathfrak{p}-1}{q-1} \left(\sum_{\mathfrak{a} \in T_{\mathfrak{p}}} n_{\mathfrak{p},\mathfrak{a}}(N\mathfrak{a}-1) \right) \equiv 0 \text{ mod. } w_\infty,$$

because $\mu(K_{\mathfrak{e}}) = \mu(H) = \{\eta(x),\ x \equiv 1 \text{ mod. } \mathfrak{p}, \text{ for all } \mathfrak{p} \in T\}$. Now Theorem 2 implies that $t \in \mathcal{E}_H$. ♡

3. Class Number formulas

We devote this paragraph to the computation of the indices $[\mathcal{O}_{K_{\mathfrak{e}}}^\times : \bar{E}_\Gamma \mathcal{O}_L^\times]$ and $[\mathcal{O}_H^\times : \bar{E}_{H,\Gamma} \mathcal{O}_L^\times]$, where $\bar{E}_\Gamma$ is the subgroup of $\mathcal{O}_{K_{\mathfrak{e}}}^\times$ generated by $\bar{E}_\tau, \tau \in \Gamma$ and $\bar{E}_{H,\Gamma} := \bar{E}_\Gamma \cap H$.

If A is an ideal of $[G]$ then we will write A^Γ for the invariants of A under the action of Γ.

Proposition 3. *Let $E := (\bar{E})^{w_k w_\infty h}$. Then we have*

$$[\mathcal{O}_L^\times : \mu(L)(E \cap L)] = (w_k w_\infty h)^{[L:k]-1} \frac{h(\mathcal{O}_L)}{|G_n|} \frac{w_k}{w_L}, \tag{3.1}$$

where G_n is the group of $n-$th roots in G and $|G_n|$ is the number of elements of G_n.

Proof. Since $l(E \cap L) = (\omega I^2)^\Gamma$, we have

$$[\mathcal{O}_L^\times : \mu(L)(E \cap L)] = [l(\mathcal{O}_L^\times) : (\omega I^2)^\Gamma] = \frac{1}{R_L} [I^\Gamma : \omega I^\Gamma] \cdot [\omega I^\Gamma : (\omega I^2)^\Gamma],$$

where $I := I_G$ and $R_L := [I^\Gamma : l(\mathcal{O}_L^\times)]$ is the regulator of the ring $\mathcal{O}_L$. But I^Γ is isomorphic to the augmentation ideal of $[G/\Gamma]$. Therefore

$$\frac{1}{R_L} \cdot [I^\Gamma : \omega I^\Gamma] = \frac{1}{R_L} \cdot \det \omega = \frac{1}{R_L} \prod_{\chi \neq 1} \left(\sum_{\tau \in G} \chi(\tau) v_\infty(\partial(\tau)) \right).$$

The product is taken over all non-trivial characters χ of G such that $\chi = 1$ on Γ. Now recall that we have $L(0,\chi) = \frac{1}{w_k w_\infty h}\sum_{\tau\in G}\chi(\tau)v_\infty(\partial(\tau))$, where $L(\cdot,\chi)$ is the L-function associated to χ, cf. Proposition 7.19 of [2]. Hence, using the analytic class number formula, cf. eq. (4.3) of [6], we obtain

$$\frac{1}{R_L}\cdot[I^\Gamma : \omega I^\Gamma] = (w_k w_\infty h)^{[L:k]-1}\frac{h(\mathcal{O}_L)}{h}\frac{w_k}{w_L}. \tag{3.2}$$

On the other hand, since $\det\omega \neq 0$ on I^Γ, we have $(\omega I^2)^\Gamma = \omega(I^2)^\Gamma$ and

$$[\omega I^\Gamma : \omega(I^2)^\Gamma] = [I^\Gamma : (I^2)^\Gamma] = |G^n| = \frac{h}{|G_n|}.$$

The formula (3.1) is now a simple consequence of the above calculus. ♡

Let I_Γ be the augmentation ideal of $[\Gamma]$, and put $s_\Gamma := \sum_{\tau\in\Gamma}\tau$; then we have the following exact sequence

$$0 \longrightarrow I_\Gamma I \longrightarrow I^2 \xrightarrow{s_\Gamma} s_\Gamma(I^2) \longrightarrow 0,$$

from which we deduce the isomorphism $I^2/(I_\Gamma I + (I^2)^\Gamma) \simeq s_\Gamma(I^2)/n(I^2)^\Gamma$, and so the formula

$$[I^2 : I_\Gamma I + (I^2)^\Gamma] = \frac{[(I^2)^\Gamma : n(I^2)^\Gamma]}{[(I^2)^\Gamma : s_\Gamma(I^2)]}.$$

But since $(I^2)^\Gamma$ is free of rank $|G/\Gamma| - 1 = [L:k]-1$, we have $[(I^2)^\Gamma : n(I^2)^\Gamma] = n^{[L:k]-1}$. On the other hand $[(I^2)^\Gamma : s_\Gamma(I^2)] = |G_n|/n$ as proved by D.Hayes in [3]. Finally we obtain

$$[I^2 : I_\Gamma I + (I^2)^\Gamma] = \frac{n^{[L:k]}}{|G_n|}. \tag{3.3}$$

Now recall that $l(E_\Gamma(E\cap L)) = \omega(I_\Gamma I + (I^2)^\Gamma)$, where $E_\Gamma := (\bar{E}_\Gamma)^{w_k w_\infty h}$. This allows us to compute the index $[\mathcal{O}_H^\times : \mu(H)E_\Gamma(E\cap L)]$. Indeed we have

$$\begin{aligned}[][\mathcal{O}_H^\times : \mu(H)E_\Gamma(E\cap L)] &= [l(\mathcal{O}_H^\times) : \omega(I_\Gamma I + (I^2)^\Gamma)] \\ &= \frac{1}{R_H}[I:\omega I]\cdot[\omega I : \omega I^2]\cdot[\omega I^2 : \omega(I_\Gamma I + (I^2)^\Gamma)] \\ &= \frac{1}{R_H}[I:\omega I]\cdot[I:I^2]\cdot[I^2 : I_\Gamma I + (I^2)^\Gamma]\end{aligned}$$

The formulas (3.2) and (3.3) give us

$$[\mathcal{O}_H^\times : \mu(H)E_\Gamma(E\cap L)] = (w_k w_\infty h)^{h-1} n^{[L:k]}\frac{h(\mathcal{O}_H)}{|G_n|}\frac{w_k}{w_\infty}. \tag{3.4}$$

Theorem 3. *We have*

$$[\mathcal{O}_{K_\mathfrak{t}}^\times : \bar{E}_\Gamma \mathcal{O}_L^\times] = n^{[L:k]} \frac{h(\mathcal{O}_H)}{h(\mathcal{O}_L)} \frac{w_L}{w_k}. \tag{3.5}$$

Proof. Let us observe first that

$$[\mathcal{O}_H^\times : E_\Gamma \mathcal{O}_L^\times] = \frac{[\mathcal{O}_H^\times : E_\Gamma(E \cap L)]}{[E_\Gamma \mathcal{O}_L^\times : E_\Gamma(E \cap L)]}.$$

But since E_Γ is the kernel of the map $N_{H/L} : E \longrightarrow L$, we have $E_\Gamma \cap L = \{1\}$ and $[E_\Gamma \mathcal{O}_L^\times : E_\Gamma(E \cap L)] = [\mathcal{O}_L^\times : E \cap L]$. Thus

$$\begin{aligned} [\mathcal{O}_H^\times : E_\Gamma \mathcal{O}_L^\times] &= \frac{[\mathcal{O}_H^\times : \mu(H) E_\Gamma(E \cap L)]}{[\mathcal{O}_L^\times : \mu(L)(E \cap L)]} \frac{w_\infty}{w_L} \\ &= (w_k w_\infty h)^{h-[L:k]} n^{[L:k]} \frac{h(\mathcal{O}_H)}{h(\mathcal{O}_L)} \end{aligned} \tag{3.6}$$

by the formulas (3.1) and (3.4). On the other hand, the isomorphism $\bar{E}_\Gamma \mathcal{O}_L^\times / E_\Gamma \mathcal{O}_L^\times \simeq M/N$, where $M := \bar{E}_\Gamma / E_\Gamma$ and $N := \mu(L) E_\Gamma / E_\Gamma$, and the fact that E_Γ is free of rank $h - [L : k]$, imply

$$[\bar{E}_\Gamma \mathcal{O}_L^\times : E_\Gamma \mathcal{O}_L^\times] = \frac{[\bar{E}_\Gamma : E_\Gamma]}{[\mu(L) E_\Gamma : E_\Gamma]} = \frac{w_\infty}{w_L} (w_k w_\infty h)^{h-[L:k]}. \tag{3.7}$$

Now the group $\mathcal{O}_{K_\mathfrak{t}}^\times / \mathcal{O}_H^\times$ contains exactly $\kappa := w_\infty / w_k$ elements. Therefore Theorem 3 is a consequence of the relation

$$[\mathcal{O}_{K_\mathfrak{t}}^\times : \bar{E}_\Gamma \mathcal{O}_L^\times] = \frac{w_\infty}{w_k} \frac{[\mathcal{O}_H^\times : E_\Gamma \mathcal{O}_L^\times]}{[\bar{E}_\Gamma \mathcal{O}_L^\times : E_\Gamma \mathcal{O}_L^\times]},$$

together with the formulas (3.6) and (3.7). ♡

Theorem 4. *Let $\bar{E}_{H,\Gamma} := \bar{E}_\Gamma \cap H$. Then we have*

$$[\mathcal{O}_H^\times : \bar{E}_{H,\Gamma} \mathcal{O}_L^\times] = n^{[L:k]} \frac{h(\mathcal{O}_H)}{h(\mathcal{O}_L)}. \tag{3.8}$$

Proof. Let $W := w_k w_\infty h$. Then the index $[\mathcal{O}_H^\times : \bar{E}_{H,\Gamma} \mathcal{O}_L^\times]$ can be expressed as follows:

$$[\mathcal{O}_H^\times : \bar{E}_{H,\Gamma} \mathcal{O}_L^\times] = [\mathcal{O}_H^\times : E_\Gamma \mathcal{O}_L^\times][E_\Gamma \mathcal{O}_L^\times : \bar{E}_{H,\Gamma}^W \mathcal{O}_L^\times] \,/\, [\bar{E}_{H,\Gamma} \mathcal{O}_L^\times : \bar{E}_{H,\Gamma}^W \mathcal{O}_L^\times]$$

But, as the indices $[\mathcal{O}_H^\times : E_\Gamma \mathcal{O}_L^\times]$ and $[\bar{E}_{H,\Gamma} \mathcal{O}_L^\times : \bar{E}_{H,\Gamma}^W \mathcal{O}_L^\times]$ are given respectively by (3.6) and (3.7), we see that we have only to compute

$[E_\Gamma \mathcal{O}_L^\times : \bar{E}_{H,\Gamma}^W \mathcal{O}_L^\times]$. Now the group $E_\Gamma \mathcal{O}_L^\times / \bar{E}_{H,\Gamma}^W \mathcal{O}_L^\times$ is isomorphic to $E_\Gamma / \bar{E}_{H,\Gamma}^W$. On the other hand, the logarithm map realizes an isomorphism from $E_\Gamma / \bar{E}_{H,\Gamma}^W$ onto $I_\Gamma I/(I_\Gamma I \cap \Delta^{(2)})$. Hence it remains to check that $[I_\Gamma I : I_\Gamma I \cap \Delta^{(2)}] = w_\infty / w_L$. Theorem 1 tells us that $I_\Gamma I/(I_\Gamma I \cap \Delta^{(2)}) \simeq \Psi^{(2)}(I_\Gamma I)$. It is well known that $\Psi^{(1)}(I_\Gamma) = m_L/\kappa$, where $m_L := w_L/w_k$. In particular we have $\Psi^{(1)}(I) = /\kappa$ and $\Psi^{(2)}(I_\Gamma I) = m_L/\kappa$, so we are done. ♡

4. Proof of Theorem 1.

Let $\widetilde{\Psi}^{(2)} : [G^2] \longrightarrow /\kappa$ be the map defined by

$$\sum_{g_1,g_2 \in G} n_{g_1,g_2}(g_1,g_2) \longmapsto \sum_{g_1,g_2 \in G} n_{g_1,g_2}\left(\frac{\psi(g_1)-1}{w_k}\right)\left(\frac{\psi(g_2)-1}{w_k}\right).$$

And let C_G be the submodule of $[G^2]$ whose elements are the sums $\sum_{g_1,g_2 \in G} n_{g_1,g_2}(g_1,g_2)$ such that $\sum_{g_1,g_2 \in G} n_{g_1,g_2}(g_1-1)(g_2-1) = 0$. Then the map $\Psi^{(2)}$ of Theorem 1 is defined if, and only if, we have $C_G \subseteq \operatorname{Ker} \widetilde{\Psi}^{(2)}$. We will prove this last condition by exhibiting a family of generators of C_G. Since $[G^2]/C_G \simeq I_G^2$ is free we see that C_G is a direct factor of $[G^2]$. In fact, if the group G is cyclic, we have

Proposition 4. *Suppose that G is a cyclic group and let g be a generator of G. Then*

$$[G^2] = C_G \oplus R_G, \tag{4.1}$$

where R_G is the -module generated by the elements $(g^i, g), i = 1, \ldots, |G|-1$. Also, we have

$$C_G = E_G \oplus F_G \tag{4.2}$$

where E_G is generated by $(1, g^i), i \in \{0, \ldots, |G|-1\}$; and F_G is generated by all the sums

$$g_{i,j} := (g^i, g^j) - (g^{i+j-1}, g) - (g^i, g^{j-1}) + (g^{j-1}, g), i, j \in \{1, \ldots, |G|-1\}.$$

Proof. We leave it as an exercise to verify that the sum $E_G + F_G + R_G$ is direct. On the other hand, we may prove by induction on j that $(g^i, g^j) \in E_G \oplus F_G \oplus R_G$, for all $i, j \in \{1, \ldots, |G|-1\}$. Moreover, since $(1, g) \in E_G$ and $g_{i,1} = (1, g) - (g^i, 1)$, we see that $(g^i, 1) \in F_G \oplus E_G \oplus R_G$ for $i = 1, \ldots, |G|-1$. Hence we obtain the decomposition

$$F_G \oplus E_G \oplus R_G = [G^2].$$

In particular $F_G \oplus E_G$ is a direct factor of C_G; but since these two -modules have the same rank, they are equal. ♡

Proposition 5. *The -module C_G is generated by the following elements of $[G^2]$:*

i) the couples $(1,d), d \in G$

ii) the differences $(c,d)-(d,c), c,d \in G-\{1\}$

iii) the sums $(c,de)-(cd,e)-(c,d)+(d,e), c,d,e \in G-\{1\}$

Proof. Let us denote by D_G the -module generated by the elements $i), ii)$, and $iii)$, and suppose that G is cyclic. Then the formula (4.2) implies $D_G = C_G$. On the other hand, suppose that $G = G_1 \times G_2$, where G_1 and G_2 are subgroups of G and let $a_1, a_2 \in G_1$ and $b_1, b_2 \in G_2$. Then the following elements of D_G

$$\begin{aligned}\alpha &:= (a_1b_1, a_2b_2) - (a_1b_1a_2, b_2) - (a_1b_1, a_2) + (a_2, b_2),\\ \beta &:= (a_1b_1, a_2) - (a_2, a_1b_1),\\ \gamma &:= (a_2, a_1b_1) - (a_2a_1, b_1) - (a_2, a_1) + (a_1, b_1),\\ \delta &:= (a_1a_2b_1, b_2) - (b_2, a_1a_2b_1),\\ \epsilon &:= (b_2, b_1a_1a_2) - (b_2b_1, a_1a_2) - (b_2, b_1) + (b_1, a_1a_2),\\ \rho &:= (a_1a_2, b_1) - (b_1, a_1a_2).\end{aligned}$$

are such that $\alpha+\beta+\gamma+\delta+\epsilon+\rho = (a_1b_1, a_2b_2)+(a_2,b_2)-(a_2,a_1)+(a_1,b_1)-(b_2,b_1)-(b_1b_2,a_1a_2)$. In particular we have

$$[G^2] \subset [G_1^2] + [G_2^2] + [G_1 \times G_2] + D_G.$$

Now, since $I_G = I_{G_1} \oplus I_{G_2} \oplus I_{G_1}I_{G_2}$, we deduce

$$C_G \subset C_{G_1} + C_{G_2} + ([G_1 \times G_2] \cap C_G) + D_G.$$

But $[G_1 \times G_2] \cap C_G = [G_1 \times \{1\}] + [\{1\} \times G_2] \subset D_G$. Thus, if $C_{G_i} = D_{G_i}$, for $i \in \{1,2\}$ then we also have $C_G = D_G$. Now, since G is a direct product of cyclic groups, we deduce by induction that $C_G = D_G$. ♡

The proposition 5 makes it clear now that $C_G \subseteq \operatorname{Ker} \widetilde{\Psi}^{(2)}$. So the morphism $\widetilde{\Psi}^{(2)}$ factors through $[G^2]/C_G \simeq I_G^2$ and yields a map $\Psi^{(2)} : I_G^2 \longrightarrow /\kappa$, as claims Theorem 1. On the other hand, since the norm of an ideal of $\mathcal{O}_k$ is a power of q, the group $G' := \psi(G)$ is cyclic. Thus, there exists $g \in G$ such that $\psi(g)$ generates G'. In particular $(\psi(g)-1)/w_k$ is prime to $\kappa := w_\infty/w_k$ (recall that w_k is the g.c.d. of $\psi(\tau)-1, \tau \in G$).

Now, let us write $\pi : [G] \longrightarrow [G']$ for the morphism $\sum_{\tau \in G} n_\tau \tau \longmapsto \sum_{\tau \in G} n_\tau \psi(\tau)$. Since the square $I_{G'}^2$ of the augmentation ideal $I_{G'}$ of $[G']$ is generated by $(\psi(g)-1)(\psi(g^i)-1), i = 1, \ldots, |G'|-1$, we see that an element $x \in I_G^2$ has the form

$$x = \sum_{i=1}^{|G'|-1} n_i (g-1)(g^i-1) + y\,, \quad \text{where } \; y \in I_G^2 \cap \operatorname{Ker} \pi.$$

It is clear that $\sum n_i(g-1)(g^i-1) \in \Delta^{(2)}$, if and only if, $\sum n_i(g^i-1) \in \Delta^{(1)}$. Thereby the inclusion $\Delta^{(2)} \subseteq I_G\Delta^{(1)}$ holds if we can prove that $I_G^2 \cap \mathrm{Ker}\,\pi \subseteq I_G\Delta^{(1)}$. But it is well known that $I_G^2 \cap \mathrm{Ker}\,\pi = I_D I_G$, where $D :=\mathrm{Ker}\,\psi$. So we are done.

References

[1] S.Galovich, M.Rosen, The class number of cyclotomic function fields, J.Number Theory **13**, (1981),363-375.

[2] B.Gross, M.Rosen, Fourier series and the special values of L-functions, Adv. Math. **69**, (1988), 1-31.

[3] D.Hayes, Elliptic units in function fields, In: Proc. of a conference on modern developments related to Fermat's last theorem, (D.Goldfeld ed.), Birkhäuser, Boston, 1982.

[4] D.Hayes, Stickelberger elements in function fields, Compositio. Math. **55** (1985), 209-239.

[5] D.Hayes, A brief introduction to Drinfeld modules, in "The Arithmetic of Function Fields" (D.Goss, D.Hayes, M.Rosen, Ed.) (1992), W.de Gruyter, Berlin, 1-32.

[6] H.Oukhaba, Groups of elliptic units in global function fields, in "The Arithmetic of Function Fields" (D.Goss, D.Hayes, M.Rosen, Ed.) (1992), W.de Gruyter, Berlin, 87-102.

[7] H.Oukhaba, Construction of Elliptic Units in Function Fields, Number Theory, edited by S.David, Cambridge University Press 1995, Printed in Great Britain, 187-207.

[8] L.Shu, Class Number formulas over global function fields, J.Number Theory **48** (1994), 133-161.

[9] L.Yin, Index-class number formulas over global function fields, Preprint (1995).

[10] L.Yin, On the index of cyclotomic units in characteristic p and its applications, Preprint (1996).

L-Series of Automorphic Cusp Forms of Drinfeld Type

Hans-Georg Rück
Institut für Experimentelle Mathematik
Universität GH Essen
Ellernstr.29, 45326 Essen (Germany)
rueck@exp-math.uni-essen.de

1 Introduction

In [Gr-Za] the authors give a relation between heights of Heegner points on modular curves and special values of the first derivative of L-series attached to cusp forms f and imaginary quadratic extensions of $\mathbb{Q}$. Their proof is divided into two parts which are more or less independent. In the first part they evaluate local and global height pairings of Heegner points. The second part is devoted to the representation of the L-series as the Petersson scalar product of f and a certain other cusp form. In the function field analogue of this theorem one works with Heegner points on Drinfeld modular curves. Their local heights are computed in [Ti]. The analogue of the L-series part is presented in this paper. We hope that we can combine both parts to formulas as in the case of number fields.

Let f be an automorphic cusp form of Drinfeld type corresponding to the rational function field $K = \mathbb{F}_q(T)$, and let $L = K(\sqrt{D})$ be an imaginary quadratic extension of K. We represent $L(f, \mathcal{A}, s)$, the L-series of f and an ideal class $\mathcal{A}$ of L, as the scalar product of f and another cusp form F_s of Drinfeld type. In Theorem 9 we evaluate the Fourier coefficients of F_s. As an example we concentrate to the case that D is an irreducible polynomial in $\mathbb{F}_q[T]$ of odd degree. The general case will follow the same pattern. Special values of these L-series or its derivatives can be obtained by specializing the "closed form" in Theorem 9.

There are a lot of calculations in this paper which are deeply influenced by the ideas in [Gr-Za]. We start with "Rankin's method" where we used the transformation rule of theta series proved in [Rü1], and we get a function Φ_s (Proposition 2) which is not necessarily a cusp form of Drinfeld type. The holomorphic projection formula (Proposition 3) using Poincaré series (cf. [Rü2]) leads us to the final result.

2 Basic Definitions

Let $\mathbb{F}_q$ be the finite field with odd characteristic p and $q = p^\alpha$ elements, and let $K = \mathbb{F}_q(T)$ be the rational function field over $\mathbb{F}_q$. For the place ∞ of K we consider the completion K_∞ with normalized valuation v_∞ and valuation ring O_∞. We fix the prime $\pi_\infty = T^{-1}$; each $x \in K_\infty$ has a power series expansion $x = \sum_{i \geq i_0} a_i \pi_\infty^i$ with $a_i \in \mathbb{F}_q$.

We define the following additive character ψ_∞ of K_∞: Take $\sigma : \mathbb{F}_q \to \mathbb{C}^*$ with $\sigma(a) = \exp(\frac{2\pi i}{p} Tr_{\mathbb{F}_q/\mathbb{F}_p}(a))$ and set $\psi_\infty(\sum a_i \pi_\infty^i) = \sigma(-a_1)$. This definition agrees with the usual one because the divisor of the differential dt equals -2∞.

The oriented edges of the Bruhat-Tits tree are given by $GL_2(K_\infty)/\Gamma_\infty K_\infty^*$, where

$$\Gamma_\infty = \{\begin{pmatrix} \alpha & \beta \\ \gamma & \delta \end{pmatrix} \in GL_2(O_\infty) \mid v_\infty(\gamma) > 0\}.$$

It can be represented by the two disjoint sets

$$\mathcal{T}_+ := \{\begin{pmatrix} \pi_\infty^m & u \\ 0 & 1 \end{pmatrix} \mid m \in \mathbb{Z}, u \in K_\infty/\pi_\infty^m O_\infty\}$$

and

$$\mathcal{T}_- := \{\begin{pmatrix} \pi_\infty^m & u \\ 0 & 1 \end{pmatrix}\begin{pmatrix} 0 & 1 \\ \pi_\infty & 0 \end{pmatrix} \mid m \in \mathbb{Z}, u \in K_\infty/\pi_\infty^m O_\infty\}.$$

We do not distinguish between matrices in $GL_2(K_\infty)$ and the corresponding classes in $GL_2(K_\infty)/\Gamma_\infty K_\infty^*$.

The groups $GL_2(\mathbb{F}_q[T])$ and $SL_2(\mathbb{F}_q[T])$ operate on $GL_2(K_\infty)/\Gamma_\infty K_\infty^*$. For $N \in \mathbb{F}_q[T]$ let

$$\Gamma_0(N) = \{\begin{pmatrix} a & b \\ c & d \end{pmatrix} \in GL_2(\mathbb{F}_q[T]) \mid c \equiv 0\,(N)\}$$

and $\Gamma_0^{(1)}(N) = \Gamma_0(N) \cap SL_2(\mathbb{F}_q[T])$.

An automorphic cusp form of Drinfeld type of level N (or harmonic cochain with values in $\mathbb{C}$) is a function $f : GL_2(K_\infty)/\Gamma_\infty K_\infty^* \to \mathbb{C}$ which satisfies the following conditions
i) f is harmonic, i.e. ,

$$f(X\begin{pmatrix} 0 & 1 \\ \pi_\infty & 0 \end{pmatrix}) = -f(X)$$

and

$$\sum_{\beta \in GL_2(O_\infty)/\Gamma_\infty} f(X\beta) = 0$$

for all $X \in GL_2(K_\infty)/\Gamma_\infty K_\infty^*$,
ii) f is invariant under $\Gamma_0(N)$, i.e. ,

$$f(AX) = f(X)$$

for all $X \in GL_2(K_\infty)/\Gamma_\infty K_\infty^*$ and $A \in \Gamma_0(N)$,
iii) f has compact support modulo $\Gamma_0(N)$, i.e. there are only finitely many elements $\bar{X}$ in $\Gamma_0(N)\backslash GL_2(K_\infty)/\Gamma_\infty K_\infty^*$ with $f(\bar{X}) \neq 0$.

Since

$$f(\begin{pmatrix} \pi_\infty^m & u \\ 0 & 1 \end{pmatrix}) = f(\begin{pmatrix} \pi_\infty^m & u+h \\ 0 & 1 \end{pmatrix})$$

for all $h \in \mathbb{F}_q[T]$, we can consider the Fourier transform f^* of f, and we get

$$f(\begin{pmatrix} \pi_\infty^m & u \\ 0 & 1 \end{pmatrix}) = \sum_{\lambda \in \mathbb{F}_q[T]} f^*(\pi_\infty^m, \lambda)\psi_\infty(\lambda u)$$

with

$$f^*(\pi_\infty^m, \lambda) = \int_{K_\infty/\mathbb{F}_q[T]} f(\begin{pmatrix} \pi_\infty & u \\ 0 & 1 \end{pmatrix})\psi_\infty(-\lambda u)\, du,$$

where du is a Haar measure with $\int_{K_\infty/\mathbb{F}_q[T]} du = 1$.

The definition of cusp forms immediately yields:

$$f^*(\pi_\infty^m, \lambda) = 0, \text{ if } \lambda = 0 \text{ or if } \deg\lambda + 2 > m$$

$$f^*(\pi_\infty^m, \lambda) = q^{-m+\deg\lambda+2} f^*(\pi_\infty^{\deg\lambda+2}, \lambda), \text{ if } \lambda \neq 0 \text{ and } \deg\lambda + 2 \leq m.$$

Hence all the Fourier coefficients of f are uniquely determined by the coefficients $f^*(\pi_\infty^{\deg\lambda+2}, \lambda)$.

Now we want to define the L-series attached to f. Let $\mathfrak{m}$ be an effective divisor of K of degree n, then $\mathfrak{m} = (\lambda)_0 + (n - \deg\lambda)\infty$ with $\lambda \in \mathbb{F}_q[T]$, $\deg\lambda \leq n$. We define $f^*(\mathfrak{m}) = f^*(\pi_\infty^{n+2}, \lambda)$. The L-series to f is then $L(f,s) = \sum_{\mathfrak{m}\geq 0} f^*(\mathfrak{m})\mathrm{N}(\mathfrak{m})^{-s}$, where $\mathrm{N}(\mathfrak{m})$ denotes the norm of the divisor $\mathfrak{m}$. The definition yields immediately (cf. [We2])

$$L(f,s) = \frac{1}{q-1}\sum_{m=2}^{\infty}(\sum_{\deg\lambda+2\leq m} f^*(\pi_\infty^m, \lambda))\, q^{-(m-2)s}.$$

If E is an elliptic curve over K with split multiplicative reduction at ∞, then the corresponding L-series is equal to the L-series $L(f,s)$ of an appropriate automorphic cusp form f (cf. [Wel], [Ge-Re]).

Let L/K be an imaginary quadratic extension (i.e. a quadratic extension where ∞ is not decomposed) in which each (finite) divisor of N is not ramified. Then there is a squarefree polynomial $D \in \mathbb{F}_q[T]$, prime to N with $L = K(\sqrt{D})$. We assume in this paper that D is irreducible and at the end in addition that its degree is odd. This has only technical reasons. In the general case the main arguments are the same, but it is more complicated to keep track of the formulas.

For a class $\mathcal{A}$ in the ideal class group of L we want to consider the partial zeta function $\sum_{\mathfrak{m}\geq 0} r_{\mathcal{A}}(\mathfrak{m})\mathrm{N}(\mathfrak{m})^{-s}$, where for a divisor $\mathfrak{m} = (\lambda)_0 + (n - \deg\lambda)\infty$ as above the number $r_{\mathcal{A}}(\mathfrak{m})$ is equal to
$\#\{\mathfrak{a} \in \mathcal{A} \mid \mathfrak{a} \text{ integral ideal in } \mathcal{A} \text{ with } \mathrm{N}_{L/K}(\mathfrak{a}) = \lambda\mathbb{F}_q[T]\}$.

Alternatively one can choose $\mathfrak{a}_0 \in \mathcal{A}^{-1}$ and $\lambda_0 \in K$ with $\mathrm{N}_{L/K}(\mathfrak{a}_0) = \lambda_0\mathbb{F}_q[T]$ and define $r_{\mathfrak{a}_0,\lambda_0}(\lambda) = \#\{\mu \in \mathfrak{a}_0 \mid \mathrm{N}_{L/K}(\mu) = \lambda_0\lambda\}$.
Then $r_{\mathcal{A}}(\mathfrak{m}) = \frac{1}{q-1}\sum_{\epsilon\in\mathbb{F}_q^*} r_{\mathfrak{a}_0,\lambda_0}(\epsilon\lambda)$ and the L-series of the theta series

$$\Theta_{\mathfrak{a}_0,\lambda_0}\left(\begin{pmatrix} \pi_\infty^m & u \\ 0 & 1 \end{pmatrix}\right) = \sum_{\deg\lambda+2\leq m} r_{\mathfrak{a}_0,\lambda_0}(\lambda)\psi_\infty(\lambda u)$$

satisfies

$$L(\Theta_{\mathfrak{a}_0,\lambda_0}, s) = \sum_{\mathfrak{m}\geq 0} r_{\mathcal{A}}(\mathfrak{m})\mathrm{N}(\mathfrak{m})^{-s}.$$

In this paper we want to study the L-series attached to an automorphic cusp form of Drinfeld type of level N and to a class $\mathcal{A}$

$$L(f,\mathcal{A},s) = \sum_{\mathfrak{m}\geq 0} f^*(\mathfrak{m})r_{\mathcal{A}}(\mathfrak{m})\mathrm{N}(\mathfrak{m})^{-s}.$$

3 Rankin's Method

The properties of the automorphic cusp form f yield

$$f^*(\pi_\infty^m,\lambda) = q^{-m+1}\sum_{u\in\pi_\infty/\pi_\infty^m} f\left(\begin{pmatrix} \pi_\infty^m & u \\ 0 & 1 \end{pmatrix}\right)\psi(-\lambda u).$$

Now we calculate

$$L(f,\mathcal{A},s) = \frac{1}{q-1}\sum_{m=2}^{\infty}\Big(\sum_{\deg\lambda+2\le m} f^*(\pi_\infty^m,\lambda)\overline{r_{\mathfrak{a}_0,\lambda_0}(\lambda)\Big)\, q^{-(m-2)s}}$$

$$= \frac{q}{q-1}\sum_{m=2}^{\infty}\sum_{u\in\pi_\infty/\pi_\infty^m} f\Big(\begin{pmatrix}\pi_\infty^m & u\\ 0 & 1\end{pmatrix}\Big)\overline{\Theta_{\mathfrak{a}_0,\lambda_0}\Big(\begin{pmatrix}\pi_\infty^m & u\\ 0 & 1\end{pmatrix}\Big)\, q^{-m(s+1)+2s}}$$

$$= \frac{q}{q-1}\int_{\mathbb{H}_\infty} f\Big(\begin{pmatrix}\pi_\infty^m & u\\ 0 & 1\end{pmatrix}\Big)\overline{\Theta_{\mathfrak{a}_0,\lambda_0}\Big(\begin{pmatrix}\pi_\infty^m & u\\ 0 & 1\end{pmatrix}\Big)}\, q^{-m(\bar{s}+1)+2\bar{s}},$$

where

$$\mathbb{H}_\infty := \begin{pmatrix}1 & \mathbb{F}_q[T]\\ 0 & 1\end{pmatrix}\backslash\begin{pmatrix}K_\infty^* & K_\infty\\ 0 & 1\end{pmatrix}/\begin{pmatrix}O_\infty^* & O_\infty\\ 0 & 1\end{pmatrix}.$$

We consider the surjective mapping

$$\mathbb{H}_\infty \to \Gamma_0^{(1)}(ND)\backslash GL_2(K_\infty)/\Gamma_\infty K_\infty^* =: X_0^{(1)}(ND)$$

and take on $X_0^{(1)}(ND)$ the measure which counts the stabilizer of an element (cf. [Ge-Re], (4.8)). Then we get $L(f,\mathcal{A},s) =$

$$= \frac{q}{2(q-1)}\int_{X_0^{(1)}(ND)} f\Big(\begin{pmatrix}\pi_\infty^m & u\\ 0 & 1\end{pmatrix}\Big)\overline{\sum_M \Theta_{\mathfrak{a}_0,\lambda_0}\Big(M\begin{pmatrix}\pi_\infty^m & u\\ 0 & 1\end{pmatrix}\Big)\, q^{-m^*(\bar{s}+1)+2\bar{s}}}$$

where the sum is taken over those $M = \begin{pmatrix}a & b\\ c & d\end{pmatrix} \in \begin{pmatrix}1 & \mathbb{F}_q[T]\\ 0 & 1\end{pmatrix}\backslash\Gamma_0^{(1)}(ND)$ with $M\begin{pmatrix}\pi_\infty^m & u\\ 0 & 1\end{pmatrix} \in \mathcal{T}_+$, and $m^* = m - 2v_\infty(cu+d)$. We remark that $M\begin{pmatrix}\pi_\infty^m & u\\ 0 & 1\end{pmatrix} \in \mathcal{T}_+$ if and only if $v_\infty(c\pi_\infty^m) > v_\infty(cu+d)$.

In ([Rü1], Theorem 6.2) we showed that for these M one has the following transformation rule

$$\Theta_{\mathfrak{a}_0,\lambda_0}\Big(M\begin{pmatrix}\pi_\infty^m & u\\ 0 & 1\end{pmatrix}\Big) = \Theta_{\mathfrak{a}_0,\lambda_0}\Big(\begin{pmatrix}\pi_\infty^m & u\\ 0 & 1\end{pmatrix}\Big)\left[\frac{d}{D}\right]\delta_{cu+d}\, q^{-v_\infty(cu+d)}, \tag{3.1}$$

where $\left[\frac{\cdot}{D}\right]$ is the Legendre symbol at D and δ denotes the local norm symbol at ∞. If one introduces the partial L-series

$$L^{(N)}(2s+1) = \frac{1}{q-1}\sum_{\substack{k\in\mathbb{F}_q[T]\\ \gcd(k,N)=1}}\left[\frac{D}{k}\right] q^{-(2s+1)\deg k},$$

some straightforward calculations yield

$$L^{(N)}(2s+1)L(f,\mathcal{A},s) = \frac{q}{2(q-1)} \int\limits_{X_0^{(1)}(ND)} f \cdot \overline{\Theta_{\mathfrak{a}_0,\lambda_0} E_{1,\bar{s}}}$$

with

$$E_{1,s}(\begin{pmatrix} \pi_\infty^m & u \\ 0 & 1 \end{pmatrix}) = q^{-m(s+1)+2s} \sum_{\substack{c,d\in\mathbb{F}_q[T] \\ c\equiv 0(ND) \\ \gcd(d,N)=1 \\ v_\infty(c\pi_\infty^m)>v_\infty(cu+d)}} \left[\frac{d}{D}\right] \delta_{cu+d}\, q^{v_\infty(cu+d)(2s+1)}.$$

In particular $\Theta_{\mathfrak{a}_0,\lambda_0} E_{1,s}$ is a function on $X_0^{(1)}(ND)$.

Let $\mu : \mathbb{F}_q[T] \to \{0,1,-1\}$ be the Moebius function with $\sum_{e|n} \mu(e) = 0$ if $n\,\mathbb{F}_q[T] \neq \mathbb{F}_q[T]$, and $\frac{1}{q-1}\sum_{e|1} \mu(e) = 1$, then $E_{1,s}(\begin{pmatrix} \pi_\infty^m & u \\ 0 & 1 \end{pmatrix}) =$

$$= \frac{q^{-m(s+1)+2s}}{q-1} \sum_{e|N} \mu(e) \left[\frac{e}{D}\right] \delta_e\, q^{-(2s+1)\deg e} E_s^{(1)}(\begin{pmatrix} \frac{N\pi_\infty^m}{e} & \frac{Nu}{e} \\ 0 & 1 \end{pmatrix})$$

with the Eisenstein series

$$E_s^{(1)}(\begin{pmatrix} \pi_\infty^m & u \\ 0 & 1 \end{pmatrix}) = \sum_{\substack{c,d\in\mathbb{F}_q[T] \\ c\equiv 0(D) \\ v_\infty(c\pi_\infty^m)>v_\infty(cu+d)}} \left[\frac{d}{D}\right] \delta_{cu+d}\, q^{v_\infty(cu+d)(2s+1)}.$$

The function

$$\Theta_{\mathfrak{a}_0,\lambda_0}(\begin{pmatrix} \pi_\infty^m & u \\ 0 & 1 \end{pmatrix})\, q^{-m(s+1)+2s} E_s^{(1)}(\begin{pmatrix} \frac{N\pi_\infty^m}{e} & \frac{Nu}{e} \\ 0 & 1 \end{pmatrix})$$

is invariant under $\Gamma_0^{(1)}(\frac{ND}{e})$. We assume that f is a newform of level N. Then f is orthogonal (with respect to the Petersson scalar product) to functions of lower level. Therefore we get

Proposition 1 *Let f be a newform of level N, then*

$$L^{(N)}(2s+1)L(f,\mathcal{A},s) = \frac{q}{2(q-1)} \int\limits_{X_0^{(1)}(ND)} f \cdot \overline{\Theta_{\mathfrak{a}_0,\lambda_0} E_{2,\bar{s}}}$$

with

$$E_{2,s}(\begin{pmatrix} \pi_\infty^m & u \\ 0 & 1 \end{pmatrix}) = q^{-m(s+1)+2s} E_s^{(1)}(\begin{pmatrix} N\pi_\infty^m & Nu \\ 0 & 1 \end{pmatrix})$$

$$= q^{-m(s+1)+2s} \sum_{\substack{c,d\in\mathbb{F}_q[T],\ c\equiv 0(D) \\ v_\infty(cN\pi_\infty^m) > v_\infty(cNu+d)}} \left[\tfrac{d}{D}\right] \delta_{cNu+d}\, q^{v_\infty(cNu+d)(2s+1)}.$$

4 Computation of the Trace

$\Theta_{\mathfrak{a}_0,\lambda_0} E_{2,s}$ is only invariant under $\Gamma_0^{(1)}(ND)$. To make it invariant under $\Gamma_0^{(1)}(N)$ we compute the trace with respect to $\Gamma_0^{(1)}(ND)\backslash\Gamma_0^{(1)}(N)$. The trace from $\Gamma_0^{(1)}(N)$ to $\Gamma_0(N)$ is easy, this will be done at the very end of the calculations.

Since N and D are relatively prime, there are $\mu_1, \mu_2 \in \mathbb{F}_q[T]$ with $1 = \mu_1 N + \mu_2 D$. The set

$$R = \{\begin{pmatrix} 1 & 0 \\ 0 & 1 \end{pmatrix}, \begin{pmatrix} 1 & 1 \\ \mu_2 D & \mu_1 N \end{pmatrix}\begin{pmatrix} 0 & -1 \\ 1 & \lambda \end{pmatrix} \lambda \bmod D\}$$

is therefore a set of representatives of $\Gamma_0^{(1)}(ND)\backslash\Gamma_0^{(1)}(N)$. Here we used the assumption that D is irreducible. We want to evaluate $\sum_{M\in R} \Theta_{\mathfrak{a}_0,\lambda_0} E_{2,s}(M\cdot)$. We treat $\Theta_{\mathfrak{a}_0,\lambda_0}$ and $E_{2,s}$ separately.

From ([Rül], Prop. 4.4) we get, if $m > v_\infty(u)$

$$\Theta_{\mathfrak{a}_0,\lambda_0}(\begin{pmatrix} \frac{\pi_\infty^m}{u^2} & \frac{1}{u} \\ 0 & 1 \end{pmatrix}) = \Theta_{\mathfrak{a}_0,\lambda_0}(\begin{pmatrix} \frac{\pi_\infty^m}{D} & \frac{u}{D} \\ 0 & 1 \end{pmatrix})\delta_u\, q^{-v_\infty(u)}\delta_{-\lambda_0} q^{-\frac{1}{2}\deg D}\epsilon_0^{-1}$$

where $\epsilon_0 = 1$ if $\deg D$ is even and $\epsilon_0 = \delta_{-t}(-1)^{\alpha+1}\gamma(p)^\alpha$ ($q = p^\alpha$; $\gamma(p) = 1$ or i if $p \equiv 1$ or $3 \bmod 4$) if $\deg D$ is odd. Then it is not difficult to evaluate

$$\Theta_{\mathfrak{a}_0,\lambda_0}(\begin{pmatrix} 0 & -1 \\ 1 & \lambda \end{pmatrix}\begin{pmatrix} \pi_\infty^m & u \\ 0 & 1 \end{pmatrix}) = \tag{4.1}$$

$$\Theta_{\mathfrak{a}_0,\lambda_0}(\begin{pmatrix} \frac{\pi_\infty^m}{D} & \frac{-(u+\lambda)}{D} \\ 0 & 1 \end{pmatrix}) \cdot \delta_{u+\lambda}\, q^{-v_\infty(u+\lambda)}\delta_{\lambda_0}\, q^{-\frac{1}{2}\deg D}\epsilon_0^{-1}.$$

Now (3.1) and (4.1) yield the operation of the matrices $M \in R$ on $\Theta_{\mathfrak{a}_0,\lambda_0}$.

The situation for the Eisenstein series $E_s^{(1)}$ is easier. Straight forward calculations (mainly transformations of the summation indices) yield:

If $\begin{pmatrix} a & b \\ c & d \end{pmatrix} \in SL_2(\mathbb{F}_q[T])$ with $\gcd(c, D) = 1$ and if $v_\infty(c\pi_\infty^m) > v_\infty(cu+d)$ we get

$$E_s^{(1)}(\begin{pmatrix} a & b \\ c & d \end{pmatrix}\begin{pmatrix} \pi_\infty^m & u \\ 0 & 1 \end{pmatrix}) = \tag{4.2}$$

$$E_s^{(D)}(\begin{pmatrix} \frac{\pi_\infty^m}{D} & \frac{u+c^*d)}{D} \\ 0 & 1 \end{pmatrix})\left[\tfrac{c}{D}\right]\delta_D\, q^{-(2s+1)\deg D}\cdot \delta_{cu+d}\, q^{-v_\infty(cu+d)(2s+1)}$$

with $c^* \equiv c^{-1}(D)$. Here $E_s^{(D)}$ is the Eisenstein series

$$E_s^{(D)}(\begin{pmatrix} \pi_\infty^m & u \\ 0 & 1 \end{pmatrix}) = \sum_{\substack{c,d\in\mathbb{F}_q[T] \\ v_\infty(c\pi_\infty^m)>v_\infty(cu+d)}} \left[\frac{c}{D}\right]\delta_{cu+d}\, q^{v_\infty(cu+d)(2s+1)}.$$

At least the application of (3.1), (4.1) and (4.2) yields

Proposition 2 *Let f be as above and let D be irreducible, then*

$$L^{(N)}(2s+1)L(f,\mathcal{A},s) = \frac{q}{2(q-1)}\int\limits_{X_0^{(1)}(N)} f\cdot\overline{\Phi_{\bar{s}}}$$

with

$$\Phi_s(\begin{pmatrix} \pi_\infty^m & u \\ 0 & 1 \end{pmatrix}) = \textstyle\sum_{M\in R}\Theta_{a_0,\lambda_0}E_{2,s}(M\begin{pmatrix} \pi_\infty^m & u \\ 0 & 1 \end{pmatrix})$$

$$= q^{-\deg D}\textstyle\sum_{\lambda \bmod D}\Theta_{a_0,\lambda_0}(\begin{pmatrix} \frac{\pi_\infty^m}{D} & \frac{-(u+\lambda)}{D} \\ 0 & 1 \end{pmatrix})\,\mathcal{E}_s(\begin{pmatrix} \frac{\pi_\infty^m}{D} & \frac{u+\lambda}{D} \\ 0 & 1 \end{pmatrix})$$

where

$$\mathcal{E}_s(\begin{pmatrix} \pi_\infty^m & u \\ 0 & 1 \end{pmatrix}) = q^{(s+1)\deg D+2s}q^{-m(s+1)}\Big[E_s^{(1)}(\begin{pmatrix} ND\pi_\infty^m & NDu \\ 0 & 1 \end{pmatrix})+$$

$$+E_s^{(D)}(\begin{pmatrix} N\pi_\infty^m & Nu \\ 0 & 1 \end{pmatrix})\,\delta_{\lambda_0 DN}\,\epsilon_0^{-1}\left[\tfrac{D}{N}\right]q^{(-\frac{1}{2}-2s)\deg D}\Big].$$

5 Holomorphic Projection

We want to evaluate the integral $\int_{X_0^{(1)}(N)} f \cdot \overline{\Phi_{\bar{s}}}$, where f is an automorphic cusp form of Drinfeld type of level N and Φ_s is a function on $X_0^{(1)}(N)$. Since the Petersson scalar product is nondegenerate on cusp forms, we find an automorphic cusp form Ψ_s of Drinfeld type to $\Gamma_0^{(1)}(N)$ (one has to modify the definition of cusp forms to $\Gamma_0^{(1)}(N)$ in an obvious way) such that

$$\int_{X_0^{(1)}(N)} g \cdot \overline{\Psi_{\bar{s}}} = \int_{X_0^{(1)}(N)} g \cdot \overline{\Phi_{\bar{s}}}$$

for all cusp forms g.
If we set $g = f$ we obtain our result. It is the aim to compute the Fourier coefficients of Ψ_s from those of Φ_s.

For this we take $g = P_\lambda$, where P_λ ($\lambda \in \mathbb{F}_q[T], \lambda \neq 0$) are the Poincaré series introduced in [Rü2], and evaluate (cf. [Rü2], Prop. 14)

$$\int_{X_0^{(1)}(N)} P_\lambda \cdot \overline{\Psi_{\bar{s}}} = \frac{4}{q-1} \overline{\Psi_{\bar{s}}^*(\pi_\infty^{\deg \lambda + 2}, \lambda)}. \tag{5.1}$$

On the other hand we calculate

$$\int_{X_0^{(1)}(N)} P_\lambda \cdot \overline{\Phi_{\bar{s}}} = 2 \int_{\mathbb{H}_\infty} g_\lambda \cdot \overline{\Phi_{\bar{s}}} - 2 \int_{\mathbb{H}_\infty} g_\lambda \cdot \overline{\widetilde{\Phi}_{\bar{s}}} \tag{5.2}$$

where

$$g_\lambda\left(\begin{pmatrix} \pi_\infty^m & u \\ 0 & 1 \end{pmatrix}\right) := \begin{cases} 0 & \text{if } \deg \lambda + 2 > m \\ q^{-m}\psi_\infty(\lambda u) & \text{if } \deg \lambda + 2 \leq m \end{cases}$$

and where

$$\widetilde{\Phi}_s\left(\begin{pmatrix} \pi_\infty^m & u \\ 0 & 1 \end{pmatrix}\right) = \Phi_s\left(\begin{pmatrix} \pi_\infty^m & u \\ 0 & 1 \end{pmatrix}\begin{pmatrix} 0 & 1 \\ \pi_\infty & 0 \end{pmatrix}\right).$$

At these calculations we used again the surjective mapping $\mathbb{H}_\infty \to X_0^{(1)}(N)$. Since $\mathbb{H}_\infty$ represents only the part $\mathcal{T}_+$ of $GL_2(K_\infty)/\Gamma_\infty K_\infty^*$ and since Φ_s is not necessarily harmonic, we have also to consider the function $\widetilde{\Phi}_s$.

Using the Fourier expansions

$$\Phi_s(\begin{pmatrix} \pi_\infty^m & u \\ 0 & 1 \end{pmatrix}) = \sum_\mu \Phi_s^*(\pi_\infty^m, \mu)\psi_\infty(\mu u);$$

$$\widetilde{\Phi}_s(\begin{pmatrix} \pi_\infty^m & u \\ 0 & 1 \end{pmatrix}) = \sum_\mu \widetilde{\Phi}_s^*(\pi_\infty^m, \mu)\psi_\infty(\mu u)$$

and the character relations for ψ_∞, (5.2) yields

$$\int_{X_0^{(1)}(N)} P_\lambda \cdot \overline{\Phi_{\bar{s}}} = \frac{2}{q} \sum_{m=\deg\lambda+2}^{\infty} \overline{(\Phi_{\bar{s}}^*(\pi_\infty^m, \lambda) - \widetilde{\Phi}_{\bar{s}}^*(\pi_\infty^m, \lambda)} \tag{5.3}$$

Finally (5.1) and (5.3) prove

Proposition 3 *Let f and D be as above, then*

$$L^{(N)}(2s+1)L(f, \mathcal{A}, s) = \frac{q}{2(q-1)} \int_{X_0^{(1)}(N)} f \cdot \overline{\Psi_{\bar{s}}},$$

with an automorphic cusp form Ψ_s of Drinfeld type to $\Gamma_0^{(1)}(N)$, whose Fourier coefficients can be evaluated by

$$\Psi_s^*(\pi_\infty^{\deg\lambda+2}, \lambda) = \frac{q-1}{2q} \sum_{m=\deg\lambda+2}^{\infty} (\Phi_s^*(\pi_\infty^m, \lambda) - \widetilde{\Phi}_s^*(\pi_\infty^m, \lambda)),$$

where Φ_s is the function of Proposition 2.

6 Fourier Expansion of Φ_s

In this section we evaluate the Fourier coefficients $\Phi_s^*(\pi_\infty^m, \lambda)$. The function $\Theta_{\mathfrak{a}_0,\lambda_0}$ is already defined by its coefficients $r_{\mathfrak{a}_0,\lambda_0}$. It remains to evaluate the coefficients of $\mathcal{E}_s$ and therefore of the Eisenstein series $E_s^{(1)}$ and $E_s^{(D)}$.

We introduce a "basic function" on $GL_2(K_\infty)/\Gamma_\infty K_\infty^*$:

$$F_s(\begin{pmatrix} \pi_\infty^m & u \\ 0 & 1 \end{pmatrix}) = \sum_\lambda F_s^*(\pi_\infty^m, \lambda)\psi_\infty(\lambda u) = \sum_{\substack{d \in \mathbb{F}_q[T] \\ m > v_\infty(u+d)}} \delta_{u+d}\, q^{v_\infty(u+d)(2s+1)}.$$

At first we express the Eisenstein series in terms of F_s.

Elementary transformations give

$$E_s^{(1)}\left(\begin{pmatrix} ND\pi_\infty^m & NDu \\ 0 & 1 \end{pmatrix}\right) = \sum_{\substack{d\in\mathbb{F}_q[T] \\ d\neq 0}} \left[\frac{D}{d}\right] q^{-(2s+1)\deg d} + \delta_D\, q^{-(2s+1)\deg D}\,.$$

$$\cdot\sum_{\mu\neq 0}\left[\sum_{\substack{c|\mu \\ c\equiv 0(D)}} F_s^*(cN\pi_\infty^m,\frac{\mu}{c}) \sum_{d \bmod D}\left[\frac{d}{D}\right]\psi_\infty(\frac{\mu}{c}\frac{d}{D})\right]\psi_\infty(\mu N u).$$

The Gauss sum can be evaluated

$$\sum_{d \bmod D}\left[\frac{d}{D}\right]\psi_\infty(\lambda\frac{d}{D}) = \left[\frac{\lambda}{D}\right]\epsilon_0^{-1} q^{\frac{1}{2}\deg D}.$$

Therefore

$$E_s^{(1)}\left(\begin{pmatrix} ND\pi_\infty^m & NDu \\ 0 & 1 \end{pmatrix}\right) = \tag{6.1}$$

$$\sum\nolimits_{d\in\mathbb{F}_q[T] d\neq 0}\left[\tfrac{D}{d}\right] q^{-(2s+1)\deg d} + \epsilon_0^{-1}\delta_D\, q^{(-2s-\frac{1}{2})\deg D}\,.$$

$$\cdot\sum\nolimits_{\mu\neq 0}\sum_{\substack{c|\mu \\ c\equiv 0(D)}}\left[\tfrac{\mu/c}{D}\right] F_s^*(cN\pi_\infty^m,\tfrac{\mu}{c})\psi_\infty(\mu N u). \tag{6.2}$$

The same transformations as above yield

$$E_s^{(D)}\left(\begin{pmatrix} N\pi_\infty^m & Nu \\ 0 & 1 \end{pmatrix}\right) = \sum_{\substack{d\in\mathbb{F}_q[T] \\ d\neq 0}}\left[\frac{d}{D}\right] F_s^*(dN\pi_\infty^m,0) +$$

$$+\sum_{\mu\neq 0}\sum_{c|\mu}\left[\frac{c}{D}\right] F_s^*(cN\pi_\infty^m,\frac{\mu}{c})\psi_\infty(\mu N u). \tag{6.3}$$

Now we have to evaluate the Fourier coefficients of the "basic function" F_s. This is not very difficult, perhaps a little tedious to write down in detail. One starts with the definition

$$F_s^*(\pi_\infty^m,\lambda) = q^{-m+1}\sum_{u\in\pi_\infty/\pi_\infty^m} F_s\left(\begin{pmatrix} \pi_\infty^m & u \\ 0 & 1 \end{pmatrix}\right)\psi_\infty(-\lambda u)$$

and uses the character relations for ψ_∞. We don't want to carry it out in detail. As an example we restrict to $\deg D$ being odd (the other case is very similar). Then L_∞/K_∞ is ramified and the local norm symbol for $z = e_z\pi_\infty^n + \dots$ is given by $\delta_z = \chi_2(e_z)\delta_T^n$ (χ_2 is the quadratic character on $\mathbb{F}_q^*$; we remind that $\pi_\infty = T^{-1}$).

We get

Lemma 4 *Let $D \in \mathbb{F}_q[T]$ be irreducible of odd degree, then*

$$F_s^*(\pi_\infty^m, \mu) = \begin{cases} 0 & , \text{if either } \mu = 0 \text{ or } \deg\mu + 2 > m \\ \epsilon_0^{-1} q^{\frac{1}{2}} \delta_\mu q^{2s(\deg\mu+1)} & , \text{if } \mu \neq 0 \text{ and } \deg\mu + 2 \leq m \end{cases}.$$

Then (6.1), (6.2) and the lemma give

Proposition 5 *Let D be irreducible and* $\deg D$ *be odd, then*

$$\mathcal{E}_s(\begin{pmatrix} \pi_\infty^m & u \\ 0 & 1 \end{pmatrix}) = \sum_{\deg(\mu N)+2 \leq m} e_s(\pi_\infty^m, \mu)\psi_\infty(\mu N u)$$

with

$$e_s(\pi_\infty^m, 0) = q^{(s+1)\deg D + 2s} q^{-m(s+1)} \sum_{\substack{d \in \mathbb{F}_q[T] \\ d \neq 0}} \left[\frac{D}{d}\right] q^{-(2s+1)\deg d}$$

and ($\mu \neq 0$)

$$\begin{aligned} e_s(\pi_\infty^m, \mu) &= q^{(-s+\frac{1}{2})\deg D + 4s + \frac{1}{2}} q^{-m(s+1)} q^{2s\deg\mu} \cdot \\ &\cdot \Big(\sum_{\substack{c|\mu \\ c \equiv 0(D)}} \left[\frac{D}{\mu/c}\right] q^{-2s\deg c} + \delta_{\lambda_0 N\mu} \left[\frac{D}{N}\right] \sum_{c|\mu} \left[\frac{D}{c}\right] q^{-2s\deg c} \Big). \end{aligned}$$

7 Fourier Expansion of $\widetilde{\Phi}_s$

The situation is more complicated than in the last section. To extend functions canonically from $\mathcal{T}_+$ to the whole of $GL_2(K_\infty)/\Gamma_\infty K_\infty^*$ we need the following proposition.

Proposition 6 *Let* $\chi_D : (\mathbb{F}_q[T]/D\,\mathbb{F}_q[T])^* \to \mathbb{C}^*$ *be a character modulo* D *and let* $\chi_\infty : K_\infty^* \to \mathbb{C}^*$ *be a character which vanishes on the subgroup of 1-units* $O_\infty^{(1)} = \{x \in K_\infty^* \mid v_\infty(x-1) > 0\}$.
Let $F : \mathcal{T}_+ \to \mathbb{C}$ *be a function which satisfies*

$$F(\begin{pmatrix} a & b \\ c & d \end{pmatrix}\begin{pmatrix} \pi_\infty^m & u \\ 0 & 1 \end{pmatrix}) = F(\begin{pmatrix} \pi_\infty^m & u \\ 0 & 1 \end{pmatrix})\,\chi_D(d)\,\chi_\infty(cu+d),$$

for all $\begin{pmatrix} a & b \\ c & d \end{pmatrix} \in \Gamma_0^{(1)}(D)$ *such that* $\begin{pmatrix} a & b \\ c & d \end{pmatrix}\begin{pmatrix} \pi_\infty^m & u \\ 0 & 1 \end{pmatrix} \in \mathcal{T}_+$ *(or equivalently* $v_\infty(c\pi_\infty^m) > v_\infty(cu+d)$*).*

Then F *can be defined on* $GL_2(K_\infty)/\Gamma_\infty K_\infty^*$ *with*

$$F(\begin{pmatrix} a & b \\ c & d \end{pmatrix}\begin{pmatrix} \pi_\infty^m & u \\ 0 & 1 \end{pmatrix}) = F(\begin{pmatrix} \pi_\infty^m & u \\ 0 & 1 \end{pmatrix})\,\chi_D(d)\cdot \\ \cdot\begin{cases} \chi_\infty(cu+d) & , \text{if } v_\infty(c\pi_\infty^m) > v_\infty(cu+d) \\ \chi_\infty(c^{-1}) & , \text{if } v_\infty(c\pi_\infty^m) \le v_\infty(cu+d) \end{cases}. \tag{7.1}$$

Proof. For each $X \in GL_2(K_\infty)/\Gamma_\infty K_\infty^*$ there is $A \in \Gamma_0^{(1)}(D)$ and $\begin{pmatrix} \pi_\infty^m & u \\ 0 & 1 \end{pmatrix} \in \mathcal{T}_+$ such that $X = A\begin{pmatrix} \pi_\infty^m & u \\ 0 & 1 \end{pmatrix}$ in $GL_2(K_\infty)/\Gamma_\infty K_\infty^*$. Then we define $F(X)$ by equation (7.1). The assumption for F guarantees that this definition is independent of the choice of A and $\begin{pmatrix} \pi_\infty^m & u \\ 0 & 1 \end{pmatrix}$. □

We apply this proposition to Θ_{a_0,λ_0} and to the Eisenstein series.

The Eisenstein series $E_s^{(i)}$ $(i = 1, D)$ satisfy

$$E_s^{(i)}(\begin{pmatrix} a & b \\ c & d \end{pmatrix}\begin{pmatrix} \pi_\infty^m & u \\ 0 & 1 \end{pmatrix}) = E_s^{(i)}(\begin{pmatrix} \pi_\infty^m & u \\ 0 & 1 \end{pmatrix})\left[\frac{d}{D}\right]\delta_{cu+d}\,q^{-v_\infty(cu+d)(2s+1)}$$

if $\begin{pmatrix} a & b \\ c & d \end{pmatrix} \in \Gamma_0^{(1)}(D)$ and $v_\infty(c\pi_\infty^m) > v_\infty(cu+d)$. We can apply Proposition 6 with $\chi_D(d) = \left[\frac{d}{D}\right]$ and $\chi_\infty(z) = \delta_z\,q^{-v_\infty(z)(2s+1)}$.

Hence

$$E_s^{(1)}(\begin{pmatrix} \pi_\infty^m & u \\ 0 & 1 \end{pmatrix}\begin{pmatrix} 0 & 1 \\ \pi_\infty & 0 \end{pmatrix}) = \sum_{\substack{c,d \in \mathbb{F}_q[T] \\ c \equiv 0(D) \\ v_\infty(c\pi_\infty^m) \le v_\infty(cu+d)}} \left[\frac{d}{D}\right]\delta_{-c}\,q^{v_\infty(c)(2s+1)}$$

and

$$E_s^{(D)}\left(\begin{pmatrix} \pi_\infty^m & u \\ 0 & 1 \end{pmatrix}\begin{pmatrix} 0 & 1 \\ \pi_\infty & 0 \end{pmatrix}\right) = \sum_{\substack{c,d\in\mathbb{F}_q[T] \\ v_\infty(c\pi_\infty^m)\le v_\infty(cu+d)}} \left[\frac{c}{D}\right]\delta_{-c}\, q^{v_\infty(c)(2s+1)}.$$

We denote these functions by $\widetilde{E}_s^{(1)}$ and $\widetilde{E}_s^{(D)}$ as above. Then straight forward calculations, starting with the definition of the Fourier coefficients, yield

$$\widetilde{E}_s^{(1)}\left(\begin{pmatrix} ND\pi_\infty^m & NDu \\ 0 & 1 \end{pmatrix}\right) = \sum_{\substack{\mu\neq 0 \\ \deg(\mu N)+2\le m}} \left[\epsilon_0\, q^{\frac{1}{2}\deg D+\deg N+1} q^{-m}\delta_\mu \cdot \sum_{\substack{c|\mu \\ c\equiv 0(D)}} \left[\frac{D}{\mu/c}\right] q^{-2s\deg c}\right] \psi_\infty(\mu N u) \quad (7.2)$$

and

$$\widetilde{E}_s^{(D)}\left(\begin{pmatrix} N\pi_\infty^m & Nu \\ 0 & 1 \end{pmatrix}\right) = q^{\deg N+1} q^{-m}\delta_{-1}\sum_{c\neq 0}\left[\frac{D}{c}\right] q^{-2s\deg c} +$$

$$+ \sum_{\substack{\mu\neq 0 \\ \deg(\mu N)+2\le m}} \left[q^{\deg N+1} q^{-m}\delta_{-1}\sum_{c|\mu}\left[\frac{D}{c}\right] q^{-2s\deg c}\right]\psi_{co}(\mu N u). \quad (7.3)$$

In addition we have $\widetilde{q^{-m(s+1)}} = q^{-(1-m)(s+1)}$, therefore (7.2) and (7.3) give

Proposition 7 *Let D be irreducible, then*

$$\widetilde{\mathcal{E}}_s\left(\begin{pmatrix} \pi_\infty^m & u \\ 0 & 1 \end{pmatrix}\right) = \mathcal{E}_s\left(\begin{pmatrix} \pi_\infty^m & u \\ 0 & 1 \end{pmatrix}\begin{pmatrix} 0 & 1 \\ \pi_\infty & 0 \end{pmatrix}\right) =$$

$$= \sum_{\deg(\mu N)+2\le m} \widetilde{e}_s(\pi_\infty^m,\mu)\psi_\infty(\mu N u) \text{ with}$$

$$\widetilde{e}_s(\pi_\infty^m, 0) = q^{(-s+\frac{1}{2})\deg D+s+\deg N} q^{ms}\epsilon_0\,\delta_{\lambda_0 DN}\left[\frac{D}{N}\right]\sum_{c\neq 0}\left[\frac{D}{c}\right] q^{-2s\deg c}$$

and $(\mu \neq 0)$

$$\widetilde{e}_s(\pi_\infty^m, \mu) = q^{(s+\frac{3}{2})\deg D + s + \deg N} q^{ms} \epsilon_0 \left[\delta_\mu \sum_{\substack{c|\mu \\ c\equiv 0(D)}} \left[\frac{D}{\mu/c}\right] q^{-2s\deg c} + \right.$$

$$\left. + q^{(-1-2s)\deg D} \delta_{\lambda_0 DN} \left[\frac{D}{N}\right] \sum_{c|\mu} \left[\frac{D}{c}\right] q^{-2s\deg c} \right].$$

For $\Theta_{\mathfrak{a}_0,\lambda_0}$ it is not just straight forward calculation. In the following we make use of the fact that the Fourier coefficients $r_{\mathfrak{a}_0,\lambda_0}(\lambda)$ of $\Theta_{\mathfrak{a}_0,\lambda_0}$ are independent of π_∞^m if $\deg \lambda + 2 \leq m$.

$\Theta_{\mathfrak{a}_0,\lambda_0}$ satisfies Proposition 6 with $\chi_D(d) = \left[\frac{d}{D}\right]$ and $\chi_\infty(z) = \delta_z\, q^{-v_\infty(z)}$ (cf. (3.1)).
Again we denote $\widetilde{\Theta}_{\mathfrak{a}_0,\lambda_0}(\cdot) = \Theta_{\mathfrak{a}_0,\lambda_0}(\cdot \begin{pmatrix} 0 & 1 \\ \pi_\infty & 0 \end{pmatrix})$.

Let $\pi_\infty^m \in K_\infty^*$ and $u \in K_\infty$, we choose $c, d \in \mathbb{F}_q[T]$ with $c \equiv 0\,(D)$,, $\gcd(c,d) = 1$ and $v_\infty(u + \frac{d}{c}) \geq m+1$. In addition we find $a, b \in \mathbb{F}_q[T]$ with $ad - bc = 1$. Then for all $k \in \mathbb{Z}$ with $k \leq m+1$ there is the following identity in $GL_2(K_\infty)/\Gamma_\infty K_\infty^*$:

$$\begin{pmatrix} \pi_\infty^k & u \\ 0 & 1 \end{pmatrix} \begin{pmatrix} 0 & 1 \\ \pi_\infty & 0 \end{pmatrix} = \begin{pmatrix} d & -b \\ -c & a \end{pmatrix} \begin{pmatrix} \frac{\pi_\infty^{1-k}}{c^2} & \frac{a}{c} \\ 0 & 1 \end{pmatrix}$$

We use this identity for $k = m$ and $k = m+1$, Proposition 6 and the Fourier expansion for $\Theta_{\mathfrak{a}_0,\lambda_0}$, and we calculate

$$\widetilde{\Theta}_{\mathfrak{a}_0,\lambda_0}(\begin{pmatrix} \pi_\infty^m & u \\ 0 & 1 \end{pmatrix}) - \widetilde{\Theta}_{\mathfrak{a}_0,\lambda_0}(\begin{pmatrix} \pi_\infty^{m+1} & u \\ 0 & 1 \end{pmatrix}) = \left[\frac{d}{D}\right] \delta_{-c}\, q^{v_\infty(c)}\,.$$

$$\cdot \sum_{\deg \mu + 2 = 1 - m + 2\deg c} r_{\mathfrak{a}_0,\lambda_0}(\mu) \psi_\infty(\mu \frac{a}{c}). \qquad (7.4)$$

On the other hand we set $u_\epsilon = -\frac{d}{c} + \epsilon \pi_\infty^m$ for $\epsilon \in \mathbb{F}_q^*$, we compare $\frac{a}{c}$ with $\frac{au_\epsilon + b}{cu_\epsilon + d}$ and sum over all ϵ:

$$(q-1)\widetilde{\Theta}_{\mathfrak{a}_0,\lambda_0}(\begin{pmatrix} \pi_\infty^m & u \\ 0 & 1 \end{pmatrix}) - \left[\frac{d}{D}\right] \delta_{-c}\, q^{v_\infty(c)} \sum_{\epsilon \in \mathbb{F}_q^*} \Theta_{\mathfrak{a}_0,\lambda_0}(\begin{pmatrix} \frac{\pi_\infty^{1-m}}{c^2} & \frac{au_\epsilon + b}{cu_\epsilon + d} \\ 0 & 1 \end{pmatrix})$$

$$= q\left[\frac{d}{D}\right]\delta_{-c}\, q^{v_\infty(c)} \sum_{\deg\mu+2=1-m+2\deg c} r_{\mathfrak{a}_0,\lambda_0}(\mu)\psi_\infty(\mu\frac{a}{c}). \tag{7.5}$$

Now

$$\begin{pmatrix} \frac{\pi_\infty^{1-m}}{c^2} & \frac{au_\epsilon+b}{cu_\epsilon+d} \\ 0 & 1 \end{pmatrix} = \begin{pmatrix} a & b \\ c & d \end{pmatrix}\begin{pmatrix} \pi_\infty^{m+1} & u_\epsilon \\ 0 & 1 \end{pmatrix}$$

in $GL_2(K_\infty)/\Gamma_\infty K_\infty^*$. We use this to evaluate the corresponding value of $\Theta_{\mathfrak{a}_0,\lambda_0}$. A combination of (7.4) and (7.5) gives therefore

$$\begin{aligned} &q\,\widetilde{\Theta}_{\mathfrak{a}_0,\lambda_0}(\begin{pmatrix} \pi_\infty^{m+1} & u \\ 0 & 1 \end{pmatrix}) - \widetilde{\Theta}_{\mathfrak{a}_0,\lambda_0}(\begin{pmatrix} \pi_\infty^{m} & u \\ 0 & 1 \end{pmatrix}) \\ &= \delta_{-\pi_\infty^m} q^{-m} \sum_{\epsilon\in\mathbb{F}_q^*} \delta_\epsilon \Theta_{\mathfrak{a}_0,\lambda_0}(\begin{pmatrix} \pi_\infty^{m+1} & u+\epsilon\pi_\infty^m \\ 0 & 1 \end{pmatrix}). \end{aligned} \tag{7.6}$$

If we evaluate in (7.6) the Fourier coefficients at λ with $\deg\lambda+2\le m$, we get the recursion formula

$$q\,\widetilde{\Theta}^*_{\mathfrak{a}_0,\lambda_0}(\pi_\infty^{m+1},\lambda) - \widetilde{\Theta}^*_{\mathfrak{a}_0,\lambda_0}(\pi_\infty^{m},\lambda) = \delta_{\pi_\infty^m} q^{-m} \sum_{\epsilon\in\mathbb{F}_q^*} \delta_\epsilon r_{\mathfrak{a}_0,\lambda_0}(\lambda). \tag{7.7}$$

The Fourier coefficient in (7.6) at λ with $\deg\lambda+2=m+1$ yields

$$q\,\widetilde{\Theta}^*_{\mathfrak{a}_0,\lambda_0}(\pi_\infty^{\deg\lambda+2},\lambda) = \delta_{\pi_\infty^{\deg\lambda+1}} q^{-\deg\lambda-1} \sum_{\epsilon\in\mathbb{F}_q^*} \delta_\epsilon \psi_\infty(-\lambda\epsilon\pi_\infty^{\deg\lambda+1}) r_{\mathfrak{a}_0,\lambda_0}(\lambda). \tag{7.8}$$

For $\lambda=0$ we calculate

$$\widetilde{\Theta}^*_{\mathfrak{a}_0,\lambda_0}(\pi_\infty,0) = q^{-\frac{1}{2}\deg D}\delta_{\lambda_0\epsilon_0^{-1}} \sum_{\deg\mu+2\le\deg D} r_{\mathfrak{a}_0,\lambda_0}(\mu). \tag{7.9}$$

Now it is obvious how one evaluates $\widetilde{\Theta}^*_{\mathfrak{a}_0,\lambda_0}(\pi_\infty^m,\lambda)$ using (7.7), (7.8) and (7.9). As an example we restrict again to the case that $\deg D$ is odd.

Proposition 8 *Let D be irreducible with odd degree, then*

$$\widetilde{\Theta}_{\mathfrak{a}_0,\lambda_0}(\begin{pmatrix} \pi_\infty^{m} & u \\ 0 & 1 \end{pmatrix}) = \sum_{\deg\lambda+2\le m} \widetilde{\Theta}^*_{\mathfrak{a}_0,\lambda_0}(\pi_\infty^m,\lambda)\psi_\infty(\lambda u)$$

with

$$\widetilde{\Theta}^{*}_{\mathfrak{a}_0,\lambda_0}(\pi_\infty^m,0) = q^{-\frac{1}{2}\deg D+1} q^{-m} \delta_{\lambda_0} \epsilon_0^{-1} \sum_{\deg \mu+2 \leq \deg D} r_{\mathfrak{a}_0,\lambda_0}(\mu)$$

and $(\lambda \neq 0)$

$$\widetilde{\Theta}^{*}_{\mathfrak{a}_0,\lambda_0}(\pi_\infty^m,\lambda) = q^{\frac{1}{2}} q^{-m} \epsilon_0^{-1} r_{\mathfrak{a}_0,\lambda_0}(\lambda)\, \delta_\lambda.$$

8 Conclusion

We start with Proposition 3. Then the definition of Φ_s in Proposition 2 shows that

$$\begin{aligned} \Phi_s^*(\pi_\infty^m,\lambda) - \widetilde{\Phi}_s^*(\pi_\infty^m,\lambda) &= \sum_\mu r_{\mathfrak{a}_0,\lambda_0}(\mu N - \lambda D) e_s(\frac{\pi_\infty^m}{D},\mu) \\ &\quad - \sum_\mu \widetilde{\Theta}^{*}_{\mathfrak{a}_0,\lambda_0}(\frac{\pi_\infty^m}{D},\mu N - \lambda D) \widetilde{e}_s(\frac{\pi_\infty^m}{D},\mu). \end{aligned}$$

In the last two sections we evaluated the coefficients e_s, $\widetilde{\Theta}^{*}_{\mathfrak{a}_0,\lambda_0}$ and $\widetilde{e}_s$. The number m occurs in all these formulas as an exponent. Hence the summation over m (which is necessary by Proposition 3) is given by the closed formula for the corresponding geometric series.

We can now collect all the previous results to the final formula. But we still have to take the trace from $\Gamma_0^{(1)}(N)$ to $\Gamma_0(N)$.. The Fourier coefficient of the resulting cusp form, which we will denote F_s is then

$$F_s^*(\pi_\infty^{\deg\lambda+2},\lambda) = \sum_{\epsilon\in\mathbb{F}_q^*} \Psi_s^*(\pi_\infty^{\deg\lambda+2},\epsilon\lambda).$$ We get

Theorem 9 *Let f be an automorphic cusp form of Drinfeld type of level N, suppose that f is a newform. Let $D \in \mathbb{F}_q[T]$ be irreducible of odd degree, then*

$$L^{(N)}(2s+1) L(f,\mathcal{A},s) = \frac{1}{4} \int_{\Gamma_0(N)\backslash GL_2(K_\infty)/\Gamma_\infty K_\infty^*} f \cdot \overline{F_{\bar{s}}},$$

where F_s is an automorphic cusp form of Drinfeld type of level N with Fourier coefficients

$$F_s^*(\pi_\infty^{\deg\lambda+2},\lambda) = \sum_{\substack{\mu\neq 0\\ \epsilon\in\mathbb{F}_q^*}} \left[\frac{(q^{-s-1})^{m(\mu)}}{1-q^{-s-1}} q^{(-2s-\frac{1}{2})\deg D+4s+\frac{1}{2}} r_{\mathfrak{a}_0,\lambda_0}(\mu N-\epsilon\lambda D)\cdot\right.$$

$$\cdot q^{2s\deg\mu}\Big(\sum_{\substack{c\mid\mu\\ c\equiv 0(D)}}\left[\frac{D}{\mu/c}\right]q^{-2s\deg c}+\delta_{\lambda_0 N\mu}\left[\frac{D}{N}\right]\sum_{c\mid\mu}\left[\frac{D}{c}\right]q^{-2s\deg c}\Big)$$

$$-\frac{(q^{s-1})^{m(\mu)}}{1-q^{s-1}}q^{(2s+\frac{1}{2})\deg D+s+\frac{1}{2}+\deg N}r_{\mathfrak{a}_0,\lambda_0}(\mu N-\epsilon\lambda D)\,\delta_{\lambda_0\mu}\cdot$$

$$\left.\cdot\Big(\sum_{\substack{c\mid\mu\\ c\equiv 0(D)}}\left[\frac{D}{\mu/c}\right]q^{-2s\deg c}+q^{(-1-2s)\deg D}\delta_{\lambda_0 DN\mu}\left[\frac{D}{N}\right]\sum_{c\mid\mu}\left[\frac{D}{c}\right]q^{-2s\deg c}\Big)\right]+$$

$$+\sum_{\epsilon\in\mathbb{F}_q^*}\left[\frac{(q^{-s-1})^{\deg\lambda+2}}{1-q^{-s-1}}q^{2s}r_{\mathfrak{a}_0,\lambda_0}(-\epsilon\lambda D)\sum_{d\neq 0}\left[\frac{D}{d}\right]q^{-(2s+1)\deg d}\right.$$

$$\left.-\frac{(q^{s-1})^{\deg\lambda+2}}{1-q^{s-1}}q^{-\frac{1}{2}\deg D+\frac{1}{2}+s+\deg N}r_{\mathfrak{a}_0,\lambda_0}(-\epsilon\lambda D)\delta_{DN}\left[\frac{D}{N}\right]\sum_{d\neq 0}\left[\frac{D}{d}\right]q^{-2s\deg d}\right]$$

$(m(\mu) := \max\{\deg\lambda+2, \deg(\mu N)-\deg D+2\})$.

Depending on the application one has in mind, the formula in the theorem can be simplified.

References

[Ge-Re] Gekeler, E.-U., Reversat M., *Jacobians of Drinfeld modular curves* , Journal reine angew. Math. 476 (1996), 27-93.

[Gr-Za] Gross, B., Zagier, D., *Heegner points and derivatives of L-series*, Invent. Math. 84 (1986), 225-320.

[Rü1] H.-G. Rück, *Theta Series of Imaginary Quadratic Function Fields*, manuscripta math. 88, 387-407, 1995.

[Rü2] H.-G. Rück, *Poincaré Series of Drinfeld Type*,
Arch. Math. 68, 190-201, 1997.

[Ti] U. Tipp, *Local height pairings of Heegner points on Drinfeld modular curves*,
Proceedings of the conference on Drinfeld modules in Alden Biesen, this book.

[We1] A. Weil, *Dirichlet Series and Automorphic Forms*,
LNM 189, Springer.

[We2] A. Weil, *On the Analogue of the Modular group in Characteristic p*,
in "Functional Analysis, etc", Proc. Conf. in honor of M. Stone, Springer 1970, 211-223 .

Hyperelliptic Drinfeld Modular Curves

Andreas Schweizer
Dept. of Mathematics, Concordia University
1455 Boulevard de Maisonneuve Ouest
Montreal, Quebec H3G 1M8, Canada
e-mail: schweiz@cicma.concordia.ca

Abstract

After showing that for $\mathfrak{n} \in \mathbb{F}_q[T]$ the Drinfeld modular curves $X_0(\mathfrak{n})$ are conservative over $\mathbb{F}_q(T)$ we determine all $\mathfrak{n}$ for which $X_0(\mathfrak{n})$ is rational, elliptic, or hyperelliptic.

0. Introduction

Using a method which is as simple as it is efficient (namely reducing modulo a suitable prime and then counting points), A. Ogg has determined all $n \in \mathbb{N}$ for which the modular curve $X_0(n)$ is hyperelliptic (compare [Ogg2],[Ogg3]). In this paper the same will be done for Drinfeld modular curves.

One interesting aspect of the classical theory is the curve $X_0(37)$, which is hyperelliptic (the genus being 2), but whose hyperelliptic involution is not "modular" (i.e., cannot be lifted to an automorphism of the upper half-plane). Moreover, this involution maps the 2 cusps to 2 rational points on $X_0(37)$. In our situation however, such an exceptional curve doesn't exist; the hyperelliptic involution always turns out to be the full Atkin-Lehner involution.

This paper is an improved and enlarged version of the first half of a manuscript with the same title. The second half of this manuscript is intended to be published in revised form under the title "Involutory elliptic curves over $\mathbb{F}_q(T)$".

I thank Professor E.-U. Gekeler for advice and encouragement and Bin Zhang and Qing Liu for pointing out the reference [L&K] to me. I also thank DFG and CICMA for financial support.

1. Basic facts

Let $\mathbb{F}_q$ be a finite field with q elements, $A := \mathbb{F}_q[T]$ the polynomial ring, $K := \mathbb{F}_q(T)$ the rational function field, $K_\infty := \mathbb{F}_q((T^{-1}))$ its completion at ∞ $(= \frac{1}{T})$, and C the completion of the algebraic closure of K_∞. The letters $\mathfrak{n}, \mathfrak{m}, \mathfrak{l}, \ldots$ will denote monic elements of A and $\mathfrak{p}$ will be a prime (i.e., a non-constant monic irreducible element) of A.

The group $GL_2(K_\infty)$ acts on the Drinfeld upper halfplane $\Omega = C - K_\infty$ by fractional linear transformations. For $\mathfrak{n} \in \mathbb{F}_q[T]$ let $\Gamma_0(\mathfrak{n})$ be the subgroup $\{\left(\begin{smallmatrix} a & b \\ c & d \end{smallmatrix}\right) \in GL_2(\mathbb{F}_q[T]) : \mathfrak{n}|c\}$. The compactification of $\Gamma_0(\mathfrak{n}) \setminus \Omega$ (which is obtained by adding to $\Gamma_0(\mathfrak{n}) \setminus \Omega$ a finite number of points, the so-called cusps) is a (connected, smooth, proper) curve over C. It is denoted by $X_0(\mathfrak{n})$. The cusps are the orbits $\Gamma_0(\mathfrak{n}) \setminus \mathbb{P}^1(K)$ of $\Gamma_0(\mathfrak{n})$, which operates from the left on the projective line $\mathbb{P}^1(K)$ over K.

Throughout this paper we will assume $\mathfrak{n} = \prod_{i=1}^{s} \mathfrak{p}_i^{e_i}$ with different primes $\mathfrak{p}_i \in A$, and d (if not an element of A) will denote the degree of $\mathfrak{n}$. The genus of $X_0(\mathfrak{n})$ is ([Ge1] p.79):

$$g(X_0(\mathfrak{n})) = 1 + \frac{\varepsilon(\mathfrak{n}) - (q+1)\kappa(\mathfrak{n}) - 2^{s-1}(q(q-1)r(\mathfrak{n}) + (q+1)(q-2))}{q^2-1}$$

where

$$\begin{aligned}
\varepsilon(\mathfrak{n}) &= \prod_{i=1}^{s} q^{(e_i-1)deg(\mathfrak{p}_i)}(q^{deg(\mathfrak{p}_i)}+1),\\
\kappa(\mathfrak{n}) &= \prod_{i=1}^{s} (q^{\lfloor \frac{e_i}{2} \rfloor deg(\mathfrak{p}_i)} + q^{\lfloor \frac{e_i-1}{2} \rfloor deg(\mathfrak{p}_i)}),\\
r(\mathfrak{n}) &= \begin{cases} 1 & if\ 2|deg(\mathfrak{p}_i)\ for\ all\ \mathfrak{p}_i|\mathfrak{n},\\ 0 & otherwise. \end{cases}
\end{aligned}$$

Especially, $X_0(\mathfrak{n})$ is rational for $d \leq 2$. For $d = 3$ the genus is q if $\mathfrak{n}$ is square-free, and $q-1$ if not.

For every $\mathfrak{m}|\mathfrak{n}$ with $(\mathfrak{m}, \frac{\mathfrak{n}}{\mathfrak{m}}) = 1$ there exists the (partial) Atkin-Lehner involution $W_{\mathfrak{m}}^{(\mathfrak{n})}$ on $X_0(\mathfrak{n})$, sometimes simply denoted by $W_{\mathfrak{m}}$ if the $\mathfrak{n}$ is clear. It is given on $\Gamma_0(\mathfrak{n}) \setminus \Omega$ (resp. $\Gamma_0(\mathfrak{n}) \setminus \mathbb{P}^1(K)$) by multiplication from the left with any matrix $\left(\begin{smallmatrix} \mathfrak{m}a & b \\ \mathfrak{n}c & \mathfrak{m}d \end{smallmatrix}\right)$ with $a, b, c, d \in A$ and determinant $\alpha\mathfrak{m}$ for some $\alpha \in \mathbb{F}_q^{\times}$. From this it is obvious that $W_{\mathfrak{m}_1} W_{\mathfrak{m}_2} = W_{\mathfrak{m}_3}$ with $\mathfrak{m}_3 = \frac{\mathfrak{m}_1\mathfrak{m}_2}{(\mathfrak{m}_1,\mathfrak{m}_2)^2}$. Consequently the $W_{\mathfrak{m}}$'s form a 2-elementary-abelian group of order 2^s.

In particular, $W_{\mathfrak{n}}^{(\mathfrak{n})}$, represented for example by $\left(\begin{smallmatrix} 0 & -1 \\ \mathfrak{n} & 0 \end{smallmatrix}\right)$, is called the full Atkin-Lehner involution and we define $X_+(\mathfrak{n}) := W_{\mathfrak{n}} \setminus X_0(\mathfrak{n})$. For $d = 3$ the curve $X_+(\mathfrak{n})$ is rational ([Ge2]). Thus for $d = 3$, apart from some elliptic exceptions, the curve $X_0(\mathfrak{n})$ is hyperelliptic (i.e., of genus ≥ 2 and a double cover of a rational curve).

If L is an extension field of K or of $A/(\mathfrak{p})$ for some prime $\mathfrak{p}$, we denote by $L\{\tau\}$ the twisted polynomial ring with the commutation rule $\tau \circ \ell = \ell^q \circ \tau$ for all $\ell \in L$.

A (rank 2) Drinfeld module ϕ over L is an $\mathbb{F}_q$-algebra homomorphism $a \mapsto \phi_a$ from A into $L\{\tau\}$ defined by $\phi_T = \Delta\tau^2 + \lambda\tau + T$ with $\Delta, \lambda \in L$ and $\Delta \neq 0$. If we interpret τ as the Frobenius homomorphism $x \mapsto x^q$, this defines a new A- module structure on the additive group $(L, +)$.

The element $j(\phi) := \frac{\lambda^{q+1}}{\Delta}$ is called the j-invariant of ϕ. A morphism from the Drinfeld module ϕ to the Drinfeld module ψ is a $u \in L\{\tau\}$ such that $u \circ \phi_T = \psi_T \circ u$. An isomorphism is a morphism $\sigma \in L\{\tau\}^* = L^\times$. Two Drinfeld modules ϕ and ψ over L are isomorphic over the algebraic closure $\bar{L}$ of L if and only if $j(\phi) = j(\psi)$.

For $u \in L\{\tau\}$ we denote by $ker(u)$ the set of elements of $\bar{L}$ annihilated by u. If $L \supseteq K$ or $L \supseteq A/(\mathfrak{p})$ with $\mathfrak{p} \not| \mathfrak{n}$, then $ker(\phi_\mathfrak{n})$ (i.e., the $\mathfrak{n}$-torsion points of ϕ) is a free rank 2 $A/(\mathfrak{n})$-module. Its $\varepsilon(\mathfrak{n})$ different free rank 1 $A/(\mathfrak{n})$-submodules are called the cyclic $\mathfrak{n}$-kernels of ϕ. Such a kernel is called L-rational if it is stable (as a set, not necessary elementwise) under the action of the absolute Galois group of L.

A cyclic $\mathfrak{n}$-isogeny of ϕ is a $u \in \bar{L}\{\tau\}$ whose degree in τ equals $deg(\mathfrak{n})$ and for which $ker(u)$ is a cyclic $\mathfrak{n}$-kernel of ϕ. Every cyclic $\mathfrak{n}$-kernel γ gives rise to a cyclic $\mathfrak{n}$-isogeny u, unique up to an isomorphism. We may choose u to be in $L\{\tau\}$ if and only if γ is L-rational. In this case we say that u is L-rational.

As for elliptic curves over $\mathbb{C}$ there is a close connection between rank 2 Drinfeld modules over C and rank 2 A-lattices contained in C (see [Gel] or any standard reference on Drinfeld modules). As in the classical situation one can use this to show that the endomorphism ring $End(\phi)$ of a Drinfeld module ϕ over C is either A or an A-order in an imaginary-quadratic extension of K. Note that $a \mapsto \phi_a$ furnishes an embedding of A into $End(\phi)$. Also, as in the classical situation the connection between lattices and Drinfeld modules leads to:

The points of $\Gamma_0(\mathfrak{n}) \setminus \Omega$ are in one-to-one correspondence with the isomorphism classes of pairs (ϕ, γ), where ϕ is a Drinfeld module over C and γ is a cyclic $\mathfrak{n}$-kernel of ϕ. Here an isomorphism between two pairs (ϕ, γ) and (ψ, γ') is an isomorphism σ from ϕ to ψ with $\sigma(\gamma) = \gamma'$. Equivalently we may say that the points of $\Gamma_0(\mathfrak{n}) \setminus \Omega$ are in one-to-one correspondence with the isomorphism classes of pairs (ϕ, u), where ϕ is a Drinfeld module over C and u is a cyclic $\mathfrak{n}$-isogeny of ϕ.

The interpretation of $W_\mathfrak{m}^{(\mathfrak{n})}$ in terms of these pairs is as follows: We write $\mathfrak{n} = \mathfrak{l}\mathfrak{m}$. Since $(\mathfrak{l}, \mathfrak{m}) = 1$, a cyclic $\mathfrak{n}$-kernel γ is the direct sum of a unique cyclic $\mathfrak{l}$-kernel $\gamma_\mathfrak{l}$ and a unique cyclic $\mathfrak{m}$-kernel $\gamma_\mathfrak{m}$. If $u_\mathfrak{m}$ is the cyclic $\mathfrak{m}$-isogeny belonging to $\gamma_\mathfrak{m}$, then $W_\mathfrak{m}^{(\mathfrak{n})}$ maps the class of (ϕ, γ) to the class of $(\psi, u_\mathfrak{m}(ker(\phi_\mathfrak{m}) \oplus \gamma_\mathfrak{l}))$ where ψ is the image of ϕ under $u_\mathfrak{m}$.

The function field of $X_0(1)$ is $C(j)$. This curve (more precisely, its "finite" part) parametrizes isomorphism classes of Drinfeld modules and $j : GL_2(A) \setminus \Omega \to C$ is just the j-invariant of the corresponding Drinfeld modules. The function field of $X_0(\mathfrak{n})$ is $C(j, j \circ W_\mathfrak{n})$, a separable extension of $C(j)$ of degree $\varepsilon(\mathfrak{n})$. The cusps of $X_0(\mathfrak{n})$ are the places of $C(j, j \circ W_\mathfrak{n})$ lying above the place $\frac{1}{j}$ of $C(j)$.

The curve $X_0(\mathfrak{n})$ is already defined over K and the corresponding function field is $F_0(\mathfrak{n}) := K(j, j \circ W_{\mathfrak{n}})$.

The curve $X_0(\mathfrak{n})$ has good reduction mod $\mathfrak{p}$ for all $\mathfrak{p} \not| \mathfrak{n}$. If L is the algebraic closure of $A/(\mathfrak{p})$, the L-rational points of the reduction that are not lying above the point $j = \infty$ of the reduction of $X_0(1)$ correspond to L-isomorphism classes of pairs (ϕ, γ), where ϕ is a Drinfeld module over L and γ is a cyclic $\mathfrak{n}$-kernel of ϕ. If ϕ is defined over a finite extension k of $A/(\mathfrak{p})$ and γ is k-rational, then the point on $X_0(\mathfrak{n})$ *mod* $\mathfrak{p}$ corresponding to the pair (ϕ, γ) is k-rational.

We will also use the fact that for $\mathfrak{p} \not| \mathfrak{n}$ the reduction map mod $\mathfrak{p}$ is injective on the set of cusps. Although this is known to the experts, there doesn't seem to exist a published proof.

By adding a finite number of cusps to $\Gamma(\mathfrak{n}) \setminus \Omega$, where $\Gamma(\mathfrak{n})$ is the full congruence subgroup $\{\left(\begin{smallmatrix} a & b \\ c & d \end{smallmatrix}\right) \in GL_2(A) : a \equiv d \equiv 1 \ mod\ \mathfrak{n},\ b \equiv c \equiv 0 \ mod\ \mathfrak{n}\}$, we obtain the curve $X(\mathfrak{n})$. It is defined over the field $K_+(\mathfrak{n})$, a finite Galois extension of K contained in K_∞ (compare [Ge2]). The function field of $X(\mathfrak{n})$ over the constant field $K_+(\mathfrak{n})$ will be denoted by $F(\mathfrak{n})$.

Proposition 1.

a) Every cusp of $X_0(\mathfrak{n})$ has a representative $\binom{x}{y}$ where $x, y \in A$ are monic, $y|\mathfrak{n}$, and $(x, \mathfrak{n}) = 1$. Two such representatives $\binom{x}{y}$ and $\binom{x'}{y'}$ represent the same cusp of $X_0(\mathfrak{n})$ if and only if $y = y'$ and $\alpha x' \equiv x$ modulo $\tilde{y} := (y, \frac{\mathfrak{n}}{y})$ for some $\alpha \in \mathbb{F}_q^\times$.

b) The cusps with the same y are conjugate over K. If $deg(\tilde{y}) \geq 1$, there are $\frac{\varphi(\tilde{y})}{q-1}$ of these, where $\varphi(\tilde{y}) = |A/(\tilde{y})^*|$. Especially, a cusp $\binom{x}{y}$ is rational over K if and only if $\varphi(\tilde{y}) \leq q - 1$, which is the case if and only if $deg(\tilde{y}) \leq 1$ or $q = 2$ and $\tilde{y} = T^2 + T$.

c) The total number of cusps of $X_0(\mathfrak{n})$ is $2^s + \frac{\kappa(\mathfrak{n}) - 2^s}{q-1}$. The number of rational cusps is at least 2^s and may be easily calculated explicitly. (We omit the formulas.)

PROOF. According to [Ge1] and [Ge2] the cusps of $X_0(\mathfrak{n})$ are in bijection with the orbits of $\Gamma_0(\mathfrak{n})/\Gamma(\mathfrak{n})$ on the pairs $\binom{x}{y}$ where $x, y \in A/(\mathfrak{n})$ are relatively prime and pairs $\binom{x}{y}$, $\binom{\alpha x}{\alpha y}$ with $\alpha \in \mathbb{F}_q^\times$ are identified. The cusps are $K_+(\mathfrak{n})$-rational and $Gal(K_+(\mathfrak{n})/K) = A/(\mathfrak{n})^*/\mathbb{F}_q^*$ operates by $\sigma\binom{x}{y} = \binom{\sigma x}{y}$. Hence we may argue as in [Ogg1, pp.222] (with obvious modifications). Thus we obtain a) and b); and c) follows by elementary calculations. (Note that φ and κ are weakly multiplicative). □

Except for the exact number of rational cusps, part c) has been known before (see for example [G&N]).

Example 2. For $q = 2$ the curve $X_0(T^4)$ has 6 cusps, namely $\binom{1}{1}$, $\binom{1}{T^4}$, $\binom{1}{T}$, $\binom{1}{T^3}$, $\binom{1}{T^2}$, $\binom{T+1}{T^2}$. The first four are rational whereas the last two are conjugate over K.

2. The function fields of the modular curves

A finite extension E/F of function fields is called geometric if both fields have the same field of constants. In this case we will also call F a geometric subfield of E. It is known that the $F_0(\mathfrak{n})$ and $F(\mathfrak{n})$ and consequently all their geometric subfields are separably generated function fields. Moreover, they are disclosed (i.e., have a prime divisor of degree 1), because the corresponding curve always has at least one rational cusp. However, since C contains inseparable extensions of K, it is (contrary to the classical situation) not a priori clear that $F_0(\mathfrak{n})$ and $C \cdot F_0(\mathfrak{n})$ have the same genus. A function field is called conservative if all constant field extensions are genus-preserving.

Proposition 3. *The function field $F(\mathfrak{n})$ is conservative.*

PROOF. There is a unique inseparable extension of degree p $(= char(\mathbb{F}_q))$ of $K_+(\mathfrak{n})$, namely $L := K_+(\mathfrak{n})(\sqrt[p]{T})$. By a result of Kimura (compare [Ki], or [St] for a generalization), conservativity of $F(\mathfrak{n})$ is equivalent to $g(L \cdot F(\mathfrak{n})) = g(F(\mathfrak{n}))$. Denote by $s_{\sqrt[p]{T}} : L \cdot F(\mathfrak{n}) \to F(\mathfrak{n})$ the Tate-trace associated to $\sqrt[p]{T}$ (compare [St] or [Ta]) and by $\mathfrak{D} = \mathfrak{D}(s_{\sqrt[p]{T}}, L \cdot F(\mathfrak{n})/F(\mathfrak{n}))$ the corresponding different. We want to show $deg\ \mathfrak{D} = 0$; then the Hurwitz formula [St p.32] implies $g(L \cdot F(\mathfrak{n})) = g(F(\mathfrak{n}))$.

If $\wp$ is a place of $F(\mathfrak{n})$, trivial on $K_+(\mathfrak{n})$, there is a unique place $\mathfrak{P}$ of $L \cdot F(\mathfrak{n})$ lying above $\wp$ and its ramification index or its inertia degree must equal p. Denote by $\mathcal{O}_\wp$ and $\mathcal{O}_\mathfrak{P}$ the corresponding valuation rings. If $B = \{1, \sqrt[p]{T}, \ldots, \sqrt[p]{T}^{p-1}\}$ is an $\mathcal{O}_\wp$-basis of $\mathcal{O}_\mathfrak{P}$, then $v_\mathfrak{P}(\mathfrak{D}) = 0$, where $v_\mathfrak{P}$ denotes the additive valuation associated to $\mathfrak{P}$. Furthermore, if $\wp$ is inert, then B is an $\mathcal{O}_\wp$-basis of $\mathcal{O}_\mathfrak{P}$, because $\sqrt[p]{T}$ generates the residue field extension. All that remains to prove is that for every ramified $\wp$ the set B is an $\mathcal{O}_\wp$-basis of $\mathcal{O}_\mathfrak{P}$.

According to [Ge1] the extension $F(\mathfrak{n})/K_+(\mathfrak{n})(j)$ is unramified outside $\{j, \frac{1}{j}\}$. The places lying above j are tamely ramified. The cusps are $K_+(\mathfrak{n})$-rational, hence the inertia degree of the places above $\frac{1}{j}$ is 1.

Now let $\wp$ be a prime of $F(\mathfrak{n})$, ramified in $L \cdot F(\mathfrak{n})$ and lying above some prime π of $K_+(\mathfrak{n})(j)$. Then wild ramification must occur above π in $L(j)/K_+(\mathfrak{n})(j)$ or in $L \cdot F(\mathfrak{n})/L(j)$. Hence π cannot be j, because j is inert in $L(j)$ and tamely ramified in $L \cdot F(\mathfrak{n})/L(j)$. Similarly, π cannot be $\frac{1}{j}$, because the inertia degree of $\frac{1}{j}$ is 1 in $F(\mathfrak{n})$ but p in $L(j)$. So π must be ramified in $L(j)$, i.e., $\pi = (f(j))^p$ with $f(j) \in L[j]$. Furthermore, $f(j)$ is a uniformizer for $\mathfrak{P}$, because $\mathfrak{P}$ is unramified in $L \cdot F(\mathfrak{n})/L(j)$. Now $\{1, f(j), \ldots, f(j)^{p-1}\}$ is an $\mathcal{O}_\mathfrak{p}$-basis of $\mathcal{O}_\mathfrak{P}$, and $f(j) = \sum_{i=0}^{p-1} a_i \sqrt[p]{T}^i$ with $a_i \in K_+(\mathfrak{n})[j]$ shows that B is an $\mathcal{O}_\mathfrak{p}$-basis of $\mathcal{O}_\mathfrak{P}$, too. □

Lemma 4. *Let F/k be a separable function field and E/F a finite geometric extension. Suppose further that E is disclosed and conservative. Then F, too, must be conservative.*

PROOF. Let k' be a finite extension of the constant field k and denote $E' = k' \cdot E$ and $F' = k' \cdot F$. Since F is disclosed, too, the effective degrees $m(E'/E)$ and $m(F'/F)$ must equal 1 ([Art], Chapter XV, Theorem 7). So $[E' : E] = [k' : k] = [F' : F]$ and hence $[E' : F'] = [E : F]$. If $s \in Hom_F(E, F)$ is non-trivial and $s' \in Hom_{F'}(E', F')$ is a prolongation of s, then $\mathfrak{D}(s', E'/F') \leq \mathfrak{D}(s, E/F)$ holds for the corresponding differents ([St], Lemma 1). Applying the Hurwitz formula ([St] p.32) to E/F and E'/F' shows $g(F') \geq g(F)$ and hence $g(F') = g(F)$. □

Proposition 5.

a) The function field $F_0(\mathfrak{n})$ is conservative.

b) Every geometric subfield F of $F_0(\mathfrak{n})$ is conservative.

PROOF. a) Applying Lemma 4 to $E = F(\mathfrak{n})$, $F = K_+(\mathfrak{n}) \cdot F_0(\mathfrak{n})$, we see that $K_+(\mathfrak{n}) \cdot F_0(\mathfrak{n})$ is conservative. Since $K_+(\mathfrak{n})$ is separable over K, we have $g(F_0(\mathfrak{n})) = g(K_+(\mathfrak{n}) \cdot F_0(\mathfrak{n}))$, so $F_0(\mathfrak{n})$ is conservative.

Statement b) follows from Lemma 4, if we set $E = F_0(\mathfrak{n})$. □

Lemma 6. *Let F be a geometric subfield of $F_0(\mathfrak{n})$. Then F is rational (resp. elliptic, resp. hyperelliptic) if and only if $C \cdot F$ is.*

PROOF. For $g(F) \leq 1$ this follows directly from the conservativity. If $g(F) \geq 2$, we choose g linear independent elements $x_1, \ldots, x_g \in F$ from the linear space of the canonical class. Since F is disclosed, we have $m(C \cdot F/F) = 1$. This is equivalent to C and F being linearly disjoint over K ([Art], Chapter XV, Theorem 6). Hence $\{x_1, \ldots, x_g\}$ also is a C-base of the linear space of the canonical class in $C \cdot F$ (a canonical divisor of F is also a canonical divisor of $C \cdot F$ [Art], Chapter XV, Theorem 21). Let $S = K(\frac{x_2}{x_1}, \dots, \frac{x_g}{x_1})$.

If F is hyperelliptic, we have $[F : S] = 2$ and S is rational ([Art], Chapter XVI, Theorem 14). Clearly, $C \cdot S$ is also rational and $[C \cdot F : C \cdot S] = 2$.

If F is not hyperelliptic, we have $S = F$ ([Art], Chapter XVI, Theorem 13) and after constant field extension $C \cdot F = C(\frac{x_2}{x_1}, \dots, \frac{x_g}{x_1})$, so $C \cdot F$ is not hyperelliptic ([Art], Chapter XVI, Theorem 14). □

3. The fixed points of the Atkin-Lehner involutions

Proposition 7. (Gekeler) *Let w be any involution of $X_0(\mathfrak{n})$. Denote by # the number of its fixed points.*

a) If q is odd, then $g(w\backslash X_0(\mathfrak{n})) = \frac{1}{4}(2g(X_0(\mathfrak{n})) + 2 - \#)$.

b) If q is even, then $g(w\backslash X_0(\mathfrak{n})) = \frac{1}{2}(g(X_0(\mathfrak{n})) + 1 - \#)$.

PROOF. This is essentially the Hurwitz formula. In case a) all ramification is tame, whereas in case b) all ramification is wild. Now it is known that the curve $X_0(\mathfrak{n})$ is ordinary. This means the following: If p is the characteristic of K, then the p-rank of the Jacobian of $X_0(\mathfrak{n})$ equals the genus g of $X_0(\mathfrak{n})$. (More explicitely, in [G&R] the Jacobian is given as $(C^\times)^g$ divided by some g-dimensional multiplicative lattice Λ.) By Proposition 1 of [Ray] this implies that in characteristic 2 the second ramification groups of the covering $X_0(\mathfrak{n}) \to w\backslash X_0(\mathfrak{n})$ are trivial. □

We mainly want to apply this proposition to the Atkin-Lehner involutions $W_{\mathfrak{m}}^{(\mathfrak{n})}$. So we have to investigate their fixed points. We first look at the cusps, writing them in the form $\binom{x}{y}$ with monic $x, y \in A$ satisfying $y|\mathfrak{n}$ and $(x, \mathfrak{n}) = 1$. Defining $y_1 := (y, \mathfrak{m})$ and $y_2 := (y, \mathfrak{l})$ (recall that $\mathfrak{n} = \mathfrak{l}\mathfrak{m}$ with $(\mathfrak{l}, \mathfrak{m}) = 1$), we obtain $y = y_1 y_2$ with monic, coprime $y_1, y_2 \in A$.

It turns out that the description of the fixed cusps is similar, although not completely analogous, to the one in the classical situation (compare [Ogg2]).

Lemma 8. *If $\binom{x}{y}$ and $\binom{x'}{y'}$ are cusps of $X_0(\mathfrak{n})$ with $\binom{x'}{y'} = W_{\mathfrak{m}}^{(\mathfrak{n})}\binom{x}{y}$, then $y_1' = \frac{\mathfrak{m}}{y_1}$ and $y_2' = y_2$.*

PROOF. $W_{\mathfrak{m}}^{(\mathfrak{n})} = \begin{pmatrix} a\mathfrak{m} & b \\ c\mathfrak{l}\mathfrak{m} & d\mathfrak{m} \end{pmatrix} = \begin{pmatrix} 1 & 0 \\ 0 & \mathfrak{m} \end{pmatrix}\begin{pmatrix} a\mathfrak{m} & b \\ c\mathfrak{l} & d \end{pmatrix}$, the last matrix being contained in $GL_2(A)$. Therefore $a\mathfrak{m}x + by$ and $c\mathfrak{l}x + dy$ are coprime (x and y being coprime) and $(a\mathfrak{m}x + by, c\mathfrak{l}\mathfrak{m}x + d\mathfrak{m}y) = (a\mathfrak{m}x + by, \mathfrak{m}) = y_1$. Consequently,

$$W_{\mathfrak{m}}^{(\mathfrak{n})}\begin{pmatrix} x \\ y \end{pmatrix} = \begin{pmatrix} a\frac{\mathfrak{m}}{y_1}x + by_2 \\ \frac{\mathfrak{m}}{y_1}y_2(c\frac{\mathfrak{l}}{y_2}x + dy_1) \end{pmatrix}$$

where the coordinates on the right side are coprime elements of A. Since $(\mathfrak{l}x, y_1) = 1$ we may conclude $y_1' = \frac{\mathfrak{m}}{y_1}$. Moreover $y_2 | y_2'$, and applying $W_{\mathfrak{m}}^{(\mathfrak{n})}$ once again, $y_2' | y_2$, too, so $y_2' = y_2$. □

Proposition 9.

a) If q is even, the cusps of $X_0(\mathfrak{n})$ fixed by $W_{\mathfrak{m}}^{(\mathfrak{n})}$ are exactly the ones with $y_1^2 = \mathfrak{m}$.

b) If q is odd, the cusps of $X_0(\mathfrak{n})$ fixed by $W_{\mathfrak{m}}^{(\mathfrak{n})}$ are exactly the ones with $y_1^2 = \mathfrak{m}$ and $(y_2, \frac{\mathfrak{l}}{y_2}) = 1$.

PROOF. Let $\mathfrak{m} = y_1^2$. According to Lemma 8 this is a necessary condition for $\binom{x}{y_1 y_2}$ to be a fixed point of $W_{\mathfrak{m}}^{(\mathfrak{n})} = \begin{pmatrix} a\mathfrak{m} & b \\ c\mathfrak{l}\mathfrak{m} & d\mathfrak{m} \end{pmatrix}$.

We may assume $det(W_{\mathfrak{m}}^{(\mathfrak{n})}) = \mathfrak{m}$. Then $1 = ady_1^2 - bc\mathfrak{l}$ and by the proof of Lemma 8 we have

$$W_{\mathfrak{m}}^{(\mathfrak{n})}\begin{pmatrix} x \\ y_1 y_2 \end{pmatrix} = \begin{pmatrix} ay_1 x + by_2 \\ y_1 y_2 (c\frac{\mathfrak{l}}{y_2} x + dy_1) \end{pmatrix} = \begin{pmatrix} (ay_1 x + by_2)(c\frac{\mathfrak{l}}{y_2} x + dy_1) \\ y_1 y_2 \end{pmatrix}$$

$$= \begin{pmatrix} ady_1^2 x + bc\mathfrak{l}x \\ y_1 y_2 \end{pmatrix} = \begin{pmatrix} -x + 2ady_1^2 x \\ y_1 y_2 \end{pmatrix} = \begin{pmatrix} x + 2bc\mathfrak{l}x \\ y_1 y_2 \end{pmatrix} =: \begin{pmatrix} x' \\ y_1 y_2 \end{pmatrix}.$$

Here we used the fact that we can reduce the first coordinate of the cusp modulo $y_1 y_2$ and modulo $y_1 \frac{\mathfrak{l}}{y_2}$.

The above chain of equalities implies at once that for even q the cusp $\binom{x}{y_1 y_2}$ is a fixed point of $W_{y_1^2}$.

If q is odd we set $z_2 := (y_2, \frac{\mathfrak{l}}{y_2})$. In this case $\binom{x}{y_1 y_2}$ is a fixed point of $W_{y_1^2}$ if and only if there exists an $\alpha \in \mathbb{F}_q^\times$ with $x' \equiv \alpha x$ modulo $y_1 z_2$. Because of $x' \equiv -x \ mod \ y_1$ and $x' \equiv x \ mod \ z_2$ this is possible only for $z_2 = 1$. □

Example 10. Let $q = 2$ and $\mathfrak{n} = (T^2 + T + 1)^2$. Then the curve $X_0(\mathfrak{n})$ has 5 cusps, namely: $\binom{1}{1}$, $\binom{1}{\mathfrak{n}}$, $\binom{1}{T^2+T+1}$, $\binom{T}{T^2+T+1}$, $\binom{T+1}{T^2+T+1}$. The full Atkin-Lehner involution $W_{\mathfrak{n}}$ exchanges the first two and fixes the other three. Since the genus of $X_0(\mathfrak{n})$ is 2, we obtain $g(X_+(\mathfrak{n})) = 0$.

Proposition 11. *If q is odd, the number of interior (i.e., non-cusp) fixed points of $W_{\mathfrak{m}}$ on $X_0(\mathfrak{n})$ is*

$$\#_i(W_{\mathfrak{m}}^{(\mathfrak{n})}) = \begin{cases} h(\sqrt{\alpha\mathfrak{m}}) \prod\limits_{\mathfrak{p}|\frac{\mathfrak{n}}{\mathfrak{m}}} (1 + (\frac{\alpha\mathfrak{m}}{\mathfrak{p}})) & if\ 2|deg(\mathfrak{m}), \\ h(\sqrt{\mathfrak{m}}) \prod\limits_{\mathfrak{p}|\frac{\mathfrak{n}}{\mathfrak{m}}} (1 + (\frac{\mathfrak{m}}{\mathfrak{p}})) + h(\sqrt{\alpha\mathfrak{m}}) \prod\limits_{\mathfrak{p}|\frac{\mathfrak{n}}{\mathfrak{m}}} (1 + (\frac{\alpha\mathfrak{m}}{\mathfrak{p}})) & if\ 2\nmid deg(\mathfrak{m}). \end{cases}$$

Here $\alpha \in \mathbb{F}_q^\times$ is a fixed non-square, $(\frac{f}{\mathfrak{p}})$ is the Legendre symbol, and $h(\sqrt{\mathfrak{f}})$ denotes the ideal class number of the (not necessarily maximal) order $\mathbb{F}_q[T, \sqrt{\mathfrak{f}}]$.

PROOF. The proof is the same as in the classical situation ([Ogg2] p.454). Compare also [Ge2] for a special case in the Drinfeld situation. □

Structurally, the proof shows that the interior fixed points of $W_{\mathfrak{m}}$ correspond to Drinfeld modules having complex multiplication by $A[\sqrt{\alpha\mathfrak{m}}]$ or $A[\sqrt{\mathfrak{m}}]$. If q is even, the fixed points will correspond to Drinfeld modules with complex multiplication by certain orders in the inseparable extension $K(\sqrt{T})$. This makes the calculation of the number of fixed points of $W_{\mathfrak{m}}$ on $X_0(\mathfrak{n})$ lying above a given fixed point on $X_0(\mathfrak{m})$ much more complicated. For lack of a complete result we derive some formulas that cover many cases.

Lemma 12. *Suppose that q is even and $\mathfrak{n} = \mathfrak{l}\mathfrak{m}$ with $(\mathfrak{l}, \mathfrak{m}) = 1$ and $deg(\mathfrak{m}) \geq 1$. Write $\mathfrak{m} = \mathfrak{m}_2^2\mathfrak{m}_1$ with $\mathfrak{m}_1, \mathfrak{m}_2 \in \mathbb{F}_q[T]$, where $\mathfrak{m}_1$ is square-free. There exist uniquely determined $\mathfrak{s}, \mathfrak{t} \in \mathbb{F}_q[T]$, not necessarily monic, with $\mathfrak{m}_1 = \mathfrak{s}^2 + T\mathfrak{t}^2$. Let*

$$\mathfrak{t}_1 = \prod_{\mathfrak{p} \nmid \mathfrak{m}_2} \mathfrak{p}^{ord_{\mathfrak{p}}(\mathfrak{t})} \text{ and } \mathfrak{t}_2 = \prod_{\mathfrak{p} \mid \mathfrak{m}_2} \mathfrak{p}^{ord_{\mathfrak{p}}(\mathfrak{t})}.$$

a) The number of interior fixed points of $W_{\mathfrak{m}}$ on $X_0(\mathfrak{m})$ is

$$\#_i(W_{\mathfrak{m}}^{(\mathfrak{m})}) = \begin{cases} 0 & \text{if } \mathfrak{m} \text{ is a square,} \\ q^{deg(\mathfrak{m}_2\mathfrak{t}_2)} \cdot \sum_{\mathfrak{f} \mid \mathfrak{t}_1} q^{deg(\mathfrak{f})} & \text{otherwise.} \end{cases}$$

b) More generally, we have $\#_i(W_{\mathfrak{m}}^{(\mathfrak{n})}) \leq \varepsilon(\mathfrak{l})\#_i(W_{\mathfrak{m}}^{(\mathfrak{m})})$, where ε is the function defined in Section 1.

In particular, $\#_i(W_{\mathfrak{m}}^{(\mathfrak{n})}) = 0$ if $\mathfrak{m}$ is a square.

c) If there exists a $\mathfrak{p}$ with $ord_{\mathfrak{p}}(\mathfrak{l}) > 2ord_{\mathfrak{p}}(\mathfrak{t}_1) + 1$, then $\#_i(W_{\mathfrak{m}}^{(\mathfrak{n})}) = 0$.

d) If $\mathfrak{l}$ is square-free and $\mathfrak{m}$ is not a square, then

$$\#_i(W_{\mathfrak{m}}^{(\mathfrak{n})}) = q^{deg(\mathfrak{m}_2\mathfrak{t}_2)} \cdot \sum_{\mathfrak{f} \mid \mathfrak{t}_1} q^{deg(\mathfrak{f})}\varepsilon(gcd(\mathfrak{l}, \frac{\mathfrak{t}_1}{\mathfrak{f}})).$$

PROOF. As in the proof of Proposition 11 the fixed points of $W_{\mathfrak{m}}$ on $X_0(\mathfrak{m})$ are in bijection with the isomorphism classes of pairs (ϕ, u), where ϕ is a Drinfeld module over C and u a cyclic $\mathfrak{m}$-isogeny of ϕ with $u^2 = \alpha\mathfrak{m}$ in $End(\phi)$ for some $\alpha \in Aut(\phi)$. Since in $Aut(\phi)$ every element is a square, in each of these classes there is a unique representative (ϕ, λ) with $\lambda^2 = \mathfrak{m}$ in $End(\phi)$.

If $\mathfrak{m}$ is a square, then λ lies in $A \subseteq End(\phi)$ and hence cannot be a cyclic $\mathfrak{m}$-isogeny. If $\mathfrak{m}$ is not a square, then $\lambda = \mathfrak{m}_2(\mathfrak{s}+\mathfrak{t}\sqrt{T})$ implies $\mathfrak{m}_2\mathfrak{t}\sqrt{T} \in End(\phi)$. On the other hand, $End(\phi)$ must be contained in $A[\mathfrak{m}_2\sqrt{T}]$; else we could write $\lambda = \mathfrak{p}b$ with some prime factor $\mathfrak{p}$ of $\mathfrak{m}_2$ and $b \in End(\phi)$, and consequently λ wouldn't be a cyclic $\mathfrak{m}$-isogeny. $End(\phi)$ must even be contained in $A[\mathfrak{m}_2\mathfrak{t}_2\sqrt{T}]$; otherwise λ would not be cyclic because of being divisible in $End(\phi)$ by a prime divisor of $\mathfrak{t}_2$. This had been overlooked in the first version of this paper.

Since $\mathfrak{s}$ and $\mathfrak{t}$ are relatively prime, we see that for non-square $\mathfrak{m}$ the fixed points of $W_{\mathfrak{m}}$ on $X_0(\mathfrak{m})$ are in bijection with the isomorphism classes of Drinfeld modules with complex multiplication by orders $A[\tilde{\mathfrak{f}}\sqrt{T}]$ where $\tilde{\mathfrak{f}}|\mathfrak{m}_2\mathfrak{t}$ and $\mathfrak{m}_2\mathfrak{t}_2|\tilde{\mathfrak{f}}$. But the isomorphism classes of Drinfeld modules whose endomorphism ring is $A[\tilde{\mathfrak{f}}\sqrt{T}]$ are in bijection with the ideal classes of $A[\tilde{\mathfrak{f}}\sqrt{T}]$ and the ideal class number of $A[\tilde{\mathfrak{f}}\sqrt{T}]$ is $q^{deg(\tilde{\mathfrak{f}})}$. This proves a).

If P is a fixed point of $W_{\mathfrak{m}}^{(\mathfrak{n})}$ on $X_0(\mathfrak{n})$, its image in $X_0(\mathfrak{m})$ is a fixed point of $W_{\mathfrak{m}}^{(\mathfrak{m})}$. As the degree of the covering $X_0(\mathfrak{n}) \to X_0(\mathfrak{m})$ is $\varepsilon(\mathfrak{l})$, there are at most $\varepsilon(\mathfrak{l})$ fixed points $X_0(\mathfrak{n})$ lying over the same fixed point on $X_0(\mathfrak{m})$. This is statement b).

Now we investigate more closely how many fixed points of $W_{\mathfrak{m}}$ on $X_0(\mathfrak{n})$ are lying above a given fixed point on $X_0(\mathfrak{m})$. In other words: Given a pair (ϕ, λ) as in a), how many cyclic $\mathfrak{l}$-kernels $\gamma_{\mathfrak{l}}$ of ϕ satisfy $\lambda(\gamma_{\mathfrak{l}}) = \gamma_{\mathfrak{l}}$? This boils down to the question, whether there exists an $a \in A$ such that $ker(a+\lambda)$ contains a cyclic $\mathfrak{l}$-kernel. In particular $\mathfrak{l}$ must divide $a^2 + \lambda^2$ in A.

If there exists a $\mathfrak{p}$ with $\mathfrak{p}^{2ord_{\mathfrak{p}}(\mathfrak{t}_1)+2}|\mathfrak{l}$, this implies $\mathfrak{p}^{ord_{\mathfrak{p}}(\mathfrak{t}_1)+1}|(a+\lambda)$ in $\mathbb{F}_q[\sqrt{T}]$ and hence $\mathfrak{p}^{ord_{\mathfrak{p}}(\mathfrak{t}_1)+1}|\mathfrak{m}_2\mathfrak{t}_2\mathfrak{t}_1$. But this contradicts $(\mathfrak{l}, \mathfrak{m}_2\mathfrak{t}_2) = 1$. So there exists no $a \in A$ with $\mathfrak{l}|(a^2+\lambda^2)$ and assertion c) is proved.

For the proof of statement d) we fix one fixed point (ϕ, λ) of $W_{\mathfrak{m}}^{(\mathfrak{m})}$ with $End(\phi) = A[\mathfrak{m}_2\mathfrak{t}_2\mathfrak{f}\sqrt{T}]$. If $\mathfrak{l}$ is square-free, modulo $\mathfrak{l}$ there exists exactly one $a \in A$ with $\mathfrak{l}|(a^2+\lambda^2)$. The number of cyclic $\mathfrak{l}$-kernels in $ker(a+\lambda)$ is the product of the numbers of cyclic $\mathfrak{p}$-kernels in $ker(a+\lambda)$ for all prime divisors $\mathfrak{p}$ of $\mathfrak{l}$.

If $\mathfrak{p}$ divides $\frac{\mathfrak{t}_1}{\mathfrak{f}}$, then $\mathfrak{p}$ divides $\mathfrak{m}_2\mathfrak{t}\sqrt{T}$ and $a+\lambda$ in $End(\phi)$. So $a+\lambda$ annihilates $ker(\phi_{\mathfrak{p}})$. In other words, $ker(a+\lambda)$ contains all $\varepsilon(\mathfrak{p})$ different $\mathfrak{p}$-kernels.

If $\mathfrak{p}$ doesn't divide $\frac{\mathfrak{t}_1}{\mathfrak{f}}$, then $\mathfrak{p}$ also doesn't divide $a+\lambda$ in $End(\phi)$. In this case $ker(a+\lambda)$ contains exactly one $\mathfrak{p}$-kernel.

All in all there are $\varepsilon(gcd(\mathfrak{l}, \frac{\mathfrak{t}_1}{\mathfrak{f}}))$ different cyclic $\mathfrak{l}$-kernels $\gamma_{\mathfrak{l}}$ with $\lambda(\gamma_{\mathfrak{l}}) = \gamma_{\mathfrak{l}}$.

□

Example 13. Let $q = 2$ and $\mathfrak{n} = T^4 + T = T(T+1)(T^2+T+1)$. Then the partial Atkin-Lehner involution W_{T^3+1} has 5 fixed points on the curve $X_0(\mathfrak{n})$. Since $g(X_0(T^4+T)) = 8$, Proposition 7 yields $g(W_{T^3+1}\backslash X_0(T^4+T)) = 2$.

4. Hyperelliptic modular curves

In accordance with [Ogg2], a curve over a field k will be called *subhyperelliptic* if it admits a k-rational function of degree 1 or 2. Since $X_0(\mathfrak{n})$ considered over an extension field K' of K always has a rational point (there is at least one K-rational cusp), it is subhyperelliptic over K' if and only if it is rational, elliptic, or hyperelliptic over this field.

Thanks to Lemma 6 we may say that $X_0(\mathfrak{n})$ is (sub)hyperelliptic or not without reference to the constant field, which might be any extension of K. This is important in view of Theorem 20. For its proof we need Lemma 16 and 17; but the proof of Lemma 16 only makes sense if the constant field is C, whereas in Lemma 17 we must work over K.

Lemma 14. *Let X be an hyperelliptic curve over some algebraically closed field k. Let w be an involution of X which is different from the hyperelliptic involution v. Then w has at most 4 fixed points. If moreover the characteristic of k is 2, then w has at most 2 fixed points.*

Proof. Since the curve $wvw^{-1}\backslash X$ is also rational, it follows from the uniqueness of v that w and v commute. Therefore w induces an involution $\tilde{w}$ on the curve $v\backslash X$. This involution cannot be the identity, for, if it were the identity, w or wv would have infinitely many fixed points on X.

Furthermore, v permutes the fixed points of w on X. If there were more than 4, then $\tilde{w}$ would have more than 2 fixed points on the rational curve $v\backslash X$. This would contradict the Hurwitz formula.

If $char(k) = 2$, the argument is the same, taking into account that the ramification is wild. □

Example 15. *For $q = 2$ the curve $X_0(T^4+T^3+1)$ is not subhyperelliptic.*

Indeed $g(X_0(T^4+T^3+1)) = 4$ and the full Atkin-Lehner involution $W_{T^4+T^3+1}$ has 3 fixed points. So $g(X_+(T^4+T^3+1)) = 1$ and we can apply Lemma 14.

Lemma 16. *If $X_0(\mathfrak{n})$ is subhyperelliptic and $\mathfrak{m}|\mathfrak{n}$ then $X_0(\mathfrak{m})$, too, is subhyperelliptic.*

PROOF. The proof is the same as in the classical situation (compare [Ne], Theorem VII.2. and Corollary VII.2.). □

Lemma 17. *If $X_0(\mathfrak{n})$ is rational (resp. elliptic, resp. hyperelliptic) and $\mathfrak{p} \nmid \mathfrak{n}$, then the reduction of $X_0(\mathfrak{n})$ modulo $\mathfrak{p}$ is again rational (resp. elliptic, resp. hyperelliptic).*

PROOF. Since $X_0(\mathfrak{n})$ has good reduction *mod* $\mathfrak{p}$ the assertion is clear for $g(X_0(\mathfrak{n})) \leq 1$. If $X_0(\mathfrak{n})$ is hyperelliptic it follows from [L&K], Theorem 5.5, Definition 5.12, and Proposition 5.14 where we take for S the localization of A at $\mathfrak{p}$. □

Now we are ready to determine all $X_0(\mathfrak{n})$ that are subhyperelliptic. We recall the conventions $\mathfrak{n} = \prod_{i=1}^{s} \mathfrak{p}_i^{e_i}$ and $d = deg(\mathfrak{n})$. The curves with $d \leq 3$ are known to be subhyperelliptic (compare Section 1). We want to show that there is only one more such curve, namely the one treated in Example 10.

Lemma 18. *Suppose $X_0(\mathfrak{n})$ is subhyperelliptic and $d \geq 4$. Then $q = 2$, $d = 4$, and $s = 1$.*

PROOF. Let $\mathfrak{n} \in \mathbb{F}_q[T]$ with $d \geq 4$ and suppose that $X_0(\mathfrak{n})$ is subhyperelliptic.

First we assume $(T^q - T) \nmid \mathfrak{n}$, so there exists an $\alpha \in \mathbb{F}_q$ with $(T-\alpha) \nmid \mathfrak{n}$. Without loss of generality (i.e., after a translation $T \mapsto T+\alpha$) we may assume $T \nmid \mathfrak{n}$. Since $X_0(\mathfrak{n})$ *mod* T is subhyperelliptic (i.e. a double cover of a rational curve), it has at most $2(q^2+1)$ rational points over $\mathbb{F}_{q^2}$ (= quadratic extension of $A/(T)$).

On the other hand consider the Drinfeld module ϕ over $A/(T)$ defined by $\phi_T = \tau^2$. The number of cyclic $\mathfrak{n}$-kernels of this Drinfeld module is $\varepsilon(\mathfrak{n})$ and all of these are rational over $\mathbb{F}_{q^2}$, because they are stable under the action of ϕ_T which is the same as the Frobenius of $\mathbb{F}_{q^2}$. The automorphism group of ϕ is $\mathbb{F}_{q^2}^\times$, but $\mathbb{F}_q^\times$ acts trivially on these kernels. Hence there are at most $q+1$ kernels on the same orbit under the action of $Aut(\phi)$. Since ϕ is defined over $\mathbb{F}_{q^2}$ and the $\mathfrak{n}$-kernels are rational over $\mathbb{F}_{q^2}$, this gives at least $\frac{\varepsilon(\mathfrak{n})}{q+1}$ $\mathbb{F}_{q^2}$-rational points on $X_0(\mathfrak{n})$ *mod* T. Together with the at least 2^s rational cusps we obtain the inequality

$$2^s + \frac{\varepsilon(\mathfrak{n})}{q+1} \leq 2(q^2+1).$$

From this one easily deduces $q^d < \varepsilon(\mathfrak{n}) \leq 2q^2(q+1) \leq 3q^3$, hence $d = 4$ and consequently $q = 2$. Furthermore, $2^s + \frac{\varepsilon(\mathfrak{n})}{3} \leq 10$ implies $s = 1$.

Now suppose $(T^q - T)|\mathfrak{n}$, but $(T^{q^2} - T) \not| \mathfrak{n}$; so there exists a quadratic prime $\mathfrak{q} \in A$ with $\mathfrak{q} \not| \mathfrak{n}$. Let ψ be the supersingular Drinfeld module over $A/(\mathfrak{q})$. It is defined by $\psi_T = \frac{-1}{\mathfrak{q}'(t)}\tau^2 + \tau + t$, where $\mathfrak{q}'(T)$ is the formal derivative of $\mathfrak{q}(T)$ and t is the image of T in $A/(\mathfrak{q})$ ([Ge3], Example 4.4). Then $\psi_{\mathfrak{q}} = \frac{\tau^4}{\delta}$ where $\delta \in \mathbb{F}_q^\times$ is the discriminant of the polynomial $\mathfrak{q}$. Furthermore, $Aut(\psi) = \mathbb{F}_q^\times$.

Counting $\mathbb{F}_{q^4}$-rational points on $X_0(\mathfrak{n})$ mod $\mathfrak{q}$ we obtain as above $2^s + \varepsilon(\mathfrak{n}) \leq 2(q^4 + 1)$ and hence $d = 4$. Since $(T^q - T)|\mathfrak{n}$ this implies $q \leq 4$, and calculation with the exact values of $\varepsilon(\mathfrak{n})$ excludes the remaining cases with $(T^q - T)|\mathfrak{n}$ and $(T^{q^2} - T) \not| \mathfrak{n}$.

Finally, if $(T^{q^2} - T)|\mathfrak{n}$, then the curves $X_0(T^{q^2} - T)$ and $X_0(T^{q^2-1} - 1)$, too, must be subhyperelliptic. But $X_0(T^{q^2-1} - 1)$ is not subhyperelliptic for $q > 2$ by the discussion above, and combining Example 13 and Lemma 14 shows that for $q = 2$, $X_0(T^4 + T)$ is not subhyperelliptic. □

Since we have already discussed the cases $\mathfrak{n} = (T^2+T+1)^2$ and $\mathfrak{n} = T^4+T^3+1$ in Example 10 and Example 15, the only values that remain to be examined are $\mathfrak{n} = T^4$, $\mathfrak{n} = (T+1)^4$, $\mathfrak{n} = T^4 + T + 1$, and $\mathfrak{n} = T^4 + T^3 + T^2 + T + 1$. But $T^4 + T^3 + T^2 + T + 1$ is obtained from $T^4 + T^3 + 1$ by translation $T \mapsto T + 1$ and hence $X_0(T^4 + T^3 + T^2 + T + 1)$ is not subhyperelliptic.

Lemma 19. *For $q = 2$, the curves $X_0(T^4)$ and $X_0(T^4 + T + 1)$ are not subhyperelliptic.*

PROOF. Since $X_0(T^4)$ has 4 rational cusps, the proof of Lemma 18 shows that $X_0(T^4)$ (and by translation also $X_0((T+1)^4)$) is not subhyperelliptic.

For $\mathfrak{n} = T^4 + T + 1$ we need some extra considerations. The Drinfeld module ϕ defined in the proof of Lemma 18 has 17 different $(T^4 + T + 1)$-isogenies, rational over $\mathbb{F}_4$. The orbits under the action of $Aut(\phi)$ on these have length 3 or 1. This gives us at least 7 $\mathbb{F}_4$-rational points (5 orbits of length 3 and 2 of length 1). Together with the 2 rational cusps this makes 9 rational points over $\mathbb{F}_4$.

Now we show that the Drinfeld module ψ over $A/(T)$ defined by $\psi_T = \tau^2 + \tau$ has two $\mathbb{F}_2$-rational (T^4+T+1)-isogenies. Indeed, $\psi_{T^4+T+1} = \tau^8 + \tau^4 + \tau^2 + \tau + 1 = (\tau^4 + \tau^3 + 1) \circ (\tau^4 + \tau^3 + \tau^2 + \tau + 1) = (\tau^4 + \tau^3 + \tau^2 + \tau + 1) \circ (\tau^4 + \tau^3 + 1)$ and it is easy to see that the zeros of $X^{16} + X^8 + X^4 + X^2 + X$ (resp. $X^{16} + X^8 + X$) form an $\mathbb{F}_2$-vectorspace which is stable under the action of ψ. Since $j(\psi) = 1$ and $Aut(\psi) = \{id\}$ this gives us two more rational points on $X_0(T^4 + T + 1)$ mod T, enough to be sure that $X_0(T^4 + T + 1)$ is not subhyperelliptic. □

Putting everything together, we have proved

Theorem 20. *Let* $\mathfrak{n} \in \mathbb{F}_q[T]$ *be of degree d. Then*

a) $X_0(\mathfrak{n})$ *is rational if and only if* $d \leq 2$.

b) $X_0(\mathfrak{n})$ *is elliptic if and only if* $q = 2$ *and* $\mathfrak{n}$ *is one of the polynomials* T^3, $T^2(T+1)$, $T(T+1)^2$, $(T+1)^3$.

c) $X_0(\mathfrak{n})$ *is hyperelliptic if and only if* $q > 2$ *and* $d = 3$ *or if* $q = 2$ *and* $\mathfrak{n}$ *is one of the polynomials* $T^3 + T + 1$, $T^3 + T^2 + 1$, $T(T^2 + T + 1)$, $(T+1)(T^2 + T + 1)$, $(T^2 + T + 1)^2$.

In all these cases the curve $X_+(\mathfrak{n})$ *is rational. In particular, for hyperelliptic* $X_0(\mathfrak{n})$ *the hyperelliptic involution is always the full Atkin-Lehner involution.*

References

[Art] E. ARTIN: *Algebraic Numbers and Algebraic Functions,* Gordon and Breach, New York, 1967

[Ge1] E.-U. GEKELER: *Drinfeld-Moduln und modulare Formen über rationalen Funktionenkörpern,* Bonner Mathematische Schriften 119 (1980)

[Ge2] E.-U. GEKELER: Über Drinfeld'sche Modulkurven vom Hecke-Typ, *Compositio Math.* **57** (1986), 219-236

[Ge3] E.-U. GEKELER: On finite Drinfeld modules, *J. Algebra* **141** (1991), 187-203

[G&N] E.-U. GEKELER AND U. NONNENGARDT: Fundamental domains of some arithmetic groups over function fields, *Internat. J. Math.* **6** (1995), 689-708

[G&R] E.-U. GEKELER AND M. REVERSAT: Jacobians of Drinfeld Modular Curves, *J. Reine Angew. Math.* **476** (1996), 27-93

[Ki] T. KIMURA: On Conservativity of Algebraic Function Fields, *Proc. Japan Acad.* **45** (1969), 595-597

[L&K] K. LØNSTED AND S. KLEIMAN: Basics on families of hyperelliptic curves, *Compositio Math.* **38** (1979), 83-111

[Ne] M. NEWMAN: Conjugacy, Genus, and Class Numbers, *Math. Ann.* **196** (1972), 198-217

[Ogg1] A. OGG: Rational points on certain elliptic modular curves, *Proceedings of Symposia in Pure Mathematics* vol. 24, AMS (1973), 221-231

Local height pairings of Heegner points on Drinfeld curves

Ulrich Tipp
Universität des Saarlandes
Fachbereich Mathematik, Postfach 151150
D-66041 Saarbrücken, GERMANY
Ulrich@math.uni-sb.de

1 Introduction

In 1986 B.Gross and D.Zagier wrote the paper "Heegner points and derivatives of L-series" [3]. Under the assumption that the L-series of a modular elliptic curve over $\mathbb{Q}$ vanishes at $s = 1$ they established a formula connecting the value of the derivative of the L-function at the point $s = 1$ to the Néron-Tate height of a Heegner point on this curve.

Now for function fields $\mathbb{F}_q(T)$ every elliptic curve E with split multiplicative reduction at ∞ is modular in the sense that there exists a morphism $X_0(N) \to E$ for some Drinfeld modular curve $X_0(N)$ [1].

As for classical modular curves one can define Heegner points on $X_0(N)$. So the question arises whether there is some relation between the height of Heegner points and special values of L-series in this case as well.

The paper [3] consists of two parts. In the first part the authors give formulas for the local height pairings by means of numbers of ideals with certain properties in a quadratic extension of $\mathbb{Q}$. In the second part the derivative of an L-series attached to some newform f at $s = 1$ is written as scalar product of f with some cusp form. The fourier coefficients of this form are shown to be more or less the global height pairings evaluated in the first part.

In [6] I obtained some formulas for local height pairings for Heegner divisors on Drinfeld modular curves. At the finite places these formulas are very similar to those of Gross and Zagier.

In this article the results of [6] are summarized with some minor extensions.

For the other part namely a treatment of L-functions in this context we refer to the paper [5] of H.-G. Rück, elsewhere in this book.

Throughout the paper we assume that K is a rational function field $\mathbb{F}_q(T)$ and that char $K \neq 2$.

2 Local height pairings on curves

Let K_v be a local field (e.g. the completion of $\mathbb{F}_q(T)$ at some place v) and X/K_v a curve. We denote by $\mathrm{Div}^0(X/K_v)$ the group of divisors of degree 0 rational over K_v. Let $Z^0(X/K_v)$ be the subgroup of $\mathrm{Div}^0(X/K_v)$ consisting of divisors whose points are rational over K_v. If $a = \sum e_i a_i$ with $a_i \in X(K_v)$ and $e_i \in \mathbb{Z}$ then $|a| := \{a_i\}$ denotes the support of a. For a function f on X let $f(a) := \prod f(a_i)^{e_i}$.

The following propositions are taken from [2] and are due to Néron and Lichtenbaum.

Proposition 1 *There exists a uniquely determined function called "local height pairing"*

$$< \cdot, \cdot >_v : (Z^0(X/K_v) \times Div^0(X/K_v))' \longrightarrow \mathbb{R},$$

where ′ denotes the restriction to pairs of divisors with disjoint support, with the following properties.

i) linearity: $< a, b >_v + < a, c >_v = < a, b + c >_v$

ii) symmetry: $< a, b >_v = < b, a >_v$ *whenever* $b \in Z^0(X/K_v)$.

iii) $< a, div(f) > = \log |f(a)|_v$ *for all functions* f.

iv) for $b \in Div^0(X/K_v)$ *fixed and a point* $x_0 \in X(K_v) - |b|$ *the function*

$$\begin{array}{rcl} X(K_v) - |b| & \longrightarrow & \mathbb{R} \\ x & \longmapsto & < (x) - (x_0), b >_v \end{array}$$

is continuous in the $v-$*adic topology.*

This pairing can be extended to a pairing on $(Div^0(X/K_v) \times Div^0(X/K_v))'$.

For non-archimedean places the pairing can be described through intersection numbers on a regular model $\mathcal{X}/\mathcal{O}_v$ where $\mathcal{O}_v$ is the valuation ring in K_v. This can also be applied to the place at ∞, which in our situation is non-archimedean, too. In section 4 however we will give another description via values of theta functions.

Proposition 2 *Let* $\mathcal{X}/\mathcal{O}_v$ *be a regular model of* X/K_v, $\mathfrak{D}$, $\mathfrak{E} \in Div^0(X/K_v)$ *with disjoint support. If* $\tilde{\mathfrak{D}}, \tilde{\mathfrak{E}}$ *are divisors on* $\mathcal{X}$ *with* $\tilde{\mathfrak{D}}|_X = \mathfrak{D}$, $\tilde{\mathfrak{E}}|_X = \mathfrak{E}$ *and* $(\tilde{\mathfrak{D}} \,.\, \mathcal{F}_i) = 0$ *for all fibre components* $\mathcal{F}_i$ *of the special fibre of* $\mathcal{X}$ *then*

$$< \mathfrak{D}, \mathfrak{E} >_v = -(\tilde{\mathfrak{D}} \,.\, \tilde{\mathfrak{E}}) \log q_v$$

where q_v *is the cardinality of the residue class field at* v.

The condition that $(\tilde{\mathfrak{D}}\,.\,\mathcal{F}_i) = 0$ for all i can be fulfilled by taking $\tilde{\mathfrak{D}} = \overline{\mathfrak{D}} + \sum \alpha_i \mathcal{F}_i$ where $\overline{\mathfrak{D}}$ is a horizontal divisor and $\alpha_i \in \mathbb{Q}$. The fact that the intersection pairing $(\mathcal{F}_i\,.\,\mathcal{F}_j)$ is up to multiples of the whole fibre non-degenerate gives the existence of the α_i's.

3 Local height pairings on $X_0(N)$ at the finite places

Let $X_0(N)$ be the Drinfeld modular curve defined over $K = \mathbb{F}_q(T)$ parameterizing pairs (ϕ, ϕ') of Drinfeld modules of rank 2 together with a cyclic isogeny $u : \phi \to \phi'$ of degree N i.e. $\ker u \simeq A/N$ with $A = \mathbb{F}_q[T]$.

Then we can define Heegner points on it.

Definition 1 *Let L/K be an imaginary quadratic extension i.e. the place at ∞ is not split in L. Let further $\mathcal{O}$ be the integral closure of A in L. If ϕ is a Drinfeld module of rank 2 with complex multiplication by $\mathcal{O}$ then from rank 1 theory we know that it is isomorphic to one defined over the Hilbert class field H of L i.e. the maximal abelian unramified extension in which ∞ is totally split. We will write $\mathcal{O}_H$ for the integral closure of A in H.*

Suppose now that all prime divisors of N are split in L/K. Then we can decompose $N = \mathfrak{n}\bar{\mathfrak{n}}$ where $\bar{\cdot}$ denotes the non-trivial automorphism of L/K.

Again from rank 1 theory we know that there is a Drinfeld module ϕ' and an isogeny $\phi_{\mathfrak{n}}$ in the notation of Hayes [4] with $\ker \phi_{\mathfrak{n}} \simeq \mathcal{O}/\mathfrak{n} \simeq A/N$. Everything is defined over H so this gives a $H-$rational point on $X_0(N)$ called a Heegner point.

Now we summarize the formulas given in [6] and give some ideas of the proof.

Proposition 3 *Let $N \in A$ be squarefree, $D \in A$ squarefree and congruent to a square modulo N. Let $L = K(\sqrt{D})$ be an imaginary quadratic extension with Hilbert class field H. Take some $\sigma \in$ Gal (H/L) and let $\mathfrak{a}$ be an ideal in the class $\mathcal{A}$ corresponding to σ under the Artin homomorphism. We fix a generator a of $N_{L/K}(\mathfrak{a})$. Further let $M \in A$ be a monic polynomial such that (M) is not the Norm of some ideal in $\mathcal{A}$. This last condition is equivalent to $x \notin T_M x^\sigma$ for a Heegner point x and a Hecke operator T_M. For an irreducible Polynomial $P \in A$ and an arbitrary Polynomial $\mu \in A$ and $\epsilon \in \mathbb{F}_q^*$ we define*

$$r_{\mathcal{A}}(NP\mu, MD) = \#\{\mathfrak{c} \in \mathcal{A}^{-1} \text{ integral}:$$

$$\exists \lambda \, with N(\mathfrak{c}) = (\lambda),\ -a\lambda \in N_{L/K}(L^*),\ (\lambda - \epsilon NP\mu) = (MD)\}.$$

Then we get the following formulas

- *if P is split then*

$$< (x) - (\infty) \ , \ T_M[(x)^\sigma - (0)] >_P = 0$$

- *if P is inert then there are $\epsilon \in \mathbb{F}_q^* - \mathbb{F}_q^2$ and $Q \in A$ irreducible and monic such that $End_{\overline{\mathbb{F}}}(x) \otimes K$ is isomorphic to the quaternion algebra $(D, \epsilon PQ)$. $\overline{\mathbb{F}}$ denotes an algebraic closure of the residue field. Let $\mathfrak{q}$ be a divisor of Q and write $[\mathfrak{qn}]$ for the class of the ideal $\mathfrak{qn}$. Then*

$$\begin{aligned} &< (x) - (\infty) \ , \ T_M[(x)^\sigma - (0)] >_P \\ &= \sum_{v \in \operatorname{Spec} \mathcal{O}_H, v|P} -(x \ . \ T_M x^\sigma) \log q_v \\ &= -u^2 \log q \sum_{\substack{\mu \in A - \{0\}/N_{L/K}(\mathcal{O}^*) \\ \deg \mu \leq \deg MD - \deg NP}} ord_P(P^2\mu) \cdot r_A(NP\mu, MD) \cdot \delta(\mu) R_{\{\mathcal{A}[\mathfrak{qn}]\}}(\mu) \deg P \end{aligned}$$

with

$$u = |\mathcal{O}^* / \mathbb{F}_q^*| = \begin{cases} q+1, & \textit{if } \deg D = 0 \\ 1, & \textit{otherwise} \end{cases}$$

and

$$\delta(\mu) := \prod_{\mathfrak{p} | gcd(\mu, D)} 2.$$

Furthermore $R_{\{\mathcal{A}[\mathfrak{qn}]\}}(\mu)$ is the number of integral ideals $\mathfrak{c}'$, which lie in a class differing from the class $\mathcal{A}[\mathfrak{qn}]$ only by a square in the class group and whose norm is the ideal (μ) , if $-aNQ^{-1}\mu \in N_{L/K}(L^*)$ and 0 otherwise. This is well defined modulo $N_{L/K}(\mathcal{O}^*)$.

- *if P is ramified then there are $\epsilon \in \mathbb{F}_q^* - \mathbb{F}_q^2$ and $Q \in A$ irreducible and monic such that $End_{\overline{\mathbb{F}}}(x) \otimes K$ is isomorphic to the quaternion algebra $(D, \epsilon Q)$. Then*

$$\begin{aligned} &< (x) - (\infty) \ , \ T_M(x)^\sigma - (0) >_P \\ &= \sum_{v \in \operatorname{Spec} \mathcal{O}_H, v|P} -(x \ . \ T_M x^\sigma) \log q_v \\ &= - \log q \sum_{\substack{\mu \in A - \{0\}/N_{L/K}(\mathcal{O}^*) \\ \deg \mu \leq \deg MD - \deg NP}} ord_P(P\mu) \cdot r_A(NP\mu, MD) \cdot \delta(P\mu) R_{\{\mathcal{A}[\mathfrak{qn}]\}}(\mu) \deg P \end{aligned}$$

For the proof of this proposition we follow the same idea as B. Gross and D. Zagier in [3]. First of all Proposition 2 implies

$$< (x) - (\infty) \; , \; T_M(x)^\sigma - (0) >_v = -(x \; . \; T_M x^\sigma) \log q_v$$

where the intersection takes place on the regular model of $X_0(N)$ over Spec A given by the moduli problem and eventually a desingularisation in points of the fiber over the prime divisors of N.

The modular interpretation of the points then yields a description of the intersection numbers in terms of homomorphisms between Drinfeld modules. More precisely we have the following statement

[6, Prop. 3.4]: Let W be the valuation ring in the maximal unramified extension of the complete local ring $\mathcal{O}_v$. Let x be a W−rational point on $X_0(N)/\mathcal{O}_v$ and remember that we assume that $x \notin T_M x^\sigma$. Then

$$(x \; . \; T_M x^\sigma) = \frac{1}{q-1} \sum_{n \geq 1} \#\mathrm{Hom}_{W/\pi^n}(x^\sigma, x)_{\deg M} \qquad (1)$$

where π is a uniformizing Parameter at v and homomorphisms $x \to y$ are defined through their representing Drinfeld modules. The index deg M means that only homomorphisms of degree M are counted.

If v lies over a prime P which splits in L/K (especially all primes dividing N) then the right hand side of equation (1) is 0.

In the other cases the Drinfeld modules representing x, have supersingular reduction, so $\mathrm{End}_{W/\pi}(x)$ is an order in a quaternion algebra over K. A theorem of M. Eichler allows to describe the order $\mathrm{End}_{W/\pi}(x)$ in this algebra explicitely. Counting ideals in this order then gives the formulas of Proposition 3.

4 Local height pairing on $X_0(N)$ at ∞

In contrast to the situation in [3] the place at ∞ is non-archimedean here, too. Therefore we can describe the local height pairing at ∞ by intersection numbers on a regular model as well. The model is given by the analytic reduction theory on Drinfelds upper half plane $\Omega = \mathbb{P}^1(C) - \mathbb{P}^1(K_\infty)$, where C is the completion of an algebraic closure of K_∞. See [6] for the details.

Another possibility to get the local height pairing at ∞ is given by values of certain theta functions. This goes back to an idea of B. Gross who used this to give a formula for local height pairing on Mumford curves [2]. The theta functions used there were defined by Manin and Drinfeld. Here we need similar functions for the group $\Gamma_0(N)$ which were defined by E.-U. Gekeler and M. Reversat in [1].

Definition 2 *Let* $\Gamma = \Gamma_0(N) < PGL_2(A)$ *be the subgroup whose elements are represented by matrices* $\begin{pmatrix} a & b \\ c & d \end{pmatrix}$ *in* $GL_2(A)$ *with* $N|c$.

Let $\omega,\ \eta \in \Omega$. *Then*

$$\theta(\omega, \eta; z) := \prod_{\gamma \in \Gamma} \frac{z - \gamma\omega}{z - \gamma\eta}$$

is a so called theta-function $\theta(\omega, \eta; \cdot) : \Omega \to C$.

We briefly recall the main properties of these functions.

Proposition 4 [1, Th. 5.4.1]

i) There exists a constant $c(\omega, \eta, \alpha)$ *such that*

$$\theta(\omega, \eta; \alpha z) = c(\omega, \eta, \alpha)\theta(\omega, \eta; z)$$

independently of z.

ii) $c(\omega, \eta, \alpha)$ *defines a group homomorphism* $\Gamma \to C^*$ *that factors through* $\overline{\Gamma} = \Gamma^{ab}/torsion$.

iii) $\theta(\omega, \eta; z)$ *is holomorphic and non-zero at the cusps. It has zeroes in the orbit* $\Gamma\omega$ *and poles in* $\Gamma\eta$ *if* $\omega \notin \Gamma\eta$ *and it is everywhere holomorphic and non-zero if* $\omega \in \Gamma\eta$.

iv) The holomorphic function

$$u_\alpha(z) = \theta(\omega, \alpha\omega; z)$$

is independent of the choice of $\omega \in \Omega$. *It depends only on the class of* α *in* $\overline{\Gamma}$.

v) For $\alpha, \beta \in \Gamma$ *we have* $u_{\alpha\beta} = u_\alpha\, u_\beta$.

vi) We have

$$c(\omega, \eta, \alpha) = \frac{u_\alpha(\eta)}{u_\alpha(\omega)}.$$

vii) $(\alpha, \beta) := \log_q |c(\omega, \alpha\omega, \beta)|$ *is a non-degenerate pairing* $\overline{\Gamma} \times \overline{\Gamma} \to \mathbb{Q}$ [1, Cor.5.7.5]

REMARKS.

i) By linearity we define theta functions also for divisors of degree 0 on Ω. Suppose that $\mathfrak{D} = \mathfrak{D}_0 - \mathfrak{D}_\infty$ for some positive divisors $\mathfrak{D}_0 = \sum \omega_i$, $\mathfrak{D}_\infty = \sum \eta_i$ of degree d. Then we define

$$\theta_{\mathfrak{D}}(z) := \prod_{i=1}^{d} \theta(\omega_i, \eta_i; z).$$

This theta function has the automorphic factor

$$u_\gamma(\mathfrak{D}) = \prod_{i=1}^{d} \frac{u_\gamma(\eta_i)}{u_\gamma(\omega_i)}.$$

ii) The divisors we are interested in have cusps in their support. This is excluded in [1]. Therefore one has to generalize the definition of the theta function to this case and has to prove a proposition similar to [1, 5.4.1]. In order to do this one has to improve the estimates of [1, 5.3]. But as this is not written down yet in all details we do not make use of it in the sequel. We only keep in mind that the following propositions should hold in this case as well.

Before we come to the main proposition we introduce the notion of rational unitary lifting.

Definition 3 *Let* $\Gamma = \Gamma_0(N)$, $\Omega = \mathbb{P}^1(C) - \mathbb{P}^1(K_\infty)$ *the Drinfeld upper half plane,* $\Omega^* = \Omega \cup \mathbb{P}^1(K)$ *and* $\pi : \Omega^* \to X_0(N)\ (\simeq \Gamma_0(N) \setminus \Omega^*)$ *the projection. Let* $\mathfrak{D}$ *be a divisor on* $X_0(N)$. *Then* $\tilde{\mathfrak{D}} = \sum_{i=1}^{r} c_i\omega_i$ *with* $c_i \in \mathbb{Q}$ *and* $\omega_i \in \Omega$ *is called a rational lifting of* $\mathfrak{D}$, *if* $\pi_*(\tilde{\mathfrak{D}}) = \mathfrak{D}$. *The lifting is called unitary if* $\log|u_\gamma(\tilde{\mathfrak{D}})| = 0$ *for all* $\gamma \in \Gamma$.

Lemma 1 *Every divisor on* $X_0(N)$ *has a rational unitary lifting on* Ω^*.

PROOF. It suffices to show the assertion for points because the function $\log|u_\gamma(\cdot)|$ is linear. So let x be a point on $X_0(N)$ and $z \in \Omega^*$ with $\pi(z) = x$. Let $\gamma_1, \ldots, \gamma_g$ be a basis of $\overline{\Gamma} = (\Gamma^{ab}/\text{tors.})$ as $\mathbb{Z}$-module. Then

$$\tilde{z} = \alpha_0 z + \alpha_1\gamma_1 z + \cdots + \alpha_g\gamma_g z$$

with $\alpha_i \in \mathbb{Q}$ and $\sum \alpha_i = 1$ is a rational lifting and

$$\begin{aligned}
&\log_q |u_{\gamma_j}(\tilde{z})| \\
&= \alpha_0 \log_q |u_{\gamma_j}(z)| + \sum_{i=1}^{g} \alpha_i \log_q |u_{\gamma_j}(\gamma_i z)| \\
&= \log_q |u_{\gamma_j}(z)| + \sum_{i=1}^{g} \alpha_i \log_q \left| \frac{u_{\gamma_j}(\gamma_i z)}{u_{\gamma_j}(z)} \right|,
\end{aligned}$$

as $\sum_i \alpha_i = 1$, so $\alpha_0 = 1 - \sum_{i \neq 0} \alpha_i$. Now $\log_q \left| \frac{u_{\gamma_j}(\gamma_i z)}{u_{\gamma_j}(z)} \right|$ is the pairing (γ_i, γ_j) from proposition 4 vii) telling us that the pairing is non-degenerate. Therefore the system of linear equations

$$\sum_{i=1}^{g} \alpha_i \log_q \left| \frac{u_{\gamma_j}(\gamma_i z)}{u_{\gamma_j}(z)} \right| = -\log_q |u_{\gamma_j}(z)|$$

with $i = 1, \ldots, g$ has a solution with $\alpha_i \in \mathbb{Q}$. Let α_i the elements of this solution, then

$$\log_q |u_{\gamma_j}(\tilde{z})| = 0$$

for all $j = 1, \ldots, g$. It follows, that $\tilde{z}$ is a rational unitary lifting of x. □

Proposition 5 *Let $\mathfrak{D}, \mathfrak{D}' \in Div^0(X_0(N))$ be disjoint divisors defined over a quadratic extension L_∞, with no cusps in their support. Let $\tilde{\mathfrak{D}}$ be a rational unitary lifting of $\mathfrak{D}$ and $\tilde{\mathfrak{D}}'$ an arbitrary lifting of $\mathfrak{D}'$. For the valuation $v|\infty$ in L_∞ we claim*

$$< \mathfrak{D}, \mathfrak{D}' >_v = \log |\theta_{\tilde{\mathfrak{D}}}(\tilde{\mathfrak{D}}')|_v = 2 \log_q |\theta_{\tilde{\mathfrak{D}}}(\tilde{\mathfrak{D}}')| \log q.$$

Remark. The remark after proposition 4 states that $u_\gamma(\tilde{\mathfrak{D}})$ is the automorphic factor of $\theta_{\tilde{\mathfrak{D}}}(z)$ i.e.

$$\log_q \left| \frac{\theta_{\tilde{\mathfrak{D}}}(\gamma z)}{\theta_{\tilde{\mathfrak{D}}}(z)} \right| = \log_q |u_\gamma(\tilde{\mathfrak{D}})|.$$

As $\tilde{\mathfrak{D}}$ is a unitary lifting this is 0. It follows in particular that $\log_q |\theta_{\tilde{\mathfrak{D}}}(\tilde{\mathfrak{D}}')|$ does not depend on the choice of the lifting $\tilde{\mathfrak{D}}'$ of $\mathfrak{D}'$.

Proof. We have to show for the pairing on the right hand side that

a) it is symmetric,

b) it is partially continuous

c) for a principal divisor $\mathfrak{D} = (f)$ we get

$$\tilde{f}(\tilde{\mathfrak{D}}') = \theta_{\tilde{\mathfrak{D}}}(\tilde{\mathfrak{D}}'),$$

where $\tilde{f} = f \circ \pi : \Omega^* \longrightarrow C$.

ad a) Without loss of generality let $\tilde{\mathfrak{D}}'$ be also a unitary lifting. Then

$$\frac{z-\gamma\omega}{z-\gamma\eta}\cdot\frac{y-\gamma\eta}{y-\gamma\omega} = \frac{z-\gamma\omega}{y-\gamma\omega}\cdot\frac{y-\gamma\eta}{z-\gamma\eta}$$

and so

$$\frac{\theta(\omega,\eta;z)}{\theta(\omega,\eta;y)} = \frac{\theta(z,y;\omega)}{\theta(z,y;\eta)}$$

showing

$$\theta_{\tilde{\mathfrak{D}}}(\tilde{\mathfrak{D}}') = \theta_{\tilde{\mathfrak{D}}'}(\tilde{\mathfrak{D}}).$$

ad b) This follows from the continuity of the theta functions.

ad c) The function $\theta_{\tilde{\mathfrak{D}}}(z)$ has in all points of $\Gamma\omega_i$ a zero of order $\#\Gamma_{\omega_i}$. In the points of $\Gamma\eta_i$ the function has poles of order $\#\Gamma_{\eta_i}$. The same holds for the function $\tilde{f}$, as the covering $\Omega^* \longrightarrow \Gamma\backslash\Omega^*$ has ramification orders $\#\Gamma_\omega$ for an element $\omega \in \Omega$.

It follows that the quotient $\delta(z) := \theta_{\tilde{\mathfrak{D}}}(z)/\tilde{f}(z)$ is a holomorphic theta function without zeroes and poles with automorphic factor $u_\gamma(\tilde{\mathfrak{D}})$. For $\Gamma = \Gamma_0(N)$ every such function is of the form $c \cdot u_{\gamma_0}(z)$ for some $c \in C^*$ and some $\gamma_0 \in \Gamma$ [1, Corollary 5.6.5.]. So assume $\delta(z) = c \cdot u_{\gamma_0}(z)$. Then

$$(\gamma,\gamma_0) = \log_q\left|\frac{u_{\gamma_0}(\gamma z)}{u_{\gamma_0}(z)}\right| = \log_q\left|\frac{\delta(\gamma z)}{\delta(z)}\right| = \log_q|u_\gamma(\tilde{\mathfrak{D}})| = 0$$

for all $\gamma \in \Gamma$ as $\tilde{\mathfrak{D}}$ is a unitary lifting. Because of the non-degeneracy of the pairing $\delta(z)$ is constant. If $\tilde{\mathfrak{D}} = \sum \alpha_i z_i$, then

$$\theta_{\tilde{\mathfrak{D}}}(\tilde{\mathfrak{D}}') = \tilde{f}(\tilde{\mathfrak{D}}')\cdot\prod_{i=1}^{s}\delta(z_i)^{\alpha_i} = \tilde{f}(\tilde{\mathfrak{D}}')\cdot(\mathrm{const.})^{\deg\mathfrak{D}}$$

so $\theta_{\tilde{\mathfrak{D}}}(\tilde{\mathfrak{D}}') = \tilde{f}(\tilde{\mathfrak{D}}')$ for $\deg\mathfrak{D} = 0$ proving the proposition. □

Before we specialize this to the situation where each of $\mathfrak{D}$ and $\mathfrak{D}'$ consist of only two points we recall some facts about functions on Ω and harmonic cochains on the Bruhat-Tits tree $\mathcal{T}$.

i) There is an exact sequence

$$1 \longrightarrow C^* \longrightarrow \mathcal{O}(\Omega)^* \xrightarrow{r} \underline{H}(\mathcal{T}) \longrightarrow 0.$$

Let $\mathcal{T}^*$ be the tree $\mathcal{T}$ if L_∞/K_∞ is inert. If it is ramified then let $\mathcal{T}^*$ be the tree $\mathcal{T}$ with one extra vertex on every edge (the midpoint in $\mathcal{T}(\mathbb{Q})$). Then the above sequence holds also for $\mathcal{T}^*$. This is because the description of r (cf. [1, 1.7.4]) shows that the value of a function in $\mathcal{O}(\Omega)^*$ at a point which goes to a midpoint of an edge under the reduction morphism is the mean value of two points going to the adjacent vertices. This is just the harmonicity at this new vertex.

ii) There is an isomorphism $\overline{\Gamma} \to \underline{H}_!(\mathcal{T}, \mathbb{Z})^\Gamma$ [1, Prop.3.4.5] which sends $\alpha \in \Gamma$ to a function ϕ_α. For the definition see [1, 3.3]. It is a special case of Definition 4.

iii) For $\alpha \in \Gamma$ we have $r(u_\alpha) = \phi_\alpha$ [1, 5.6.1]. From this it follows that if z, z_0 reduce to vertices v, v_0 of $\mathcal{T}^*$ respectively

$$\log_q \left| \frac{u_\alpha(z)}{u_\alpha(z_0)} \right| = \int_{v_0}^{v} \phi_\alpha$$

where the integral is just the sum of the values of the function on the edges of the unique path without backtracking from v_0 to v.

Corollary 1 *Let x, x_0, x', x'_0 be four disjoint non-cuspidal points on $X_0(N)$ defined over L_∞. Let z, z_0, z', z'_0 be points in Ω lying over the corresponding points and v, v_0, v', v'_0 be the vertices to which the points reduce. Let $\tilde{z} = \alpha_0 z + \alpha_1\gamma_1 z + \cdots + \alpha_g\gamma_g z$ and $\tilde{z}_0 = \beta_0 z_0 + \beta_1\gamma_1 z_0 + \cdots + \beta_g\gamma_g z_0$ be some rational unitary lifting as described in lemma 1.*

Then

$$\begin{aligned} &< (x) - (x_0)\,,\, (x') - (x'_0) > \\ &= \; 2\log q\Big(\log_q|\theta(z', z'_0; z)| - \log_q|\theta(z', z'_0; z_0)| - \sum_{i=1}^{g}(\alpha_i - \beta_i)\int_{v'_0}^{v'} \phi_{\gamma_i}\Big). \end{aligned}$$

PROOF. Set $\gamma_0 := 1$. Proposition 5 yields

$$\begin{aligned}
&< (x) - (x_0)\,,\,(x') - (x'_0) > \frac{1}{2\log q} \\
&= \log_q |\theta_{\tilde{\mathfrak{D}}}(\tilde{\mathfrak{D}}')| = \log_q |\theta_{\tilde{\mathfrak{D}}'}(\tilde{\mathfrak{D}})| \\
&= \sum_{i=0}^{g} \alpha_i \log_q |\theta(z', z'_0; \gamma_i z)| - \beta_i \log_q |\theta(z', z'_0; \gamma_i z_0)| \\
&= \log_q |\theta(z', z'_0; z)| - \log_q |\theta(z', z'_0; z_0)| \\
&\quad + \sum_{i=1}^{g} \alpha_i \log_q \left|\frac{\theta(z', z'_0; \gamma_i z)}{\theta(z', z'_0; z)}\right| - \beta_i \log_q \left|\frac{\theta(z', z'_0; \gamma_i z_0)}{\theta(z', z'_0; z_0)}\right| \\
&\quad \text{as } \alpha_0 = 1 - \textstyle\sum_{i\neq 0} \alpha_i \text{ and } \beta_0 = 1 - \textstyle\sum_{i\neq 0} \beta_i \\
&= \log_q |\theta(z', z'_0; z)| - \log_q |\theta(z', z'_0; z_0)| - \sum_{i=1}^{g} (\alpha_i - \beta_i) \int_{v'_0}^{v'} \phi_{\gamma_i}. \ \square
\end{aligned}$$

At the end we give an interpretation of the integral in the formula of Corollary 1 as scalar product with some "holomorphic projection".

We consider the space

$$\begin{aligned}
C_!(\mathcal{T}^*, \mathbb{Q})^\Gamma \ := \ & \{f : E(\mathcal{T}^*) \to \mathbb{Q} : f(\gamma e) = f(e)\, \forall \gamma \in \Gamma, \\
& f(e^-) = -f(e),\ f \text{ has compact support modulo } \Gamma\}
\end{aligned}$$

where e^- is the opposite oriented edge to e. Let G be the graph $\mathcal{T}^*/\Gamma$. Let $E(G)$ be the set of oriented edges of G. Set $C_!(G) = \{f : E(G) \to \mathbb{Q},\ f \text{ has compact support}\}$. Then there is a bijection

$$\begin{aligned}
C_!(\mathcal{T}^*, \mathbb{Q})^\Gamma &\longrightarrow C_!(G) \\
f &\longmapsto \overline{f} : \overline{f}(\overline{e}) = f(e) \frac{1}{|\mathrm{stab}(e)|}
\end{aligned}$$

where $\overline{e}$ is an edge of G represented by $e \in E(\mathcal{T})$. The order of the stabilizer does not depend on the chosen e, so this is well-defined.

For pairs v_1, v_2 of vertices of $\mathcal{T}^*$ we construct some special functions in $C_!(\mathcal{T}^*, \mathbb{Q})^\Gamma$.

Definition 4 *Denote by $(v_1 \to v_2)$ the set of oriented edges of the path without backtracking from v_1 to v_2. Then we define*

$$i_{v_1,v_2}(\gamma, e) := \begin{cases} 1 & \text{if } \gamma e \in (v_1 \to v_2) \\ -1 & \text{if } \gamma e \in (v_2 \to v_1) \\ 0 & \text{otherwise} \end{cases}$$

and

$$f_{v_1,v_2}(e) := \sum_{\gamma\in\Gamma} i_{v_1,v_2}(\gamma, e).$$

In particular if $v_2 = \alpha v_1$ *for some* $\alpha \in \Gamma$ *we get* $f_{v_1,v_2} = \phi_\alpha$.

From the construction of the functions f_{v_1,v_2} it is clear that they belong to the space $C_!(\mathcal{T}^*, \mathbb{Q})^\Gamma$.

Now we define a scalar product on this space by

$$(f,g) := \sum_{\bar{e}\in E(G)} f(e)\cdot g(e)\cdot |\mathrm{stab}(e)|^{-1} = \sum_{\bar{e}\in E(G)} \bar{f}(\bar{e})\cdot \bar{g}(\bar{e})\cdot |\mathrm{stab}(e)|.$$

This is well-defined and restricted to the space $\underline{H}_!(\mathcal{T}^*, \mathbb{Q})^\Gamma$ of $\Gamma-$ invariant harmonic cochains we get the Petersson product [1, 4.8].

If f is a function in the space $C_!(\mathcal{T}^*, \mathbb{Q})^\Gamma$ we denote by f^{hol} the uniquely determined function of $\underline{H}_!(\mathcal{T}^*, \mathbb{Q})^\Gamma$ which has the same scalar product with all functions in $\underline{H}_!(\mathcal{T}^*, \mathbb{Q})^\Gamma$ as f and call it the holomorphic projection.

Corollary 2 *With the same notations as in Corollary 1 we have*

$$\begin{aligned} &< (x) - (x_0)\,,\, (x') - (x'_0) > \\ =\ & 2\log q\big(\log_q|\theta(z', z'_0; z)| - \log_q|\theta(z', z'_0; z_0)| - \frac{1}{2}(f^{\mathrm{hol}}_{v_0,v}, f_{v'_0,v'})\big). \end{aligned}$$

PROOF. The values α_i, β_i describing the unitary liftings come from the system of linear equations

$$\sum_{i=1}^{g}(\alpha_i - \beta_i)\,(\gamma_i, \gamma_j) = -\log_q\left|\frac{u_{\gamma_j}(z)}{u_{\gamma_j}(z_0)}\right| = -\int_{v_0}^{v}\phi_\gamma.$$

Now [1, Th. 5.7.1] states that

$$2\cdot(\gamma_i, \gamma_j) = (\phi_{\gamma_i}, \phi_{\gamma_j}).$$

In the same way it is easy to see that

$$2\cdot\int_{v_0}^{v}\phi_{\gamma_j} = (f_{v_0,v}, \phi_{\gamma_j})$$

and so we get

$$(\sum(\alpha_i - \beta_i)\phi_{\gamma_i}, \phi_{\gamma_j}) = -2\cdot\int_{v_0}^{v}\phi_{\gamma_j} = (f_{v_0,v}, \phi_{\gamma_j})$$

for all j which implies $\sum(\alpha_i - \beta_i)\phi_{\gamma_i} = f^{\mathrm{hol}}_{v_0,v}$.

Now the Corollary is a direct consequence of Corollary 1. □

References

[1] E.-U. Gekeler, M. Reversat: *Jacobians of Drinfeld Modular curves*, Crelle Journal **476** (1996), 27-93

[2] B. Gross: *Local heights on curves* in "Arithmetic geometry", (ed. Cornell/Silverman) Springer 1986

[3] B. Gross, D. Zagier: *Heegner points and derivatives of L−series*, Invent.Math. **84** (1986), 225-320

[4] D. R. Hayes: *Explicit class field theory in global function fields*, Studies in Algebra and Number Theory, Advances in Mathematics supplementary studies vol.6, 173-217, Academic Press (1979)

[5] H.-G. Rück: *L−Series of Automorphic Cusp Forms of Drinfeld Type*, this book

[6] U. Tipp: *Höhenberechnungen von Heegnerpunkten auf Drinfeldschen Modulkurven*, Dissertation, Preprint No.10 (1996), Inst. f. exp. Math., Essen

List of Participants

Participants of the workshop Drinfeld modules, modular schemes and applications, Alden-Biesen, September 9-14, 1996

ALVAREZ ARTURO
Universidad de Salamanca
Departamento de Matematicas
Plaza de la Merced 1-4
E-37008 Salamanca, SPAIN
aalvarez@gugu.usal.es

ANGLES BRUNO
Université de Toulouse III
Laboratoir de Mathématiques
Emile Picard UMR 5580
F-31062 Toulouse Cedex, FRANCE
angles@picard.ups-tlse.fr

BROWN MARTIN L.
Université Joseph Fourier
Institut Fourier B.P. 74
URA 188 du CNRS,
UFR de Mathématiques
F-38402 Saint-Martin-d' Hères cedex
FRANCE
Martin.Brown@ujf-grenoble.fr

CORNELISSEN GUNTHER
Universiteit Gent
Dept. of Pure Mathematics
and Computer Algebra
Galglaan 2,B-9000, Gent, BELGIUM
gc@cage.rug.ac.be

DENEF JAN
University of Louvain KUL
Dept. of Algebra and Topology
Celestijnenlaan 200B
B-3001 Heverlee, BELGIUM
Jan.Denef@wis.kuleuven.ac.be

GEKELER ERNST-ULRICH
Universität des Saarlandes, Fachbereich Mathematik, Postfach 151150
D-66041 Saarbrücken, GERMANY
gekeler@math.uni-sb.de

HELLEGOUARCH YVES
Université de CAEN
Mathématiques & Mécanique
F-14032 Caen Cedex, FRANCE

HOEPPNER SABINE
Rijksuniversiteit Groningen
Vakgroep Wiskunde,Postbus 800
NL-9700 AV Groningen,
THE NETHERLANDS
sabinek@math.rug.nl

LOPEZ BARTOLOME
Universidad Complutense de Madrid
Dpto. ALGEBRA,Ciudad Universitaria
E-28040 Madrid, SPAIN
bartolo@sunal1.mat.ucm.es

MATZAT HEINRICH
Univerisität Heidelberg
Interdisziplinares Zentrum für
Wissenschaftliches Rechnen
Im Neuenheimer Feld, D-69120 Heidelberg, GERMANY
matzat@clio.iwr.uni-heidelberg.de

MAUDUIT VERONIQUE
Université de CAEN
Mathématiques & Mécanique
F-14032 Caen Cedex,FRANCE
mauduit@math.unicaen.fr

MICHAUX CHRISTIAN
Université de Mons-Hainaut
Avenue Maistriau 15, B-7000
Mons, BELGIUM
chrmich@sun1.umh.ac.be

OUKHABA HASSAN
Université de Franche-Comté
Laboratoire de Mathématiques
et Mécanique Théorique
16, Route de Gray
F-25030 Besangon Cedex, FRANCE
hassan@math.univ-fcomte.fr

REBINGUET NADJA
Université de Toulouse III
Laboratoire de Mathématiques
Emile Picard UMR 5580
F-31062 Toulouse Cedex, FRANCE
rebingue@picard.ups-tlse.fr

RECHER FRANCOIS
Université Lille 1
UFR Mathématiques
F-59655 Villeneuve d'Ascq, FRANCE
recher@univ-lille1.fr

REVERSAT MARC
Université de Toulouse III
Laboratoire de Mathématiques
Emile Picard UMR 5580
F-31062 Toulouse Cedex, FRANCE
reversat@picard.ups-tlse.fr

RÜCK HANS-GEORG
Universität GH Essen
Institut für Experimentelle Mathematik
D-45326 Essen, GERMANY
rueck@exp-math.uni-essen.de

RUST IMKE
Universität des Saarlandes
Fachbereich Mathematik
Postfach 151150
D-66041 Saarbrücken, GERMANY
turtle@math.uni-sb.de

SAIDI MOHAMED
Univerisität Heidelberg
Interdisziplinares Zentrum
fur Wissenschaftliches Rechnen
Im Neuenheimer Feld
D-69120 Heidelberg, GERMANY
SAIDI@harmless.mathi.
uni-heidelberg.de

SCHEJA ORTWIN
Universität des Saarlandes
Fachbereich Mathematik
Postfach 151150
D-66041 Saarbrücken, GERMANY
scheja@math.uni-sb.de

SCHOLTEN JASPER
Rijksuniversiteit Groningen
Vakgroep Wiskunde, Postbus 800
NL-9700 AV Groningen,
THE NETHERLANDS
J.Scholten@math.rug.nl

SNYDER BRIAN
Ohio State University
100 Mathematics tower
231 W. 18th Avenue
Columbus, Ohio 43210-1174, USA
snyder@math.ohio-state.edu

TIPP ULRICH
Universität des Saarlandes
Fachbereich Mathematik
Postfach 151150
D-66041 Saarbrücken, GERMANY
Ulrich@math.uni-sb.de

TOEN BERTRAND
Université de Toulouse III
Laboratoir de Mathématiques
Emile Picard UMR 5580
F-31062 Toulouse Cedex, FRANCE

TOP JAAP
Rijksuniversiteit Groningen
Vakgroep Wiskunde, Postbus 800
NL-9700 AV Groningen,
THE NETHERLANDS
J.Top@math.rug.nl

ULMER DOUGLAS
University of Arizona
Dept. of mathematics
Tucson AZ 85721,Arizona, USA
email: ulmer@math.arizona.edu

VAN DER PUT MARIUS
Universiteit Groningen
Mathematisch Instituut, P.O. Box 800
NL-9700 AV Groningen,
THE NETHERLANDS
mvdput@math.rug.nl

VAN GEEL JAN
Universiteit Gent
Dept. of Pure Mathematics
and Computer Algebra
Galglaan 2, B-9000 Gent, BELGIUM
jvg@cage.rug.ac.be

VAN HAMME LUCIEN
Vrije Universiteit Brussel VUB
Toegepaste Wetenschappen
Pleinlaan 2, B-1050 Brussel, BELGIUM
lvhamme@vnet3.vub.ac.be

VAN STEEN GUY
University of Antwerp RUCA
Dept. Wiskunde
Groenenborgerlaan 171
B- 2020 Antwerpen, BELGIUM
vansteen@ruca.ua.ac.be

VIDUNAS RAIMUNDAS
Rijksuniversiteit Groningen
Vakgroep Wiskunde, Postbus 800
NL-9700 AV Groningen,
THE NETHERLANDS
R.Vidunas@math.rug.nl

WACK BODO
Universität des Saarlandes
Fachbereich Mathematik
Postfach 151150
D-66041 Saarbrücken,
GERMANY

ZAGIER DON
Max-Panck-Institut
für Mathematik
Gottfried-Claren-Strasse 26
D-53225 Bonn, GERMANY
don@mpim-bonn.mpg.de

www.ingramcontent.com/pod-product-compliance
Ingram Content Group UK Ltd.
Pitfield, Milton Keynes, MK11 3LW, UK
UKHW020400150926
8624IPUK00004B/266

9 789810 230678